Broadening the Genetic Base of Grain Cereals

Mohar Singh • Sandeep Kumar
Editors

Broadening the Genetic Base of Grain Cereals

Springer

Editors
Mohar Singh
ICAR-National Bureau of Plant
Genetic Resources Regional Station
Shimla, HP, India

Sandeep Kumar
ICAR-National Bureau of Plant
Genetic Resources
New Delhi, India

ISBN 978-81-322-3611-5 ISBN 978-81-322-3613-9 (eBook)
DOI 10.1007/978-81-322-3613-9

Library of Congress Control Number: 2016950437

Printed on acid-free paper

This Springer imprint is published by Springer Nature
The registered company is Springer (India) Pvt. Ltd.

Preface

Worldwide, majority of grain cereals have intrinsically a narrow genetic base which has been exploited to the maximum levels of productivity. Further breakthroughs in enhancing yield and improving stability in future crop cultivars require the identification of useful materials with desirable traits of interest and their incorporation into the elite genetic background. Knowledge, access, and use of genetic diversity available in cultivated and wild relatives are essential for broadening the genetic base of cultivated species. In view of this context, an effort has been made to bring together the rather scattered research work done in this useful area in the form of an edited collection, a compilation that should be of great value to the researchers across the world. The book comprises a total of ten chapters on different cereals and millets, contributed by eminent researchers from various reputed institutions of the world. The introductory chapter describes some key issues linked to bottlenecks and wild species utilization on current trends of interspecific hybridization. The subsequent chapters deal with different aspects related to widening the genetic base of cultivated gene pool. Each chapter provides detailed information on crop genepool, evolutionary relationships and systematic assessment of gene flow, gene flow constraints, level of diversity in crop germplasm, production related problems, traits of importance for widening the genetic base, wide hybridization, barriers to wide hybridization, methods of wide hybridization, and genome mapping and genomics status of each candidate crop. The editors are extremely thankful to all authors for their significant contributions to this book. The entire process of preparing the manuscript was marked by cordial collegiality. Thanks are also due to Ms. Megha Bakshi working as Project Assistant with editors for her assistance during the preparation of this multi-authored edited book in terms of compilation, edited chapter processing, and typographical scientific work. We are also indebted to Springer India for their excellent professional support in the completion of this project. Despite several rounds of proofreading and our best efforts, the book may still have some scientific, technical, and printing errors. We will appreciate if these omissions are brought to our notice, so that they may be rectified in future editions. Finally, we hope this book will be very useful to researchers and other readers across the world.

Shimla, India — Mohar Singh
New Delhi, India — Sandeep Kumar

Contents

About the Editors

Mohar Singh is currently working as Senior Scientist (Plant Breeding) at the National Bureau of Plant Genetic Resources Regional Station in Shimla, India. Dr. Singh received his doctoral degree in Plant Breeding from the Himachal Pradesh Agricultural University in Palampur, India. He has been working on the genetic and genomic resources of pulses, cereals, and potential crops for the last several years. He has identified several useful gene sources for various traits of interest in wild *Cicer* and *Lens* species, some of which have since been introgressed into the cultivated background for diversification of the cultivated gene pool. Dr. Singh has published more than 50 research papers in journals of national and international repute. He also holds two textbooks and three edited books to his credit.

Sandeep Kumar is presently working as Senior Scientist (Biochemistry) at the Germplasm Evaluation Division, National Bureau of Plant Genetic Resources (NBPGR), an institute of the Indian Council of Agricultural Research (ICAR), New Delhi, since December 22, 2010. Earlier, he worked as a Scientist (Biochemistry) at the Directorate of Rapeseed-Mustard Research, Sewar, Bharatpur, which is also an institute of the ICAR. He received his master's and doctoral degrees in Biochemistry from Chaudhary Charan Singh Haryana Agricultural University, Hisar, Haryana. He has substantial experience in teaching various graduate and postgraduate courses at SBS PGI, Balawala, Dehradun, and graduate courses at D.N. College, Hisar. He has also researched biochemical aspects in various groups of crops including cereals, oilseeds, and forages. Presently, his research focuses on the nutritional and other biochemical aspects of cereals and oilseeds. He has to his credit one book as co-editor and more than 45 scientific publications published in journals of national and international repute.

Introduction

1

Mohar Singh and Sandeep Kumar

Abstract

The human population growth has passed seven billion figures and is assumed to reach nine billion by 2050 (UN 2013). With such a higher population increasing rate, the amount of food we produce today will not be sufficient in the coming years to keep all alive, healthy and wealthy, which develops substantial pressure on the agricultural production system. Agricultural production demands are expected to increase in the twenty-first century with demands for cereals in particular expected to rise by 70 % from 2007 to 2050 (FAO 2012). Besides, the need of biomass for bioenergy and biorefineries is also expected to increase considerably. There are, however, fewer possibilities to increase the area under agriculture substantially for enhancing production systems in the world. Global climate change is also going to affect the production potential more adversely mainly due to increase in temperatures which in terms reduce the photosynthetic capacity of important agricultural crop plants. Further, water is expected to become a limiting factor in most of areas, and the pressure from devastating pathogens and pests is also assumed to increase, substantially. However, challenges of meeting future cereal production demands as mentioned above will be entirely dependent on the development of potential cultivars well adapted to the altered growing conditions. The tailoring of genetically potential crop varieties with high and stable yield, adequate quality attributes and low environment impact will in turn be dependent on the available invaluable wild/weedy genetic resources and their subsequent selection and utilization for diversification of culti-

M. Singh (✉)
ICAR-National Bureau of Plant Genetic Resources
Regional Station, Shimla, HP 171 004, India
e-mail: singhmohar_2003@yahoo.com

S. Kumar
ICAR-National Bureau of Plant Genetic Resources,
Pusa Campus, New Delhi 110 012, India
e-mail: kumarsandeep_biochem@rediffmail.com

M. Singh, S. Kumar (eds.), *Broadening the Genetic Base of Grain Cereals*,
DOI 10.1007/978-81-322-3613-9_1

vated gene pool. Gene banks across the globe are primarily focusing on plant genetic resources management activities of elite germplasm. But there is an urgent need for active involvement with all stakeholders to enhance the utilization of crop wild relatives. There is considerable germplasm availability of cereals including collections of adapted varieties and genetic stocks carrying useful traits of interest through to landraces and crop wild relatives. However, this richness of plant genetic diversity is greatly underutilized, with important traits not being exploited into the cultivated genetic background. The conservation of genetic resources must be linked to their enhanced and sustainable use if they carry potential traits to play a pivotal role under adverse situations. Pre-breeding approach provides us a useful tool that can be used for the introgression of demanding traits of interest into suitable, agronomically adapted genetic backgrounds.

Keywords
Base broadening • Yield enhancement • Wild species • Grain cereals

The human population growth has passed seven billion figures and is assumed to reach nine billion by 2050 (UN 2013). With such a higher population increasing rate, the amount of food we produce today will not be sufficient in the coming years to keep all alive, healthy and wealthy, which develops substantial pressure on the agricultural production system. Agricultural production demands are expected to increase in the twenty-first century with demands for cereals in particular expected to rise by 70 % from 2007 to 2050 (FAO 2012). Besides, the need of biomass for bioenergy and biorefineries is also expected to increase considerably. There are, however, fewer possibilities to increase the area under agriculture substantially for enhancing production systems in the world. Global climate change is also going to affect the production potential more adversely mainly due to increase in temperatures which in terms reduce the photosynthetic capacity of important agricultural crop plants. Further, water is expected to become a limiting factor in most of areas, and the pressure from devastating pathogens and pests is also assumed to increase, substantially. However, challenges of meeting future cereal production demands as mentioned above will be entirely dependent on the development of potential cultivars well adapted to the altered growing conditions. The tailoring of genetically potential crop varieties with high and stable yield, adequate quality attributes and low environment impact will in turn be dependent on the available invaluable wild/weedy genetic resources and their subsequent selection and utilization for diversification of cultivated gene pool. Gene banks across the globe are primarily focusing on plant genetic resources management activities of elite germplasm. But there is an urgent need for active involvement with all stakeholders to enhance the utilization of crop wild relatives. There is considerable germplasm availability of cereals including collections of adapted varieties and genetic stocks carrying useful traits of interest through to landraces and crop wild relatives. However, this richness of plant genetic diversity is greatly underutilized, with important traits not being exploited into the cultivated genetic background. The conservation of genetic resources must be linked to their enhanced and sustainable use if they carry potential traits to play a pivotal role under adverse situations. Pre-breeding approach provides us a useful tool that can be used for the introgression of demanding traits of interest into suitable, agronomically adapted genetic backgrounds.

1.1 Traits of Importance for Base Broadening

1.1.1 Abiotic Stresses

Tolerance to prevailing abiotic stresses such as drought, cold, salinity, heat, water logging, nutrient use efficiency and mineral toxicity has traditionally enabled cereal crops to cope with the major stresses with a balance of useful characters. In view of climate changing scenario, shifts in the prevalence of some abiotic stresses with drought being increasingly important, given the effect of climate change on the amount of precipitation and its distribution over the growing seasons in most parts of the world. Therefore, it is expected that the agricultural areas will experience enhanced or novel abiotic stresses making it increasingly urgent to develop cereals that can withstand such environmental changes, in order to increase or even just to maintain current yield levels.

1.1.2 Biotic Stresses

Plant pathogens cause considerable yield losses in cereal production, reducing crop quality and threatening food safety. Disease prevention and control are thus a prerequisite for competitive cereal production with the breeding of genetically disease-resistant crops being one of the most environmentally and economically desirable ways to manage plant diseases. The prevalence of different plant diseases is changing due to changing environmental conditions, including global climate change, but also changes in agricultural production with trends towards larger areas planted to fewer and/or genetically more uniform varieties, reduced crop rotation and soil tillage, loss of biodiversity, changing use of pesticides, and global trade. In such a changing environmental and economical context, plant diseases will inevitably appear and compromise crop production in regions where they did not represent a problem before, as is already being seen with the emergence of a new strain of heat-tolerant wheat yellow rust and increasing problems with *Fusarium* head blight and *Ramularia* leaf spot. Understanding the host-pathogen biology is the first step towards minimizing the risks represented by plant diseases. Durable, both race non-specific and race-specific, resistance incorporated into high-yielding genotypes is the main method to manage diseases of cereals.

1.1.3 Quality Parameters

Cereal grains provide carbohydrates, protein, minerals, dietary fibre, vitamins and various functional molecules with medicinal properties. Increasing the future production potential of cereals and increasing/maintaining the nutritional traits make the task more challenging for the breeders. Wild species in general had higher content of micronutrients and vitamins which can be used for biofortification of the cultivated species. Further, high fibre content and functional biomolecules or nutraceuticals are other biochemical components where wild species can play a major role in development of materials containing higher concentration of these components.

New durable and efficient sources of resistance/nutritional traits will have to be sought, in the case of cereals, from landraces and from wild relatives of the crops. This appears to be an insignificant part of cereal germplasm accessions that are conserved in various gene banks across the world. Knowledge, access, and use of diversity available in the cultivated and wild relatives are essential for broadening the genetic base of cultivars to sustain crop improvement. An overview of the existing level of diversity and the genetic base of major grain cereals is presented in this chapter based on the contributions made by respective crop experts collated in this book.

1.2 Wheat

Wheat is one of the most important staple food crops for 35 % of the world's population and is becoming important in the developing countries (CIMMYT 2011). To meet the demand for tailoring high-yielding and stress-resistant crop varieties, it is a prerequisite to widen the genetic base of cultivated gene pool (Khalighi et al. 2008). The narrow genetic base of cultivated genotypes has limited the plant breeders' progress towards further improvement (Rejesus et al. 1996). Therefore, there is an urgent need to exploit new sources of variation, and incorporating them into cultivated backgrounds is one of the basic and fundamental ways of most of the wheat breeding programmes (Baghizadeh and Khosravi 2011). However, landraces and their wild relatives are base genetic materials for diversity maintenance and improvement (Baghizadeh and Khosravi 2011). The wild relatives of cultivated wheat are considered as potential sources of useful alleles for genetic base widening (Khalighi et al. 2008), because of high polymorphism found in wild gene resources, and that can be useful for wheat crop improvement (Cenkci et al. 2008). Introduction of useful genes from *Aegilops* can contribute to traits like major yield contributing component characters, quality and resistance to various biotic and abiotic stresses (Schneider et al. 2008). Likewise, *Aegilops* species are also reservoirs of invaluable genes for resistance against heat and drought (Zaharieva et al. 2001a; Molnár et al. 2004), salinity (Colmer et al. 2006) and cold (Monneveux et al. 2000) stresses. Many of these wild species contain valuable genes for resistance to various pathogens (Aghaee-Sarbarzeh et al. 2001; Schneider et al. 2008). Besides, *Aegilops* can also contribute to yield through modification of physiology of cultivated gene pool by increasing the size of flag leaf and thus biomass production (Monneveux et al. 2000); *Ae. kotschyi* and *Ae. tauschii* also possess high iron and zinc content (Chhuneja et al. 2006).

1.3 Rice

The worldwide demand for rice is estimated to increase from 676 million tons in 2010 to 852 million tons in 2035 with an overall increase of 26% in the next 25 years (Khush 2013). To achieve this target, new rice varieties with higher yield potential and greater yield stability are vital to relieve the growing pressure of greater rice production. Rice is bestowed with tremendous genetic diversity of which only a small fraction has been used for its improvement. Today, yield stagnation in rice is a proven fact rather than mere perception, and further yield jump does not seem to be forthcoming with present set of genetic material being used in most of the cross-breeding programmes. In order to achieve high yield potential and breaking yield plateau, utilization of wild gene pool can widen the genetic base of cultivated varieties. The genus *Oryza* of the Gramineae family consisted of 24 species. Among these, two species *O. sativa* and *O. glaberrima* are cultivated and the remaining 22 are wild species. The wild species have either $2n=24$ or $2n=48$ chromosomes representing AA, BB, CC, BBCC, CCDD, EE, FF, GG and HHJJ genomes. The genes from the cultivated rice (*O. sativa*) and its closely related wild species, having AA genome (*O. nivara*, *O. rufipogon*, *O. glumaepatula*, *O. longistaminata* and *O. meridionalis*), can be easily transferred to cultivated species by conventional hybridization and backcrossing procedures (Brar and Khush 1997).

1.4 Maize

Maize originated in Mexico and Central America and belongs to the tribe Maydae of family Poaceae. Genetic resources especially crop wild relatives and landraces harbour novel alleles/genes for imparting resistance/tolerance to various agro-morphological traits and biotic and abiotic stresses including nutritional quality traits. Among wild relatives, teosintes and *Tripsacum* are native to Mexico and Central America, while *Coix*, *Chionachne*, *Trilobachne* and *Polytoca* originated in Southeast Asia. Likewise, interspecific compatibility between maize and teosinte develops hybrids. While generation of crosses between maize and *Tripsacum* species is difficult, Asiatic genera possess high degree of cross-incompatibility with maize. However, landrace accessions possessing unexplored alleles/genes serve as valuable donors for important traits of interest. These invaluable germplasm resources can be systematically and effectively utilized in genetic enhancement and to widen the genetic base of cultivated varieties to develop high-yielding and nutritious germplasm resilience to desirable traits.

1.5 Barley

Barley being the earliest domesticated cereal is one of today's major crops. Wide adaptation of barley to ecogeographic conditions and higher response to low input and stressful environments make it cultivable globally. It is self-pollinating and a true diploid ($2n=2x=14$) and therefore is considered as an ideal genetic material for basic research purposes. Besides other sources of genetic variation, barley landraces continue to be an important tool of genetic diversity conserved *ex situ* in gene banks. Gene flow involving wild progenitors and domesticated cultivars has attracted breeders' interest in order to broaden the genetic base of barley. Biotic and abiotic factors affecting barley yield have to be determined specifically. There is little or no opportunity for interspecific gene transfer in nature between the other species of genus *Hordeum* as there is no chromosome pairing at meiosis with few exceptions. Wheat-barley hybridization has the potential to incorporate favourable characteristics of stem strength and winter hardiness of wheat into barley. Use of DNA markers and high-throughput techniques in barley has enhanced the precision for introgressing specific traits of importance for diversification of cultivated gene pool. Consensus maps have allowed comparative mapping to explore information available in other crop species. Recent approaches like advanced backcross QTL (AB-QTL) analysis and association mapping (AM) studies have aided in understanding of the complex agronomic traits in barley. Further, interspecific and intergeneric hybridization in barley can be beneficial for introgression of useful traits in the barley genetic background. Owing to health benefits of barley, industrial interest for introduction of different barley products in human diets has increased considerably. Further, efforts are in progress to sequence the barley genome by the International Barley Genome Sequencing Consortium. The future barley breeding programme should combine the application of new tools and techniques with traditional and efficient plant breeding methods to achieve desired goals rapidly.

1.6 Oats

Oats rank sixth in world cereal production and are being grown as multipurpose crop for grain, pasture and forage or as a rotation crop in several segments of the world. The collection of germplasm and the conservation of gene pools of cultivated and related wild/weedy species are essential for crop improvement. The main focus is to summarize recent knowledge and significant achievements on genetic resources activities including plant taxonomy, diversity, collection, conservation, characterization, evaluation, and utilization in breeding of oats for diversification of cultivated varieties. Wild species of *Avena* are grouped into three gene pools depending on their interfertility with cultivated hexaploid oat. As far as wide hybridization is concerned, resistance to powdery mildew was introduced to breeding materials with the use of germplasm from *A. sterilis* (Hayes and Jones 1966), *A. barbata* (Aung et al. 1977; Thomas et al. 1980) and diploids *A. pilosa* (Hoppe and Kummer 1991; Sebesta et al. 1986), *A. hirtula* (Thomas 1968), *A. ventricosa* (Thomas 1970) and *A. prostrata* (Griffiths 1984; Morikawa 1995). In more recent findings of Herrmann (Yu and Herrmann 2006), the resistance was transferred from the tetraploid perennial oat *A. macrostachya*.

1.7 Sorghum

Sorghum [*Sorghum bicolor* (L.) Moench] is the world's fifth most important multipurpose cereal crop, valued for its grain, stalks and leaves. It is considered as drought, heat and insect resistant. Its small diploid genome ($2n=20$) and phenotypic diversity make it an ideal C_4 grass model (Mace et al. 2013). The *Sorghum* genus consists of 31 cultivated species and 17 related wild species (Snowden 1936, 1955), but as per USDA ARS (2007), it includes 25 species. The number of species constituting the genus varies in different scientific publications confirming the dynamic nature of the classification of cultivated sorghum and its wild relatives. Gene flow between cultivated types and their wild/weedy relatives plays an important role in structuring the genetic variability within and among populations. The consequences of gene flow can substantially contribute to the scientific basis for managing agricultural systems, understanding evolutionary processes and designing in situ conservation measures for genetic resources and using these resources to secure current and future crop breeding programmes. In the case of sorghum, efforts have been made towards development of interspecific hybrids. More recently, an interspecific hybrid was produced between a genetic male-sterile *S. bicolor* and *S. halepense*. The chromosome number was likely generated from fertilization by a rare monohaploid *S. halepense* pollen grain (Dweikat 2005). This plant was fertile and showed normal segregation of polymorphic SSR markers in the F_2 generation. Interspecific hybrids between tetraploid sorghum ($2n=40$) and *S. halepense* have been developed easily and used in attempts to create perennial grain sorghums (Piper and Kulakow 1994; Cox et al. 2002). None of the efforts to date have succeeded in producing agronomically useful germplasm through wide hybridization approaches.

1.8 Pearl Millet

Pearl millet is an important staple crop in the semiarid tracts of Asia and Africa due to its high nutritive value and adaptation to varying stress conditions. Worldwide, *Pennisetum* spp. germplasm is conserved as. 66,682 accessions in 97 gene banks, with ICRISAT in India holding the largest collection. Enormous genetic diversity has been reported in cultivated gene pool for agro-morphological traits and resistance to abiotic and biotic stresses, including nutritional traits. In pear millet, there has been a general lack of interest in using wild species because of the large amount of genetic variability already available in landraces. However, new sources for different traits such as *P. glaucum* subsp. *monodii* for cytoplasmic-nuclear male sterility (CMS); *P. purpureum* for forage, stiff stalk and restorer genes of the A_1 CMS system; *P. mezianum* for drought tolerance; *P. orientale* for drought tolerance and forage; *P. schweinfurthii* for large seeds; *P. pedicellatum* and *P. polystachion* for downy mildew resistance; and *P. squamulatum* for apomictic gene have been identified (Rai et al. 1997). Further, inter-subspecific, interspecific and intergeneric hybridization has been attempted to widen the cultigen gene pool in cultivated pearl millet. Inter-subspecific crosses involving subsp. *glaucum* (cultivated), *monodii* (wild relative) and *stenostachyum* (weedy relative) have been successful to transfer desirable traits like rust resistance, male sterility, and alleles for enhancing yield component traits. However, interspecific hybridization between *P. glaucum* and *P. purpureum* has led to development of forage hybrids with high biomass and better quality.

1.9 Finger Millet

Finger millet is highly nutritious with diverse usage, well adapted to marginal lands, and mostly grown by resource-poor farmers worldwide. There are indications that wild and weedy ancestor *E. africana* in Africa has contributed to the cultivated gene pool enhancement. Moreover, the African germplasm appears to possess a higher level of resistance to blast and other prevailing stresses. Identification of molecular markers/QTLs for important agro-morphological traits can accelerate the crop improvement of finger millet with more emphasis on introgression of genes into locally well-adapted elite germplasm for developing stress-tolerant, high-yielding cultivars with improved nutritional traits. Till now, little efforts have been made to enhance finger millet germplasm at DNA level. Both of conventional and contemporary breeding approaches will be useful for rapid development of high-yielding finger millet cultivars.

1.10 Foxtail and Barnyard Millets

Foxtail millet [*Setaria italica* (L.) P. Beauv.] is one of the oldest domesticated diploids ($2n=2x=18$) and cultivated in 26 countries in the world. Its genome structure is highly conserved relative to the ancestral grass lineage making it a suitable model species for genetic and molecular studies (Devos et al. 1998; Jayaraman et al. 2008). On the other hand, barnyard millet [*Echinochloa frumentacea* (Roxb.) Link.] is valued for its drought tolerance, early maturity, good yield and superior nutritional traits. Its grain contains 55 % carbohydrate, 11.2 % protein, 3.9 % fat and 13.6 % crude fibre with significant amounts of both calcium and iron (Saleh et al. 2013). In foxtail millet, different barriers to interspecific hybridization include morphological barriers and genus complexity, whereas in the case of barnyard millet, no information is available on genome organization, and some morphological and physiological barriers have been identified. Some efforts have been made towards development of interspecific hybrids in both the crops. The wild relatives of *S. italica* have been found to donate valuable traits to the cultivated gene pool of foxtail millet, e.g. resistance to certain herbicides such as triazine, sethoxydim, dinitroaniline and trifluralin (Darmency and Pernes 1985; Wang et al. 1996; Wang and Darmency 1997; Darmency et al. 2011). The information on interspecific hybridization in the case of barnyard millet is very meagre. The transfer of resistance to grain smut from Japanese to Indian barnyard millet is hampered by sterility in the hybrids between the two species (Sood et al. 2014). There is immense scope for utilization of wild species/landraces for development of stress tolerance cultivars with enhanced nutritional value through pre-breeding. Further, the availability of genome-wide sequences in foxtail millet has led to the development and identification of different molecular markers linked to desired traits and can be used for introgression of genes into cultivated varieties. In barnyard millet, comparative genomics can play an important role. The identified markers in foxtail millet can be utilized in barnyard millet for genetic diversity studies and in linkage mapping.

References

Aghaee-Sarbarzeh M, Harjit S, Dhaliwal HS (2001) A microsatellite marker linked to leaf rust resistance transferred from *Aegilops triuncialis* into hexaploid wheat. Plant Breed 120:259–261

Aung T, Thomas H, Jones I (1977) The transfer of the gene for mildew resistance from *Avena barbata* (4x) into the cultivated oat *A. sativa* by an induced translocation. Euphytica 26(3):623–632. doi:10.1007/BF00021687

Baghizadeh A, Khosravi S (2011) Genetic diversity assessment of *Aegilops* germplasm by RAPD molecular markers. Agric Biol J 2(2):197–202

Brar DS, Khush GS (1997) Alien introgression in rice. In: *Oryza*: from molecule to plant. Springer, Dordrecht, pp 35–47

Cenkci S, Yildiz M, Konuk M, Eren Y (2008) RAPD analyses of some wild *Triticum* and *Aegilops* species and wheat cultivars in Turkey. Acta Biol 50(1):35–42

Chhuneja P, Dhaliwal HS, Bains NS, Singh K (2006) *Aegilops kotschyi* and *Aegilops tauschii* as sources for higher levels of grain Iron and Zinc. Plant Breed 125:529–531

CIMMYT (2011) Wheat –global alliance for improving food security and the livelihoods of the resource-poor in the developing world. Proposal submitted by CIMMYT and ICARDA to the CGIAR Consortium Board, 30 August 2011

Colmer TD, Flowers TJ, Munns R (2006) Use of wild relatives to improve salt tolerance in wheat. J Exp Bot 57:1059–1078

Cox TS, Bender M, Picone C et al (2002) Breeding perennial grain crops. Crit Rev Plant Sci 21:59–91

Darmency H, Pernes J (1985) Use of wild *Setariaviridis* (L.) Beauv. to improve triazine resistance in cultivated *S. italica* (L.) by hybridization. Weed Res 25:175–179

Darmency H, Picard JC, Wang T (2011) Fitness costs linked to dinitroaniline resistance mutation in *Setaria*. Heredity 107:80–86

Devos KM, Wang ZM, Beales J, Sasaki T, Gale M (1998) Comparative genetic maps of foxtail millet (*Setaria italica*) and rice (*Oryza sativa*). Theor Appl Genet 96:63–68

Dweikat I (2005) A diploid, interspecific, fertile hybrid from cultivated sorghum, *Sorghum bicolor*, and the common Johnson grass weed *Sorghum halepense*. Mol Breed 16:93–101

FAO (2012) World agriculture towards 2030/2050: the 2012 revision. In: Alexandratos N, Bruinsma J (eds) ESA Working Paper No. 12–03, June, Rome (Available at http://typo3.fao.org/fileadmin/templates/esa/Global_persepctives/world_ag_2030_50_20 12_rev.pdf)

Griffiths N (1984) Studies on chromosome manipulation in Avena. (Ph.D. thesis), University of Wales, Aberystwyth, UK

Hayes J, Jones I (1966) Variation in the pathogenicity of *Erysiphe graminis* D.C. F. Sp. Avenae, and its relation to the development of mildew-resistant oat cultivars. Euphytica 15(1):80–86. doi:10.1007/BF00024081

Hoppe H, Kummer M (1991) New productive hexaploid derivatives after introgression of *Avena pilosa* features. Vortr Pflanzenzucht 20:56–61

Jayaraman A, Puranik S, Rai NK, Vidapu S, Sahu PP, Lata C, Prasad M (2008) cDNA- AFLP analysis reveals differential gene expression in response to salt stress in foxtail millet (*Setaria italica*). Mol Biotechnol 40:241–251

Khalighi M, Arzani A, Poursiahbidi MA (2008) Assessment of genetic diversity in *Triticum* spp. and *Aegilops* spp. using AFLP markers. African J Biotechnol 7(5):546–552

Khush GS (2013) Strategies for increasing the yield potential of cereals: case of rice as an example. Plant Breed 132:433–436

Mace ES, Tai S, Gildin EK et al (2013) Whole-genome sequencing reveals untapped genetic potential in Africa's indigenous cereal crop sorghum. Nat Comm 4:2320. doi:10.1038/ncomms3320

Molnár I, Gáspár L, Sárvári E, Dulai S, Hoffmann B, Molnár-Láng M, Galiba G (2004) Physiological and morphological responses to water stress in *Aegilops biuncialis* and *Triticum aestivum* genotypes with differing tolerance to drought. Funct Plant Biol 31(12):1149–1159

Monneveux P, Zaharieva M, Rekika D (2000) The utilization of *Triticum* and *Aegilops* species for the improvement of durum wheat. Options Mediterran 40:71–82

Morikawa T (1995) Transfer of mildew resistance from the wild oat Avena prostrata into the cultivated oat. bulletin of the university of osaka prefecture. Ser B Agric Life Sci 47:1–10

Piper JK, Kulakow PA (1994) Seed yield and biomass allocation in *Sorghum bicolor* and F_1 backcross generations of *S. bicolor* x *S. halepense* hybrids. Can J Bot 72:468–474

Rai KN, Appa Rao S, Reddy KN (1997) Pearl millet. In: Fuciillo D, Sears L, Stapleton P (eds) Biodiversity in trust: conservation and use of plant genetic resources in CGIAR centers. Cambridge University Press, Cambridge, pp 243–258

Rejesus RM, van Ginkel M, Smale M (1996) Wheat breeder's perspectives of genetic diversity and germplasm use. CIMMYT, Mexico, p 21

Saleh ASM, Zhang Q, Chen J, Shen Q (2013) Millet grains: nutritional quality: processing and potential health benefits. Comp Rev Food Sci Food Saf 12(3):281–295

Schneider A, Molnar I, Mornar-Lang M (2008) Utilisation of *Aegilops* species to widen the genetic diversity of cultivated wheat. Euphytica 163:1–19

Sebesta J, Harder D, Jones L, Kummer M, Clifford B, Zwatz B (1986) Pathogenicity of crown rust, stem rust and powdery mildew on oats in Europe and sources of resistance. Paper presented at the 2nd international oat conference, Martinus

Snowden JD (1936) The cultivated races of sorghum. Adlard, London, pp 1–274

Snowden JD (1955) The wild fodder sorghums of the section Eu-sorghum. J Linn Soc Lond 55:191

Sood S, Khulbe RK, Saini N, Gupta A, Agrawal PK (2014) Interspecific hybrid between *Echinochloa esculenta* (Japanese barnyard millet) and *E. frumentacea* (Indian barnyard millet) – a new avenue for genetic enhancement of barnyard millet. Elect J Plant Breed 5:248–253

Thomas H (1968) The addition of single chromosomes of avena hirtula to the cultivated hexaploid oat. *A Sativa*. Can J Genet Cytol 10(3):551–563

Thomas H (1970) Chromosome relationships between the cultivated oat *Avena sativa* (6x) and *A. ventricosa* (2x). Can J Genet Cytol 12(1):36–43

Thomas H, Powell W, Aung T (1980) Interfering with regular meiotic behaviour in *Avena sativa* as a method of incorporating the gene for mildew resistance from *A. barbata*. Euphytica 29(3):635–640

UN (2013) World population projected to reach 9.6 billion by 2050. http://www.un.org/en/development/desa/news/population/un-report-world-population-projected-to-reach-9-6-billion-by-2050.html

USDA ARS (2007) National genetic resources program. Germplasm Resources Information Network (GRIN) [Online Database]

Wang T, Darmency H (1997) Inheritance of sethoxydim resistance in foxtail millet. *Setaria italica* (L.) Beauv. Euphytica 94:69–73

Wang T, Fleury A, Ma J, Darmency H (1996) Genetic control of dinitroaniline resistance in foxtail millet (*Setaria italica*). J Hered 87:423–426

Yu J, Herrmann M (2006) Inheritance and mapping of a powdery mildew resistance gene introgressed from Avena macrostachya in cultivated oat. Theor Appl Genet 113(3):429–437

Zaharieva M, Gaulin E, Havaux M, Acevedo E, Monneveux P (2001) Drought and heat responses in the wild wheat relative *Aegilops geniculata* roth: potential interest for wheat improvement. Crop Sci 41:1321–1329

Wheat

2

W. Tadesse, S. Rajaram, F.C. Ogbonnaya,
M. Sanchez-Garcia, Q. Sohail, and M. Baum

Abstract

Wheat is one of the most important staple crops of global food and nutritional security. The demand for wheat has been increasing substantially with the increasing human population pressure. Wheat genetic resources including crop wild relatives (CWRs) have played a significant role in wheat genetic improvement by contributing useful gene sources for yield potential, wide adaptation, dwarf plant height, improved grain quality, and resistance/tolerance to major abiotic and biotic stresses. In view of the threat of genetic erosion associated to many natural and anthropogenic factors including climate change and the rapid expansion and domination of mega wheat cultivars across the major wheat agroecologies, efforts have been made to collect and conserve wheat genetic resources including wild relatives *ex situ* in several gene banks of the world. It is therefore imperative to introgress these unadapted gene sources for diversification of cultivated gene pool. Further, adoption of modern tools and techniques such as focused identification of germplasm strategy (FIGS) and effective gene introgression procedures and genomics are essential in improving genetic resource utilization and breeding efficiency.

Keywords

Wheat • Wild relatives • Base widening • Production problems

W. Tadesse (✉) • M. Sanchez-Garcia • Q. Sohail
M. Baum
International Center for Agricultural Research in the Dry Areas (ICARDA), Beirut, Lebanon
e-mail: w.tadesse@cgiar.org

S. Rajaram
International Center for Agricultural Research in the Dry Areas (ICARDA), Beirut, Lebanon

Resource Seed Mexicana, Mexico City, Mexico

F.C. Ogbonnaya
Grains Research and Development Corporation (GRDC), Barton, Australia

M. Singh, S. Kumar (eds.), *Broadening the Genetic Base of Grain Cereals*,
DOI 10.1007/978-81-322-3613-9_2

2.1 Introduction

Wheat is believed to be one of the first grains domesticated by humans since the Neolithic period ~10,000 years B.C., and to date it remains one of the most important cereal grains feeding the increasing world population (Feldman 2001). All the cultivated wheats belong to the genus *Triticum*, which in turn was divided into three major taxonomic groups, einkorn, emmer, and dinkel, by Schultz (1913). The diploid einkorn wheat, *Triticum monococcum* ($2n=2x=14$, A^mA^m), has been domesticated directly from its wild form, *T. aegilopoides* ($2n=2x=14$, A^mA^m). Similarly, the cultivated emmer wheat, *Triticum dicoccum* ($2n=2x=28$, AABB), is adapted from the wild emmer, *Triticum dicoccoides* ($2n=2x=28$, AABB), which is an allopolyploid, arose by amphiploidy between *Triticum urartu* ($2n=2x=14$, AA) and *Aegilops speltoides* ($2n=2x=14$, BB) (Johnson and Dhaliwal 1976; Feldman and Levy 2005). There is also another tetraploid wheat, *Triticum timopheevii* ($2n=4x=28$, A^tA^t GG), which, however, has been cultivated in a very limited extent. It is believed to be domesticated from wild emmer wheat, *T. dicoccoides* ssp. *armeniacum* (Feldman 2001). According to Naranjo (1990) and Jiang and Gill (1994), a species-specific translocation involving chromosomes $6A^t$, 1G, and 4G distinguishes *T. timopheevii* from *T. turgidum*, which contains a translocation involving chromosomes 4A, 5B, and 7B (Devos et al. 1995).

The evolution of the common wheat, *Triticum aestivum* ($2n=6x=42$, AABBDD), has been the subject of many investigations and intense discussions for several decades. It is now considered certain that hexaploid wheat was formed from a hybrid between the tetraploid wheat species *T. turgidum* ($2n=4x=28$, AABB) and the diploid species *Ae. tauschii* var. *strangulata* ($2n=2x=14$, DD) (McFadden and Sears 1946; Riley et al. 1958; Dvorak et al. 1998). Molecular evidences clearly showed that *T. urartu* is the A genome donor of both tetraploid and hexaploid wheats (Dvorak et al. 1993). Archaeological and botanical studies of both wild and cultivated forms have indicated that the Fertile Crescent is the birth place of cultivated wheats about 8000–10,000 years ago (Gill and Friebe 2001). Among diploid wheats, einkorn wheat (*Triticum monococcum* L.) is still cultivated to a limited extent, and its wild form, *T. aegilopoides*, is widely distributed in the Middle East (Johnson 1975; Heun et al. 1997). The tetraploid hulled wheat, *T. turgidum* ssp. *dicoccum* (emmer wheat), was one of the ancient cultivated wheats. However, it is the free-threshing macaroni or durum wheat, which arose by few mutations from primitive emmer wheats, that are widely cultivated in the present times (Gill and Friebe 2001). The farming communities of the Hauran plateau in Southern Syria, similarly to those in other parts of the Fertile Crescent, have contributed for millennia to the evolution and in situ conservation of the durum landrace Haurani. According to Vavilov (1951) and Harlan (1992), the landrace Haurani can be considered as an evolutionary link between wild emmer wheat (*T. dicoccoides*), the wild progenitor of all domesticated wheats, and, through science-based breeding, modern cultivars. The remains of the cultivated emmer (*Triticum turgidum* ssp. *dicoccum*) have been discovered at several archaeological sites in Syria dating to 7500 B.C. (Zohary and Hopf 1993; Zohary 1999). The other cultivated tetraploid wheat, *Triticum timopheevii* ($2n=4x=28$, AAGG), is of little economic importance. The wild forms of both tetraploid wheats, *Triticum turgidum* ssp. *dicoccoides* and *T. timopheevii* ssp. *armeniacum*, are widely distributed in the Fertile Crescent. *T. dicoccoides* is found exclusively in Israel, Syria, and Lebanon, while *T. armeniacum* is predominantly found in Azerbaijan and Armenia, and yet both overlap in Turkey, northern Iraq, and possibly Iran (Gill and Friebe 2001; Feldman et al. 1995; Feldman 2001). The hexaploid species, *T. aestivum* ($2n=6x=42$, AABBDD) and *T. zhukovskyi* ($2n=6x=42$, AABBGG), have no wild progenitors and are only found in cultivated forms in farmers' fields (Feldman 2001). Currently, at global level, 95 % of the cultivated wheat is the hexaploid bread wheat ($2n=6x=42$, AABBDD). Currently, wheat is the most widely grown cereal crop in the world and one of the central pillars of global food security. World trade for wheat is greater than for all other crops combined. The wheat flour is important not only for making bread, biscuit, and pastry products but also for the production of commercial starch and gluten. The raised bread loaf is possible because the wheat kernel contains gluten, an elastic form of protein that traps minute bubbles of carbon dioxide, when fermentation occurs in leavened dough, causing the dough to rise. It is the best of the cereal foods and provides more nourishment for humans than any other food source (Hanson et al. 1982).

2.2 Trends in Wheat Production

As indicated in Fig. 2.1, wheat production at global level has increased dramatically from the 1960s through the 2000s without much change in production area.

The adoption and utilization of green revolution packages such as improved high-yielding varieties and irrigation and improved moisture utilization, fertilizers, pesticides, and associated management skills with conducive policy structures and implementations have been instrumental for such drastic increase in wheat production both vertically and horizontally. According to FAO (2014), in 2012, about 651 million tons of wheat was produced on an average of 217 million ha with productivity level of 3.1 t ha^{-1}, which is a highly significant increment from the wheat production and productivity levels of about 222 million tons and 1.2 t/ha, respectively, in 1961.

Though tremendous progress was made, wheat productivity varies from country to country, and almost in all developing countries, there is a big gap between potential yield and national average yield. The 20 top wheat-producing countries in the last 10 years (2002–2012) are indicated in Table 2.1. During such periods, China produced annually about 106 million tons of wheat on a total area of 23.4 million ha at a yield level of 4.5 t/ha, while India produced 77 million tons on an average of 27.5 million ha at productivity level of 2.8 t/ha.

At regional/continental level, Asia has produced the largest amount of wheat followed by Europe and America during the periods 2002–2012 (Table 2.2). At the same period, Africa produced an average of about 22 million tons of wheat annually, which is the least as compared to the rest of the continents/regions. However, because of increasing populations and change in lifestyle and increasing trends toward urbanization, it is expected that wheat production in Africa will enhance significantly in the near future.

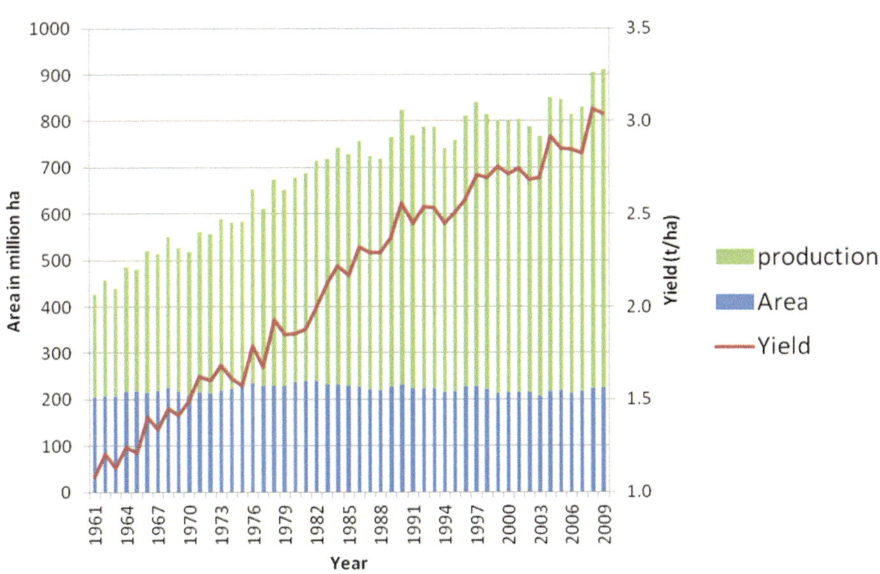

Fig. 2.1 World wheat yield, production, and area from 1961 to 2009 (FAO 2014)

Table 2.1 Wheat average production, area, and yield of the top 20 countries in the world during the last 10 years (2002–2012)

Country	Production (Mt)	Area (Mha)	Yield (t/ha)
China	105.88	23.42	4.51
India	76.94	27.47	2.79
USA	57.55	20.04	2.87
Russia	48.46	23.55	2.04
France	36.91	5.35	6.91
Canada	24.16	9.26	2.61
Germany	23.00	3.13	7.34
Pakistan	21.83	8.54	2.55
Australia	20.93	12.97	1.59
Turkey	19.84	8.50	2.34
Ukraine	17.27	5.92	2.82
UK	14.84	1.92	7.73
Kazakhstan	13.37	12.51	1.06
Iran	13.31	6.54	2.03
Argentina	13.02	4.91	2.68
Poland	8.88	2.26	3.92
Egypt	7.76	1.21	6.41
Italy	7.38	2.07	3.59
Uzbekistan	6.06	1.41	4.30
Spain	5.93	2.02	2.94

Table 2.2 Global and regional average wheat production, area, and yield in the last 10 years (2002–2012)

Region	Production (Mt)	Area (M ha)	Yield (t/ha)
Africa	21.99	9.54	2.30
Northern Africa	16.73	6.97	2.39
Central Africa	0.02	0.01	1.32
Eastern Africa	3.20	1.75	1.81
Western Africa	0.10	0.07	1.46
Southern Africa	1.94	0.73	2.71
America	106.79	38.45	2.78
Northern America	81.72	29.30	2.79
Central America	3.36	0.67	5.02
South America	21.71	8.48	2.57
Asia	279.59	98.58	2.83
Central Asia	23.40	15.45	1.51
Easter Asia	107.11	23.90	4.47
Southeast Asia	0.16	0.10	1.59
Southern Asia	118.36	46.03	2.57
Western Asia	30.56	13.10	2.34
Europe	206.65	57.25	3.60
Oceania	21.29	13.02	1.62
World	636.30	216.84	2.93

Source: FAOSTAT (2014)

2.3 Wheat Production Environments and Challenges

Wheat is a widely adapted C3 crop growing from temperate, irrigated to dry, and high-rain-fall areas and from warm, humid to dry, and cold environments. It grows suitably at the latitude range of 30° and 60°N and 27° and 40°S (Nuttonson 1955), but can grow from the Arctic Circle to the equator. In terms of altitude, it grows from the sea level to areas as high as 4570 m. amsl in Tibet (Percival 1921). The optimum growing temperature is about 25 °C, with minimum and maximum growth temperatures of 3–4 °C and 30–32 °C, respectively. It is believed that this wide adaptation and plasticity of bread wheat is associated with its large and complex genome nature. According to Rajaram et al. (1995), a mega-environment (ME) is a region with similar climate, moisture regimes, soil types, growth habit, plant performance, prevalent diseases, and insect pests. Thus, based on moisture availability, cropping systems, and temperature regimes, the wheat production area in the world can be classified under five distinct large mega-environments, namely, favorable irrigated and high-rainfall spring wheat environment (E1), semiarid (Mediterranean rain-fed) spring wheat environment (E2), favorable irrigated winter/facultative wheat environment (E3), semiarid rain-fed winter/facultative wheat environment (E4), and high-latitude (spring planted) spring wheat environment (E5). Because of these complexities, it is prudent that international breeding programs take the mega-environment approach and use the shuttle breeding and multilocation yield trial strategy to develop high-yielding and widely adapted genotypes for each mega-environment. Most of the wheat produced in the developed world is rain-fed, while in the developing world, especially in the large producers such as India and China, more than half of the wheat area is in irrigated condition. Wheat productivity varies not only between irrigated and rain-fed production systems but also between countries applying similar agronomic practices. Compared to the global average, the difference in productivity of wheat in the developing world in general and in the Central and West Asia and North Africa (CWANA) region in particular is due to abiotic stresses (drought, cold, heat, salinity) and biotic stresses (yellow rust, leaf rust, stem rust, root rots, Russian wheat aphid, Barley yellow dwarf virus, sunn pest, and Hessian fly). Principally, drought and yellow rust are the most important wheat yield limiting factors in CWANA. With the current climate change, it is anticipated that new pests and diseases will emerge as already exemplified in the recent epidemics of yellow rust across the CWANA region and Ug99 epidemic in East African countries (Solh et al. 2012). The effect of climate change is also evident on the quality of wheat as increased heat results in shriveled wheat grains. In general, after the quantum leap of the green revolution, wheat yields have been rising by only 1.1 % per year, a level that falls far short of the demand of a population that is growing 1.5 % or more annually. According to some estimates, the global wheat production must increase at least by 1.6 % annually to meet a projected yearly wheat demand of 760 million tons by 2020. In the year 2050, the world population is estimated to be nine billion, and the demand for wheat reaches more than 900 million tons. Meeting this demand poses considerable challenges with the apparent effect of climate change, increasing drought/water shortage, soil degradation, and reduced supply and increasing cost of fertilizers, increasing demand for biofuel, and emergence of new virulent diseases and pests. Offsetting these challenges requires understanding of the drivers of past trends and future changes in wheat production and designing an effective strategy with the application of new technologies, enabling policies and establishment of new networks and collaborations.

2.4 Wheat Breeding: The Case at ICARDA

The International Center for Agricultural Research in the Dry Areas (ICARDA) was established at Aleppo, Syria, in 1977 to undertake multidisciplinary agricultural research in order to generate agricultural technologies for the non-tropical dry areas of the world. The center started its wheat (spring bread wheat, facultative/winter bread wheat, and durum wheat) improvement program since its inception targeting West Asia and North Africa (WANA) region and, more recently, including Central Asia and Caucasus (CAC) countries, collectively known as the CWANA region. The spring bread wheat and durum wheat improvement program has been managed as a joint activity with the International Center for Maize and Wheat Improvement (CIMMYT) as the CIMMYT/ICARDA wheat improvement program from the 1970s until 2003 with major emphasis on the semiarid rain-fed environments of the CWANA region. In 2005, a joint program was reestablished as the ICARDA/CIMMYT Wheat Improvement Program (ICWIP) with ICARDA taking the lead for the CWANA region. ICARDA is part of the International Winter Wheat Improvement Program (IWWIP) which is managed by Turkey-CIMMYT-ICARDA (TCI) (Tadesse et al. 2013). ICARDA is also part of the International Wheat Improvement Network (IWIN), which is an alliance of international research centers (CIMMYT and ICARDA), national agricultural research systems (NARSs), and advanced research institutes, which has been established with a general objective of enhancing the productivity, yield stability, and end-use quality of wheat production at global level with major emphasis in the developing world (Baum et al. 2013). IWIN has contributed significantly in the development of germplasm that has made major contributions to the green revolution and to improving food security and farmers' livelihoods in many developing countries (Dixon et al. 2009).

In this specific instance, it is noteworthy to mention Dr. Norman Borlaug, who has developed the semidwarf input-responsive wheat cultivars for the green revolution in Mexico, India, Pakistan, and Turkey and many other developing countries. Starting from 2011, CIMMYT and ICARDA are implementing the WHEAT Consortium Research Program (CRP 3.1) which is part of a concerted effort of the Consultative Group on International Agricultural Research (CGIAR) to implement a new, results-oriented strategy through a series of CRPs that fully exploit the potential of international agricultural research for development to enhance global food security and environmental sustainability. WHEAT draws on and potentiates the capacities and commitment of two leading international centers (CIMMYT and ICARDA), in partnership with farming communities, national research systems, advanced research institutes, private companies, policymakers, and diverse development organizations.

2.5 Wheat Breeding Approaches

The wheat breeding program at ICARDA applies both conventional and molecular breeding approaches and techniques in order to develop high-yielding and widely adapted germplasm with resistance/tolerance to the major biotic and abiotic constraints prevailing in the developing world. Some of these strategies and techniques include classification of mega-environments (MEs) and assembling of targeted crossing blocks, shuttle breeding, utilization of doubled haploids (DH) and marker-assisted selection (MAS), key location yield trials, distribution of germplasm to NARS through international nurseries, and partnership and capacity building of NARS (Rajaram et al. 1995; Ortiz Ferrara et al. 1987; van Ginkel et al. 2002; Tadesse et al. 2012a). As water is becoming scarce even in the irrigated areas, IWIN's germplasm development approach is to identify genotypes with disease resistance, high yield potential, and water use efficiency so that wheat genotypes targeted for irrigated areas can cope with temporary drought periods. Similarly, this approach enables to maximize grain yield during drought and good seasons, respectively, for the rain-fed production system.

2.5.1 Assembling of Crossing Blocks Targeting Mega-Environments

Principally, high-yielding and adapted CWANA hallmark wheat varieties representing each ME, synthetic wheats, and elite lines from ICARDA are used as parents. The parents are characterized for adaptation zones, quality, and resistance/tolerance to biotic and abiotic stresses, photoperiod sensitivity, dwarfing genes, vernalization requirements, and other regionally important traits both at the phenotypic and molecular levels. Annually, 1000–2000 crosses (simple, top, and back crosses), representing the major MEs, are made for spring bread wheat, winter bread wheat, and durum wheat both under plastic house and field conditions. It is anticipated that germplasm developed for a given ME withstands the major stresses present within that ME, but not always the significant secondary stresses (Rajaram et al. 1995). It is however important to include genetic diversity of the germplasm for additional traits of importance within that particular mega-environment. Such diversity is also important to counter the threat of genetic vulnerability. The national programs are basically responsible to address these specific niches of environments within the given mega-environment.

2.5.2 Evaluation of Segregating Populations Through Shuttle Breeding

The shuttle breeding methodology proposed 50 years ago and implemented by Borlaug (1968) has been responsible for the production of photoperiod-insensitive and otherwise widely adapted germplasm. The shuttle breeding basically involves evaluation of segregating generations in contrasting locations in terms of latitude, altitude, and rainfall during the summer and winter cycles in order to shorten the breeding cycle. The methodology has been also very effective to introduce and select the *Ppd1* and *Ppd2* photoperiod-insensitive genes, which along with the dwarfing genes, *Rht1* and *Rht2*, resulted in widely adapted, high-yielding, lodging-tolerant, and input-responsive semidwarf wheat varieties. The shuttle breeding methodology can be applied using the different selection methods. In the past, different selection methods including pedigree, single seed descent, and modified bulk have been employed for evaluation of segregating generations (Ortiz Ferrara et al. 1987; van Ginkel et al. 2002). Currently, as indicated in Fig. 2.2, segregating populations from F_2 to F_6 are selected using modified pedigree-bulk selection method whereby on an average 2000 F_2 plants are space planted and individual plants are selected and harvested separately and then planted in the F_3–F_4 as entries with the selected plants in each entry harvested in bulk. In the F_5, selected plants or heads are harvested separately and grown in the F_6 as individual plant head rows and with selected plants harvested in bulk. The crosses and populations targeting irrigated and rain-fed conditions are handled jointly until F_6 under optimum irrigation (favorable environment) in order for resistant plants to be easily identified under heavy artificial epidemics. For spring bread wheat, a shuttle breeding approach involving the winter-summer cycle at Terbol station in Lebanon, winter cycle at the Marchouch station in Morocco and the Sids station in Egypt, and summer cycle at the Kulumsa station in Ethiopia is being followed (Fig. 2.2).

Such shuttle enables the identification of disease-resistant, photoperiod-insensitive, high-yielding, and widely adapted genotypes. It also shortens the breeding cycle by half as two generations can be grown per year. In the case of winter wheat, advancement of two field-planted generations per year is very difficult, if not impossible, and hence evaluation of segregating generations has been undertaken both in Lebanon and Turkey only during the winter cycle.

2.5.3 Doubled Haploid

In vitro haploid production followed by chromosome doubling reduces the time needed for the production of completely homozygous lines into a single generation and increases the precision and efficiency of selection process in wheat breeding (Tadesse et al. 2012b). It also facilitates the detection of linkage and gene interactions and estimation of genetic variance and the number of genes for quantitative characteristics; produces genetic translocations, substitutions, and chromosome addition lines; and facilitates genetic transformation and mutation studies. Wheat varieties developed from doubled haploids using

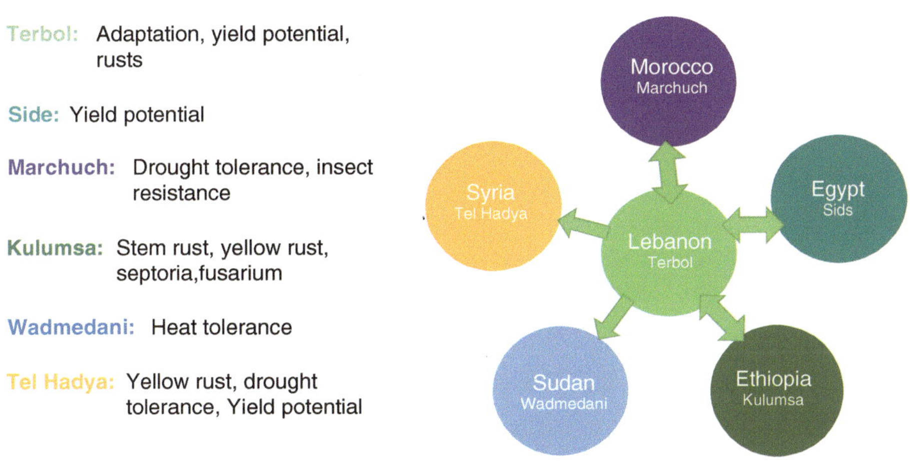

Fig. 2.2 Key locations and shuttle breeding sites of ICARDA's wheat breeding program; (Note: the *arrow* directions indicate the direction of shipment of subsequent segregating populations and fixed germplasm)

anther culture or maize induction systems have been released for cultivation in both developed and developing countries. At ICARDA, DH lines are produced using anther-isolated microspore culture. Between 1000 and 5000 DH lines are produced each year. We undertake DH production from F_1 plants originating from elite × elite crosses instead of F_2 plants to increase the gain in time rather than the genetic gain which could be obtained by having one more generation of recombination. This is specifically important for winter wheat as only one field cycle of generation advancement per year is possible compared to two generations per year for spring wheat using shuttle breeding under diverse irrigated and rainfed schemes in winter and summer seasons or using the single seed descent (SSD) method under greenhouse conditions.

2.5.4 Marker-Assisted Selection

The wheat breeding program at ICARDA also uses physiological and molecular screening techniques in order to increase the rates of genetic gain through (a) strategic trait crossing to combine complementary traits in the progeny, (b) high-throughput phenotyping to enhance the accumulation of desirable alleles in intermediate generations, and (c) exploration of genetic resources to broaden the genetic base for hybridization (Reynolds and Tuberosa 2008). MAS using recommended diagnostic markers is used to characterize new parental materials for disease resistance (stripe rust, leaf rust, stem rust, nematodes), insect resistance (Hessian fly and Russian wheat aphid), phenological traits such as photoperiodism (PPd), vernalization requirement (Vrn), plant height (Rht), grain hardness, and other desired genes (Tadesse et al. 2012a). Diagnostic markers are also used for gene pyramiding in the F_2, F_1top, and BC_1F_1 populations (William et al. 2007). Association mapping studies using elite bread wheat genotypes from ICARDA's wheat breeding program has led to the identification of novel sources of genes for drought tolerance, heat tolerance, and stripe rust resistance (Tadesse et al. 2014). Molecular screening and rigorous field screening of spring bread wheat genotypes at the stem rust-prone locations of Njoro, Kenya, and Debrezeit, Ethiopia, have led to the identification of Ug99-resistant, widely adapted, and high-yielding genotypes. These genotypes have been distributed to the CWANA NARS through ICARDA's international nursery system. Recently, three Ug99-resistant bread wheat varieties have been released in Ethiopia from this program.

2.5.5 Multilocation Yield Trials

A total of 2000–3000 F_7 generation genotypes selected from the F_6 nursery are tested annually in preliminary yield trials (PYTs) using nonreplicated augmented designs at the Terbol and Kheferdan (Lebanon), Sids (Egypt), Marchouch (Morocco), Wad Madani (Sudan), and Kulumsa (Ethiopia) research stations. The Tel Hadya and Breda sites in Syria were also used until 2012, which were then stopped due to security concerns. On an annual basis, about 400–500 lines are selected from the PYTs based on yield potential data from the Sids and Terbol stations; disease (rusts, fusarium, septoria, and tan spot) resistance data from the Kulumsa station; heat tolerance data from the Wad Madani, Terbol, and Marchouch (late planting) stations; and drought tolerance data from Kheferdan and Marchouch stations, and insect resistance (Hessian fly and sunn pest) data are yield tested in advanced yield trials (AYTs) using alpha-lattice designs in two replications. They are grown at the same locations as the PYT, both under irrigation and rainfed conditions, to screen for phenological traits, disease and insect pest resistance, yield and yield components, and grain quality traits. Such key location yield trial testing enables to identify germplasm with wide adaptation targeting the major mega-environments.

2.6 Germplasm Distribution and Capacity Development

Based on request, different forms of germplasm such as genetic stocks for crossing blocks, segregating generations, and fixed genotypes have been distributed to the CWANA NARS and others. The genetic stocks and the segregating generations are sent with the objective of decentralizing the breeding program and creating genetic variability to be selected under alternative conditions, while the fixed materials are sent with the objective of being considered for release of adapted genotypes as varieties. The experience obtained so far is that most NARSs have released more varieties from directly introduced, essentially finished material than from early segregating populations. Research infrastructure, budget availability, and overall strength of NARSs are the main factors accounting for these differences (Byerlee and Moya 1993). At present, unless specifically requested, most of the germplasm distribution to NARS consists of essentially finished materials. Since 2005, the spring bread wheat program has distributed germplasms for heat tolerance, Hessian fly resistance, Ug99 resistance, drought tolerance, and high yield potential for the CWANA region and beyond. On an annual basis, the spring bread wheat composes and distributes the following nurseries and yield trials:

1. Spring bread wheat observation nursery for CWANA (CWANA SBWON)
2. Spring bread wheat observation nursery for heat tolerance (SBWON HT)
3. Spring bread wheat yield trial for irrigated environments (CWANA ISBWYT)
4. Spring bread wheat yield trial for dryland rain-fed environments (CWANA DSBWYT)
5. Elite spring bread wheat yield trial (ESBWYT)

Each of these observation nurseries is composed of 120–150 genotypes targeted for the respective environments, while the respective yield trials consist of 24 genotypes with the 24th slot reserved for the local check commercial varieties. For winter bread wheat, selected genotypes from ICARDA are included in the Facultative and Winter Wheat Observation Nursery (FAWWON) and International Winter Wheat Yield Trial (IWWYT) which are distributed to more than 50 countries on an annual basis by IWWIP from Konya, Turkey. In general, the observation nurseries are targeted to identify adapted varieties for release by the respective NARSs, while the yield trials are primarily targeted for identifying widely adapted genotypes to be used in future crosses at ICARDA. All nurseries are distributed on an annual basis, based on requests from the national research programs across CWANA and other regions of the world.

Germplasm distribution to the NARS by its own will not bring the expected results of releasing and adoption of improved varieties unless it is handled and managed by trained and qualified breeders. The wheat breeding program at ICARDA organizes both short- and long-term training courses in wheat breeding and has trained hundreds of wheat breeders from all over the wheat-growing NARS. This has helped in the promotion of the fundamental concepts of international wheat breeding and development and release of many new and superior wheat varieties. In the last decade, many new technologies such as molecular markers have been developed and utilized in the public and private crop breeding programs of the developed world. Such rapid advances in biotechnology and molecular genetics not only provide unprecedented opportunities to enhance breeding efficiency but also create new challenges in training breeders with skills integrating both conventional and molecular breeding approaches and techniques. To this end, the wheat breeding program at ICARDA has revised its training program to accommodate classical and molecular approaches for wheat improvement. Junior and mid-career scientists experience a comprehensive hands-on course on breeding for durable disease resistance, high-yield potential and stability, drought and heat tolerance, end-use seed quality, and seed health issues using conventional and molecular tools. ICARDA believes that well-trained young scientists are the key to the future of national research programs in addition to improved wheat genetic materials.

2.7 Variety Release, Adoption, and Impacts

The success of wheat improvement within the CGIAR has been remarkable, and today more than 70 % of all spring wheat cultivars grown in developing countries are CIMMYT and ICARDA derived, reaching 90 % in South Asia, parts of West Asia, and North Africa (Byerlee and Moya 1993; Lantican et al. 2005). The impact of WHEAT has been witnessed not only by farmers, governments, policymakers, and professionals but also by donors such as the World Bank. According to the World Bank (2008), for no other major crop is the percentage of improved cultivars in farmers' fields in developing countries higher than for wheat. From the CIMMYT/ICARDA wheat breeding, it has been reported that more than 1500 wheat varieties have been released during the period of 1966–1990 with an average of 65 varieties released annually. More than 80 % of the varieties released are spring bread wheats, which account for 77 % of the wheat area in the developing world outside China (Byerlee and Moya 1993). Among the many crosses developed to date at CIMMYT/ICARDA, the VEERY cross (KVZ/BUHO//KAL/BB) and its derivatives have been by far the most successful cross enabling the release of at least 65 varieties in more than 30 countries. Among the VEERYs, VEERY 5 (CM33027-F-15M-500Y-0M) was the most popular and released in many countries with different local names such as SERI 82 in Lebanon, Turkey, and Mexico; Dashen in Ethiopia; Tilila in Morocco; Aziz in Yemen; Pirsabak 85 in Pakistan; Loerie in Zambia; MACS2496 in India; SCW101 in Zimbabwe; Rassol in Iran; GIZA 164 in Egypt; SASARAIB in Sudan; and TAUSI in Tanzania. After the VEERYs, Cham-6 (W3918A/JUP), also called Neser, was released and grown in Syria, Lebanon, Jordan, Iraq, and Algeria. Similarly, Kauz (JUP/BJY//URES) has been released in many countries with different local names such as Cham-8 in Syria, Tanur in Lebanon, Mehdia in Morocco, Atrak in Iran, Bacanora in Mexico, and WH 542 in India and with other different names in many other countries. After Kauz, many successful varieties have been originated from the Attila cross (ND/VG9144//KAL/BB/3/YACO/4/VEE#5) and grown in many countries as mega-cultivars such as Kubsa in Ethiopia; Imam in Sudan; Utique 96 in Tunisia; PBW343 in India; Chamran, Gaher, and Shiroodi in Iran; Ziyabey 98 in Jordan; and MH-97 in Pakistan and with different names in many other countries (Tadesse et al. 2010). Recently, from the CIMMYT/ICARDA wheat breeding, Ug99-resistant wheat varieties have been released in many countries (Table 2.3).

Adoption studies of the modern varieties (MVs) originated from the CIMMYT/ICARDA have indicated that in the 1990s, MVs covered close to 50 million hectares, or 70 % of the wheat area in the developing world, excluding China (Byerlee and Moya 1993). According to their report, spring bread wheat varieties have been the most successful with MVs occupying an estimated 85 % of spring bread wheat area and account for 93 % of production. Since spring bread wheat dominates developing country wheat production (about 70 % of the total), success in spring bread wheat accounts for the overall large area sown to semidwarfs. Though estimating the economic impact of the CIMMYT/ICARDA international wheat breeding program is very difficult given the diversity of environments and number of countries and research involved, Byerlee and Moya (1993) reported that the adoption of MVs of spring bread wheat in 1977–1990 resulted in about 15.5 million tons of additional wheat production in 1990, valued at about US$ three billion. For the spring wheat areas under consideration, this amounts to a production increase of 16 % (an increase of about 1.1 % annually over the period). Nearly 80 % of all durum wheat cultivars in developing countries are CIMMYT/ICARDA selections.

An adoption study conducted in Syria by the Farm Resource Management Program (FRMP) of ICARDA during 1991 showed that modern high-yielding varieties (HYVs) such as Bohouth 2, Bohouth 4, and Bohouth 6 and Cham 2, Cham 4, Cham 6, Cham 8, and Cham 10 account for 87 % of the area planted and were grown by 86 % of the farmers surveyed (Tutwiler and Mazid 1991). The area planted by Mexipak 65 has dramatically reduced in the year 2000, and since 2005 Cham 8 covers more than 50 % of the spring bread wheat production area in Syria. Semidwarf durum wheats became available in the early 1970s, and now over half of the durum wheat

Table 2.3 List of Ug99-resistant bread wheat varieties released recently by the CWANA NARS from ICARDA's wheat breeding program

Variety name	Type	Cross/pedigree	Selection history	Year of release	Country
SIDS-13	SBW	KAUZ//TSI/SNB	ICW94-0375-4AP-2AP-030AP-0APS-3AP-0APS-050AP-0AP	2010	Egypt
EGSeed-7	SBW	KAUZ//TSI/SNB	ICW94-0375-5AP-1AP-030AP-0APS-11AP-0APS-050AP-0AP	2010	Egypt
TIJABAN-2010	SBW	W3918/JUP	CM40096-8M-7Y-0M-0AP-0JOR-0PAK	2010	Pakistan
HOGGANA	SBW	PYN/BAU//MILAN (= ETBW 5780	CMSW94WM00188S-0300M-0100Y-0100M-15Y-8M-0Y-0IAP-0QTAP-0YT	2011	Ethiopia
SHORIMA	SBW	UTQE96/3/PYN/BAU//MILAN	ICW02-00330-4AP/0TS-0AP-030AP-10KUL-030KUL-0AP/0KUL-0DZ/0AP	2011	Ethiopia
HULUKA	SBW	UTQE96/3/PYN/BAU//MILAN	ICW02-00330-9AP/0TS-0AP-030AP-1KUL-030KUL-0AP/0KUL-0DZ/0AP	2011	Ethiopia
KARIM	SBW	T.AEST/SPRW//CA8055/3/BACANORA86	ICW92-0477-1AP-1AP-4AP-1AP-0AP	2011	Iran
GOUMRIA-3	SBW	VEE#7/KAUZ	ICW94-0029-0L-1AP-1AP-7AP-0APS-0AP-0SD	2013	Sudan
NORMAN	FWW	OR F1.158/FDL//BLO/3/SHI4414/CROW	ICWH860291-3AP-1AP-0AP-1AP-0AP	2007	Tajikistan
GIZIL BUGDA	FWW	SAULESKU41/SADOVO1	TCI950295-3AP-0AP-0E-1YE-0YE-1YM-0YM	2009	Azerbaijan
MTSHETSKAYA 1	FWW	TAST/SPRW//ZAR	ICWH840048		Georgia
CHUMON	FWW	CADET/6/YUMAI13/5/NAI60/3/14.53/ODIN//CI13441/CANON	TCI-02-417-0AP-0AP-1AP-0AP-5A-0AP	2012	Tajikistan

area is sown to MVs. MVs of durum wheat such as Bohouth 1, Cham 1, Cham 3, and Cham 5, which were originated from the CIMMYT/ICARDA wheat breeding program, have been released and widely grown in Syria (Nachit 1992). The utilization and adoption of such MVs of bread and durum wheat coupled by supplementary irrigation and other inputs such as fertilizers and herbicides by the Syrian farmers have increased the wheat production significantly without a change in the area of wheat production as indicated in Fig. 2.3.

International breeding efforts focusing only on winter wheat started since 1986 through the establishment of IWWIP. To date more than 55 winter/facultative wheat varieties of IWWIP origin have been released in 10 countries of Central and West Asia including Afghanistan, Iran, and Turkey (Tadesse et al. 2013). These new varieties have broad genetic diversity as they are developed from parents of diverse sources originated from ICARDA, CIMMYT, and a wide range of genetically unrelated winter wheats from Turkey, Iran, Russia, Ukraine, Romania, Bulgaria, Hungary, and the USA. Some lines such as OK82282//BOW/NKT and YMH/TOB//MCD/3/LIRA have been identified and released under different names in different countries indicating their broad adaptation. The former (?) is released in Afghanistan and Kyrgyzstan, while the latter (Kinaci 97) has been released in Turkey, Afghanistan, and Uzbekistan. However, MVs of winter wheat still cover a small area. Old cultivars such as Bezostaya and Gereek are dominantly cultivated in Turkey.

2.8 Breeding Progress

Development and identification of high-yielding genotypes with wide adaptation and resistance to biotic and abiotic stresses remain the top priorities of the wheat breeding programs. There are different approaches of determining the breeding progress or the rate of genetic gain for grain yield and other traits. Some studies use yield of histori-

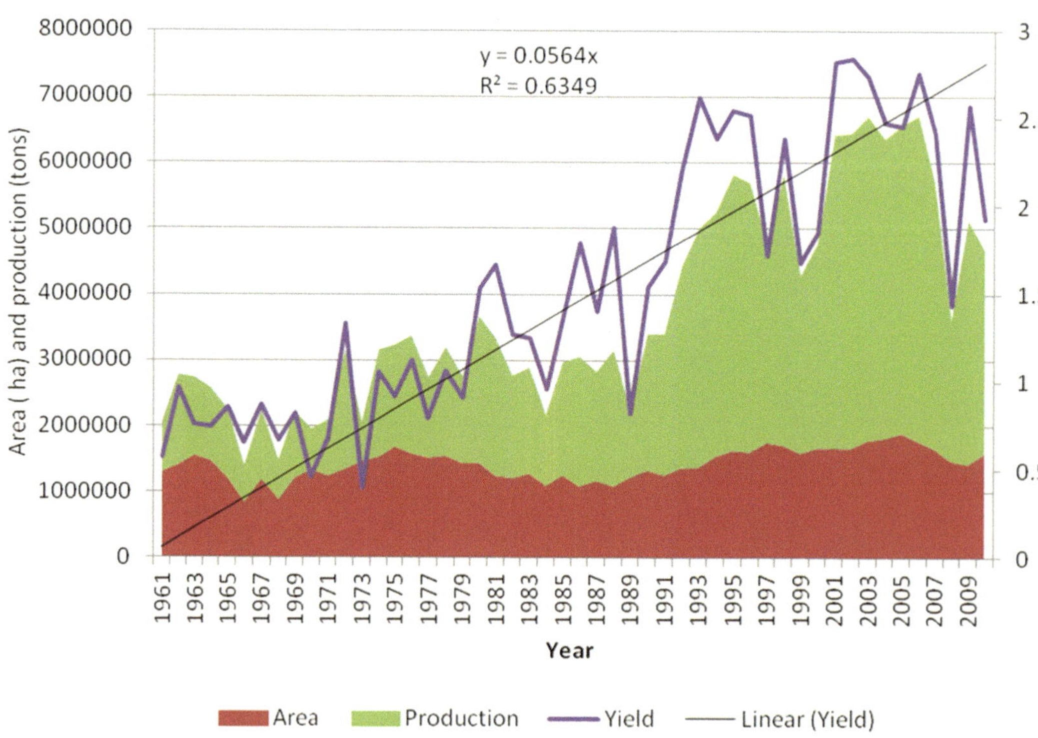

Fig. 2.3 Area, production, and yield of wheat in Syria, 1961–2010 (FAOSTAT 2014)

cal genotypes grown in the same environment, while others have used mean yield to examine progress over time in highly productive environments. Trethowan et al. (2002) have used regression analysis using the mean of the five highest-yielding entries expressed as percent of the trial mean across years to determine the rate of breeding progress in elite spring wheat yield trial (ESWYT) and semiarid wheat yield trial (SAWYT). Tadesse et al. (2010) have used success rate analysis of best lines for the high-rainfall wheat yield trials (HRWYT) of CIMMYT to demonstrate yield gains over years or trials. Genetic gain studies for the CIMMYT/ ICARDA wheat breeding program have shown continuous progress in yield and other traits (Sayre et al. 1997; Trethowan et al. 2002; Sharma et al. 2012; Tadesse et al. 2010). Recently, Tadesse et al. (2013) have determined the breeding progress for the International Winter Wheat Improvement Program (IWWIP) and reported that the grain yield of the best line (BL) increased at a rate of 110 kg/ha/year (R^2=0.66; P=0.001), while the trial mean (TM) increased at a rate of 91.9 kg/ha/year (R^2=0.53; P=0.007) indicating a continuous yield improvement at IWWIP.

In addition to grain yield, significant progress has been made in developing resistant wheat germplasm to diseases and pests ensuring that developing and deploying genetically resistant varieties adapted to target growing environments are the best economical and environmentally friendly strategy for controlling rust diseases of wheat particularly for resource poor farmers. However, because of the coevolution of the host and pathogen, the deployment of individual resistance genes leads to the emergence of new virulent pathogen mutants, and hence the "boom and bust cycle" of cultivar performance continues. Recently, a new stem rust race Ug99 (TTKS) has been first detected in Uganda in 1999 and then spread to Kenya, Ethiopia, Yemen, Sudan, and Iran and became a global threat to the wheat industry of the world for the very fact that it overcomes many of the known and most common stem rust resistance genes such as *Sr31*, *Sr24*, and *Sr36* (Jin and Singh 2006). Similarly, the breakdown of yellow rust resistance genes *Yr9* in cultivars derived from "VEERY" in the 1980s and *Yr27* in 2000s in major mega-cultivars derived from "Attila" cross such as PBW343 (India), Inquilab-91 (Pakistan), Kubsa (Ethiopia), and others such as Achtar in Morocco, Hidab in Algeria, and many other cultivars in the CWANA region (Solh et al. 2012) has caused significant wheat production loss. Through a coordinated international effort, many wheat varieties resistant to Ug99 and yellow rust have been released which replaced the susceptible cultivars. *Septoria tritici* blotch, caused by *Septoria tritici* (Rob. ex Desm) [sexual stage *Mycosphaerella graminicola* (Fuckel) Schroeter], is an increasingly important wheat disease all over the world. In the CWANA region, the disease became more important especially in the high-rainfall areas. The semidwarf MVs were in general susceptible to *Septoria*, which led to their low adoption in the high-rainfall areas. With this understanding, in the 1970s and 1980s, the wheat breeding at CIMMYT incorporated sources of resistance from South America and from winter wheat into many spring wheat lines, especially those targeted for high-rainfall zones (van Ginkel et al. 2002). Without this resistance, losses to *Septoria* diseases in these environments may average as much as 10–20 % (Brennan and Murray 1988). The wheat breeding program at ICARDA routinely undertakes evaluation of genotypes for major diseases and insect pests in selected hot spot locations of the region. As indicated earlier, the Kulumsa station in Ethiopia is the key location for screening the major wheat diseases including *Septoria*. The Marchouch station is also an important location for screening of germplasm against *Septoria* and enables to detect regional race variability. With such key location testing approach, progress has been made to identify high-yielding genotypes with high levels of resistance to *Septoria* that are effective across CWANA.

In most developing countries, apart from grain yield and disease resistance, grain quality was not a strong criterion of variety selection. However, things have changed through time, and some developing NARSs are critically looking for better quality varieties suiting for preparation of different end products. Varieties such as Bezostaya, Achtar, VEERY, HD1220, and Pavon-76 are known for their excellent grain quality. These varieties are still dominantly grown in some countries not only because of their wide adaptation, high yield potential, and stability but also because of their high protein content and quality. With this understanding, the wheat breeding at CIMMYT and ICARDA undertakes evaluation of germplasm for quality traits following standard grain quality procedures. Most of the currently available elite genotypes for both irrigated and rain-fed environments are excellent in quality with protein levels of 12–16 %. Most of these genotypes have the 5 + 10 (Glu-D1), 7 + 8 (Glu-B1), and 2* (Glu-A1) alleles. These alleles, especially the 5 + 10 Glu-D1 allele, have been reported to be highly correlated with protein quality and are being used intensively as a selection criterion in wheat breeding for improving end-use quality.

2.9 Future Directions and Prospects

According to some estimates (Fischer 2009), the global wheat production must increase at least by 1.6 % annually during 2005–2020 to meet a projected wheat demand of 760 million tons by 2020. In the year 2050, the world population is estimated to be nine billion (Alexandratos 2009), and the demand for wheat reaches more than 900 million tons. Fulfilling this increasing demand for wheat is very challenging with the current scenario of climate change (IPCC 2007; Battisti and Naylor 2009), increasing drought/water shortage, soil degradation, reduced supply and increasing cost of fertilizers, increasing demand for biofuel, and emergence of new virulent diseases and pests. Offsetting these challenges requires understanding of the drivers of past trends and future changes in wheat production and designing an effective research strategy for gene mining, introgression, and deployment with the application of new technologies and tools.

2.9.1 Strengthening of Gene Mining and Deployment Activities

Located in the heart of the Fertile Crescent, ICARDA houses more than 41,000 wheat accessions including rich collections of landraces, primitive wheat, *Aegilops*, and wild *Triticum* species. Synthetic wheats have been developed by crossing different tetraploid ancestors of wheat such as *Triticum dicoccum*, *Triticum dicoccoides*, or *Triticum carthlicum* ($2n = 4x = 28$, AABB) (Lange and Jochemsen 1992; Niwa et al. 2010). The most common way of producing hexaploid synthetic wheat which is also called primary synthetics ($2n = 6x = 42$, AABBDD) is by crossing the *T. turgidum* ssp. *durum* ($2n = 4x = 28$, AABB) with *Ae. tauschii* ($2n$ $2x = 14$, DD) (Mujeeb-Kazi and Hettel 1995). Although most of the synthetic wheat lines are agronomically inferior to modern cultivars, they possess superior genes for both monogenic and polygenic traits. These germplasm are novel sources of resistance genes against biotic and abiotic stresses for wheat production (van Ginkel and Ogbonnaya 2007). However, despite the existence of this promising resource of new genes, there has been limited deployment and/or effective use in cultivated bread wheat mainly due to the high cost of screening of such huge number of accessions and the potential simultaneous transfer of deleterious genes.

The development of focused identification of germplasm strategy (FIGS) and the availability of new molecular tools such as genotyping by sequencing (GBS) would hasten and facilitate the characterization and mining of novel genes and alleles effectively and rapidly from such gene bank accessions (Ogbonnaya et al. 2013). It is also important to apply modern tools including genome-wide selection and advanced statistical analysis of multilocation evaluation data for wheat breeding in order to allow faster integration of desirable traits and improve breeding efficiency, especially for complex traits such as grain yield under optimum, drought, and heat conditions (Ortiz Ferrara et al. 1987; Braun et al.

2010). Synthetic hexaploid wheat is now being used to transfer genes from wild relatives to cultivated wheat for resistance to biotic and abiotic stresses and yield potential per se (Ma et al. 1995; Ogbonnaya et al. 2013; Tadesse et al. 2006; Sohail et al. 2011). According to Ogbonnaya et al. (2013), synthetic wheats are being utilized as parents both at ICARDA and CIMMYT breeding extensively. It is anticipated that the availability of modern genetic tools such as genotype by sequencing will greatly enhance the utilization of synthetic wheats in the wheat improvement at global level.

2.9.2 Genetically Modified Wheat

Unlike many other major crops (notably maize, soybean, cotton, and canola) that now account for more than 100 million ha of commercial transgenic crop production across many countries, there is no genetically modified (GM) wheat production in any country. The GM approach would be particularly valuable for traits for which there is limited or no genetic variation available within the *Triticum* species. This would include herbicide resistance, *Fusarium* resistance, novel quality traits, and technologies for creating hybrid cultivars. In addition, GM technologies hold promise for enhancing drought and heat tolerance, as well as disease and pest resistance. Major efforts are needed to break yield barrier in wheat to increase wheat yield potential by 50 % in order to cope with its growing demand. Increasing the radiation use efficiency of wheat through modification of key enzymes (e.g., RuBisCO) and biochemical pathways to increase photosynthesis, ear size, and lodging resistance is a key area of wheat research through integration of physiological and molecular breeding methodologies to increase wheat yield potential. Further increase in yield potential would be achieved through the development of hybrid wheat systems based on native and transgenic interventions in collaborative approach, leveraging private sector technologies for the benefit of partners and stakeholders in the developing world. There have been substantial commercial concerns regarding the effect of consumer resistance to GM products in some countries. However, more recently, there has been a resurgence of interest in GM wheat, and it is very likely that GM wheat cultivars will be released within the next 10 years.

2.9.3 Hybrid Wheat

In the past, hybrid wheat production has been discouraged for its poor economic feasibility emanated from the very limited heterotic advantage (10 %); lack of clear advantage in terms of agronomic, disease, and quality traits; and expensive seed production costs and due to the argument that heterosis can be fixed and consequently hybrids would have no biological advantage over inbred lines (Pickett and Galwey 1997). However, through time it became apparent that hybrid wheats provide higher grain yield, higher thousand grain weight, more tillers, higher biomass, deeper roots, and better resistance to both biotic and abiotic stresses as compared to their parents. Currently, hybrid wheat is dominantly produced in Europe using the chemical hybridization agent (CHA) CROISOR® 100. Other countries such as the USA, China, India, and Australia use the cytoplasmic male sterility (CMS) method. It is anticipated that the application of biotechnological methods will enable to capture increased heterosis by direct selection of favorable alleles and development of new genetically based systems to control male sterility.

2.9.4 Establishment of Efficient Networking

The International Wheat Improvement Network (IWIN) coordinated by CIMMYT and ICARDA has been the most successful and efficient network for making available and widespread distribution of new wheat genotypes globally (Payne 2004; Reynolds and Borlaug 2006; Dixon et al. 2009; Byerlee and Dubin 2010). Such a network needs to be strengthened through the establishment of other networks and collaborations in order to develop, disseminate, and market more productive, stress-tolerant, and nutritive wheat varieties and to perfect and promote production practices based on the principles of conservation agriculture that boost yields while conserving or enhancing critical resources like soil and water. During the main green revolution in Asia between 1961 and 1981, one-third of the increase in food production was attributable to crop improvement; the other two-thirds arose from a variety of crop management and institutional factors, in particular increased fertilizer use and better weed control, water management, and market access (Evenson and Gollin 2003). Furthermore, there is scope for exploiting the positive interactions between genotype and cropping systems management. One of the proven crop management routes for improving the productivity of sustainable agriculture is the application of conservation agriculture systems (including reduced tillage, which saves resources, slashes costs, reduces greenhouse gases, and stabilizes production) while creating the management conditions for the expression of a greater proportion of genetic yield potential than in degraded, infertile conditions. Key elements of such an approach include effective weed control, using herbicides as appropriate, and soil fertility management. Thus, without improved and profitable crop management, the full benefits of improved wheat germplasm will not be realized. Within the context of choices open to farmers, the specific outcomes for national food security and wheat-based farming systems from agricultural and rural policies will depend on the particular combinations of resource policies (e.g., land and water regulations and pricing), agricultural market policies (e.g., seed, fertilizer, and machinery subsidies), and grain procurement, subsidies, and consumer policies (e.g., food price controls and subsidies). Policymakers, researchers, and farmers generally depend on different types of information from different sources. Scientists developing new wheat varieties need to have access to data from genomics laboratories linked with phenotypic data from field trials in various environments representative of target regions. Farmers will often benefit as a result of extension systems having faster access to new information and technologies from laboratory and experimental fields, which will allow them to make more efficient and sustainable decisions at the farm level. Similarly, policymakers and farmers need to access up-to-date global and local market information to guide decision-making. To address global challenges in wheat science and production, the exchange of both wheat genetic material and associated knowledge through existing networks and new partnerships (e.g., IWIN) will be a critically important international public good that must remain freely available to achieve impact.

References

Alexandratos N (2009) Expert meeting on "How to feed the world in 2050". Critical evaluation of selected projections. FAO, Rome 24–26 June 2009

Battisti DS, Naylor RL (2009) Historical warnings of future food insecurity with unprecedented seasonal heat. Science 323(5911):240–244

Baum M, Tadesse W, Singh R, Payne T, Braun HJ (2013) Global crop improvement networks to bridge technology gaps. In: The 12th international wheat genetic symposium, Yakohama, Japan

Borlaug NE (1968) Wheat breeding and its impact on world food supply. In: Finlay KW, Shephard KW (eds) Proceedings of the 3rd international wheat genetics symposium. Australian Academy of Sciences, Canberra

Braun HJ, Atlin G, Payne T (2010) Multi-location testing as a tool to identify plant response to global climate change. In: Reynolds MP (ed) Climate change and crop production. CABI Publishers, Wallingford

Brennan JP, Murray GM (1988) Australian wheat diseases-assessing their economic importance. Agri Sci 1:26–35

Byerlee D, Dubin HJ (2010) Crop Improvement in the CGIAR as a global success story of open access and international sharing. Intern J Comm 4:452–480

Byerlee D, Moya P (1993) Impacts of international wheat breeding research in the developing world, 1966–1990. CIMMYT, Mexico

Devos KM, Dubcovsky J, Dvorak J, Chinoy CN, Gale MD (1995) Structural evolution of wheat chromosomes 4A, 5A, and 7B and its impact on recombination. Theor Appl Genet 91:282–288

Dixon J, Braun HJ, Crouch J (2009) Transitioning wheat research to serve the future needs of the developing world. In: Dixon J, Braun HJ, Kosina P (eds) Wheat facts and futures. CIMMYT, Mexico, pp 1–19

Dvorak J, di Terlizzi P, Zhang HB, Resta P (1993) The evolution of polyploid wheat: identification of the A genome donor species. Genome 36:21–31

Dvorak J, Luo MC, Yang ZL, Zhang HB (1998) The structure of the Aegilops tauschii gene pool and the evolution of hexaploid wheat. Theor Appl Genet 97:657–670

Evenson RE, Gollin D (2003) Assessing the impact of the green revolution: 1960 to 2000. Science 300:758–761

FAO (2014) FAOSTAT. FAO, Rome, Italy. http://faostat.fao.org. Accessed 8 July 2014

Feldman M (2001) Origin of cultivated wheat. In: Bonjean AP, Angus WJ (eds) The world wheat book. A history of wheat breeding. Lavoiser Publishing, Paris

Feldman M, Levy AA (2005) Allopolyploidy – a shaping force in the evolution of wheat genomes. Cytogenetics 109:250–258

Feldman M, Lupton FGH, Miller TE (1995) Wheats. In: Smart J, Simonds NW (eds) Evolution of crop plants. Longman Group Ltd., London

Fischer G (2009) World Food and Agriculture to 2030/50: how do climate change and bioenergy alter the long-term outlook for food, agriculture and resource availability? Paper for the expert meeting "How to feed the world in 2050", FAO, Rome

Gill BS, Friebe B (2001) Cytogenetics, phylogeny and evolution of cultivated wheats. In: Bonjean AP, Angus WJ (eds) The world wheat book. A history of wheat breeding. Lavoiser Publishing, Paris

Hanson H, Borlaug NE, Anderson RG (1982) Wheat in the third world. Westview Press, Boulder

Harlan JR (1992) Crops and man, 2nd edn. CSSA, Madison

Heun M, Schafer-Pregl R, Klawan D, Castagna R, Accerbi M, Borghi B, Salamini F (1997) Site of Einkorn wheat domestication identified by DNA fingerprinting. Science 278:1312–1314

IPCC (2007) Climate change: the physical science basis. Contribution of working group I to the fourth assessment report of the Intergovernmental Panel on Climate Change. In: Salomon S, Quin D, Manning M et al (eds) Cambridge University Press, Cambridge/New York

Jiang J, Gill BS (1994) Different species-specific chromosome translocations in *Triticum timopheevii* and *T. turgidum* support the *diphyletic* origin of polyploid wheats. Chrom Res 2:59–64

Jin Y, Singh RP (2006) Resistance in U.S. wheat to recent African isolates of *Puccinia graminis* f. sp. *tritici* with virulence to resistance gene *Sr31*. Plant Dis 90:476–480

Johnson BL (1975) Identification of the apparent B genome donor of wheat. Can J Genet Cytol 17:21–39

Johnson BL, Dhaliwal HS (1976) Reproductive isolation of Triticum boeoticum and Triticum urartu and the origin of the tetraploid wheat. Am J Bot 63:1088–1094

Lange W, Jochemsen G (1992) Use of the gene pools of Triticum turgidum ssp. dicoccoides and Aegilops squarrosa for the breeding of common wheat (T. aestivum), through chromosome-doubled hybrids I. Two strategies for the production of the amphiploids. Euphytica 59:197–212

Lantican MA, Dubin HJ, Morris ML (2005) Impacts of international wheat breeding research in the developing world, 1988–2002. CIMMYT, Mexico

Ma H, Singh RP, Mujeeb-Kazi A (1995) Resistance to stripe rust in Triticum turgidum, T. tauschii and their synthetic hexaploids. Euphytica 82:117–124

McFadden ES, Sears ER (1946) The origin of *Triticum spelta* and its free-threshing hexaploid relatives. Heredity 37:81–89

Mujeeb-Kazi A, Hettel GP (1995) Utilizing wild grass biodiversity in wheat improvement: 15 years of wide cross research at CIMMYT, CIMMYT research report no. 2. CIMMYT, Mexico

Nachit MM (1992) Durum breeding for Mediterranean drylands of North Africa and West Asia. In: Rajaram S, Saari EE, Hettel GP (eds) Durum wheats: challenges and opportunities, Wheat special report no. 9. CIMMYT, Mexico

Naranjo T (1990) Chromosome structure of durum wheat. Theor Appl Genet 79:397–400

Niwa K, Aihara H, Yamada A, Motohashi T (2010) Chromosome number variations in newly synthesized hexaploid wheats spontaneously derived from self-fertilization of *Triticum carthlicum* Nevski *Aegilops tauschii* Coss. F1 hybrids. Cereal Res Comm 38:449–458

Nuttonson MY (1955) Wheat-climatic relationships and the use of phenology in ascertaining The thermal and photo thermal requirements of wheat. American Institute of Crop Ecology, Washington, DC

Ogbonnaya FC, Abdalla O, Mujeeb-Kazi A, Kazi AG, Gosnian N, Lagudah ES (2013) Synthetic hexaploids: harnessing species of the primary gene pool for wheat improvement. Plant Breed Rev 37:35–122

Ortiz Ferrara G, Mulitze D, Yau SK (1987) Bread wheat breeding for tolerance to thermal stresses occurring in West Asia and North Africa. In: Acevedo E, Fereres E, Gimenez C, Srivastava JP (eds) Improvement and management of winter cereals under temperature, drought and salinity stresses. INIA, Madrid, pp 267–282

Payne T (2004) The International Wheat Improvement Network (IWIN) at CIMMYT. www.cimmyt.org

Percival J (1921) The wheat plant. A monograph. E.P. Dutton & Company, New York

Pickett AA, Galwey NW (1997) A further evaluation of hybrid wheat. Plant Var Seeds 10:15–32

Rajaram S, van Ginkel M, Fischer RA (1995) CIMMYT's wheat breeding mega-environments (ME). In: Proceedings of the 8th international wheat genetic symposium, Beijing, China

Reynolds M, Borlaug NE (2006) Impacts of breeding on international collaborative wheat improvement. J Agric Sci 144:3–17

Reynolds M, Tuberosa R (2008) Translational research impacting on crop productivity in drought-prone environments. Curr Opin Plant Biol 11:171–179

Riley R, Unrau J, Chapman V (1958) Evidence on the origin of the B genome of wheat. Heredity 49:91–98

Sayre KD, Rajaram S, Fischer RA (1997) Yield potential progress in short bread wheats in northwest Mexico. Crop Sci 37:36–42

Schultz A (1913) Die Geschichte der kultivierten Getreide. Nebert, Halle

Sharma RC, Crossa J, Velu G, Huerta-Espino J, Vargas M, Payne TS, Singh RP (2012) Genetic gains for grain yield in CIMMYT spring bread wheat across international environments. Crop Sci 52:1522–1533

Sohail Q, Inoue T, Tanaka H, Eltayeb AE, Matsuoka Y, Tsujimoto H (2011) Applicability of *Aegilops tauschii* drought tolerance traits to breeding of hexaploid wheat. Breed Sci 61:347–357

Solh M, Nazari K, Tadesse W, Wellings CR (2012) The growing threat of stripe rust worldwide. Borlaug Global Rust Initiative (BGRI) conference, Beijing

Tadesse W, Hsam SLK, Wenzel G, Zeller FJ (2006) Identification and monosomic analysis of tan spot resistance genes in synthetic wheat lines (*Triticum turgidum* L. x *Aegilops tauschii* Coss.). Crop Sci 46:1212–1217

Tadesse W, Manes Y, Singh R, Payne T, Crossa J, Braun HJ (2010) Adaptation and performance of CIMMYT spring wheat genotypes targeted to high rain fall areas of the world. Crop Sci 50:240–248

Tadesse W, Abdalla O, Ogbonnaya F, Nazari K, Tahir I, Baum M (2012a) Agronomic performance of elite stem rust resistant spring wheat genotypes and association among trial sites in the CWANA region. Crop Sci 52:1105–1114

Tadesse W, Inagaki M, Tawkaz S, Baum M, van Ginkel M (2012b) Recent advances and application of doubled haploids in wheat breeding. Afr J Biotech 89:15484–15492

Tadesse W, Morgounov AI, Braun HJ et al (2013) Breeding progress for yield and adaptation of winter wheat targeted to irrigated environments at the International Winter Wheat Improvement Program (IWWIP). Euphytica 194:177–185

Tadesse W, Ogbonnaya FC, Jighly A, Nazari K, Rajaram S, Baum M (2014) Association mapping of resistance to yellow rust in winter wheat cultivars and elite genotypes. Crop Sci 54:607–616

Trethowan RM, van Ginkel M, Rajaram S (2002) Progress in breeding wheat for yield and adaptation in global drought affected environments. Crop Sci 42:1441–1446

Tutwiler R, Mazid A (1991) Impact of modem wheat technology in Syria. Part one: the adoption of new technologies. Farm Resource Management Program: annual report 1991. ICARDA, pp 176–209

van Ginkel M, Ogbonnaya F (2007) Novel genetic diversity from synthetic wheats in breeding cultivars for changing production conditions. Field Crops Res 104:86–94

van Ginkel M, Trethowan RM, Ammar K, Jiankang Wang, Lillemo M (2002) Guide to bread wheat breeding at CIMMYT. Wheat special report no. 5 (revised edition), CIMMYT, Mexico, DF

Vavilov NI (1951) The origin, variation, immunity and breeding of cultivated plants. Soil Sci 72:482

William M, Trethowan R, Crosby-Galvan EM (2007) Wheat breeding assisted by markers: CIMMYT's experience. Euphytica 157:307–319

World Bank (2008) World development report. The World Bank, Washington, DC

Zohary D (1999) Monophyletic vs. polyphyletic origin of the crops on which agriculture was founded in the near East. Genet Resour Crop Evol 46:133–142

Zohary D, Hopf M (1993) Domestication of plants in the old world, 2nd edn. Clarendon, Oxford

Rice

3

Kuldeep Singh, Kumari Neelam, Amanpreet Kaur, and Karminderbir Kaur

Abstract

Rice crop is affected by numerous biotic and abiotic stresses at various stages of its life cycle, which together cause 60 % losses in yield. To deal with such losses, new resistant varieties are required. However, there is limited genetic diversity available in cultivated rice due to selection of desirable traits during the course of evolution. The number of alleles in cultivated rice is reduced by 50–60 % compared to wild rice. Thus, there is a need to explore the wild germplasm and landraces for desirable genes and transfer these into elite cultivars to raise rice yields and augment stress resistance. The genus *Oryza* contains two cultivated and 22 wild species which together constitute three gene pools. The genes from wild species can be transferred to cultivated varieties through crossing and backcrossing procedures. Advanced mapping populations like backcross inbred lines (BILs), near isogenic lines (NILs), and chromosome segment substitution lines (CSSLs), MAGIC population, and association mapping panels are powerful tools for identifying the naturally occurring, favorable alleles in unadapted germplasm. But this success is limited largely to the primary gene pool of rice. The pre- and post-zygotic barriers like low crossability, hybrid sterility, hybrid inviability, and hybrid breakdown mar the transfer of useful traits from secondary and tertiary gene pool to the cultivated rice. However, the advancements in tissue culture have made it possible to overcome these barriers. A combination of conventional plant breeding methods with tissue culture and molecular approaches helps in precise transfer of genes from wild to cultivated species. Several improved varieties developed through wide hybridization and MAS have been released including Dhanrasi (blast resistance), Co-31 (drought tolerance), and improved Pusa Basmati 1 (bacterial blight resistance).

K. Singh (✉) • K. Neelam • A. Kaur • K. Kaur
Punjab Agricultural University,
Ludhiana, Punjab 141 004, India
e-mail: kuldeep35@pau.edu

M. Singh, S. Kumar (eds.), *Broadening the Genetic Base of Grain Cereals*,
DOI 10.1007/978-81-322-3613-9_3

Keywords
Wide hybridization • Pre and post zygotic barriers • Mapping populations • Abiotic and biotic stresses • Marker assisted selection

3.1 Introduction

With the global population expected to cross the nine billion mark by 2050, the major challenge that lies ahead is to produce enough food to feed the increased population. Rice, being the most widely consumed cereal and staple food for more than half of the world's population, needs an increase in productivity to avoid food scarcity in the future. The global demand for rice is estimated to rise from 676 million tons in 2010 to 852 million tons in 2035 with an overall increase of 26 % in the next 25 years (Khush 2013). Achieving this increase from the present land or lesser land demands an increase in the inherent potential of rice to give higher yield. New rice varieties with higher yield potential and greater yield stability are vital to relieve the growing pressure of greater rice production. Besides land, water, and climate, the other factors deterring rice productivity are the biotic and abiotic stresses. Abiotic stresses, including drought, salinity, high or low temperatures, flooding, low nutrient availability, mineral deficiency, heavy metals, and pollutants, represent a severe threat to rice yield. Biotic stresses, brought about by biological agents such as viruses, bacteria, fungi, nematodes, insects, or herbivores, further reduce rice productivity.

During the course of domestication of cultivated rice from wild relatives, there has been a significant reduction in genetic diversity as desirable agronomic traits like non-shattering of seeds, efficient seed production, and determinate growth habit were selected (Morishima 2001). Sun et al. (2001) reported that the number of alleles in cultivated rice had been reduced by 50–60 % compared to wild rice. Exploring wild germplasm and landraces for desirable genes and transferring these into elite cultivars have shown to be a feasible path for raising rice yields and augmenting stress resistance (Price et al. 2002; McCouch et al. 2007). The genus *Oryza* of the Gramineae family consisted of 24 species. Among these, two species *O. sativa* and *O. glaberrima* are cultivated and the remaining 22 are wild species. The wild species have either $2n=24$ or $2n=48$ chromosomes representing AA, BB, CC, BBCC, CCDD, EE, FF, GG, and HHJJ genomes. The genes from the cultivated rice (*O. sativa*) and its closely related wild species, having AA genome (*O. nivara*, *O. rufipogon*, *O. glumaepatula*, *O. longistaminata*, and *O. meridionalis*), can be easily transferred to cultivated rice by conventional crossing and backcrossing procedures (Brar and Khush 1997). Several genes and QTLs have been mined from wild species of rice for resistance to biotic and abiotic stresses and for enhancing the productivity of modern cultivars (Khush et al. 1977, 1990; Xiao et al. 1996, 1998; Zhang et al. 2000; Moncada et al. 2001; Aluko et al. 2004; Linh et al. 2008; Rangel et al. 2008; Chen et al. 2009). But this success is limited largely to the primary gene pool of rice. The pre- and post-zygotic barriers like low crossability, hybrid sterility, hybrid inviability, and hybrid breakdown mar the transfer of useful traits from secondary and tertiary gene pool to the cultivated rice. In order to combat stress epidemics along with main focus on enhancing yield potential, diversification and broadening of genetic base of cultivated rice are much urgently required. Therefore, pre-breeding efforts are required involving particularly those wild species from distant gene pool with potential to improve yield, quality, and stress resistance. In this chapter, we review the information on wild gene pool of rice, their systematics, gene flow, crossability barriers and means to overcome them, conventional and contemporary approaches to interspecific gene transfer, and utilization of wild gene pool for crop improvement.

3.2 Origin and Domestication

Civilization as we know today largely owes its existence to domestication of poaceous cereal crops, rice being one of them. Rice (genus *Oryza*) belongs to the tribe Oryzae which originated 130 million years ago in Gondwanaland (Chang 1976a, b; Khush 1997). The drifting of the super-continent led to distribution of genus *Oryza* in the humid tropics of Africa, South America, South and Southeast Asia, and Oceania (Chang 1976b). Domestication of rice began ~9000 years ago, in Yangtze valley in China. Cultivation of the wild rice preceded domestication. The alternating periods of drought and variations in temperature during the early neothermal age led to development of annuals in parts Southeast Asia and Indian subcontinent (Whyte 1972). It has been hypothesized that cultivation of Asian rice races has occurred independently and concurrently at many sites extending from the Ganges plains below through upper Burma, Northern Thailand, Laos, and North Vietnam to Southwest and South China (Roschevicz 1931; Ramiah 1937; Chatterjee 1951; Chang 1964, 1976a; Morinaga 1968). The cultivation of African rice races is more recent.

Prehistoric people initially gathered and consumed rice grains. These people also hunted, fished, and gathered other edible plant parts as food. Unknowingly, they exercised selection by searching for plants bearing heavier panicles (Hawkes 1970; Whyte 1972). The cultivated species (*O. glaberrima*) was most likely domesticated from the wild annual *O. barthii* (Harlan et al. 1973). Rice cultivation is probably not more than 3500 years old. During the process of cultivation and domestication of *O. sativa*, a number of morphological and physiological changes occurred. Larger leaves, longer and thicker culms, and longer panicles resulted in increased plant size. Concurrently, there were decreases in (or losses of) pigmentation, rhizome formation, ability to float in deep water, awning, shattering, duration of grain dormancy, photoperiod response, and sensitivity to low temperatures. The frequency of cross-pollination also declined, so that the crop became more inbred than the wild races (Chang 1976a).

It is generally agreed that river valleys of Yangtze and Mekong rivers could be the primary center of origin of *O. sativa* while delta of Niger River in Africa as the primary center of origin of *O. glaberrima* (OECD 1999). The foothills of the Himalayas, Chhattisgarh, Jeypore tract of Orissa, Northeastern India, Northern parts of Myanmar and Thailand, Yunnan Province of China, etc. are some of the center of diversity for Asian cultigens. The inner delta of Niger River and some areas around Guinean coast of Africa are considered to be center of diversity of the African species of *O. glaberrima* (Chang 1976b; Oka 1988).

The *O. sativa* and *O. glaberrima* are believed to have evolved independently from two different progenitors, viz., *O. nivara* and *O. barthii*, and they are believed to be domesticated in South or Southeast Asia and tropical West Africa, respectively (Fig. 3.1). The progenitors of *O. sativa* are considered to be the Asian AA genome diploid species and those of *O. glaberrima* to be African AA genome diploid species *O. barthii* and *O. longistaminata* (Chang 1976b; Siddiq 2000).

3.3 Levels of Diversity

3.3.1 Wild Relatives

The genus *Oryza* consists of two cultivated and 22 wild species representing ten genome types. The cultivated species, *O. sativa* and *O. glaberrima*, are diploid $2n=24$ and have AA genome. The wild species have either $2n=24$ or $2n=48$ chromosomes representing AA, BB, CC, BBCC, CCDD, EE, FF, GG, and HHJJ genomes. The *Oryza* species have been divided into four complexes: (1) *O. sativa* complex, (2) *O. officinalis* complex, (3) *O. meyeriana* complex, and (4) *O. ridleyi* complex. However, *O. brachyantha*, *O. schlechteri*, *O. coarctata*, and *Leersia perrieri* are not placed in any of these groups (Vaughan 1989, 1994; Brar and Singh 2011; Sanchez et al. 2013).

The *sativa* complex has two cultivated species (*O. sativa* and *O. glaberrima*) and six wild species (*O. nivara*, *O. rufipogon*, *O. barthii*, *O. longistaminata*, *O. meridionalis*, and *O. glumaepatula)* with the AA genome. Asian rice, *O.*

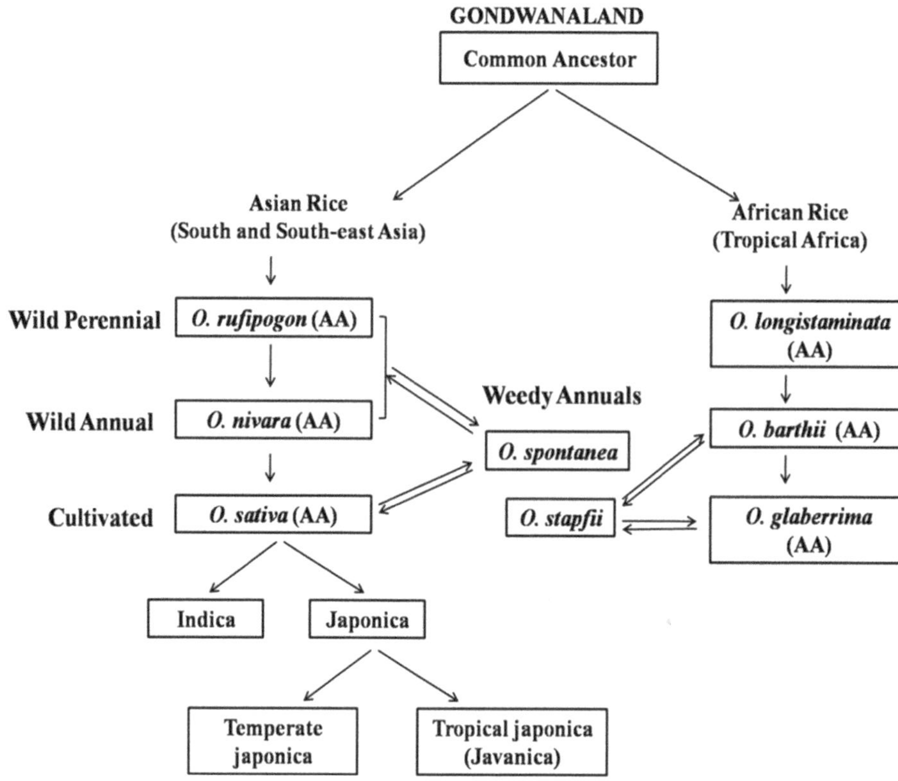

Fig. 3.1 Evolutionary pathways of Asian and African cultivated rice (Adapted from Khush 1997 and Chang 1976b)

sativa, is a major cereal crop and consumed as staple food in most parts of Asia, whereas the African cultivated rice, *O. glaberrima*, is grown in small areas in West Africa. Of the two Asian wild species, the perennial *O. rufipogon* is distributed throughout tropical Asia and Oceania, whereas the annual *O. nivara* is restricted to tropical continental Asia. The *O. longistaminata* perennial and rhizomatous wild species is endemic to Africa. Two other perennial wild species, *O. meridionalis* and *O. glumaepatula*, are found in tropical Australia and South and Central America, respectively.

The *O. officinalis* complex, also known as *O. latifolia* complex (Tateoka 1962), is the largest complex in the genus *Oryza*, with nine wild species comprising five diploid (*O. punctata* (BB, BBCC), *O. officinalis* (CC), *O. rhizomatis* (CC), *O. eichingeri* (CC), and *O. australiensis* (EE)) and four allotetraploid species (*O. minuta* (BBCC), *O. latifolia* (CCDD), *O. alta* (CCDD), and *O. grandiglumis* (CCDD)). The *O. officinalis* is widely distributed in South and Southeast Asia and South and Southwest China. The *O. rhizomatis* is found in Sri Lanka. The two African species of *O. officinalis* complex are *O. punctata* and *O. eichingeri*. *O. australiensis* is a species of tropical Australia. The *O. minuta* has been found only in the Philippines and Papua New Guinea. The *O. latifolia* is widely distributed in Central and South America as well as in the Caribbean islands. *O. alta* and *O. grandiglumis* grow only in South America, primarily in the Amazon basin, except for one population of *O. alta* reported from Belize (Oka 1961).

O. meyeriana complex has two diploid species with GG genome: *O. meyeriana* and *O. granulata*. These species are cross incompatible with *O. sativa*. *O. meyeriana* is found in Southeast Asia, and *O. granulata* is distributed across South Asia, Southeast Asia, and Southwest China.

O. ridleyi complex contains two tetraploid species, viz., *O. longiglumis* and *O. ridleyi* with HHJJ genome. *O. ridleyi* is found in continental

Southeast Asia and across the Malay Archipelago to New Guinea, while *O. longiglumis* is restricted to New Guinea.

3.3.2 Cultivated Species

The *sativa* complex has two cultivated species, viz., *O. sativa* and *O. glaberrima*. *O. sativa*, the Asian cultivated rice, is grown all over the world. The African cultivated rice, *O. glaberrima*, is grown on a small scale in West Africa. Chinese scientists have recognized two rice varietal groups, "hsien" in the south and "keng" in the temperate or northern region in China. The two types were first proposed as two subspecies, namely, *indica* and *japonica*, by Kato et al. (1928). The *indica* and *japonica* types are each characterized by an association of certain diagnostic characters, such as $KClO_3$ resistance, cold tolerance, apiculus hair length, and phenol reaction (Oka 1958).

In addition to *indica* and *japonica* types, another term "javanica" was introduced by Morinaga (1954) to include varieties from Indonesia. Glaszmann (1987), on the basis of genetic affinity using isozyme analysis, reported that javanica varieties fall within the *japonica* group and are now referred to as tropical *japonicas*, and the so-called typical *japonicas* are referred to as temperate *japonicas*.

3.3.3 Landraces

Rice landraces or regional ecotypes have evolved from their wild progenitor under natural and human selection, leading to the maintenance of high genetic diversity. According to Harlan (1975), landraces are "balanced populations in equilibrium with both the environment and pathogens, and are genetically dynamic." They are local varieties of a domesticated plant species which were adapted to the natural and cultural environment in which they live. India is one of the major centers of diversity of rice and is home to several landraces. These indigenous farmer landraces retain huge genetic diversity and posses characteristics like early maturity, adaptation to particular soil types, and resistance or tolerance to biotic and abiotic stresses. Different ecotypes found in Asia are listed below.

- Aus: Early maturing, photo-insensitive types, summer rice, drought tolerant, grown during April–July.
- Aman: Late types mostly photoperiod sensitive and flower during specific time regardless of when they are sown or transplanted.
  - Transplanted aman: Autumn rice, transplanted, grown during July to December.
  - Broadcast aman: deepwater rice, broadcast, grown during April to November.
- Boro: Winter rice, broadcast, cold tolerant, grown during December–May.
- Ashina: Deepwater, broadcast sown, grown during April to August.
- Rayada: Deepwater, photoperiod sensitive, very long duration, grown during March to December.
- Aromatic (basmati and sadri): Possessing extremely valuable quality traits like elongation, aroma, flavor, etc.
- Bulu and tjereh: Cultivated in Indonesia. Bulu have awns while tjereh are awnless.
- Japanese lowland and upland rice: Grown during March to October.

In India and Bangladesh, aus, aman, and boro have been distinguished by their cropping time. Aus is sown in March or April and harvested during the summer and is thus called the summer rice. Transplanted aman is sown in June or July, transplanted in July or August, and harvested between October and January. Broadcast aman is sown by broadcast method during April and is harvested in November. Winter rice, boro, sown during December can tolerate cold temperature in the early vegetative stage. Rayada and ashina are floating rice of Bangladesh and India, respectively. Aromatic rice such as basmati from Pakistan, Nepal, and India and sadri from Iran has a distinctive popcorn-like aroma and is highly prized for their quality (Garris et al. 2005). Bulu and tjereh are grown in Indonesia and differ from each other in physiological and morphological traits (Morinaga 1951). Generally, the cultivars which have awn are bulu, while those without are tjereh (Ueno et al. 1990). Tjereh was introduced to Indonesia from India and China and was generally cultivated in lowland places. Bulu has been cultivated in Indonesia for a long time and has many of the characteristics of upland rice, such as long culms and a small number of long panicles.

The aus ecotype of West Bengal developed directly from the southeast Indian hill. The spread of cultivation of aus types to the lower Gangetic valley resulted in the introgression of *O. rufipogon* genes into these types and led to evolution of aman ecotypes. Subsequently, the aman types were carried by Indian colonizers to Indochina and Indonesia during the ninth century A.D. The tjereh ecotype was developed from the aman ecotype in Indonesia.

Ueno et al. (1990) examined different rice cultivars belonging to five ecotypes (aus, aman, boro, bulu, and tjereh) and to two groups of Japanese rice (lowland and upland) with respect to $KC1O_3$ resistance, phenol reaction, and apiculus hair length. They suggested that aman, boro, and tjereh ecotypes should be classified as typical *indica* and that the Japanese lowland rice cultivars are mainly typical *japonica*. Some of the aus, bulu, and Japanese upland rice cultivars differ from typical *indica* and typical *japonica*.

3.4 Rice Gene Pools

Gene pool concept is very useful in explaining genetic relationships in cultivated and wild species. Harlan and de Wet (1971) proposed a concept involving three levels of the gene pool: primary, secondary, and tertiary (Fig. 3.2).

1. Primary gene pool: It includes crop species, and its wild ancestors which are cross compatible show homologous chromosome pairing and produce fertile hybrids between them (Jena 2010). The AA genome species belonging to the *O. sativa* complex form the primary gene pool for rice improvement. Hybrids between *O. sativa* and *O. rufipogon* are partially fertile; however, *O. sativa*×*O. glaberrima* and *O. sativa*×*O. longistaminata* F_1s are highly sterile (Brar and Singh 2011). Although partial F_1 sterility can occur between popula-

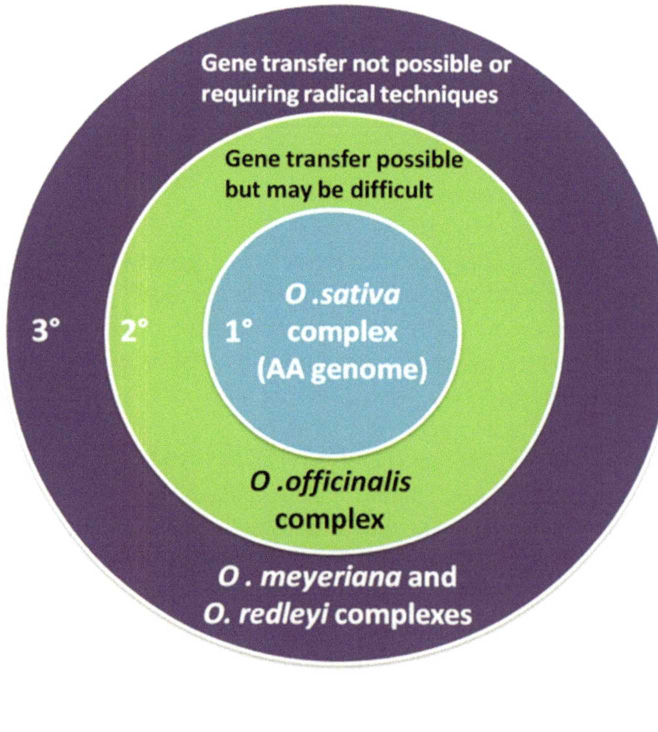

Fig. 3.2 Schematic diagram of three gene pools in rice (Edit the figure – *O. sativa*, *O. officinalis*, *O. meyeriana*)

tions within this gene pool, it does not prevent gene flow (Morishima et al. 1992).

2. Secondary gene pool: This pool includes wild taxa that exchange genes with the crop to a limited degree. These species are cross incompatible with the cultivated species, *O. sativa*, and show nonhomologous chromosome pairing making gene transfer into cultivated rice difficult (Jena 2010). *O. officinalis* complex, comprising five diploid (*O. punctata*, *O. officinalis*, *O. rhizomatis*, *O. eichingeri*, and *O. australiensis*) and four allotetraploid species (*O. minuta*, *O. latifolia*, *O. alta*, and *O. grandiglumis*), constitutes the secondary gene pool.
3. Tertiary gene pool: It includes wild taxa that hardly exchange genes with the crop. Gene transfer is either not possible with known techniques or else rather extreme or radical measures are required, e.g., embryo culture to obtain hybrids, doubling chromosome number or using bridging species to obtain some fertility. *O. meyeriana* and *O. ridleyi* complexes and unclassified wild species collectively form the tertiary gene pool.

3.5 Gene Flow

Gene flow may be defined as movement of genes or alleles from one population to another. Gene flow across the distinct gene pools appears to have been crucial to the domestication process. Rice domestication is amalgamation of both genetic exchange and genetic containment. It involved episodes of hybridization and introgression between ancient gene pools against a backdrop of emerging inbreeding and sterility barriers that progressively restricted gene flow between subpopulations. These competing evolutionary forces, sculpted by human selection, gave rise to an array of interconnected, but well-differentiated, subpopulations of *O. sativa*. Gene flow assessment is necessary for domestication studies as various domestication events are result of gene flow. Molecular markers, particularly the simple sequence repeat (SSR) markers, are commonly used for the detection of crop-to-crop and crop-to-wild gene flow frequencies under different conditions and have been proved to be very effective (Messeguer et al. 2001; Reboud 2002; Tranel et al. 2002; Song et al. 2003; Chen et al. 2004).

Gene flow in rice occurs through means of pollination and seed dispersal. Dissemination of seeds is necessary for species and gene migration. It also plays a critical role in creation of species and varietal diversity. The seed dispersal occurring as a result leads to gene flow between different genotypes that otherwise was not possible due to distance that separated them. This seems to have played important role in rice domestication also. Cultivated rice is a strictly self-pollinating crop, cleistogamy being major outcrossing barrier. The amount of pollen produced by rice plants is quite low compared to other members of grass family. Also, the viability of rice pollens decreases in a matter of few minutes after anthesis. Cultivated rice pollen longevity in air is around 10 min though pollen from wild and weedy relatives can remain viable for longer duration (Lu 2008; Song et al. 2001). Therefore, pollen-mediated gene flow frequency is limited in cultivated rice. Even most of the wild relatives of rice are self-pollinating with exception of *O. longistaminata*. The outcrossing frequency from cultivated rice to weedy and wild relatives is limited (1–2 %) but significant (Chen et al. 2004). As the relatives of cultivated rice, weedy and wild rice species (e.g., *O. rufipogon*, *O. nivara*, and *O. sativa* f. *spontanea*) are commonly found and even coexist in rice farming systems in many Asian, African, and American countries (Baki et al. 2000; Noldin 2000; Chen et al. 2001). Many of these wild relatives contain the same AA genome and are highly compatible sexually with the cultivated rice (Lu et al. 2003). Thus, prediction of gene flow frequency between rice and its weedy and wild relatives becomes important for crop improvement. Biotechnology-based crop improvement in rice focuses currently on development of transgenics, mutants, and development of hybrids for study of various traits.

3.6 Interspecific Hybridization and Its Barriers

In order to widen the genetic pool of present-day cultivars, it becomes necessary to harness ultimate variability present in wild species of rice. The wide hybridization refers to the crosses involving cultivated rice and related wild species of *Oryza*. During the last few decades, improved varieties of rice have been developed by combining and reshuffling of the genes primarily from primary gene pools and landraces of rice for enhanced agronomic performances; insect, pest, disease resistance; and high yields, though several incompatibility barriers have limited the transfer of useful genes from the secondary and tertiary gene pool into cultivated species (Table 3.1). The barriers can be classified into pre-zygotic and post-zygotic barriers:

- Pre-zygotic barriers include divergence in spatial and ecological habitats, flowering time, abnormalities in pollen germination, and pollen tube growth.
- Post-zygotic barriers include hybrid sterility, hybrid inviability, hybrid breakdown, and lack of recombination.

3.6.1 Pre-zygotic Barriers

Pre-zygotic barriers have been reported between *indica* and *japonica*, although only a few studies have been carried out on this subject. Xu (1995) showed that the number of pollen grains adhering to the stigmas during inter-subspecific pollination (a *japonica* variety hand-pollinated with pollen from an *indica* variety and vice versa) is much lesser than the number that do so in other pollinations (*indica* on *indica*, *japonica* on *japonica*, and a wide compatibility variety on either *indica* or *japonica*). The interspecific crosses also encounter difficulties in pollen tube growth. Liu et al. (2004) observed that more than 50 % of the microspores produced in *indica-japonica* hybrid could not develop into functional pollen and the hybrids encountered difficulties in pollen tube growth. Mizuta et al. (2010) identified hybrid

Table 3.1 Barriers to interspecific hybridization and techniques for overcoming it

Barriers	Reasons	Techniques for overcoming barriers
(I) Pre-zygotic barriers		
(a) Failure of pollen germination	Flowering time differences, autoimmune response, pollen-pistil incompatibilities, habitat difference	Mechanical removal of the pistil and use of recognition pollen
(b) Slow pollen tube growth		In vitro fertilization, use of growth hormones and of immunosuppressant
(c) Arresting of pollen tube in style, ovary, and ovule		In vitro fertilization
(d) Failure in sexual hybrid formation		Protoplast fusion
(e) Differences in ploidy level		Chromosome doubling, use of bridging species
(II) Post-zygotic barrier		
(a) Hybrid sterility	Structural differences between chromosomes, cytoplasmic differences, hybrid sterility genes	Chromosome doubling, backcrossing, silencing the sterility genes through RNAi or microRNA technology
(b) Hybrid inviability	Embryo abortion, growth of embryo arrested at young stage, autoimmunity	Embryo rescue, in vivo/vitro embryo culture (embryo implantation), anther culture, ovary culture
(c) Hybrid breakdown	Cryptic structural hybridity, digenic interactions, nuclear-cytoplasmic interaction	Growing larger F_2 populations
(d) Lack of recombination	–	Inducing chromosomal exchange through tissue culture or irradiation

incompatibility genes, *DOPPELGANGER1* (*DPL1*) and *DOPPELGANGER2* (*DPL2*), which are expressed in mature anther. Independent disruptions of *DPL1* and *DPL2* occurred in *indica* and *japonica*, respectively. Pollen carrying two defective *DPL* alleles became nonfunctional and did not germinate, suggesting an essential role for *DPL*s in pollen germination.

3.6.2 Post-zygotic Barriers

3.6.2.1 Hybrid Sterility

Hybrid sterility has been extensively investigated in rice. It can be either genic, chromosomal, or cytoplasmic (Grant 1981). Chromosomal sterility can result from structural differences between chromosomes, which reduce their degree of homology and consequently their pairing and which cause deficiency duplication and other disharmonious combinations of chromosome segments to be distributed to the gametes. Hybrid sterility is also caused by cytoplasmic difference, as in the case of *O. rufipogon* where the cytoplasm frequently induced male sterility (Shinjo 1984). Genes causing hybrid sterility are either gametophytic or sporophytic in action, which could affect the development of gametes produced by male or female or both. A number of loci conferring hybrid male or female sterility (or, in a few cases, both) have been identified such as *S1*, *S3*, *S5*, *S13*, *S18*, *S19*, *S20*, *S21*, *S22(t)*, *S23(t)*, *S27(t)*, *S28(t)*, *S29(t)*, and *S44(t)* (Doi et al. 1998, 1999; Sano 1983, 1986; Taguchi et al. 1999; Sobrizal et al. 2000, 2001, 2002). Recently, three new interspecific hybrid sterile loci, designated as *S37(t)*, *S38(t)*, and *S39(t)*, were detected between *O. sativa* and *O. glaberrima* (Xu et al. 2014).

Several genetic models have been proposed to explain the genetic mechanism of F_1 sterility including the one-locus allelic interaction and the two-locus epistatic interaction. The one-locus allelic interaction model proposes that an allelic interaction of a hybrid sterility gene in a heterozygote causes selective abortion of gametes carrying a given allele. This type of interaction has been found in *O. sativa* and *O. longistaminata* hybrids due to a sterility gene *S44(t)* located on chromosome 6. The interaction between *S44(t)-sativa* and *S44(t)-O. longistaminata* leads to the partial abortion of male gametes carrying the allele of *S44(t)-sativa* in the heterozygotes (Zhao et al. 2012).

3.6.2.2 Hybrid Breakdown

The weakness and sterility occurring in F_2 and later inbred generations are referred to as hybrid breakdown (Stebbins 1958). Although the F_1 is fertile, the recombinants in F_2 or later generations tend to be weak or lethal and are gradually eliminated, resulting in only parental types. Hybrid breakdown may occur due to cryptic structural hybridity, digenic interactions, and unfavorable nuclear-cytoplasmic interactions. The F_1 hybrids, being in heterozygous condition, tolerate chromosome duplications or deficiencies. However, F_2 progenies homozygous for these abnormalities are lethal (Stebbins 1958). In rice, the occurrence of weak F_2 segregants was first reported by Oka (1957) in an *indica-japonica* cross, Ac451 × Ac521. Analysis of the underlying genes revealed that this may be due to some duplicate genes whose double recessive combination lowers the fertility through some physiological effect upon the carrier. Hybrid breakdown has been described in inter-subspecific (*O. sativa* ssp. *japonica*×ssp. *indica*) and interspecific crosses (Fukuoka et al. 1998, 2005; Kubo and Yoshimura 2002, 2005; Matsubara et al. 2007; Miura et al. 2008; Yamamoto et al. 2010; Ichitani et al. 2012). Miura et al. (2008) identified a locus *hbd1(t)* on short arm of chromosome 2 of *O. nivara* that induced hybrid breakdown in the Koshihikari genetic background.

3.6.2.3 Hybrid Inviability

Hybrid inviability (or weakness) has frequently been described in plants. The products of intra- and interspecific hybridization may fail to reach reproductive maturity due to a failure of seed development or an aberrant vegetative growth. In some cases, the growth of hybrid embryo gets arrested at early stages and may lead to subsequent abortion of the embryos such as in hybrids of *O. sativa* and *O. meyeriana* (Fu et al. 2009). The embryo abortion may be attributed to the abnormal development of endosperm in interspecific hybrids that leads to nutritional starvation of young embryos. Embryo abortion can also occur due to action of a set of complementary dominant lethal genes. D1 and D2 genes present in *O. longistaminata* and *O. sativa*, respectively, induce differential levels of albumen deterioration according to their dosage in the triploid endosperm and cause abortion of embryo in interspecific hybrid (Chu and Oka 1970). The analysis of backcross progenies indicates that the D1 gene becomes inactivated in F_1 and is no longer complemented by D2 (Ghesquiere 1991). This indicates that eliminating the undesirable traits from wild germplasm requires repeated backcrossing to female cultivated parent. Imbalanced interactions in an interspecific cross between *indica* and the wild rice progenitors (*O. sativa* and *O. rufipogon*) also result in hybrid incompatibility or hybrid weakness (Chen et al. 2014). Interaction between leucine-rich receptor-like kinase protein (LRR-RLK) gene products originating from the *Hwi1* gene of wild rice and a secreted putative subtilisin-like protease encoded by the cultivated *Hwi2* locus activated autoimmune responses and impaired growth in hybrids and their progeny.

3.7 Conventional and Contemporary Approaches for Interspecific Gene Transfer

The major consideration in alien gene transfer is to selectively transfer agronomically important genes from wild species while avoiding linkage drag. To achieve precise transfer of genes from wild species, a combination of conventional plant breeding methods along with molecular approaches, tissue culture, and genetic engineering techniques has become important. Hybrids have been produced through direct crosses between rice and AA genome wild species. However, embryo rescue is required to produce hybrids and backcross progenies (alien introgression lines, AILs) between rice and all the wild species of *Oryza* (Brar et al. 1991; Brar and Khush 2002, 2006). Interspecific hybridization combined with embryo rescue has been used to transfer several valuable genes, such as resistance genes for brown plant hopper, white-backed plant hopper, bacterial blight, and blast, from non-AA genome wild species including *O. officinalis* (Jena and Khush 1990; Jena et al. 1992; Huang et al. 2001), *O. minuta* (Amante-Bordeos et al. 1992; Mariam et al. 1996), *O. brachyantha* (Brar et al. 1996), *O. latifolia* (Multani et al. 2003), *O. australiensis* (Ishii et al. 1994; Multani et al. 1994), and *O. eichingeri* (Liu et al. 2001a, b).

3.7.1 Conventional Approaches

Conventional methods include hybridization of cultivated rice with wild species, and further two to three backcrossing are required with recurrent parent for transfer of the desired traits. Several types of mapping populations have been generated from interspecific crosses between *Oryza* species and *O. sativa* to transfer favorable alleles from wild and unadapted sources into elite cultivars. Conventional mapping populations including F_2, backcross, recombinant inbred lines (RILs), and doubled haploids (DH) have been used for mapping both major genes and quantitative trait loci. RIL and DH are immortal and can be replicated over years and locations. While in case of interspecific crosses, especially with wild and unadapted germplasm, conventional mapping populations are not very beneficial because of hybrid sterility and high background noise (Jacquemin et al. 2013). Advanced mapping populations like backcross inbred lines (BILs), near isogenic lines (NILs), and chromosome segment substitution lines (CSSLs), MAGIC population, and association mapping panels are powerful tools for identifying the naturally occurring, favorable alleles in unadapted germplasm (Fig. 3.3) (Ali et al. 2010; Jacquemin et al. 2013). Many QTLs have been fine mapped and cloned using NIL populations (Ashikari et al. 2005; Xing et al. 2008; Xue et al. 2008; He et al. 2010; Balkunde et al. 2013). Chromosome segment substitution lines (CSSLs) and backcross inbred lines (BILs) have also been widely used for mapping and fine mapping of QTLs (Ahamadi et al. 2008; Ando et al. 2008; Hao et al. 2009; Lee et al.

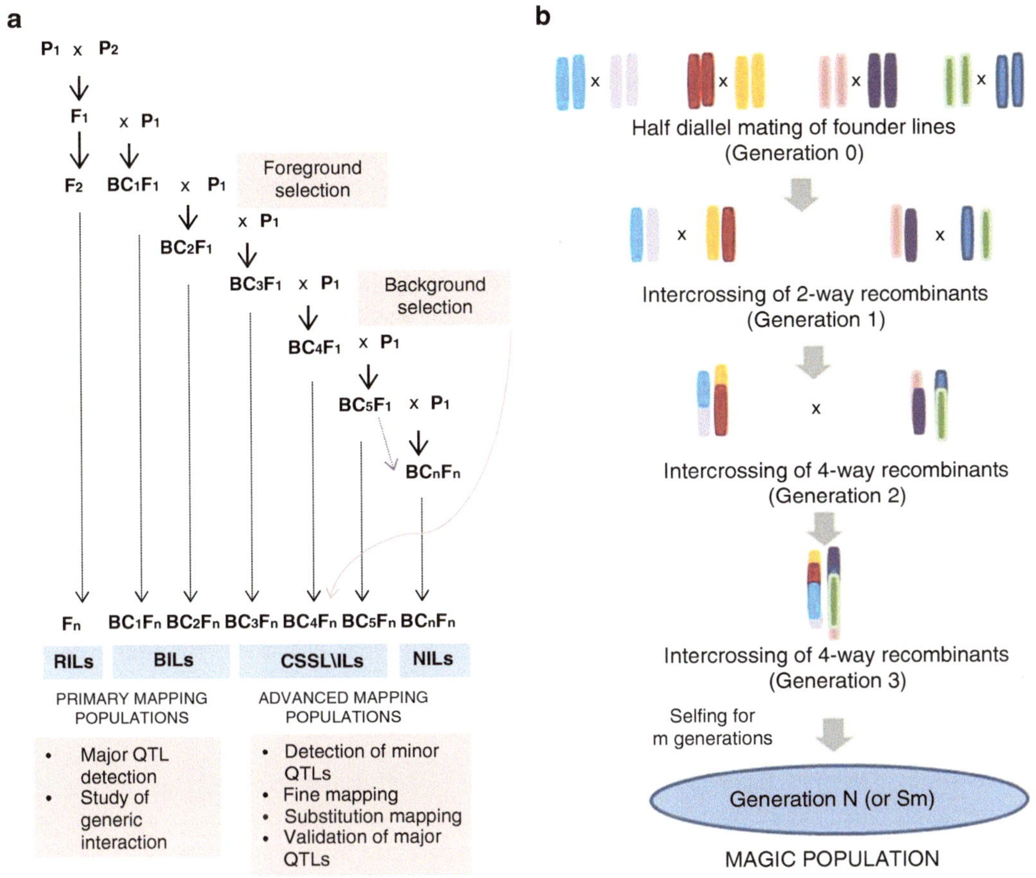

Fig. 3.3 Different approaches for mapping alien introgression

2012; Lu et al. 2013). A mapping population of 96 backcross inbred lines (BILs) derived from the cross between Habataki (*indica* variety) and Koshihikari (*japonica* variety) was used by Ashikari et al. (2005) to identify five QTLs for increasing grain number and four QTLs for plant height. *Gn1a* on chromosome 1 was identified as most effective QTL for increasing grain number. *GS3*, a major QTL for grain length and weight, was identified in near isogenic lines (NILs) derived from a cross between Minghui 63 and Chuan7 (*indica* varieties) (Fan et al. 2006). Balkunde et al. (2013) used NILs derived from a cross between the Korean *japonica* cultivar "Hwaseongbyeo" and *O. minuta* to fine map the QTL *qSPP7* across a 28.6 kb region. Ando et al. (2008) detected 38 QTLs for panicle architecture using a set of 39 CSSLs derived from the backcross progeny of a high-yielding *indica* cultivar, Habataki, with large panicles and a *japonica* cultivar, Sasanishiki, with a relatively small panicle. Lee et al. (2012) observed variations of mesocotyl elongation in BILs developed from a cross between Kasalath and Nipponbare and identified a total of five QTLs controlling mesocotyl length.

3.7.2 Genetic Engineering Approaches

Genetic engineering has become a versatile platform for cultivar improvement as well as for studying functions of genes in plants. The combination of genetic engineering and conventional breeding programs allows useful traits to be introduced into commercial crops with a lesser amount of time. Genetic transformation could be defined as the directed desirable gene transfer from one organism to other and the subsequent stable integration and expression of a foreign gene into the genome. Thus, genetic transformation facilitates introduction of only specifically desirable genes without cotransfer of any undesirable genes from donor species which normally occurs by conventional breeding methods.

3.7.2.1 Prerequisites for Plant Transformation

The tissue culture is required in most current transformation protocols to ultimately recover plants. Indeed, it is the totipotency of plant cells that underlies most plant transformation systems. Plants are regenerated from cell culture via two methods, somatic embryogenesis and organogenesis. Somatic embryogenesis is the generation of embryos from somatic tissues, such as embryos, microspores, or leaves. Organogenesis is the generation of organs, usually shoots, from a variety of tissues. Except for monocot leaf explants that contain meristems only at the leaf base, cotyledons, leaf fragments, hypocotyls, and scutella from embryos generally have the potential to generate shoots when placed on medium containing a shoot-inducing hormone, usually a cytokinin, such as 6-benzylaminopurine. The prerequisites of transformation systems are (a) target tissues competent for propagation or regeneration; (b) an efficient DNA delivery method; (c) agents to select for transgenic tissues; (d) the ability to recover fertile transgenic plants at a reasonable frequency; (e) a simple, efficient, reproducible, genotype-independent, and cost-effective process; and (f) a tight time frame in culture to avoid somaclonal variation and possible sterility.

3.7.2.2 Genetic Engineering Methods

Several plant transformation techniques have been developed out of which protoplast transformation, electroporation, particle bombardment, and *Agrobacterium*-mediated gene transfer have been used widely for rice transformation.

3.7.2.2.1 Protoplast Transformation

Briefly, in this method, protoplasts are isolated either by a mechanical or by an enzymatic process to remove the cell wall. This results in the production of a suspension containing millions of individual cells. Protoplasts are frequently obtained from an established suspension cell line of callus initiated from immature embryos, immature inflorescences, mesocotyls, immature leaf bases, and anthers. Protoplasts can either be transformed by *Agrobacterium* or by direct DNA uptake methods, facilitated by polyethylene glycol treatment, electroporation, or liposomes. This system was considered as most efficient system for monocots transformation.

3.7.2.2.2 Biolistic Gun/Particle Gun Method

The method was first used by Klein et al. (1987) for transient assay in onion epidermis and then extended by Sanford in 1988 to transient assays in wheat, maize, and rice. This method can be used to transform almost all plant species and virtually all types of tissues. Tungsten or gold particles (0.6 μm) coated with DNA are accelerated to velocities which enable their entry into plant cells using a machine called particle gun/gene gun which varies considerably in design and function. The most successful machine accelerates particles by using helium gas.

3.7.2.2.3 *Agrobacterium*-Mediated Transformation

The natural ability of the soil microorganism *Agrobacterium* to transform plants is utilized in the *Agrobacterium*-mediated transformation method. The process that involves a specific segment of the vector, T-DNA, which can be engineered to have a selectable marker and/or genes of interest, is transferred from the bacterium to the host plant cells and inserted into the nuclear genome. These functions are mediated by a set of virulence genes with optimal expression occurring at acidic pH and in the presence of phenolic inducers, such as acetosyringone, that are released by wounded plant cells (Hansen and Wright 1999). This system is basically used to transform dicot plants.

The summary of various methods used for transformation of plants with desired genes is given below in Table 3.2 and suitable examples are discussed under Sect. 3.11.

Table 3.2 Four plant transformation systems are being used these days

Plant transformation approaches			
(a) Physical	(b) Chemical	(c) Biological	(d) In planta
Microinjection	PEG (polyethylene glycol)	*A. tumefaciens*	Floral dip method
Pressure	DEAE-dextran	*A. rhizogenes*	Meristem transformation
Biolistics – gene gun/particle bombardment	Calcium phosphate	Virus mediated	Pollen transformation
Electroporation	Artificial lipids		
Microinjection	Proteins		
Silica/carbon fibers	Dendrimers		
Laser mediated			
SAT			

3.8 Molecular Markers, Genome Mapping, and Genomics as an Adjunct to Breeding

The consistent objective of all plant breeding programs is to improve existing cultivars through either conventional breeding methods or with the aid of marker technology and other biotechnological tools. Conventional breeding programs involve several crosses spanning several generations and careful phenotypic selection. Time intensive as it is, this approach is also not fool proof as tight linkage between desired traits and certain undesirable characters make it difficult for breeders to obtain desired end results. Invention of molecular markers over the last few decades played an important role in speeding up rice breeding programs. Molecular markers offer numerous advantages over conventional methods as they are consistent, not affected by the environment, cost-effective, and amenable to high-throughput genotyping. It also enables precise transfer of small segments of wild species or landraces into elite varieties, cultivar identity, assessment of genetic diversity, study of heterosis, and early generation selection. Basic marker techniques can be classified into two categories: (1) non-PCR-based techniques or hybridization-based techniques and (2) PCR-based techniques. Hybridization-based markers refer to the restriction fragment length polymorphic (RFLP) markers which are visualized by hybridization of restriction enzyme-digested DNA to a labeled probe of known sequence. However, these markers are not used nowadays as they are time consuming and unsafe. The PCR-based markers involve in vitro amplification of specific DNA sequence by using either specifically designed or arbitrarily chosen random primers. Among the PCR-based DNA markers, randomly amplified polymorphic DNAs (RAPDs), amplified fragment length polymorphisms (AFLPs), inter-simple sequence repeats (ISSRs), simple sequence repeats (SSRs), sequence-tagged sites (STS), sequence-characterized amplified regions (SCARs), and single nucleotide polymorphism (SNP)-based markers are commonly used markers in tagging and mapping genes in rice. Among all markers, microsatellites or simple sequence repeats (SSRs) have been the major choice in rice breeding, especially due to the abundance of these markers.

In rice, a research group at Cornell University first applied the RFLP technique to construct a rice linkage map (McCouch et al. 1988), and the group subsequently developed a fine linkage map with 726 DNA markers (Causse et al. 1994). Further, this map was supplemented with additional markers like RAPD, SSR, STS, and AFLP markers (McCouch et al. 2002). The availability of genomic sequences of the two rice subspecies, *O. sativa* ssp. *japonica* (cv. Nipponbare) and *O. sativa* ssp. *indica* 93-11, paved the way for generation of more markers (Goff et al. 2002; Yu et al. 2002; International Rice Genome Sequencing Project 2005). More than 20,000 of SSR markers have been developed in rice so far, and their chromosomal location can be obtained at Gramene database (http://www.gramene.org). Several landmarks like construction of a comprehensive genetic and physical map of the rice genome, construction of a high-density molecular map for gene mapping and map-based gene cloning (Temnykh et al. 2001; Coburn et al. 2002; International Rice Genome Sequencing Project 2005; Orjuela et al. 2010), development of the genetic maps of chloroplast and mitochondrial genomes, development of Oryza Map Alignment Project (OMAP), several user-friendly databases such as Rice Annotation Project Database (Ohyanagi et al. 2006; Tanaka et al. 2008), and the MSU/TIGR Rice Genome Annotation Database (Ouyang et al. 2007; http://rice.plantbiology.msu.edu/) have helped in making rice as model crop for genetic studies. Nowadays, single nucleotide polymorphic (SNPs) markers are gaining importance in plant breeding and genetics as they are more abundant, amenable to automation, and cost-effective and help in high-throughput profiling of large populations (Edward and Batley 2010; McCouch et al. 2010). SNPs are the most abundant form of genetic variation in eukaryotic genomes, and they occur in both coding and noncoding regions. SNP markers have wide applicability in marker-assisted selection, QTL mapping, genome-wide association mapping, positional cloning of genes, germplasm characterization, pedigree analysis, and many others. High-throughput single nucleotide polymorphisms (Affymetrix SNP chips, 96 Plex set, 384 Plex OPA set, 1536 plex) are being actively used for diversity analysis, DNA fingerprinting, QTL mapping, and MAS (McNally et al. 2009; Thomson et al. 2011; Huang et al. 2010; Zhao et al. 2011; www.oryza-snp.org).

The advances made in molecular marker techniques have proven extremely useful in speeding up the process of selection of complex traits and thus shorten the time taken in the development of a variety. It is more efficient, reliable, and cost-effective than phenotypic selection. Using MAS, a large number of genes and QTLs controlling agronomic traits, nutritional quality, and abiotic and biotic stresses have been identified and tagged (Jena and Mackill 2008; http://www.knowledgebank.irri.org/, http://www.rkmp.co.in/). Marker-assisted backcrossing (MAB) is an effective method for developing improved versions of widely grown "mega" varieties. Some of the successful examples of pyramiding of the genes from wild species of rice using MAS for the improvement of modern-day cultivars are given in Table 3.3.

3.9 Limiting Factors to Crop Productivity

Numerous environmental factors affect a crop plant at various stages of its life cycle. These stresses can be either biotic or abiotic. Rice is best suited to the regions of high temperature, high humidity, prolonged sunshine, and assured water supply. A temperature range of 20–37.5 °C is required for its optimum growth. The crop requires a higher temperature at tillering than that during early growth. The loamy soils are the best for rice cultivation. Whenever these conditions are not met or some biotic stresses such as insect pests or diseases attack the rice plant, the crop productivity will be compromised. Together, biotic and abiotic stresses can cause a yield loss of up to 60 %. The short growth duration of modern rice varieties has modified existing cropping patterns and increased farming intensity from single crop to double or triple crops, thereby inducing a favorable environment for insect pests and pathogen multiplication.

3.9.1 Biotic Stresses

3.9.1.1 Insect Pests

3.9.1.1.1 Brown Plant Hopper

The brown plant hoppers are most serious pests of paddy. Nymphs and adults congregate at the base of plants, above water level, sucking the plant sap. The leaves turn yellow then brown and finally the plants dry and die. Sudden slumping of crop is the first sign of damage, and affected patches give a scorched appearance called "hopperburn." It is also the carrier and transmitter of

Table 3.3 Some successful examples of MAS in rice improvement (limited to the genes coming from wild species of rice)

Traits and related gene(s)/QTL(s)	Marker(s) used	Target variety	Remarks	Reference
1. Blast resistance				
Pi1, Pi2, Pi33	SSR markers – RM224, RM527, RM3215	Jin23B	Marker-assisted gene pyramiding	Chen et al. (2008)
Pi1, Piz-5, Pi2, Pita	RFLP markers for *Pi1*, *Pi2*, and *Pita* and a PCR-based SAP marker for *Piz-5*	Co-39	Marker-assisted gene pyramiding	Hittalmani et al. (2000)
Pi9(t)	pB8	Luhui17	Introgressed *Pi-9(t)* from the donor parent P2 into hybrid restorer Luhui17 by using MAS	Wen and Gao (2011)
Pi9	STS marker (195R-1/195 F-1) derived from the Nbs2-Pi9 candidate gene	–	Identification of rice blast resistance gene Pi9 from Indian rice landraces using MAS	Imam et al. (2013)
Pi40, Pib	STS marker for *Pi40* and SSR marker for *Pib*	–	Selected plants pyramided with genes *Pi40* and *Pib* using molecular markers	Wang et al. (2012)
2. Bacterial blight				
Xa21	STS (pTA248)	Minghui 63	Marker-assisted backcross breeding	Chen et al. (2000)
		–	MAS applied for early generation selection for BB resistance	Reddy et al. (1997)
Xa4, xa5, xa13, and Xa21	STS for *Xa4*, CAPS for *xa5* (RG556+DraI), CAPS for *xa13* (RG136+HinfI), STS for *Xa21* (pTA248)	–	Marker-assisted gene pyramiding	Huang et al. (1997), Dokku et al. (2013)
xa5, xa13, and Xa21	CAPS for *xa5* (RG556+DraI), CAPS for *xa13* (RG136+HinfI), STS for *Xa21* (pTA248)	PR106	Marker-assisted backcross breeding	Singh et al. (2001)
Xa4, xa5, Xa21	Gene-specific markers, MP1+MP2 for *Xa4*, 10603.T10Dw for *xa5*, and U1/I1 for *Xa21*	Mangeumbyeo	Marker-assisted backcross breeding	Suh et al. (2013)
xa13 and Xa21	CAPS for *xa13* (RG136+HinfI), STS for *Xa21* (pTA248)	Taraori Basmati and Basmati 386	Marker-assisted backcross breeding	Pandey et al. (2013)

(continued)

Table 3.3 (continued)

Traits and related gene(s)/QTL(s)	Marker(s) used	Target variety	Remarks	Reference
Xa7, Xa21, Xa22, Xa23	SSR for *Xa7*(RM20593), STS for *Xa21* (pTA248), SSR (RM224) for *Xa22*, STS (03STS1) for *Xa23*	Huahui 1035	Introgressed four BB resistance genes into hybrid rice restorer lines through MAS	Huang et al. (2012)
3. Brown plant hopper resistance				
Bph18(t)	STS (7312.T4A)	–	Validated STS marker for BPH resistance using MAS	Jena et al. (2006)
Bph18	STS (7312.T4A)	Junambyeo	Marker-assisted backcross breeding	Suh et al. (2011)
Bph14, Bph15	SSR markers (MRG2329 and MS5)	Restorer lines 9311 and 1826	Marker-assisted gene pyramiding	Li et al. (2006)
Bph14, Bph15	SSR markers (MRG2684, MRG2346, MRG2329) and one InDel marker 76–2 for *Bph14*, SSR markers (RM5953 and RM261) and one InDel marker 15-6 for *Bph15*	Minghui 63, Shanyou63	Marker-assisted gene pyramiding	Hu et al. (2012)
4. Yellow mottle virus				
QTL_{12}, QTL_7	RFLP and SSR markers	IR64	Marker-assisted backcross breeding	Ahmadi et al. (2001)
5. Yield				
yld1.1, yld1.2	SSR markers	9311	MAS for yield-enhancing genes	Liang et al. (2004)

two viruses, rice ragged stunt virus, and rice grassy stunt virus.

3.9.1.1.2 White-Backed Plant Hopper

The white-backed plant hopper, *Sogatella furcifera*, is another serious insect pest of rice in Asia (Chen et al. 2010). Severely attacked seedlings do not grow. They are stunted, wilt, and eventually die (Dale 1994). The nymphs and adults suck cell sap at the base of the rice plant and the leaf surface. The attacked plants turn yellow and later acquire a rust-red appearance, spreading from the leaf tips to the rest of the plants (Atwal et al. 1967; Dale 1994). The tillers dry up and turn brown as a result of excessive removal of plant sap (Khan and Saxena 1985; Reissig et al. 1986). Gravid females cause additional damage by making oviposition punctures in leaf sheaths. The honeydew produced by the hoppers serves as a medium for mold growth (Dale 1994). The number of grains and the panicle length decreases (Dale 1994). During the heading stage, damaged glumes become brown and some remain unfilled (Noda 1986).

3.9.1.1.3 Green Leafhopper

Affected plants become pale yellow in color and get stunted in growth. If the plants are tapped, a large number of leafhoppers may be seen jumping to water. Both nymphs and adults suck the plant sap from the leaf and leaf sheath. The xylem and phloem vessels are plugged with their stylet sheath that causes disruption in the transport of food substances in the vessels. Mild infestation reduces the vigor of the plant and the number of reproductive tillers. Heavy infestation causes withering and complete drying of the crop. Plants are predisposed to fungal and bacterial infection through feeding and ovipositional punctures. It also transmits plant diseases such as dwarf, transitory yellowing, yellow dwarf, and rice *Tungrovirus*. Two species of green leafhoppers (GLH) can spread tungro: *Nephotettix malayanus* and *Nephotettix virescens*.

Various other insect pests – rice gall midge, rice stem borer, green leaf folder, and rice hispa – also infest rice crop (Table 3.4).

3.9.1.2 Diseases

3.9.1.2.1 Bacterial Blight

Xanthomonas oryzae pv. *oryzae* causes bacterial blight (BB) of rice which is one of the most important diseases of rice in most of the rice-growing countries (Hopkins et al. 1992). It causes wilting of seedlings and yellowing and drying of leaves. In general, the disease favors temperatures at 25–34 °C, with relative humidity above 70 %. It is commonly observed when strong winds and continuous heavy rains occur, allowing the disease-causing bacteria to easily spread through ooze droplets on lesions of infected plants.

3.9.1.2.2 Rice Blast

Blast is caused by the fungus *Pyricularia oryzae*. Young seedlings, leaves, panicles, and other aerial parts of the adult plant are affected and so often called as leaf blast, rotten neck, or panicle blast. Lesions and spots can be seen on leaves, nodes, panicles, and collar of the flag leaves. Leaf spots are of spindle-shaped with brown or reddish-brown margins, ashy centers, and pointed ends. Infection of panicle base causes neck rot and causes the panicle to fall off (igmoris.nic.in/Files2%5CBiologyDocuments%5CBiology_of_Rice.pdf).

3.9.1.2.3 Sheath Blight of Rice

This disease is caused by *Rhizoctonia solani* Kuhn. In addition to *R. solani*, *R. oryzae* causing rice sheath spot and *R. oryzae-sativae* causing aggregate sheath spot have been found to be associated with this disease. All the three pathogens may occur concurrently and sometimes referred to as rice sheath blight disease complex. Symptoms become apparent at filtering or flowering stage. Spots or lesions first develop near the water level (in flooded fields) or soil (in upland fields) and spots initially appear on the leaf sheath. Spots may be oral or ellipsoidal and measure 1–3 cm long. Lesions on the leaf blade are usually irregular and banded with green, brown, and orange coloration (igmoris.nic.in/Files2%5C BiologyDocuments%5CBiology_of_ Rice.pdf).

3.9.1.2.4 Stem Rot of Rice

The stem rot is caused by *Sclerotium oryzae* Cattaneo. The infection bodies or sclerotia are found in the upper soil layer. They survive in air-dry soil, buried moist rice soil, and in tap water. They can also survive on straw, which is buried in the soil. The sclerotia float on irrigation water and infect newly planted rice during land preparation.

3.9.1.2.5 Rice Grassy Stunt

Rice grassy stunt is a viral disease caused by RGSV transmitted by brown plant hopper *Nilaparvata lugens*. It causes severely stunted plants, excessive tillering, very upright plant growth, grassy and rosette appearance of plants, yellowish green leaves that are shorter and narrower than normal, and leaves that remain yellow even after application of sufficient nitrogen fertilizers. Numerous small rusty spots or patches emerge on leaves, which merge into blotches. Severely affected plants are unable to produce panicles.

3.9.1.2.6 Rice Tungro Disease

Rice tungro bacilliform virus (RTBV) is a plant pararetrovirus of the family Caulimoviridae. Tungro means "degenerated growth" and was first observed in the Philippines. Rice tungro bacilliform virus was first described by Saito in 1975. Tungro disease viruses are transmitted from one plant to another by leafhoppers that feed on tungro-infected plants. The most efficient vector is the green leafhopper. Tungro-infected plants also show symptoms of stunting, delayed flowering which may delay maturity, reduced number of tillers, small and not completely exserted panicles, as well as a higher than normal

Table 3.4 Rice insect pest, causal organism, and their symptoms

I. Insect/pests	*Causal organism*	*Symptoms*
Rice stem borer	*Scirpophaga incertulas*	Death of central stem known as "dead hearts"
		"White earhead" at milky stage
Green leaf folder	*Cnaphalocrocis medinalis*	Chlorophyll removal
		Characteristic small scratches on leaves
Rice grain bug or stink bug	*Leptocorisa oratorius*	Chaffy grains
		Grain break at milking stage
Brown plant hopper	*Nilaparvata lugens*	Slumping of crop
		Hopperburn
Rice gall midge	*Orseolia oryzae*	Formation of a tubular sheath gall called silver shoot
		Tiller is rendered sterile
Rice hispa	*Dicladispa armigera*	Tunneling of lower and upper epidermis results in regular translucent white patches
		White parallel streaks are visible
		Formation of blotches on the leaves
Rice case worm	*Nymphula depunctalis guen*	Larvae cut the leaf tips and roll by spinning both margins to make tubular case
		Leaves are skeletonized and appear whitish in color
Rice root aphid	*Tetraneura nigriabdominalis* Sasaki	Seedling growth is stunted
		Yellowing of leaves
White-backed plant hopper	*Sogatella furcifera*	Plants turn yellow and later acquire a rust-red appearance
		Number of grains and the panicle length decreases
Green leafhopper	*Nephotettix* sp.	Affected plants become pale yellow
		Disruption in the transport of food substances
		Drying of the crop
II. Diseases		
(a) Fungal diseases		
Rice blast	Pyricularia oryzae	Spots or lesions on leaves, nodes, panicles, and collar of the flag leaves
		Rotting of panicle neck causing panicle to fall off
Sheath blight	*Rhizoctonia solani*	Spots appear on the leaf sheath
		Symptoms become apparent at filtering or flowering stage
Brown spot	*Bipolaris oryzae*	Brown, circular to oval spots appear on the coleoptile leaves of the seedlings
False smut	*Ustilaginoidea virens*	Large orange to brown-green fruiting structures appear on one or more grains of the mature panicle
Black sheath rot	*Gaeumannomyces graminis*	Dark brown to black discoloration of the leaf sheaths
		Reduced tillering, poor grain fill, and lodging

(continued)

Table 3.4 (continued)

(b) Bacterial diseases		
Bacterial blight	*Xanthomonas oryzae* pv. *oryzae*	Water-soaked lesion on the edges of the leaf blades
		Entire plant wilt completely
		Leaves wilt and roll up
Bacterial leaf streak	*Xanthomonas oryzae* pv. *oryzicola*	Dark green and water-soaked streaks appear initially on interveins
		Patches on leaves become translucent

percentage of sterile panicles or partially filled grains, covered with dark brown blotches. The tungro virus is known to have at least two strains – S and M. The "S" strain in these varieties produces conspicuous interveinal chlorosis, giving an appearance of yellow stripe and sometimes irregular chlorotic specks on younger leaves. On the other hand, the "M" strain produces only mottling.

3.9.2 Abiotic Stresses

3.9.2.1 Cold Stress

Rice is a tropical plant and does not have inbuilt cold tolerance; cold at time of flowering is especially damaging. Low temperatures during the rice-growing season cause poor germination, slow growth, withering, and anther injury (Andaya and Tai 2007; Hu et al. 2008). On exposure to cold temperatures, the quantity of invertase enzyme is decreased leading to non-transport of sugars across tapetum. Thereby pollen grains are unable to accumulate starch (Oliver et al. 2005). Cold temperature stress at the young microspore stage enhances and induces partial degradation of proteins in the rice anthers at the trinucleate stage (Imin et al. 2004). Low temperatures during the rice-growing season cause poor germination, slow growth, withering, and anther injury (Andaya and Tai 2007; Hu et al. 2008). The optimum temperature range for germination and early seedling growth is 25–30 °C, and temperatures below 15–17 °C delay plant establishment, reduce plant competitive ability against weeds, delay plant maturity, and decrease grain yield.

3.9.2.2 Drought Stress

Anthesis and fertilization are particularly sensitive to drought. Water stress during flowering may reduce the harvest index by as much as 60 %, largely as a result of a reduction in grain set (Ekanayake et al. 1989; Garrity and O'Toole 1994). Panicle exertion and anther dehiscence are also effected (O'Toole and Namuco 1983; Ekanayake et al. 1990). The failure of panicle exertion alone accounts for approximately 25–30 % of spikelet sterility because the unexerted spikelets cannot complete anthesis and shed pollen (Matsui et al. 2001; Matsui and Omasa 2002; Prasad et al. 2006). Water deficiency, which may also be caused by dry winds during or immediately prior to rice flowering, leads to a substantial reduction in seed set. The anthers may shrivel up, so that insufficient pollen is available for fertilization. These reproductive abnormalities may prevent fertilization completely (Ekanayake et al. 1993). Grain abortion at the early stages following fertilization also accounts for part of the reduction in grain number in rice (O'Toole and Namuco 1983). High-temperature stress at flowering reduces spikelet fertility in rice (*O. sativa* L.). Water stress during the grain-filling period reduces photosynthesis, induces early senescence, and shortens the grain-filling period.

3.9.2.3 Aluminum Toxicity and Acidic Soils

Aluminum toxicity in soil inhibits the growth of plant shoots by causing nutrient deficiencies in magnesium, calcium, and phosphorus. It also depresses uptake of manganese. It can also lead to drought stress and plant hormone imbalance. Al toxicity is relatively rare in irrigated rice systems. Al toxicity occurs on acid upland soils (ultisols, oxisols) with large exchangeable Al content. These soils are characterized by pyrite (FeS_2) which produces high acidity (soil pH < 3.5) when they are exposed to the atmosphere due to drainage, resulting in the release of high amounts of Al^{3+} and Fe^{2+} into the environment (Shamshuddin et al. 2004), affecting rice growth (Elisa et al. 2011). It is known that Al^{3+} is attracted to the roots due to their negatively charged cell walls, causing toxicity (Yang et al. 2009).

3.9.2.4 Phosphorus Deficiency

P deficiency is widespread in all major rice ecosystems and is the major growth-limiting factor in acid upland soils where soil P-fixation capacity is often large. Soils particularly prone to P deficiency include coarse-textured soils containing small amounts of organic matter and small P reserves, highly weathered, clayey, acid upland soils with high P-fixation capacity. It leads to stunted growth of plants and reduced tillering. Older leaves are narrow, short, and very erect and have a "dirty" dark green color. Stems are thin and spindly and plant development is retarded.

3.10 Base Broadening Using Wild Species Through Conventional Approaches

The wild species of genus *Oryza* are phenotypically less desirable than modern varieties in their overall appearance and performance. These are, in fact, grasslike plants which possess weedy traits such as poor plant type, poor grain type, grain shattering, and low grain yield (Brar and Singh 2011; Sanchez et al. 2013). Despite their inferior phenotype, wild species constitute numerous valuable genes for resistance to biotic and abiotic stresses and increase yield which can be utilized to improve present-day cultivars as discussed below (Table 3.5).

3.10.1 Resistance to Biotic Stresses

3.10.1.1 Disease Resistance

3.10.1.1.1 Bacterial Blight Resistance Gene(s)

Bacterial blight (BB) caused by *Xanthomonas oryzae* pv. *oryzae* is one of the most destructive diseases of rice in Asia. Several genes for resistance to bacterial blight, viz., *Xa21*, *Xa23*, *Xa38*, *Xa27*, and *Xa29(t)*, have been introgressed from *O. longistaminata*, *O. rufipogon*, *O. nivara*, *O. minuta*, and *O. officinalis*, respectively. BB resistance was also identified in *O. latifolia* (CCDD), *O. australiensis* (EE), and *O. brachyantha* (FF). *Xa34(t)*, exhibiting broad spectrum resistance, was identified in *O. brachyantha* (IRGC101232) as a single dominant gene after examining the crossed progeny of two resistant introgression lines derived from IR56 (recurrent parent) and *O. brachyantha* (Ram et al. 2010b).

3.10.1.1.2 Rice Blast Resistance Gene(s)

Rice is also affected by a destructive fungal disease called blast. To date, 41 blast resistance genes have been reported; however, there are only two genes, *Pi9* and *Pi40(t)*, that have been identified in wild *Oryza* species, *O. minuta* and *O. australiensis*, respectively (Shakiba and Eizenga 2014). Screening of 99 *O. glaberrima* accessions by Silué et al. (1992) for blast resistance revealed that nine accessions were resistant to all ten *M. oryzae* strains and 32 accessions were moderately resistant, suggesting these accessions may be the source of novel resistance genes. Both bacterial blight and blast resistance were identified in *O. minuta* (IRGC101141). To transfer these resistance genes into the background of diploid, cultivated rice, Amante-Bordeos et al. (1992) used embryo rescue and

Table 3.5 Genes transferred from wild species of *Oryza* into rice

Trait	Donor wild *Oryza* species	Gene	Genome	References
Grassy stunt resistance	*O. nivara*	*GS*	AA	Khush et al. (1977)
Bacterial blight resistance	*O. rufipogon*	*Xa23(t)*	AA	Zhang et al. (1998), Jin et al. (2007)
	O. longistaminata	*Xa21*	AA	Khush et al. (1990)
	O. nivara	*Xa38*	AA	Cheema et al. (2008), Bhasin et al. (2012)
	O. officinalis	*Xa29 (t)*	CC	Tan et al. (2004a)
	O. minuta	*Xa27(t)*, *Xa35(t)*	BBCC	Gu et al. (2004), Guo et al. (2010)
	O. latifolia	Unknown	CCDD	Zhang et al. (1994), Multani et al. (2003)
	O. australiensis	*Xa32(t)*	EE	Zheng et al. (2009)
	O. brachyantha	*Xa34(t)*	FF	Ram et al. (2010b)
Blast resistance	*O. glaberrima*	Unknown	AA	Silue and Notteghem (1991)
	O. rufipogon	*Pirf2-1(t)*	AA	Utami et al. (2008)
	O. minuta	*Pi9*	BBCC	Amante-Bordeos et al. (1992)
	O. rhizomatis	*Pi54rh*	CC	Das et al. (2012)
	O. australiensis	*Pi40*	EE	Jeung et al. (2007)
Brown plant hopper resistance	*O. rufipogon*	*bph18(t)*, *bph19(t)*, *bph22(t)*, *bph23(t)*, *bph24(t)*, *Bph27*	AA	Li et al. (2010), Hou et al. (2011), Deen et al. (2010), Huang et al. (2013)
	O. glaberrima	*Bph22(t)*	AA	Ram et al. (2010a)
	O. officinalis	*bph11*, *bph12/bph16*, *Bph13(t)*, *Bph14*, *Bph15*, and *Bph17*	CC	Hirabayashi et al. (1998), Renganayaki et al. (2002), Huang et al. (2001), Yang et al. (2004), Sun et al. (2005)
	O. eichingeri	*Bph13(t)*	CC	Liu et al. (2001a, b)
	O. minuta	*Bph20(t)*, *Bph21(t)*	BBCC	Rahman et al. (2009)
	O. latifolia	*Bph12(t)*	CCDD	Yang et al. (2002)
	O. australiensis	*Bph10(t)*, *Bph18(t)*	EE	Ishii et al. (1994), Jena et al. (2006)
White-backed plant hopper resistance	*O. rufipogon*	QTLs – *qWbph2*, *qWbph5*, and *qWbph9*	AA	Chen et al. (2010)
	O. officinalis	*Wbph7* (t), *Wbph8* (t)	CC	Tan et al. (2004b)
	O. latifolia	Unknown	CCDD	Multani et al. (2003)
Yellow stem borer (larval mortality)	*O. longistaminata*	QTL	AA	Brar and Khush (2006)

(continued)

Table 3.5 (continued)

Trait	Donor wild *Oryza* species	Gene	Genome	References
Cytoplasmic male sterility	*O. sativa* f. *spontanea*	Unknown	AA	Lin and Yuan (1980)
	O. perennis	Unknown	AA	Dalmacio et al. (1995)
	O. glumaepatula	Unknown	AA	Dalmacio et al. (1996)
	O. rufipogon	Unknown	AA	Hoan et al. (1997)
Tungro tolerance	*O. rufipogon*	Unknown	AA	Kobayashi et al. (1993)
Tolerance to iron toxicity	*O. rufipogon*	Unknown	AA	Brar and Khush (2006)
	O. glaberrima	QTLs	AA	
Heat and/or drought-related traits	*O. australiensis*	Unknown	EE	Sanchez et al. (2013)
	O. barthii	Unknown	AA	Sanchez et al. (2013)
	O. glaberrima	QTLs	AA	Bocco et al. (2012)
	O. glumaepatula	Unknown	AA	Sanchez et al. (2013)
	O. meridionalis	Unknown	AA	Scafaro et al. (2009)
	O. officinalis	Unknown	CC	Sanchez et al. (2013)
	O. punctata	Unknown	BB	Sanchez et al. (2013)
	O. rhizomatis	Unknown	CC	Sanchez et al. (2013)
Cold tolerance	*O. rufipogon*	QTLs – *qCtss 3*, *qCtss 10*, *qCtss 11*, *qCTS4-1*, *qCTS12a*, *qCTS12b*	AA	Koseki et al. (2010), Andaya and Mackill (2003)
Tolerance to aluminum toxicity	*O. rufipogon*	QTL	AA	Nguyen et al. (2003)
Tolerance to acidic conditions	*O. glaberrima*	Unknown	AA	Brar and Khush (2006)
	O. rufipogon	Unknown	AA	
Tolerance to P deficiency	*O. glaberrima*	Unknown	AA	Brar and Khush (2006)
	O. rufipogon	Unknown	AA	
		QTLs		
Yield-enhancing loci	*O. rufipogon*	*yld1.1*, *yld2.1*, *qGY2-1*, *gw8.1*, *gw9.1*, *gpa7*, *spd6*	AA	Xiao et al. (1996), He et al. (2006), Xie et al. (2006, 2008), Tian et al. (2006b), Shan et al. (2009)
	O. minuta	*qSPP7*	BBCC	Balkunde et al. (2013)
	O. grandiglumis	QTL	CCDD	Yoon et al. (2006)
Increased elongation ability	*O. rufipogon*	Unknown	AA	Brar and Khush (2006)

backcrossing to produce interspecific hybrids between the elite *O. sativa* line, IR31917-45-3-2, (recurrent parent) and *O. minuta* (donor parent).

3.10.1.1.3 Sheath Blight Resistance Gene(s)

Several *Oryza* species accessions were screened for sheath blight resistance at the International Rice Research Institute (IRRI) in the Philippines with resistance being identified in accessions of *O. minuta*, *O. rufipogon* (Amante et al. 1990), and *O. officinalis* (Lakshmanan and Velusamy 1991). Prasad and Eizenga (2008) identified seven *Oryza* spp. accessions as moderately resistant to sheath blight disease with three accessions classified as *O. nivara* and one each as *O. barthii*, *O. meridionalis*, *O. nivara*/*O. sativa*, and *O. officinalis*. Eizenga et al. (2013) used two accessions of these *O. nivara* accessions as donor parents to develop two advanced backcross populations with the US rice cultivar, Bengal, as the recurrent parent and identified a major QTL *qShB6* for resistance to sheath blight.

3.10.1.2 Insect Resistance

3.10.1.2.1 Grassy Stunt Virus

The first example of transfer of a useful gene from wild species is the introgression of a gene for grassy stunt virus resistance from *O. nivara* to cultivated rice varieties (Khush et al. 1977). The first set of grassy stunt-resistant varieties, IR28, IR29, and IR30, was released for cultivation in 1974. Subsequently, many such varieties, for example, IR34, IR36, IR38, IR40, IR48, IR50, IR56, and IR58, have been released, some developed at the International Rice Research Institute, and others by national agricultural research and extension system. Rice tungro virus disease (RTD) is another serious viral disease in South and Southeast Asia. It is caused by rice tungro spherical virus (RTSV) and rice tungro bacilliform virus (RTBV). The wild species *O. longistaminata* and *O. rufipogon* show tolerance to RTBV. IR20, IR26, and IR30 were the first tungro-resistant cultivars released by the IRRI followed by IR28, IR29, IR34, IR36, IR38, and IR40 during the 1970s (Khush et al. 2004).

3.10.1.2.2 Plant Hopper Resistance Gene(s)

The brown plant hopper (BPH) and white-backed plant hopper (WBPH) are most economically important insect pests on rice crops in Asian countries (Su et al. 2013). A total of 27 brown leafhopper resistance genes have been reported. Of these, 12 genes, *Bph1*, *bph2*, *Bph3*, *bph4*, *bph5*, *Bph6*, *bph7*, *bph8*, *Bph9*, *bph19*, *Bph25*, and *Bph26*, are found in *O. sativa*; two genes, *Bph10* and *Bph18*, in *O. australiensis*; six genes, *bph11*, *bph12/bph16*, *Bph13*, *Bph14*, *Bph15*, and *Bph17*, in *O. officinalis*; one gene, *Bph13*, in *O. eichingeri*; one gene, *Bph12*, in *O. latifolia*; two genes, *Bph20(t)* and *Bph21(t)*, in *O. minuta*; one gene, *bph22(t)*, in *O. glaberrima*; and six genes, *bph18(t)*, *bph19(t)*, *bph22(t)*, *bph23(t)*, *bph24(t)*, and *Bph27*, in *O. rufipogon*.

Six major genes conferring WBPH resistance in rice have been identified, including *Wbph1* (Sidhu and Khush 1979), *Wbph2* (Angeles et al. 1981; Ravinder et al. 1982), *Wbph3* (Hernandez and Khush 1981), *Wbph4* (Hernandez and Khush 1981), *Wbph5* (Wu and Khush 1985), and *Wbph6* (Min et al. 1991). Tan et al. (2004b) located two QTLs, *Wbph7(t)* and *Wbph8(t)*, on chromosomes 3 and 4, respectively, using an introgression line of *O. officinalis* as the resistance donor. Chen et al. (2010) detected three QTLs, viz., *qWbph2*, *qWbph5*, and *qWbph9*, for WBPH resistance with favorable allele from *O. rufipogon* in a BIL population of *O. sativa*×*O. rufipogon*, of which *qWph9* having the most stable effect.

3.10.2 Resistance to Abiotic Stresses

3.10.2.1 Cold Tolerance

Studies involving large number of cultivars belonging to *indica* and *japonica* subspecies have shown that *japonica* genotypes have higher degree of cold tolerance at the germination stage (Lee 2001; Mertz et al. 2009) as well as at the vegetative and reproductive stages (Li et al. 1981; Mackill and Lei 1997). Nagamine (1991) suggested that tolerance to chilling injury at the seedling stage in rice is controlled by a single dominant gene (Nagamine 1991). However, segregation analysis performed with the same F_2 population showed that there was no linkage relationship between chilling injury and low-temperature chlorosis, indicating that tolerance to both stresses is controlled by different loci (Nagamine 1991). Saito et al. (2001) detected two closely linked QTLs (*Ctb-1* and *Ctb-2*) for cold tolerance and suggested their association with anther length.

Wild rice species, such as *O. rufipogon*, contain QTLs that can be integrated into cultivated rice to improve cold tolerance (Koseki et al. 2010). A mapping population of F_2 plants derived from a cold-tolerant wild rice, W1943 (*O. rufipogon*), and a sensitive *indica* cultivar, Guang-lu-ai 4, was used to identify QTLs associated with cold tolerance at the seedling stage. Three QTLs were detected on chromosomes 3 (*qCtss 3*), 10 (*qCtss 10*), and 11 (*qCtss 11*) with *qCtss11* explaining about 40 % of the phenotypic variation.

Andaya and Mackill (2003) developed recombinant inbred lines (RIL) from a cross between cold-tolerant temperate *japonica* cultivar (M-202) and cold-sensitive *indica* cultivar (IR50) and identified fifteen QTLs. The QTL detected on chromosome 4 (*qCTS4-1*) accounted for 21 % of the phenotypic variation for general cold tolerance. A major QTL was identified on chromosome 12, named *qCTS12a*, which accounted for 41 % of the phenotypic variation in tolerance to cold-induced wilting tolerance. Another QTL on chromosome 12, *qCTS12b*, explained 42 % of the cold-induced necrosis tolerance.

3.10.2.2 Heat and Drought Tolerance

Tian et al. (2006a) developed a set of 159 introgression lines derived from the cross between an *indica* cultivar Guichao 2 and an accession of common wild rice collected from Dongxiang County, Jiangxi province, China. Zhang et al. (2006) evaluated these 159 ILs for drought tolerance and mapped a total of 12 QTLs. Among these, four QTLs, *qSDT2-1* near SSR marker RM279 on chromosome 2, *qSDT6-1* near RM253 on chromosome 6, *qSDT6-2* near RM217 on chromosome 6, and *qSDT12-2* near RM17 on chromosome 12, had strong positive additive effect, indicating the alleles from Dongxiang wild rice could increase drought tolerance.

Heat tolerance in the wild rice *O. meridionalis* was established by comparison of seedling growth and photosynthetic rates at optimal and high temperatures, using *O. sativa* ssp. *japonica* (cv. Amaroo) as a domesticated control cultivar (Scafaro et al. 2009). Bocco et al. (2012) evaluated the drought tolerance potential of 54 introgression lines derived from *O. glaberrima* (TOG 5681) and *O. sativa* (IR 64). Some of the lines showed higher drought tolerance and yielding potential than their parents indicating the successful transfer of the beneficial genes controlling tolerance to drought and giving high productivity to the descendants from IR 64 and TOG 5681.

3.10.2.3 Tolerance to Aluminum

Wu et al. (2000) identified several QTLs conferring aluminum tolerance in a random inbred mapping population derived from Azucena and IR1152. Nguyen et al. (2001) also detected five QTLs for Al tolerance scattered on five chromosomes, with a major QTL located on chromosome 1. Nguyen et al. (2003) used *O. rufipogon* as a source for transfer of aluminum tolerance. They evaluated 171 F_6 RILs derived from IR64 (*indica*, susceptible) x *O. rufipogon* (IRGC106424, tolerant) for aluminum tolerance. Nine QTLs were identified including one for root length under non-stress conditions (CRL), three for root length under Al stress (SRL), and five for relative root length (RRL). *O. rufipogon* contributed favorable alleles for each of the five QTLs for RRL, which is a primary parameter for Al tolerance.

3.10.2.4 Tolerance to Low Phosphorus

Wissuwa et al. (2002) identified a major QTL for phosphorus-deficiency tolerance, *Pup1*, on long arm of chromosome 12 in the traditional aus-type rice variety Kasalath. Sequencing of the *Pup1* locus in Kasalath showed the presence of a *Pup1*-specific protein kinase gene, *PSTOL1* (Gamuyao et al. 2012). This gene was absent from the Nipponbare reference genome and other P-starvation-intolerant rice varieties. *PSTOL1* enables the plants to acquire more phosphorus and other nutrients by enhancing early root growth.

3.11 Base Broadening Using Genetic Engineering Approaches

3.11.1 Transgenic Rice for Disease Resistance

Till date, more than 30 BB resistance genes (R genes) have been identified in rice. Introducing R genes in rice is a direct and convenient way to obtain disease resistance. Zhang et al. (1998) introduced a broad spectrum R gene *Xa21* into elite *indica* varieties IR64, IR72, hybrid restorer line Minghui 63, and BG90-2 using particle bombardment. The transgenic plants showed enhanced resistance to bacterial blight. Rice is also affected by fungal diseases like blast and sheath blight. Overexpression of pathogenesis-related proteins (PRs), including chitinase, β-1,3-glucanases, and thaumatin-like proteins, is a common strategy to develop fungus-resistant rice (Nishizawa et al. 1999, 2003; Datta et al. 1999, 2000, 2002; Xu et al. 2003; Nandakumar et al. 2007). Helliwell et al. (2013) reported that pathogen-inducible production of ethylene in transgenic rice can enhance resistance to blast and sheath blight in rice.

3.11.2 Transgenic Rice for Insect Resistance

Bt genes derived from *Bacillus thuringiensis* have been successfully transferred in rice to confer resistance to several insects including striped stem borer, yellow stem borer, and leaf folder. The most commonly used Bt genes in rice are *cry1Ab*, *cry1Ac*, and *cry1Ab/Ac* fusion gene. Fujimoto et al. (1993) developed insect-resistant rice by introducing a truncated δ-endotoxin gene, *cryIA(b)*. Transgenic plants expressing the *CryIA(b)* protein were highly resistant to striped stem borer and leaf folder. Wunn et al. (1996) and Nayak et al. (1997) introduced *cry1A(b)* and *cry1A(c)* genes, respectively, into cultivated rice through particle bombardment for resistance to yellow stem borer. Maqbool et al. (1998) transformed Basmati 370 and M7 cultivars using a novel δ-endotoxin *cry2A Bacillus thuringiensis* gene. Tu et al. (2000) reported that transgenic *indica* rice Minghui 63 (T51-1) and its derived hybrid plant, expressing a Bt fusion protein derived from *CryIA(b)* and *CryIA(c)* sequences, exhibited excellent protection against yellow stem borer and leaf folder. Chen et al. (2005) and Tang et al. (2006) developed transgenic rice highly resistant against lepidopteran rice pests with synthetic *cry2A** and *cry1C**, respectively.

Besides *Bacillus thuringiensis* genes, insect resistance in rice has been obtained from plant- or animal-derived genes. Plant lectin genes have a high insecticidal activity, among which *Galanthus nivalis agglutinin* (*GNA*) gene has been widely used. GNA genes have insecticidal effects on plant hoppers, leafhoppers, and aphids (Rao et al. 1998; Tang et al. 1999; Tinjuangjun et al. 2000; Foissac et al. 2000; Sun et al. 2002; Nagadhara et al. 2003, 2004; Lee et al. 2012; Nakasu et al. 2014). Another gene, *Allium sativum agglutinin*, from leaf (*ASAL*) exhibits resistance to brown plant hopper and leafhopper (Saha et al. 2006; Bala et al. 2013; Chandrashekhar et al. 2014). There are very few studies reporting animal-derived insect-resistant genes. Huang et al. (2001) developed transgenic insect-resistant rice against striped stem borer and leaf folder by introducing an insecticidal gene *SpI* from spider into rice.

3.11.3 Transgenic Rice for Tolerance to Abiotic Stresses

Genetic engineering approaches offer great potential to develop rice varieties showing tolerance to abiotic stresses. Mohanty et al. (2002) reported that transgenic *indica* rice generated by *Agrobacterium*-mediated transformation with choline oxidase (*codA*) gene from *Arthrobacter globiformis* could survive salt stress. Sarangi et al. (2011) developed transgenic plants by introducing Am-SOD gene isolated from *Avicennia marina* for achieving salinity tolerance in rice. Hu et al. (2006) reported a drought-tolerance transcription factor gene SNAC1. Overexpressing SNAC1 significantly enhanced drought resistance in transgenic rice under severe drought stress conditions without showing any phenotypic changes or yield penalty. Yu et al. (2013) reported that transgenic rice plants overexpressing the *Arabidopsis* (*Arabidopsis thaliana*) homeodomain-leucine zipper transcription factor EDT1/HDG11 (enhanced drought tolerance/HOMEODOMAIN GLABROUS11) showed increased drought tolerance and improved grain yield.

3.12 Yield-Related QTLs Transferred from Wild

Grain yield is one of the most important indexes in rice breeding, which is governed by quantitative trait loci (QTLs). Grain yield is determined by three major yield components, panicles per plant, spikelets per panicle, grain weight and grain number, etc. (Yoshida 1983; Sakamoto and Matsuoka 2008). Indirect components such as plant height, tiller number, and heading date also have an impact on yield. Each component trait has varying degrees of contribution toward yield increment. The QTLs for yield and its components are expected to include structural or regulatory genes essential for inflorescence development, fertility of male and female gametes, embryogenesis, and seed development. At least some of the QTL alleles mapped for yield or related traits could also be considered as loci for fertility or positive alleles of sterility loci. Primary populations such as F_2 and recombinant inbred line populations have been widely used to discover QTLs in rice genome, with hundreds of yield-related QTLs being detected.

Among wild species of rice, *O. rufipogon* has been widely exploited for improving agronomically important traits in rice, including yield. Two *O. rufipogon* alleles at chromosome 1 (*yld1.1*) and chromosome 2 (*yld2.1*) were found to be associated with increase in the grain number (Xiao et al. 1996). In another experiment, Xiao et al. (1998) detected a total of 68 QTLs for 12 traits using a backcross population derived from *O. rufipogon* and cultivated rice. Moncada et al. (2001) followed an advanced backcross breeding strategy and analyzed BC_2F_2 populations derived from the cross involving an upland *japonica* rice cultivar, Caiapo, from Brazil and an accession of *O. rufipogon* from Malaysia. They identified two putative *O. rufipogon*-derived QTLs for yield, 13 for yield components, four for maturity, and six for plant height. A grain yield QTL, *qGY2–1*, was identified in a population derived from the cross between Guichao 2 and Dongxiang wild rice (*O. rufipogon*), and its wild rice allele increased grain yield (He et al. 2006). Xie et al. (2006, 2008) fine mapped two QTLs, *gw8.1* and *gw9.1*, in the population derived from cross between Korean *japonica* cultivar Hwaseongbyeo and *O. rufipogon*. The alleles from low-yielding parent *O. rufipogon* were found to be beneficial in the Hwaseongbyeo background. Yoon et al. (2006) mapped a QTL for yield components introgressed from *O. grandiglumis*. The QTLs, *gpa7* and *spd6*, were also identified in the populations derived from a cross between cultivated rice and O. *rufipogon* (Tian et al. 2006b; Shan et al. 2009). Zha et al. (2009) have reported map-based cloning of gene cluster including eight leucine-rich repeat receptor-like kinase (*LRK*) genes in Dongxiang wild rice (*O. rufipogon*), which increased the grain yield by 16 %. Balkunde et al. (2013) constructed a high-resolution physical map for *qSPP7* QTL for spikelet number per panicle on rice chromosome 7 using BC_7F_4 NILs derived from cross between Hwaseongbyeo and *O. minuta*.

3.12.1 Cloned QTLs for Yield Component Traits

In 1990s, many QTLs related to grain yield and its components were primarily mapped based on the various primary populations (www.gramene.org). Advanced populations such as near isogenic lines (NILs) are efficient to further fine map and clone target QTLs. NILs for primarily identified QTLs have been proposed and confirmed to be the ideal population for map-based cloning. A number of QTLs for yield and yield-related traits have been mapped and cloned from cultivated species (Table 3.6). To date, 21 QTLs directly affecting rice grain yield and its components have been cloned with NIL-based F_2 populations, and 14 new grain yield QTLs have been validated in the NILs (Bai et al. 2012).

In 2005, Ashikari et al. cloned *Gn1a* – a gene shown to produce cytokinin oxidase (OsCKX2) – and 11 other putative OSCKXs were predicted. In total, six QTLs have been found responsible for grain number. Five QTLs with pleiotropic effects on grain weight and grain number have been mapped on five different chromosomes, and three did not significantly increase grain yield in the corresponding NILs due to their counteract-

Table 3.6 Summary of QTLs cloned by map-based cloning approach in rice

QTLs	Phenotype	Marker interval	Reference
MOC1	Grain number and tiller number	17-2-12-2	Li et al. (2003)
Gn1a	Grain productivity	3A28-3A20	Ashikari et al. (2005)
GS3	Grain weight and length	GS63-SF19	Fan et al. (2006)
APO1	Harvest index and grain number	3628-41-3628-56 R3819-C11635	Ikeda et al. (2007), Terao et al. (2010)
GW2	Grain width and weight	W024-W004	Song et al. (2007)
qSW5(5)	Grain width and weight	MS40671-M16 CW5-CW6	Shomura et al. (2008), Weng et al. (2008)
Ghd7	Grain number, plant height, and heading date	RM5436-C39	Xue et al. (2008)
GIF1	Grain weight and grain filling	SSLP1-CAPS8 CAPS4-CAPS8	Wang et al. (2008)
PROG1	Grain number and tiller number	S3204-P71 pr5-pr7	Jin et al. (2008), Tan et al. (2008)
DEP1	Panicle number and grain number	S2-S11-2	Huang et al. (2009)
EP3	Grain number and tiller number	STS5803-5-STS5803-7	Piao et al. (2009)
LRK1	Grain number and tiller number	RM279-RM5654	Zha et al. (2009)
SP1	Grain size and grain number	M7-M8	Li et al. (2009)
DEP2	Grain weight and grain size	P5-P7 M2-M15 25285As-25340As	Zhu et al. (2010), Li et al. (2010), Abe et al. (2010)
FLO(a)/FLO2FLO2	Grain weight and grain filling	P18-P22 218042-218787	Qiao et al. (2010), She et al. (2010)
OsSPL14	Panicle number and grain productivity		Jiao et al. (2010), Miura et al. (2010)
SRS3	Grain weight and grain size	3130-3151	Kitagawa et al. (2010)
DEP3	Panicle number and grain number	P21-P23	Qiao et al. (2011)
Ghd8/DTH8	Grain number, plant height, and heading date	SEQ3-1-SEQ5-1 Ind8-47-Ind8-15	Yan et al. (2011), Wei et al. (2010)
GS5	Grain size	S2-RM574	Li et al. (2011)
NAL1	Number of spikelet	Ind4-Ind12	Fujita et al. (2013)

*Modified from Bai et al. (2012)

ing effects on grain weight and number. These yield QTLs (*Gn1a*, *GS3*, *GW2*, *qSW5*, *Ghd7*, *DEP1*, and *OsSPL14*) may serve to elucidate the molecular genetics mechanisms for yield traits in crops and provide useful tools for improving crop yield. Using mapping populations based on a cross between erect-type and high-yield variety, Shennong 265, and non-erect-type and low-yield variety, Nipponbare, Huang et al. (2009) detected a QTL, *DEP1*, for panicle density and grain yield. *DEP1* narrowed to an 85 kb segment, and it was confirmed that *DEP1* encoded a PEBP (phosphatidylethanolamine-binding protein)-like domain protein by sequence comparisons. N terminus of *DEP1* is partially homologous with *GS3* mentioned above. Transgene and other experiments revealed that *dep1* allele enhances meristematic activity and promotes cell proliferation and then acts as a dominant negative regulator of panicle architecture and grain number. Two parents, TN1 (*indica*) and SNJ1 (*japonica*), having different plant architectures, such as tiller and panicle number, were used by Jiao et al. (2010) to identify *IPA1* – a QTL influencing rice shoot branching and grain yield. They constructed 110 BC_2F_2 plant lines and selected 5500 BC_2F_2 plants and narrowed the candidate region to ~78 kb. At the same time, Miura et al. (2010) also detected this QTL and named it *WFP*, but they used two different parents, Nipponbare and ST-12. Further analysis revealed that this QTL encodes a plant-specific transcription factor OsSPL14, and *OsSPL14* mRNA is targeted and regulated by microRNA (miRNA) OsmiR156.

3.13 Conclusions and Future Prospects

The information provided in this chapter comprehensively describes the utilization of wild species of rice in crop improvement using interspecific hybridization. Increased intervention of humans to the natural climate leads to the simultaneous evolution of new insect pest and disease of rice. So, everyday there is a need for broadening of the genetic base of the cultivated gene pool. Wide hybridization coupled with cytogenetic manipulation of chromosome pairing helped to introduce desirable traits from wild relatives to crop plants. Later, with the advent of genomic era, the process of plant improvement was further accelerated. It became possible to improve the present-day cultivars with the aid of molecular markers for agronomically important traits such as yield, nutritional quality, and biotic and abiotic stresses. Improved version of Samba Mahsuri, Pusa Basmati 1, Punjab Basmati 3, and many others are the successful examples of gene(s) pyramiding using molecular markers and are being grown at various parts of the country. A large number of introgression lines carrying desirable traits from *O. rufipogon*, *O. glaberrima*, *O. nivara*, *O. minuta*, *O. longistaminata*, and many others are being used as a donor parent in second-generation cross. Some of the varieties released involving wide hybridization and marker-assisted selection are given in Table 3.7.

Table 3.7 Varieties released through wide hybridization and MAS

S. no	Traits	Wild species	Varieties released	Country
Disease				
1	Bacterial blight resistance	*O. longistaminata*	NSICRc112, NSICRc142, NSICRc154	Philippines
			Improved Samba Mahsuri (RP Bio-226), improved Pusa Basmati 1	India
			Guodao 1, Xieyou 218, Neizyou	China
2	Blast resistance	*O. rufipogon*	Dhanrasi	India
		O. glaberrima	Yun Dao	China
3	Insect and pest resistance			
	BPH resistance	*O. officinalis*	MTL 98, MTL 103, MTL 105, MTL 114	Vietnam
	Grassy stunt virus resistance	*O. nivara*	Many rice varieties	Rice growing countries in Asia
	Tungro resistance	*O. rufipogon*	Matatag 9	Philippines
4	Abiotic stress			
	Tolerance to heat	*O. meridionalis*	Arizona Rice-1, Arizona Rice-2	USA
	Salinity tolerance	*O. rufipogon*	BRRIdhan55	Bangladesh
	Drought	*O. perennis*	Co-31	India
	Acid sulfate tolerance	*O. rufipogon*	AS996	Vietnam
5	High-yield earliness, weed competitiveness	*O. rufipogon*	Jarava	India
		O. glaberrima	Many NERICA rices	African countries
		O. nivara	DRR Dhan 40	India

Adopted from Brar and Singh (2011)

The development of new rice for Africa "NERICA" is one of the best examples of the utilization of *O. glaberrima-O. sativa* interspecifics for sustainable agriculture. In 2000, the first seven NERICA varieties were released on wide scale for cultivation in Côte d'Ivoire and Guinea. NERICA varieties have high yield potential, short growth cycle, weed competitiveness, early vigor trait, and resistance to African insect and pest along with higher protein content and amino acid balance. NERICA varieties were also tested for adaptation in Asia, France, and China and were found suitable for cultivation, thus indicating wider adoptability of these varieties.

Future priorities in the utilization of wild relatives of *Oryza* in rice improvement should focus on continuous collection, identification, and characterization of new wild germplasm from different countries and unexplored regions of the world. Emphasis should be given on the utilization of secondary and tertiary gene pool for getting new genes or alleles for biotic, abiotic, and yield component traits. Though genetic transformation offers direct access to the unadapted gene pools, the successful deployment of the transgenic plant to the farmer's field is still questionable and requires thorough policy formation and adequate knowledge dissemination. The advancement made in biotechnology and genomics is only a compliment to the conventional plant breeding. Thus, new generation technologies should go hand in hand with traditional plant breeding for the accelerated alien gene introgression to the present-day cultivars.

References

Abe Y, Mieda K, Ando T, Kono I, Yano M, Kitano H, Iwasaki Y (2010) The *SMALL AND ROUND SEED1* (*SRS1/DEP2*) gene is involved in the regulation of seed size in rice. Genes Genet Syst 85:327–339

Ahamadi J, Fotokian MH, Fabriki-Orang S (2008) Detection of QTLs influencing panicle length, panicle grain number and panicle grain sterility in rice (*Oryza sativa* L.). J Crop Sci Biotechnol 11:163–170

Ahmadi N, Albar L, Pressoir G, Pinel A, Fargette D, Ghesquière A (2001) Genetic basis and mapping of the resistance to rice yellow mottle virus. III. Analysis of QTL efficiency in introgressed progenies confirmed the hypothesis of complementary epistasis between two resistance QTLs. Theor Appl Genet 103:1084–1092

Ali ML, Sanchez PL, Yu S, Lorieux M, Eizenga GC (2010) Chromosome segment substitution lines: a powerful tool for the introgression of valuable genes from *Oryza* wild species into cultivated rice (*O. sativa* L.). Rice 3:218–234

Aluko G, Martinez C, Tohme J, Castano C, Bergman C, Oard JH (2004) QTL mapping of grain quality traits from interspecific cross *Oryza sativa* x *O. glaberrima*. Theor Appl Genet 109:630–639

Amante A, De L, Sitch L, Leung H, Mew T (1990) Sheath blight (ShB) resistance in wild rices. Int Rice Res Newsl 15:5

Amante-Bordeos A, Sitch L, Nelson R, Dalmacio R, Oliva N, Aswidinnoor H, Leung H (1992) Transfer of bacterial blight and blast resistance from the tetraploid wild rice *Oryza minuta* to cultivated rice, *Oryza sativa*. Theor Appl Genet 84:345–354

Andaya VC, Mackill DJ (2003) Mapping of QTLs associated with cold tolerance during the vegetative stage in rice. J Exp Bot 54:2579–2585

Andaya VC, Tai TH (2007) Fine mapping of the *qCTS4* locus associated with seedling cold tolerance in rice (*Oryza sativa* L.). Mol Breed 20:349–358

Ando T, Yamamoto T, Shimizu T et al (2008) Genetic dissection and pyramiding of quantitative traits for panicle architecture by using chromosomal segment substitution lines in rice. Theor Appl Genet 116:881–890

Angeles E, Khush G, Heinrichs E (1981) New genes for resistance to white backed planthopper (*Sogatella furcifera*) in rice. Crop Sci 21:47–50

Ashikari M, Sakakibara H, Lin S et al (2005) Cytokinin oxidase regulates rice grain production. Science 309:741–745

Atwal AS, Chaudhary JP, Sohi BS (1967) Studies on the biology and control of *Sogatella furcifera* Horvath (Delphacidae: Homoptera) in Punjab. J Res Punjab Agric Univ 4:547–555

Bai XF, Wu B, Xing YZ (2012) Yield related QTLs and their applications in rice genetic improvement. J Integr Plant Biol 54:300–311

Baki BB, Chin DV, Mortimer AM (2000) Wild and weedy rice in rice ecosystems in Asia-a review. International Rice Research Institute, Philippines

Bala A, Roy A, Behura N, Hess D, Das S (2013) Insight to the mode of action of *Allium sativum* leaf agglutinin (ASAL) expressing in T3 rice lines on brown plant hopper. Am J Plant Sci 4:400–407

Balkunde S, Le HL, Lee HS, Kim DM, Kang JW, Ahn SN (2013) Fine mapping of a QTL for the number of spikelets per panicle by using near-isogenic lines derived from an inter-specific cross between *Oryza sativa* and *Oryza minuta*. Plant Breed 132:70–76

Bhasin H, Bhatia D, Raghuvanshi S et al (2012) New PCR-based sequence-tagged site marker for bacterial blight resistance gene *Xa38* of rice. Mol Breed 30:607–611

Bocco R, Lorieux M, Seck PA, Futakuchi K, Manneh B, Baimey H, Ndjiondjop MN (2012) Agro-morphological characterization of a population of introgression lines derived from crosses between IR64 (*Oryza sativa Indica*) and TOG5681 (*Oryza glaberrima*) for drought tolerance. Plant Sci 183:65–66

Brar DS, Khush GS (1997) Alien introgression in rice. In: Oryza : from molecule to plant. Springer, Netherlands, pp 35–47

Brar DS, Khush GS (2002) Transferring genes from wild species into rice. In: Kang MS (ed) Quantitative genetics, genomics and plant breeding. CABI, Wallingford, pp 197–217

Brar DS, Khush GS (2006) Cytogenetic manipulation and germplasm enhancement of rice (*Oryza sativa* L.). In: Singh RJ, Jauhar PP (eds) Genetic resources, chromosome engineering and crop improvement. CRC Press, Boca Raton, pp 115–158

Brar DS, Singh K (2011) Oryza. In: Kole C (ed) Wild crop relatives: genomics and breeding resources cereals. Springer, Berlin/Heidelberg, pp 321–365

Brar DS, Elloran R, Khush GS (1991) Inter-specific hybrids produced through embryo rescue between cultivated and eight wild species of rice. Rice Genet Newsl 8:91–93

Brar DS, Dalmacio R, Elloran R, Aggarwal R, Angeles R, Khush GS (1996) Gene transfer and molecular characterization of introgression from wild *Oryza* species into rice. In: Khush GS (ed) Rice genetics, 3rd edn. International Rice Research Institute, Manila, pp 477–485

Causse MA, Fulton TM, Cho YG et al (1994) Saturated molecular map of rice genome based on interspecific backcross population. Genetics 138:1251–1274

Chandrashekhar K, Vijayalakshmi M, Vani K, Kaul T, Reddy MK (2014) Phloem–specific expression of lectin gene from *Allium sativum* confers resistance to sap-sucker *Nilaparvata lugens*. Biotechnol Lett 36:1059–1067

Chang TT (1964) Present knowledge of rice genetics and cytogenetics. Int Rice Res Inst Tech Bull 1:96

Chang TT (1976a) Rice. In: Simmonds NW (ed) Evolution of crop plants. Longman, London

Chang TT (1976b) The origin, evolution, cultivation, dissemination, and diversification of Asian and African rices. Euphytica 25:425–441

Chatterjee D (1951) Note on the origin and distribution of wild and cultivated rice. Indian J Agric Sci 18:185–192

Cheema KK, Grewal NK, Vikal Y et al (2008) A novel bacterial blight resistance gene from *Oryza nivara* mapped to 38 Kbp region on chromosome 4L and transferred to *O. sativa* L. Genet Res 90:397–407

Chen S, Lin XH, Xu CG, Zhang QF (2000) Improvement of bacterial blight resistance of Minghui 63 an elite restorer line of hybrid rice, by molecular marker-assisted selection. Crop Sci 40:239–244

Chen LJ, Suh HS, Lee DS (2001) Evolutionary significance of Chinese weedy rice 'Lu-tao'. SABRAO J Breed Genet 33:99–109

Chen LJ, Lee DS, Song ZP, Suh HS, Lu BR (2004) Gene flow from cultivated rice (*Oryza sativa*) to its weedy and wild relatives. Ann Bot 93:67–73

Chen H, Tang W, Xu CG, Li XH, Lin YZ, Zhang QF (2005) Transgenic *indica* rice plants harbouring a synthetic *cry2A** gene of *Bacillus thuringiensis* exhibit enhanced resistance against lepidopteran rice pests. Theor Appl Genet 111:1330–1337

Chen H, Chen ZX, Ni S, Zuo SM, Pan XB, Zhu XD (2008) Pyramiding three genes with resistance to blast by marker assisted selection to improve rice blast resistance of Jin 23B. Chn J Rice Sci 22:23–27

Chen ZW, Hu FY, Xu P et al (2009) QTL analysis for hybrid sterility and plant height in interspecific populations derived from a wild rice relative, *Oryza longistaminata*. Breed Sci 59:441–445

Chen J, Huang DR, Wang L, Liu GJ, Zhuang JY (2010) Identification of quantitative trait loci for resistance to white backed planthopper, *Sogatella furcifera*, from an inter-specific cross *Oryza sativa*×*O. rufipogon*. Breed Sci 60:153–159

Chen C, Chen H, Lin Y et al (2014) A two-locus interaction causes interspecific hybrid weakness in rice. Nat Commun 5:3357

Chu YE, Oka HI (1970) Introgression across isolating barriers in wild and cultivated *Oryza* species. Evolution 24:344–355

Coburn JR, Temnykh SV, Paul EM, McCouch SR (2002) Design and application of microsatellite marker panels for semiautomated genotyping of rice (*Oryza sativa* L.). Crop Sci 42:2092

Dale D (1994) Insect pest of rice plant-their biology and ecology. In: Heinrichs EA (ed) Biology and management of rice insects. Wiley, New York, pp 363–485

Dalmacio R, Brar DS, Ishii T, Sitch LA, Virmani SS, Khush GS (1995) Identification and transfer of a new cytoplasmic male sterility source from *Oryza perennis* into indica rice (*O. sativa*). Euphytica 82:221–225

Dalmacio RD, Brar DS, Virmani SS, Khush GS (1996) Male sterile line in rice (*Oryza sativa*) developed with *O. glumaepatula* cytoplasm. IRRN 21:22–23

Das A, Soubam D, Singh PK, Thakur S, Singh NK, Sharma TR (2012) A novel blast resistance gene, *Pi54rh* cloned from wild species of rice, *Oryza rhizomatis* confers broad spectrum resistance to *Magnaporthe oryzae*. Funct Integr Genomics 12:215–228

Datta K, Velazhahan R, Oliva N et al (1999) Over-expression of the cloned rice thaumatin-like protein (PR-5) gene in transgenic rice plants enhances environmental friendly resistance to Rhizoctonia solani causing sheath blight disease. Theor Appl Genet 98:1138–1145

Datta K, Koukolikova-Nicola Z, Baisakh N, Oliva N, Datta SK (2000) Agrobacterium-mediated engineering for sheath blight resistance of indica rice cultivars from different ecosystems. Theor Appl Genet 100:832–839

Datta K, Baisakh N, Thet KM, Tu J, Datta S (2002) Pyramiding transgenes for multiple resistance in rice against bacterial blight, yellow stem borer and sheath blight. Theor Appl Genet 106:1–8

Deen R, Ramesh K, Gautam SK et al (2010) Identification of new gene for BPH resistance introgressed from *O. rufipogon*. Rice Genet Newsl 25:70–71

Doi K, Yoshimura A, Iwata N (1998) RFLP mapping and QTL analysis of heading date and pollen sterility using backcross population between *Oryza sativa* L. and *Oryza glaberrima* Steud. Breed Sci 48:395–399

Doi K, Taguchi K, Yoshimura A (1999) RFLP mapping of *S20* and *S21* for F_1 pollen semi-sterility found in backcross progeny of *Oryza sativa* and *O. glaberrima*. Rice Genet Newsl 16:65–68

Dokku P, Das KM, Rao GJN (2013) Pyramiding of four resistance genes of bacterial blight in Tapaswini, an elite rice cultivar, through marker-assisted selection. Euphytica 192:87–96

Edward D, Batley J (2010) Plant genome sequencing: applications for crop improvement. Plant Biotechnol J 8:2–9

Eizenga GC, Prasad B, Jackson AK, Jia MH (2013) Identification of rice sheath blight and blast quantitative trait loci in two different *O. sativa/O. nivara* advanced backcross populations. Mol Breed 31:889–907

Ekanayake IJ, Steponkus PL, Datta SK (1989) Spikelet sterility and flowering response of rice to water stress at anthesis. Ann Bot 63:257–264

Ekanayake IJ, Steponkus PL, de Datta SK (1990) Sensitivity of pollination to water deficits at anthesis in upland rice. Crop Sci 30:310–315

Ekanayake IJ, De Datta SK, Steponkus PL (1993) Effect of water deficit stress on diffusive resistance, transpiration and spikelet desiccation of rice (*Oryza sativa* L.). Ann Bot 72:73–80

Elisa AA, Shamshuddin J, Fauziah CI (2011) Root elongation, root surface area and organic acid exudation by rice seedling under Al^{3+} and/or H^{+} stress. Am J Agric Bio Sci 6:324–331

Fan CC, Xing YZ, Mao HL et al (2006) *GS3*, a major QTL for grain length and weight and minor QTL for grain width and thickness in rice, encodes a putative transmembrane protein. Theor Appl Genet 112:1164–1171

Foissac X, Loc NT, Christou P, Gatehouse AM, Gatehouse JA (2000) Resistance to green leafhopper (Nephotettix virescens) and brown planthopper (Nilaparvata lugens) in transgenic rice expressing snowdrop lectin (Galanthus nivalis agglutinin; GNA). J Insect Physiol 46:573–583

Fu XL, Lu YG, Liu XD, Li JQ (2009) Crossability barriers in the inter-specific hybridization between *Oryza sativa* and *O. meyeriana*. J Integr Plant Biol 51:21–28

Fujimoto H, Itoh K, Yamamoto M, Kyozuka J, Shimamoto K (1993) Insect resistant rice generated by introduction of a modified δ-endotoxin gene of Bacillus thuringiensis. Nat Biotechnol 11:1151–1155

Fujita D, Trijatmiko KR, Tagle AG, Sapasap MV, Koide Y, Sasaki K, Kobayashi N (2013) *NAL1* allele from a rice landrace greatly increases yield in modern indica cultivars. Proc Natl Acad Sci U S A 110:20431–20436

Fukuoka S, Namai H, Okuno K (1998) RFLP mapping of the genes controlling hybrid breakdown in rice (Oryza sativa L.). Theor Appl Genet 97:446–449

Fukuoka S, Newingham MCV, Ishtiaq M, Nagamine T, Kawase M, Okuno K (2005) Identification and mapping of two new loci for hybrid breakdown in cultivated rice. Rice Genet Newsl 22:29–31

Gamuyao R, Chin JH, Pariasca-Tanaka J et al (2012) The protein kinase *Pstol1* from traditional rice confers tolerance of phosphorus deficiency. Nature 488:535–539

Garris AJ, Tai TH, Coburn J, Kresovich S, McCouch S (2005) Genetic structure and diversity in *Oryza sativa* L. Genetics 169:1631–1638

Garrity DP, O'Toole JC (1994) Screening rice for drought resistance at the reproductive phase. Field Crops Res 39:99–110

Ghesquiere A (1991) Re-examination of genetic control of the reproductive barrier between Oryza longistaminata and O. sativa and relationship with rhizome expression. In: Khush GS (ed) Rice genetics II. International Rice Research Institute (IRRI), Los Banos, pp 729–730

Glaszmann JC (1987) Isozymes and classification of Asian rice varieties. Theor Appl Genet 74:21–30

Goff SA, Ricke D, Lan TH et al (2002) A draft sequence of the rice genome (*Oryza sativa* L. ssp. japonica). Science 296:92–100

Grant V (1981) Plant speciation, 2nd edn. Columbia University Press, New York

Gu K, Tian D, Yang F et al (2004) High-resolution genetic mapping of *Xa27(t)*, a new bacterial blight resistance gene in rice, *Oryza sativa* L. Theor Appl Genet 108:800–807

Guo SB, Zhang DP, Lin XH (2010) Identification and mapping of a novel bacterial blight resistance gene *Xa35(t)* originated from *Oryza minuta*. Sci Agric Sin 43:2611–2618 (Chinese with English abstract)

Hansen G, Wright MS (1999) Recent advances in the transformation of plants. Trends Plant Sci 4:226–231

Hao W, Zhu MZ, Gao JP, Sun SY, Lin HX (2009) Identification of quantitative trait loci for rice quality in a population of chromosome segment substitution lines. J Integr Plant Biol 51:500–512

Harlan JR (1975) Crops and man. American Society of Agronomy, Madison

Harlan JR, de Wet JMJ (1971) Towards a rational classification of cultivated plants. Taxon 20:509–517

Harlan JR, DeWet JMJ, Price EG (1973) Comparative evolution of cereals. Evolution 27:311–325

Hawkes JG (1970) The origins of agriculture. Econ Bot 24:131–133

He GM, Luo XJ, Tian F et al (2006) Haplotype variation in structure and expression of a gene cluster associated with a quantitative trait locus for improved yield in rice. Genome Res 16:618–626

He Q, Zhang KX, Xu CG, Xing YZ (2010) Additive and additive×additive interaction make important contributions to spikelets per panicle in rice near isogenic *(Oryza sativa* L.) lines. J Genet Genomics 37:795–803

Helliwell EE, Wang Q, Yang Y (2013) Transgenic rice with inducible ethylene production exhibits broad-spectrum disease resistance to the fungal pathogens Magnaporthe oryzae and Rhizoctonia solani. Plant Biotechnol J 11:33–42

Hernandez JE, Khush GS (1981) Genetics of resistance to white-backed planthopper in some rice (*Oryza sativa* L.) varieties. Oryza 18:44–50

Hirabayashi H, Kaji R, Okamoto M, Ogawa T, Brar DS, Angeles ER, Khush GS (1998) Mapping QTLs for brown planthopper (BPH) resistance introgressed from *Oryza officinalis* in rice. In: Khush GS, Brar DS, Hardy B (eds) Advances in rice genetics. IRRI, Philippines, pp 268–270

Hittalmani S, Parco A, Mew TV, Zeigler RS, Huang N (2000) Fine mapping and DNA marker-assisted pyramiding of the three major genes for blast resistance in rice. Theor Appl Genet 100:1121–1128

Hoan NT, Sarma NP, Siddiq EA (1997) Identification and characterization of new sources of cytoplasmic male sterility in rice. Plant Breed 116:547–551

Hopkins CM, White FF, Choi SH, Guo A, Leach JE (1992) Identification of a family of avirulence genes from *Xanthomonas oryzae* pv. *oryzae*. Mol Plant-Microbe Interact 5:451–459

Hou LY, Ping YU, Qun XU et al (2011) Genetic analysis and preliminary mapping of two recessive resistance genes to Brown Planthopper, *Nilaparvata lugens* Stål in rice. Rice Sci 18:238–242

Hu H, Dai M, Yao J, Xiao B, Li X, Zhang Q, Xiong L (2006) Overexpressing a NAM, ATAF, and CUC (NAC) transcription factor enhances drought resistance and salt tolerance in rice. Proc Natl Acad Sci U S A 103:12987–12992

Hu H, You J, Fang Y, Zhu X, Qi Z, Xiong L (2008) Characterization of transcription factor gene *SNAC2* conferring cold and salt tolerance in rice. Plant Mol Biol 67:169–181

Hu J, Li X, Wu C et al (2012) Pyramiding and evaluation of the brown planthopper resistance genes Bph14 and Bph15 in hybrid rice. Mol Breed 29:61–69

Huang N, Angeles ER, Domingo J et al (1997) Pyramiding of bacterial blight resistance genes in rice: marker-assisted selection using RFLP and PCR. Theor Appl Genet 95:313–320

Huang N, He G, Shu L, Li X, Zhang Q (2001) Identification and mapping of two brown planthopper genes in rice. Theor Appl Genet 102:929–934

Huang X, Qian Q, Liu Z et al (2009) Natural variation at the *DEP1* locus enhances grain yield in rice. Nat Genet 41:494–497

Huang XH, Wei XH, Sang T et al (2010) Genome-wide association studies of 14 agronomic traits in rice landraces. Nat Genet 42:961–967

Huang B, Xu JY, Hou MS, Ali J, Mou TM (2012) Introgression of bacterial blight resistance genes *Xa7, Xa21, Xa22* and *Xa23* into hybrid rice restorer lines by molecular marker-assisted selection. Euphytica 187:449–459

Huang D, Qiu Y, Zhang Y et al (2013) Fine mapping and characterization of *BPH27*, a brown planthopper resistance gene from wild rice (*Oryza rufipogon* Griff.). Theor Appl Genet 126:219–229

Ichitani K, Takemoto Y, Iiyama K, Taura S, Sato M (2012) Chromosomal location of HCA1 and HCA2, hybrid chlorosis genes in rice. Int J Plant Genom 2012:649081

Ikeda K, Ito M, Nagasawa N, Kyozuka J, Nagato Y (2007) Rice *ABERRANT PANICLE ORGANIZATION 1*, encoding an F-box protein, regulates meristem fate. Plant J 51:1030–1040

Imam J, Alam S, Variar M, Shukla P (2013) Identification of rice blast resistance gene *Pi9* from Indian rice land races with STS marker and its verification by virulence analysis. Proc Natl Acad Sci India Sect B Biol Sci 83:499–504

Imin N, Kerim T, Rolfe BG, Weinman JJ (2004) Effect of early cold stress on the maturation of rice anthers. Proteomics 4:1873–1882

International Rice Genome Sequencing Project (2005) The map based sequence of the rice genome. Nature 436:793–800

Ishii T, Brar DS, Multani DS, Khush GS (1994) Molecular tagging of genes for brown planthopper resistance and earliness introgressed from *Oryza australiensis* into cultivated rice *O. sativa*.. Genome 37:217–221

Jacquemin J, Bhatia D, Singh K, Wing RA (2013) The international *Oryza* ,map alignment project: development of a genus-wide comparative genomics platform to help solve the 9 billion-people question. Curr Opin Plant Biol 16:147–156

Jena KK (2010) The species of the genus *Oryza* and transfer of useful genes from wild species into cultivated rice, *O. sativa*. Breed Sci 60:518–523

Jena KK, Khush GS (1990) Introgression of genes from *Oryza officinalis* Well ex Watt to cultivated rice, *O. sativa* L. Theor Appl Genet 80:737–745

Jena KK, Mackill DJ (2008) Molecular markers and their use in marker-assisted selection in rice. Crop Sci 48:1266–1276

Jena KK, Khush GS, Kochert G (1992) RFLP analysis of rice (*Oryza sativa*) introgression lines. Theor Appl Genet 84:608–616

Jena KK, Jeung JU, Lee JH, Choi HC, Brar DS (2006) High-resolution mapping of a new brown planthopper (BPH) resistance gene, *Bph18* (t), and marker-assisted selection for BPH resistance in rice (*Oryza sativa* L.). Theor Appl Genet 112:288–297

Jeung JU, Kim BR, Cho YC, Han SS, Moon HP, Lee YT, Jena KK (2007) A novel gene, *Pi40(t)*, linked to the DNA markers derived from NBS-LRR motifs confers broad spectrum of blast resistance in rice. Theor Appl Genet 115:1163–1177

Jiao YQ, Wang YH, Xue DW et al (2010) Regulation of *OsSPL14* by *OsmiR156* defines ideal plant architecture in rice. Nat Genet 42:541–544

Jin XW, Wang CL, Yang Q, Jiang QX, Fan YL, Liu GC, Zhao KJ (2007) Breeding of near-isogenic line CBB30 and molecular mapping of *Xa30(t)*, a new resistance gene to bacterial blight in rice. Sci Agric Sin 40:1094–1100 (Chinese with English abstract)

Jin J, Huang W, Gao JP et al (2008) Genetic control of rice plant architecture under domestication. Nat Genet 40:1365–1369

Kato S, Kosaka H, Hara S (1928) On the affinity of rice varieties as shown by fertility of hybrid plants. Bull Sci Fac Agric Kyushu Univ 3:132–147

Khan ZR, Saxena RC (1985) A selected bibliography of the white backed planthopper *Sogatella furcifera* (Horvath) (Homoptera: Delphacidae). Insect Sci Appl 2:115–134

Khush GS (1997) Origin, dispersal, cultivation and variation of rice. In: Oryza: from molecule to plant. Springer, Netherlands, pp 25–34

Khush GS (2013) Strategies for increasing the yield potential of cereals: case of rice as an example. Plant Breed 132:433–436

Khush GS, Ling KC, Aquino RC, Aguiero VM (1977) Breeding for resistance to grassy stunt in rice. In: Proceedings of the 3rd International Congress SABRAO, Canberra, Australia, pp 3–9

Khush GS, Bacalangco E, Ogawa T (1990) A new gene for resistance to bacterial blight from *O. longistaminata*. Rice Genet Newsl 7:121–122

Khush GS, Angeles R, Virk PS, Brar DS (2004) Breeding rice for resistance to tungro virus at IRRI. SABRAO J 36:101–106

Kitagawa K, Kurinami S, Oki K et al (2010) A novel Kinesin 13 protein regulating rice seed length. Plant Cell Physiol 51:1315–1329

Klein TM, Wolf ED, Wu R, Sanford JC (1987) High-velocity microprojectiles for delivering nucleic acids into living cells. Nature 327:70–73

Kobayashi N, Ikeda R, Domingo IT, Vaughan DA (1993) Resistance to infection of rice tungro viruses and vector resistance in wild species of rice (Oryza spp.). Jpn J Breed 43:377–387

Koseki M, Kitazawa N, Yonebayashi S, Maehara Y, Wang ZX, Minobe Y (2010) Identification and fine mapping of a major quantitative trait locus originating from wild rice, controlling cold tolerance at the seedling stage. Mol Genet Genomics 284:45–54

Kubo T, Yoshimura A (2002) Genetic basis of HB in a Japonica/Indica cross of rice, Oryza sativa L. Theor Appl Genet 105:906–911

Kubo T, Yoshimura A (2005) Epistasis underlying female sterility detected in hybrid breakdown in a Japonica–Indica cross of rice (Oryza sativa L.). Theor Appl Genet 110:346–355

Lakshmanan P, Velusamy R (1991) Resistance to sheath blight (ShB) and brown spot (BS) in lines derived from *Oryza officinalis*. Int Rice Res Newsl 16:8

Lee MH (2001) Low temperature tolerance in rice: the Korean experience. In: Fukai S, Basnayake J (eds) Increased lowland rice production in the Mekong region. Proceedings of an international workshop, Vientiane, Laos, ACIAR, Canberra, Australia, pp 109–117

Lee HS, Sasaki K, Higashitani A, Ahn SN, Sato T (2012) Mapping and characterization of quantitative trait loci for mesocotyl elongation in rice (*Oryza sativa* L.). Rice 5:1–10

Li TG, Visperas RM, Vergara BS (1981) Correlation of cold tolerance at different growth stages in rice. Acta Bot Sin 23:203–207

Li XY, Qian Q, Fu ZM et al (2003) Control of tillering in rice. Nature 422:618–620

Li JB, Xia MY, Qi HX, He GC, Wan BL, Zha ZP (2006) Marker-assisted selection for Brown Planthopper (*Nilaparvata lugens* Stal) resistance genes *Bph14* and *Bph15* in rice. Sci Agric Sin 10:28

Li R, Li L, Wei S et al (2010) The evaluation and utilization of new genes for brown planthopper resistance in common wild rice (*Oryza rufipogon* Griff.). Mol Entomol 1:1

Li SB, Qian Q, Fu ZM et al (2009) *Short panicle 1* encodes a putative PTR family transporter and determines rice panicle size. Plant J 58:592–605

Li F, Liu WB, Tang JY et al (2010) Rice *DENSE AND ERECT PANICLE 2* is essential for determining panicle outgrowth and elongation. Cell Res 20:838–849

Li YB, Fan CC, Xing YZ et al (2011) Natural variation in *GS5* plays an important role in regulating grain size and yield in rice. Nat Genet 43:1266–1269

Liang F, Deng Q, Wang Y, Xiong Y, Jin D, Li J, Wang B (2004) Molecular marker-assisted selection for yield-enhancing genes in the progeny of "9311× *O. rufipogon*" using SSR. Euphytica 139:159–165

Lin SC, Yuan LP (1980) Hybrid rice breeding in China. In: Innovative approaches to rice breeding. IRRI, Manila, pp 35–51

Linh LH, Hang NT, Jin FX, Kang KH, Lee YT, Kwon SJ, Ahn SN (2008) Introgression of a quantitative trait locus for spikelets per panicle from Oryza minuta to the O. Sativa cultivar Hwaseongbyeo. Plant Breed 127:262–267

Liu GQ, Yan HH, Fu Q, Qian Q, Zhang ZT, Zhai WX, Zhu LH (2001a) Mapping of a new gene for brown planthopper resistance in cultivated rice introgressed from *Oryza eichingeri*. Chin Sci Bull 46:1459–1462

Liu GQ, Yan H, Fu Q, Qian Q, Zhang ZT, Zhai WX, Zhu LH (2001b) Genetic mapping of dominant genes of brown planthopper resistance from *O. eichingeri*. Chin Sci Bull 46:738–742

Liu HY, Xu CG, Zhang Q (2004) Male and female gamete abortions, and reduced affinity between the uniting gametes as the causes for sterility in an indica/japonica hybrid in rice. Sex Plant Reprod 17:55–62

Lu BR (2008) Transgene escape from GM crops and potential biosafety consequences: an environmental perspective. International Centre for Genetic

Engineering and Biotechnology (ICGEB). Collect Biosaf Rev 4:66–141

Lu BR, Song ZP, Chen JK (2003) Can transgenic rice cause ecological risks through transgene escape? Prog Natl Sci 13:17–24

Lu B, Yang C, Xie K et al (2013) Quantitative trait loci for grain-quality traits across a rice F2 population and backcross inbred lines. Euphytica 192:25–35

Mackill DJ, Lei X (1997) Genetic variation for traits related to temperate adaptation of rice cultivars. Crop Sci 37:1340–1346

Maqbool SB, Husnain T, Riazuddin S, Masson L, Christou P (1998) Effective control of yellow stem borer and rice leaf folder in transgenic rice indica varieties Basmati 370 and M7 using the novel δ-endotoxin cry 2A Bacillus thuringiensis gene. Mol Breed 4:1–7

Mariam AL, Zakri AH, Mahani MC (1996) Inter-specific hybridization of cultivated rice, *Oryza sativa* L. with the wild rice, *O. minuta* Presl. Theor Appl Genet 93:664–671

Matsubara K, Ando T, Mizubayashi T, Ito S, Yano M (2007) Identification and linkage mapping of complementary recessive genes causing hybrid breakdown in an intraspecific rice cross. Theor Appl Genet 115:179–186

Matsui T, Omasa K (2002) Rice (*Oryza sativa* L.) cultivars tolerant to high temperature at flowering: anther characteristics. Ann Bot 89:683–687

Matsui T, Omasa K, Horie T (2001) The difference in sterility due to high temperatures during the flowering period among *japonica*-rice varieties. Plant Prod Sci 4:90–93

McCouch SR, Kochert G, Yu ZH, Wang ZY, Khush GS, Coffman WR, Tanksley SD (1988) Molecular mapping of rice chromosomes. Theor Appl Genet 76:815–829

McCouch SR, Teytelman L, Xu Y et al (2002) Development and mapping of 2240 new SSR markers for rice (Oryza sativa L.). DNA Res 9:199–207

McCouch SR, Sweeney M, Li J et al (2007) Through the genetic bottleneck: *O. rufipogon* as a source of trait-enhancing alleles for *O. sativa*. Euphytica 154:317–339

McCouch SR, Zhao K, Wright M et al (2010) Development of genome-wide SNP assays for rice. Breed Sci 60:524–535

McNally K, Childs K, Bohnert R et al (2009) Genome wide SNP variation reveals relationships among landraces and modern varieties of rice. Proc Natl Acad Sci U S A 106:12273–12278

Mertz LM, Henning FA, Soares RC, Baldiga RF, Peske FB, Moraes DM (2009) Physiological changes in rice seeds exposed to cold in the germination phase. Rev Bras Sement 31:254–262

Messeguer J, Fogher C, Guiderdoni E, Marfa V, Catala MM, Baldi G, Mele E (2001) Field assessment of gene flow from transgenic to cultivated rices (*Oryza sativa* L.) using a herbicide resistance genes as tracer marker. Theor Appl Genet 103:1151–1159

Min S, Li X, Xiong Z, Hu G (1991) Genetic analysis of resistance to whitebacked planthopper *Sogatella furcifera* Horvath in rice. In: Rice genetics II. IRRI, Manila, pp 263–267

Miura K, Yamamoto E, Morinaka Y, Takashi T, Kitano H, Matsuoka M, Ashikari M (2008) The hybrid breakdown 1(t) locus induces interspecific hybrid breakdown between rice Oryza sativa cv. Koshihikari and its wild relative O. nivara. Breed Sci 58:99–105

Miura K, Ikeda M, Matsubara A et al (2010) *OsSPL14* promotes panicle branching and higher grain productivity in rice. Nat Genet 42:545–549

Mizuta Y, Harushima Y, Kurata N (2010) Rice pollen hybrid incompatibility caused by reciprocal gene loss of duplicated genes. Proc Natl Acad Sci 107:20417–20422

Mohanty A, Kathuria H, Ferjani A, Sakamoto A, Mohanty P, Murata N, Tyagi A (2002) Transgenics of an elite indica rice variety Pusa Basmati 1 harbouring the codA gene are highly tolerant to salt stress. Theor Appl Genet 106:51–57

Moncada P, Martinez CP, Borrero J et al (2001) Quantitative trait loci for yield and yield components in an *Oryza sativa* x *Oryza rufipogon* BC_2F_2 population evaluated in an upland environment. Theor Appl Genet 102:41–52

Morinaga T (1951) Rice cultivars and breeding method in Indonesia (in Japanese). Nogyo oyobi Engei 26:623–625

Morinaga T (1954) Classification of rice varieties on the basis of affinity. In: Reports for the 5th meeting of international rice commission's working party on rice breeding, Ministry of Agriculture and Forestry, Tokyo, pp 1–14

Morinaga T (1968) Origin and geographical distribution of Japanese rice. Jpn Agric Res Q 3:1–5

Morishima H (2001) Evolution and domestication of rice. In: Khush GS, Brar DS, Hardy B (eds) Rice genetics IV. Science Publishers Inc, New Delhi, pp 63–77

Morishima H, Sano Y, Oka HI (1992) Evolutionary studies in cultivated rice and its wild relatives. Oxf Surv Evol Biol 8:135–184

Multani DS, Jena KK, Brar DS, de los Reyes BG, Angeles ER, Khush GS (1994) Development of monosomic alien addition lines and introgression of genes from *Oryza australiensis* Domin to cultivated rice of *O. sativa* L. Theor Appl Genet 88:102–109

Multani DS, Khush GS, Delos Reyes BG, Brar DS (2003) Alien genes introgression and development of monosomic alien addition lines from *Oryza latifolia* Desv. to rice, *Oryza sativa* L. Theor Appl Genet 107:395–405

Nagadhara D, Ramesh S, Pasalu IC et al (2003) Transgenic indica rice plants resistant to sap-sucking insects. Plant Biotechnol J 1:231–240

Nagadhara D, Ramesh S, Pasalu IC, Rao YK, Sarma NP, Reddy VD, Rao KV (2004) Transgenic rice plants expressing the snowdrop lectin gene (gna) exhibit high-level resistance to the whitebacked planthopper

(Sogatella furcifera). Theor Appl Genet 109:1399–1405

Nagamine T (1991) Genetic control of tolerance to chilling injury at seedling in rice, *Oryza sativa* L. Jpn J Breed 41:35–40

Nakasu EY, Edwards MG, Fitches E, Gatehouse JA, Gatehouse AM (2014) Transgenic plants expressing ω-ACTX-Hv1a and snowdrop lectin (GNA) fusion protein show enhanced resistance to aphids. Front Plant Sci 5:673

Nandakumar R, Babu S, Kalpana K, Raguchander T, Balasubramanian P, Samiyappan R (2007) Agrobacterium-mediated transformation of indica rice with chitinase gene for enhanced sheath blight resistance. Biol Plant 51:142–148

Nayak P, Basu D, Das S et al (1997) Transgenic elite indica rice plants expressing CryIA(c) δ-endotoxin of Bacillus thuringiensis are resistant against yellow stem borer (Scirpophaga incertulas). Proc Natl Acad Sci U S A 94:2111–2116

Nguyen VT, Burow MD, Nguyen HT, Le BT, Le TD, Paterson AH (2001) Molecular mapping of genes conferring aluminum tolerance in rice (*Oryza sativa* L.). Theor Appl Genet 102:1002–1010

Nguyen BD, Brar DS, Bui BC, Nguyen TV, Pham LN, Nguyen HT (2003) Identification and mapping of the QTL for aluminum tolerance introgressed from new source, *Oryza rufipogon* Griff. into *indica* rice, (*Oryza sativa* L.). Theor Appl Genet 106:583–593

Nishizawa Y, Nishio Z, Nakazono K, Soma M, Nakajima E, Ugaki M, Hibi T (1999) Enhanced resistance to blast (Magnaporthe grisea) in transgenic Japonica rice by constitutive expression of rice chitinase. Theor Appl Genet 99:383–390

Nishizawa Y, Saruta M, Nakazono K et al (2003) Characterization of transgenic rice plants over-expressing the stress-inducible β-glucanase gene Gns1. Plant Mol Biol 51:143–152

Noda H (1986) Damage to ears of rice plants caused by the white-backed planthopper, *Sogatella furcifera* (Homoptera: Delphacidae). Appl Entomol Zool 21:474–476

Noldin JA (2000) Red rice status and management in America. In: Baki BB, Chin DV, Mortimer M (eds) Wild and weedy rice in rice ecosystems in Asia – a review. International Rice Research Institute, Los BanÄos, pp 21–24

OECD (1999) Trends in international migration 1999 continuous reporting system on migration. OECD Publishing, France

Ohyanagi H, Tanaka T, Sakai H et al (2006) The Rice Annotation Project Database (RAP-DB): hub for *Oryza sativa* ssp. japonica genome information. Nucl Acid Res 34:741–744

Oka HI (1957) Genetic analysis for the sterility of hybrids between distantly related varieties of cultivated rice. J Genet 55:397–409

Oka HI (1958) Intervarietal variation and classification of cultivated rice. Ind J Genet Plant Breed 18:79–89

Oka HI (1961) Report of trip for investigation of rice in Latin American countries. National Institute of Genetics, Japan, pp 1–40

Oka HI (1988) Weedy form of rice. In: Origin of cultivated rice. Elsevier/Japan Sci Soc Press, Amsterdam/Tokyo, pp 107–114

Oliver SN, Van Dongen JT, Alfred SC et al (2005) Cold induced repression of the rice anther-specific cell wall invertase gene *OSINV4* is correlated with sucrose accumulation and pollen sterility. Plant Cell Environ 28:1534–1551

Orjuela J, Garavito A, Bouniol M et al (2010) A universal core genetic map for rice. Theor Appl Genet 120:563–572

O'Toole JC, Namuco OS (1983) Role of panicle exertion in water-stress induced sterility. Crop Sci 23:1093–1097

Ouyang S, Zhu W, Hamilton J et al (2007) The TIGR rice genome annotation resource: improvements and new features. Nucleic Acids Res 35:883–887

Pandey MK, Rani NS, Sundaram RM et al (2013) Improvement of two traditional Basmati rice varieties for bacterial blight resistance and plant stature through morphological and marker-assisted selection. Mol Breed 31:239–246

Piao RH, Jiang WZ, Ham TH et al (2009) Map-based cloning of the *ERECT PANICLE 3* gene in rice. Theor Appl Genet 119:1497–1506

Prasad B, Eizenga GC (2008) Rice sheath blight disease resistance identified in *Oryza* spp. accessions. Plant Dis 92:1503–1509

Prasad PVV, Boote KJ, Allen LH, Sheehy JE, Thomas JMG (2006) Species, ecotype and cultivar differences in spikelet fertility and harvest index of rice in response to high temperature stress. Field Crop Res 95:398–411

Price AH, Townend J, Jones MP, Audebert A, Courtois B (2002) Mapping QTLs associated with drought avoidance in upland rice grown in the Philippines and West Africa. Plant Mol Biol 48:683–695

Qiao YL, Lee SI, Piao RH et al (2010) Fine mapping and candidate gene analysis of the floury endosperm gene, *FLO(a)*, in rice. Mol Cells 29:167–174

Qiao YL, Piao RH, Shi JX et al (2011) Fine mapping and candidate gene analysis of *dense and erect panicle 3, DEP3*, which confers high grain yield in rice (*Oryza sativa* L.). Theor Appl Genet 122:1439–1449

Rahman ML, Jiang W, Chu SH et al (2009) High-resolution mapping of two rice brown planthopper resistance genes, *Bph20(t)* and *Bph21(t)*, originating from *Oryza minuta*. Theor Appl Genet 119:1237–1246

Ram T, Deen R, Gautam SK, Ramesh K, Rao YK, Brar DS (2010a) Identification of new genes for brown planthopper resistance in rice introgressed from *O. glaberrima* and *O. minuta*. Rice Genet Newsl 25:67–69

Ram T, Laha GS, Gautam SK, Deen R, Madhav MS, Brar DS, Viraktamath BC (2010b) Identification of new gene introgressed from *Oryza brachyantha* with broad-spectrum resistance to bacterial blight of rice in India. Rice Genet Newsl 25:57

Ramiah K (1937) Rice in Madras a popular handbook. Government Press, Madras

Rangel PN, Brondani RP, Rangel PH, Brondani C (2008) Agronomic and molecular characterization of introgression lines from the interspecific cross *Oryza sativa* (BG90-2)×*Oryza glumaepatula* (RS-16). Genet Mol Res 7:184–195

Rao KV, Rathore KS, Hodges TK et al (1998) Expression of snowdrop lectin (GNA) in transgenic rice plants confers resistance to rice brown planthopper. Plant J 15:469–477

Ravinder SS, Khush GS, Heinrichs EA (1982) Genetic analysis of resistance to whitebacked planthopper, *Sogatella furcifera* (Horvath), in some rice varieties. Crop Prot 1:289–297

Reboud X (2002) Effect of a gap on gene flow between otherwise adjacent transgenic *Brassica napus* crops. Theor Appl Genet 106:1048–1058

Reddy JN, Baraoidan MR, Bernardo MA, George MLC, Sridhar R (1997) Application of marker-assisted selection in rice for bacterial blight resistance gene, Xa21. Curr Sci 73:873–875

Reissig WH, Heinrichs EA, Litsinger JA, Moody K, Fiedler L, Mew TW, Barrion AT (1986) Illustrated guide to integrated pest management in rice in tropical Asia. International Rice Research Institute, Laguna

Renganayaki K, Fritz AK, Sadasivam S et al (2002) Mapping and progress toward map-based cloning of brown planthopper biotype-4 resistance gene introgressed from *Oryza officinalis* into cultivated rice, *O. sativa*. Crop Sci 42:2112–2117

Roschevicz RJ (1931) A contribution to the knowledge of rice. Bull Appl Bot Genet Plant Breed 27:31–33

Saha P, Majumder P, Dutta I, Ray T, Roy SC, Das S (2006) Transgenic rice expressing Allium sativum leaf lectin with enhanced resistance against sap-sucking insect pests. Planta 223:1329–1343

Saito K, Miura K, Nagano K, Hayano-Saito Y, Araki H, Kato A (2001) Identification of two closely linked quantitative trait loci for cold tolerance on chromosome 4 of rice and their association with anther length. Theor Appl Genet 103:862–868

Sakamoto T, Matsuoka M (2008) Identifying and exploiting grain yield genes in rice. Curr Opin Plant Biol 11:209–214

Sanchez PL, Wing RA, Brar DS (2013) The wild relative of rice: genomes and genomics. In: Zhang Q, Wing RA (eds) Genetics and genomics of rice, plant genetics and genomics: crops and models 5. Springer Science + Business Media, New York, pp 9–25

Sano Y (1983) A new gene controlling sterility in F_1 hybrids of two cultivated rice species. J Hered 74:435–439

Sano Y (1986) Sterility barriers between *Oryza sativa* and *O. glaberrima*. In: International Rice Research Institute-IRRI (ed) Rice genetics. IRRI, Manila, pp 109–118

Sarangi S, Ghosh J, Bora A, Das S, Mandal AB (2011) Agrobacterium-mediated genetic transformation of indica rice varieties involving Am-SOD gene. Indian J Biotechnol 10:9–18

Scafaro AP, Haynes PA, Atwell BJ (2009) Physiological and molecular changes in *Oryza meridionalis* Ng, a heat-tolerant species of wild rice. J Exp Bot 61(1):191–202. doi:10.1093/jxb/erp294

Shakiba E, Eizenga GC (2014) Unraveling the secrets of rice wild species. In: Yan W, Bao J (eds) Rice – germplasm, genetics and improvement. In Tech, Croatia. doi:10.5772/58393

Shamshuddin J, Muhrizal S, CheFauziah I, Van Ranst E (2004) A laboratory study of pyrite oxidation in an acid sulfate soils. Commun Soil Sci Plant Anal 35:117–129

Shan JX, Zhu MZ, Shi M, Gao JP, Lin HX (2009) Fine mapping and candidate gene analysis of *spd6*, responsible for small panicle and dwarfness in wild rice (*Oryza rufipogon Griff.*). Theor Appl Genet 119:827–836

She KC, Kusano H, Koizumi K (2010) A novel factor *FLOURY ENDOSPERM2* is involved in regulation of rice grain size and starch quality. Plant Cell 22:3280–3294

Shinjo C (1984) Cytoplasmic male sterility and fertility restoration in rice having genome A. In: Tsunoda S, Takahashi N (eds) Biology of rice. Elsevier, Amsterdam, pp 321–338

Shomura A, Izawa T, Ebana K, Ebitani T, Kanegae H, Konishi S, Yano M (2008) Deletion in a gene associated with grain size increased yields during rice domestication. Nat Genet 40:1023–1028

Siddiq EA (2000) Bridging the rice yield gap in India. In: Proceedings of expert conference on bridging the rice yield gap in the Asia-Pacific Region, Bangkok, Thailand, 5–7 Oct 1999 (In press)

Sidhu G, Khush G (1979) Linkage relationships of some genes for disease and insect resistance and semidwarf stature in rice. Euphytica 28:233–237

Silue D, Notteghem J (1991) Resistance of 99 *Oryza glaberrima* varieties to blast. Int Rice Res Newsl 16:13–14

Silué D, Nottéghem JL, Chaume J (1992) Etude de la résistance spécifique d' *Oryza glaberrima* Steud à la pyriculariose du riz (translation Study of the specific resistance of *Oryza glaberrima* Steud to rice blast). Agronomie Tropicale 42:121–129

Singh S, Sidhu JS, Huang N et al (2001) Pyramiding three bacterial blight resistance genes (*xa5*, *xa13* and *Xa21*) using marker-assisted selection into *indica* rice cultivar PR106. Theor Appl Genet 102:1011–1015

Sobrizal YM, Sanchez PL, Ikeda K, Yoshimura A (2000) Identification of a gene for male gamete abortion in backcross progeny of *Oryza sativa* and *Oryza glumaepatula*. Rice Genet Newsl 17:59–61

Sobrizal YM, Matsuzaki Y, Yoshimura A (2001) Mapping of a gene for pollen semi-sterility on chromosome 8 of rice. Rice Genet Newsl 18:59–61

Sobrizal YM, Matsuzaki Y, Yoshimura A (2002) Mapping of pollen semi-sterility gene, *S28(t)*, on rice chromosome 4. Rice Genet Newsl 19:80–81

Song ZP, Lu BR, Chen JK (2001) A study of pollen viability and longevity in *Oryza rufipogon*, *O. sativa* and their hybrid. Int Rice Res Notes 26:31–32

Song ZP, Lu BR, Zhu YG, Chen JK (2003) Gene flow from cultivated rice to the wild species *Oryza rufipogon* under experimental field conditions. New Phytol 157:657–665

Song XJ, Huang W, Shi M, Zhu MZ, Lin HX (2007) A QTL for rice grain width and weight encodes a previously unknown RING-type E3 ubiquitin ligase. Nat Genet 39:623–630

Stebbins GL (1958) The inviability, weakness, and sterility of inter-specific hybrids. Adv Genet 9:147–215

Su J, Wang Z, Zhang K et al (2013) Status of insecticide resistance of the whitebacked planthopper, *Sogatella furcifera* (Hemiptera: Delphacidae). Fla Entomol 96:948–956

Suh JP, Yang SJ, Jeung JU et al (2011) Development of elite breeding lines conferring *Bph18* gene-derived resistance to brown planthopper (BPH) by marker-assisted selection and genome-wide background analysis in japonica rice (*Oryza sativa* L.). Field Crops Res 120:215–222

Suh JP, Jeung JU, Noh TH et al (2013) Development of breeding lines with three pyramided resistance genes that confer broad-spectrum bacterial blight resistance and their molecular analysis in rice. Rice 6:5

Sun CQ, Wang XK, Yoshimura A, Iwata N (2001) Comparison of the genetic diversity of common wild rice (*Oryza rufipogon* Griff.) and cultivated rice (*Oryza sativa* L.) using RFLP markers. Theor Appl Genet 102:157–162

Sun X, Wu A, Tang K (2002) Transgenic rice lines with enhanced resistance to the small brown plant hopper. Crop Prot 21:511–514

Sun L, Su C, Wang C, Zhai H, Wan J (2005) Mapping of a major resistance gene to the brown planthopper in the rice cultivar Rathu Heenati. Breed Sci 55:391–396

Taguchi K, Doi K, Yoshimura A (1999) RFLP mapping of *S19*, a gene for F1 pollen semi-sterility found in backcross progeny of *Oryza sativa* and *O. glaberrima*. Rice Genet Newsl 16:70–71

Tan GX, Ren X, Weng QM, Shi ZY, Zhu LL, He GC (2004a) Mapping of a new resistance gene to bacterial blight in rice line introgressed from *Oryza officinalis*. Acta Genet Sin 31:724–729 (Chinese with English abstract)

Tan GX, Weng QM, Ren X, Huang Z, Zhu LL, He GC (2004b) Two whitebacked planthopper resistance genes in rice share the same loci with those for brown planthopper resistance. Heredity 92:212–217

Tan LB, Li XR, Liu FX et al (2008) Control of a key transition from prostrate to erect growth in rice domestication. Nat Genet 40:1360–1364

Tanaka T, Antonio BA, Kikuchi S et al (2008) The Rice Annotation Project Database (RAP-DB): 2008 update. Nucleic Acids Res 36:1028–1033

Tang K, Tinjuangjun P, Xu Y et al (1999) Particle-bombardment-mediated co-transformation of elite Chinese rice cultivars with genes conferring resistance to bacterial blight and sap-sucking insect pests. Planta 208:552–563

Tang W, Chen H, Xu C, Li X, Lin Y, Zhang Q (2006) Development of insect-resistant transgenic *indica* rice with a synthetic *cry1C** gene. Mol Breed 18:1–10

Tateoka T (1962) Taxonomic studies of *Oryza*. I. *O. latifolia* complex. II. Bot Mag 75:418–427

Temnykh S, DeClerck G, Lukashova A, Lipovich L, Cartinhour S, McCouch S (2001) Computational and experimental analysis of microsatellites in rice (*Oryza sativa* L.): frequency, length variation, transposon associations, and genetic marker potential. Genome Res 11:1441

Terao T, Nagata K, Morino K, Hirose T (2010) A gene controlling the number of primary rachis branches also controls the vascular bundle formation and hence is responsible to increase the harvest index and grain yield in rice. Theor Appl Genet 120:875–893

Thomson MJ, Zhao K, Wright M, McNally KL, Leung H, McCouch SR (2011) Development and application of 96- and 384-plex single nucleotide polymorphism (SNP) marker sets for diversity analysis, mapping and marker-assisted selection in rice. Second Africa Rice Congress, Bamako, Mali, 22–26 March 2010: Innovation and Partnerships to Realize Africa's Rice Potential

Tian F, Li DJ, Fu Q, Zhu ZF, Fu YC, Wang XK, Sun CQ (2006a) Construction of introgression lines carrying wild rice (*Oryza rufipogon* Griff.) segments in cultivated rice (*Oryza sativa* L.) background and characterization of introgressed segments associated with yield-related traits. Theor Appl Genet 112:570–580

Tian F, Zhu ZF, Zhang BS, Tan LB, Fu YC, Wang XK, Sun CQ (2006b) Fine mapping of a quantitative trait locus for grain number per panicle from wild rice (*Oryza rufipogon* Griff.). Theor Appl Genet 113:619–629

Tinjuangjun P, Loc NT, Gatehouse AMR, Gatehouse JA, Christou P (2000) Enhanced insect resistance in Thai rice varieties generated by particle bombardment. Mol Breed 6:391–399

Tranel PJ, Wassom JJ, Jeschke MR, Rayburn AL (2002) Transmission of herbicide resistance from a monoecious to dioecious weedy *Amaranthus* species. Theor Appl Genet 105:674–679

Tu J, Zhang G, Datta K et al (2000) Field performance of transgenic elite commercial hybrid rice expressing *Bacillus thuringiensis* δ-endotoxin. Nat Biotechnol 18:1101–1104

Ueno K, Sato T, Takahashi N (1990) The *indica-japonica* classification of Asian rice ecotypes and Japanese lowland and upland rice (*Oryza sativa* L.). Euphytica 46:161–164

Utami DW, Moeljopawiro S, Aswidinnoor H, Setiawan A, Hanarida I (2008) Blast resistance genes in wild rice *Oryza rufipogon* and rice cultivar IR64. Indones J Agr 1:71–76

Vaughan DA (1989) The genus *Oryza* L.: current status of taxonomy. International Rice Research Institute, Manila, Philippines, IRRI research paper series vol 138, p 21

Vaughan DA (1994) The wild relatives of rice: a genetic resources handbook. International Rice Research Institute, Manila, p 137

Wang ET, Wang JJ, Zhu XD et al (2008) Control of rice grain-filling and yield by a gene with a potential signature of domestication. Nat Genet 40:1370–1374

Wang JM, Lin XY, Liu XM, Sun Q, Li PZ, Zhang SY (2012) Selecting the pyramids of blast resistance genes *Pi40* and *Pib* in rice assisted by molecular marker. Acta Agric boreali-sinica 27:218–221

Wei XJ, Xu JF, Guo HN et al (2010) *DTH8* suppresses flowering in rice, influencing plant height and yield potential simultaneously. Plant Physiol 153:1747–1758

Wen S, Gao B (2011) Introgressing blast resistant gene *Pi-9(t)* into elite rice restorer Luhui17 by marker-assisted selection. Rice Genome Genet 2:31–36

Weng JF, Gu SH, Wan XY et al (2008) Isolation and initial characterization of *GW5*, a major QTL associated with rice grain width and weight. Cell Res 18:1199–1209

Whyte RO (1972) The Gramineae, wild and cultivated of monsoonal and equatorial Asia. I. Southeast Asia. Asian Perspect 15:127–151

Wissuwa M, Wegner J, Ae N, Yano M (2002) Substitution mapping of *Pup1*: a major QTL increasing phosphorus uptake of rice from a phosphorus-deficient soil. Theor Appl Genet 105:890–897

Wu C, Khush GS (1985) A new dominant gene for resistance to whitebacked planthopper in rice. Crop Sci 25:505–509

Wu P, Liao CY, Hu B, Yi KK, Jin WZ, Ni JJ, He C (2000) QTLs and epistasis for aluminum tolerance in rice (*Oryza sativa* L.) at different seedling stages. Theor Appl Genet 100:1295–1303

Wunn J, Kloti A, Burkhardt PK, Ghosh Biswas GC, Launis K, Iglesias VA, Potrykus I (1996) Transgenic indica rice breeding line IR58 expressing a synthetic Cry1A(b) gene from Bacillus thuringiensis provides effective insect pest control. Biotechnology (NY) 14:171–176

Xiao J, Grandillo S, Ahn SN, McCouch S, Tanksley SD (1996) Genes from wild rice improves yield. Nature 384:223–224

Xiao J, Li J, Grandillo S, Yuan L, Tanksley SD, Mc Couch SR (1998) Identification of traits improving quantitative trait loci alleles from wild rice relative *Oryza rufipogon*. Genetics 150:899–909

Xie XB, Song MH, Jin FX, Ahn SN, Suh JP, Hwang HG, McCouch SR (2006) Fine mapping of a grain weight quantitative trait locus on rice chromosome 8 using near-isogenic lines derived from a cross between *Oryza sativa* and *Oryza rufipogon*. Theor Appl Genet 113:885–894

Xie XB, Jin FX, Song MH et al (2008) Fine mapping of a yield-enhancing QTL cluster associated with transgressive variation in an *Oryza sativa×O.rufipogon* cross. Theor Appl Genet 116:613–622

Xing YZ, Tang WJ, Xue WY, Xu CG, Zhang QF (2008) Fine mapping of a major quantitative trait loci, *qSSP7*, controlling the number of spikelets per panicle as a single Mendelian factor in rice. Theor Appl Genet 116:789–796

Xu CG (1995) Adsorption and germination of pollen in crosses within and between *indica* and *japonica* rice. J Huazhong Agric Univ 14:421–424 (In Chinese with English abstract)

Xu M, Tang Z, Tan Y et al (2003) A study on introduction of chitinase gene and β-1,3-glucanase gene into restorer line of dian-type hybrid rice (Oryza sativa L.) and enhanced resistance to blast (Magnaporthe grisea). Acta Genet Sin 30:330–334

Xu P, Zhou J, Li J et al (2014) Mapping three new interspecific hybrid sterile loci between *Oryza sativa* and *O. glaberrima*. Breed Sci 63:476

Xue WY, Xing YZ, Weng XY et al (2008) Natural variation in *Ghd7* is an important regulator of heading date and yield potential in rice. Nat Genet 40:761–767

Yamamoto E, Takashi T, Morinaka Y et al (2010) Gain of deleterious function causes an autoimmune response and Bateson–Dobzhansky–Muller incompatibility in rice. Mol Genet Genomics 283:305–315

Yan WH, Wang P, Chen HX et al (2011) A major QTL, *Ghd8* plays pleiotropic roles in regulating grain productivity, plant height, and heading date in rice. Mol Plant 4:319–330

Yang H, Ren X, Weng Q, Zhu L, He G (2002) Molecular mapping and genetic analysis of a rice brown planthopper (*Nilaparvata lugens* Stål) resistance gene. Hereditas 136:39–43

Yang H, You A, Yang Z, Zhang F, He R, Zhu L, He G (2004) High-resolution genetic mapping at the *Bph15* locus for brown planthopper resistance in rice (*Oryza sativa* L.). Theor Appl Genet 110:182–191

Yang J, Li Y, Zhang Y, Zheng S (2009) Possible involvement of cell wall pectic polysaccharides in Al resistance of some plant species. In: Liao H, Yan X, Kochian L (eds) Proceedings of the 7th international symposium on the plant-soil interaction at low pH… South China University of Technology, pp 57–58

Yoon DB, Kang KH, Kim HJ et al (2006) Mapping quantitative trait loci for yield components and morphological traits in an advanced backcross population between *Oryza grandiglumis* and the *O. sativa japonica* cultivar Hwaseongbyeo. Theor Appl Genet 112:1052–1062

Yoshida S (1983) Rice. In: Smith WH, Banta SJ (eds) Potential productivity of field crops under different environments'. International Rice Research Institute Publishing, Los Baños, pp 103–127

Yu J, Hu SN, Wang J et al (2002) A draft sequence of the rice genome (*Oryza sativa* L. ssp *indica*). Science 296:79–92

Yu L, Chen X, Wang Z et al (2013) Arabidopsis Enhanced Drought Tolerance1/HOMEODOMAIN GLABROUS11 confers drought tolerance in transgenic rice without yield penalty. Plant Physiol 162:1378–1391

Zha XJ, Luo XJ, Qian XY, He GM, Yang MF, Li Y, Yang JS (2009) Over-expression of the rice *LRK1* gene improves quantitative yield components. Plant Biotechnol J 7:611–620

Zhang Q, Wang CL, Shi AN, Bai JF, Ling SC, Li DY, Chen CB, Pang HH (1994) Evaluation of resistance to bacterial blight (*Xanthomonas oryzae* pv. *oryzae*) in wild rice species. Sci Agric Sin 27:1–9

Zhang Q, Lin SC, Zhao BY et al (1998) Identification and tagging a new gene for resistance to bacterial blight (*Xanthomonas oryzae* pv. *oryzae*) from *O. rufipogon*. Rice Genet Newsl 15:138–142

Zhang Q, Zhao BY, Zhao KJ et al (2000) Identifying and mapping a new gene *Xa23*(t) for resistance to bacterial blight (*Xanthomonas oryzae* pv. *oryzae*) from *O. rufipogon*. Acta Agron Sin 26:536–542

Zhang X, Zhou S, Fu Y, Su Z, Wang X, Sun C (2006) Identification of a drought tolerant introgression line derived from Dongxiang common wild rice (*O. rufipogon* Griff.). Plant Mol Biol 62:247–259

Zhao K, Tung CW, Eizenga GC et al (2011) Genome-wide association mapping reveals a rich genetic architecture of complex traits in *Oryza sativa*. Nat Commun 2:467

Zhao J, Li J, Xu P et al (2012) A new gene controlling hybrid sterility between *Oryza sativa* and *Oryza longistaminata*. Euphytica 187:339–344

Zheng CK, Chun-Lian W, Yuan-Jie YU, Liang YT, Kai-Jun Z (2009) Identification and molecular mapping of *Xa32*(t), a novel resistance gene for bacterial blight (*Xanthomonas oryzae* pv. *oryzae*) in rice. Acta Agron Sin 35:1173–1180

Zhu KM, Tang D, Yan CJ et al (2010) Erect panicle 2 encodes a novel protein that regulates panicle erectness in *indica* rice. Genetics 184:343–350

Maize

4

Firoz Hossain, Vignesh Muthusamy, Jayant S. Bhat, Shailendra K. Jha, Rajkumar Zunjare, Abhijit Das, Konsam Sarika, and Rajesh Kumar

Abstract

Maize (*Zea mays* ssp. *mays*) originated in Mexico and Central America and belongs to the tribe Maydae of the family Poaceae. It possesses over 32,000 genes on ten chromosomes with a genome size of 2.3 gigabase. Among various theories, teosinte (*Z. mays* ssp. *parviglumis*) is the most likely progenitor of maize. Genetic loci such as *teosinte branched 1* (*tb1*) and *teosinte glume architecture 1* (*tga1*) have played pivotal role in transforming teosinte to modern maize. From the centre of origin, maize later diffused to different parts of the world including America, Europe, Africa and Asia. Genetic resources especially wild relatives and landraces harbour novel alleles/genes for imparting resistance/tolerance to several biotic/abiotic stresses and improvement of productivity and nutritional quality traits. Among wild relatives, teosintes and *Tripsacum* are native to Mexico and Central America, while *Coix, Chionachne, Sclerachne, Trilobachne* and *Polytoca* originated in Southeast Asia. Cross compatibility between maize and teosinte generates hybrids, although the extent of success depends on teosinte species. While generation of crosses between maize and *Tripsacum* species is difficult, Asiatic genera possess high degree of cross incompatibility with maize. Several factors or loci that regulate cross compatibility between different genera or species have been identified and play pivotal role in introgression of genes from wild relatives. Landrace accessions possessing unexplored alleles/genes serve as valuable donors for important traits. The chapter deals with different theories of maize origin and recent understanding on diffusion of maize worldwide, besides providing classifications of wild relatives and their cross compatibility with maize. The importance of wild relatives and maize

F. Hossain (✉) • V. Muthusamy • J.S. Bhat • S.K. Jha
R. Zunjare • A. Das • K. Sarika • R. Kumar
Division of Genetics, ICAR-Indian Agricultural Research Institute, New Delhi 110012, India
e-mail: fh_gpb@yahoo.com

M. Singh, S. Kumar (eds.), *Broadening the Genetic Base of Grain Cereals*,
DOI 10.1007/978-81-322-3613-9_4

landrace accessions in providing valuable genes for various important traits has been discussed.

Keywords
Characterization • Landrace • Teosinte • *Tripsacum* • Wild relatives • *Zea mays*

4.1 Introduction

Maize (*Zea mays* ssp. *mays*) belongs to the tribe Maydae, family Poaceae and was originated in Mexico and Central America. It possesses somatic chromosome number of 20, a genome size of 2.3 gigabase and more than 32,000 genes (Schnable et al. 2009). Maize grows well in various agroecologies and is unparalleled to any other crop due to its ability to adapt in diverse environments. It has emerged as a crop of global importance owing to its multiple end uses as a human food and livestock feed and serves as an important component for varied industrial products. Besides, maize serves as a model organism for biological research worldwide. Globally, about 1016.73 million metric tonnes of maize is produced every year – the highest among major staple cereals (FAOSTAT 2013). A major portion of maize produced worldwide is used for animal consumption as it serves as a vital source of proteins and calories to billions of people in developing countries, particularly in Africa, Mesoamerica and Asia (Shiferaw et al. 2011). Further, it is a source of important vitamins and minerals to the human body. Along with rice and wheat, maize provides at least 30 % of the food calories to more than 4.5 billion people in 94 developing countries. Maize provides over 20 % of total calories in human diets in 21 countries and over 30 % in 12 countries that are home to a total of more than 310 million people (Shiferaw et al. 2011). At present, the developed world uses more maize than the developing world, but forecasts indicate that by the year 2050, the demand for maize in the developing countries will double owing to the rapid growth in poultry industry, the biggest driver of growth in maize production (Rosegrant et al. 2009; Prasanna 2014). Improved maize hybrids with substantial increase in production per unit area are required to feed the ever-growing population. Further, with changing climatic conditions, several new biotic stresses have emerged and minor disease and insect pests have become more prevalent and started inflicting more damages. Among abiotic stresses drought, heat and water logging are the major one and their simultaneous occurrence are now more frequent than ever. Malnutrition caused by deficiency of minerals and vitamins, especially iron, zinc and vitamin A, has been identified as one of the most important problems that require urgent attention worldwide (Bouis et al. 2011; Gupta et al. 2015a, b, c). Germplasm including wild relatives and landraces possess enormous potential as genetic resource for harbouring important and novel alleles/genes. These valuable germplasm can be systematically and effectively utilized in the crop improvement programmes worldwide, to develop high yielding and nutritious maize with resilience to biotic and abiotic stresses. Here we present origin and spread of maize, importance of wild relatives and landraces in the maize improvement programme.

4.2 Origin of Maize

The term ‘maize’ seems to be derived from the word ‘mahiz’ of Taino language of the Caribbean islands, which became ‘maiz’ in Spanish (Oxford dictionary 2015). Based on this common name, Linnaeus included the name as species in the botanical classification of *Zea*. Maize is also popularly known as ‘corn’ in English-speaking countries. In some countries, ‘corn’ means the ‘local staple’, while in some others it is used for any ‘cereal’. The ear of maize is unique among cereals, and morphologically similar wild progenitor of maize could not be found. Therefore, its evolution has been a great scientific challenge and of great interest for both biologists and archaeologists. Many hypotheses/theories have been proposed by different scientists to explain the origin of maize. Among them, (1) tripartite hypothesis, (2) catastrophic sexual transmutation theory, (3) *Tripsacum-Zea diploperennis* hypothesis, and, (4) teosinte hypothesis were debated and discussed in detail by different scientists.

4.2.1 Tripartite Hypothesis

Tripartite hypothesis, proposed by Mangelsdorf and Reeves (1938, 1939), and later elaborated by Mangelsdorf (1974), states that maize was domesticated from some unknown wild, now extinct maize plant that had structures similar to the ear of modern maize. The hypothesis comprised three parts; (1) the progenitor of maize was a wild maize prototype from South America, which has become extinct or remained undiscovered; (2) teosinte is the offspring of a cross between maize and *Tripsacum*; and, (3) sections of *Tripsacum* chromosomes had contaminated maize germplasm. Thus, Mangelsdorf and Reeves explained the extreme morphological differences between maize and teosinte by imagining a missing ancestor, while relied on *Tripsacum* to explain their similarities. The evidence/validation for this explanation came from their own experiment. Mangelsdorf and Reeves obtained few sterile maize-*Tripsacum* hybrids from the artificial crossing. They could identify a minimum of four factors with large effect that controlled defined morphological syndrome. They believed that each of these four factors with large effects represented a block of many linked genes. This interpretation was tied to their hypothesis that maize evolved in nature from a now extinct form of wild maize over a long evolutionary period. They believed that the differences between maize and teosinte were very large, and hence it is not possible that teosinte was transformed into maize in the few 1000 years during which virtually all crops were domesticated. Until the 1960s, the tripartite hypothesis was widely accepted. However, data were not sufficient and their hypothesis could not withstand the test of time (Doebley 2001).

4.2.2 Catastrophic Sexual Transmutation Theory

Iltis (1983) proposed that maize was originated due to a sudden sexual transmutation that condensed the branches of teosinte and placed them in the female expression area of the plant. It states that the ear of maize was derived from the central spike of the tassel of teosinte. According to Iltis, this has happened due to a phenomenon known as 'genetic assimilation'. This resulted in substantial alterations in the nutrient distribution of the plant and led to drastic morphological changes. Morphogenetic and structural imbalance possibly had led to the transformation into primitive maize. During the late 1980s, teosinte hypothesis started gaining importance and the catastrophic sexual transmutation theory became less convincing.

4.2.3 *Tripsacum-Zea diploperennis* Hypothesis

Tripsacum-Z. diploperennis hypothesis can be considered as a modern version of the tripartite hypothesis and was given by Eubanks (1995). It proposed that maize arose from the progeny of a cross between *Z. diploperennis* and *Tripsacum dactyloides* (Eubanks 2001). This proposal was put forward with the observations on two putative hybrids viz. 'Tripsacorn' and 'Sundance' originated from these two grasses. The rudimentary ear of these putative hybrids had exposed kernels attached to a central rachis or cob. If such hybrids occurred naturally, then at least – according to proponents of the hypothesis – the evolutionary puzzle of the origin of maize and its unparalleled architecture is solved. Thus, this hypothesis challenged the idea that maize is a domesticated form of teosinte. However, *Tripsacum* and *Z. diploperennis* could not be hybridized successfully. The chromosome number of both 'Tripsacorn' and 'Sundance' is $2n=20$. These hybrids would be expected to have 28 or 46 chromosomes if *Tripsacum* (2n=36 or 72) had indeed been one of the parents. Of the polymorphisms identified by RFLP data (Eubanks 1997), 'Tripsacorn' and 'Sundance' shared four times as many bands with *Z. diploperennis* as with *Tripsacum*, indicating a much closer relationship with teosinte than with *Tripsacum*. Besides, 23 % of the molecular markers surveyed were not found in either of the parents.

4.2.4 Teosinte Hypothesis

The teosinte hypothesis states that teosinte is the sole progenitor of maize. As proposed by Beadle (1939, 1972, 1978), the teosinte hypothesis states that (1) teosinte provided a useful food source and ancient people cultivated it; (2) during the cultivation of teosinte, mutations that improved teosinte's usefulness to humans arose and were selected by people; (3) as few as five major mutations would be sufficient to convert teosinte into a primitive form of maize; (4) different mutations controlled different traits, viz. one mutation would have converted the disarticulating ear-type of teosinte into the solid ear type of maize; and, (5) over the period of time, humans selected additional major mutations coupled with many minor ones. Beadle (1939) believed that missing ancestor is not needed to explain the origin. He could obtain completely fertile hybrids between maize and teosinte. He interpreted that these two species were conspecific and that they had diverged only recently. Beadle actually used Mangelsdorf and Reeves's own data against them and suggested that their four factors might correspond to four major genes. Each of these major genes controlled a single trait that transformed teosinte into maize. In modern form of teosinte hypothesis, *Z. mays* ssp. *parviglumis* (wild Mexican grass teosinte) has been pinpointed as the likely progenitor. Further, maize arose through large changes in *parviglumis* – through artificial selection for specific traits. Most maize geneticists and evolutionists have now accepted that maize is a domesticated derivative of *parviglumis* (Bennetzen et al. 2001). *Z. mays* ssp. *parviglumis* grows as a wild plant alongside of the Balsas river and hence commonly known as Balsas teosinte.

4.3 Characteristics of Teosinte and Maize

Though it is accepted that teosinte is the progenitor of maize, extreme morphological differences persist between teosinte and maize (Doebley and Stec 1991) (Table 4.1). Some reported that the morphological differences between maize and

Table 4.1 Differences among teosinte and maize for various morphological characteristics

Teosinte	Maize
1. Teosinte plants are branched and produce many ears	1. Maize plants produce a single upright stem with one or few ears
2. Terminal position of primary branch bears a tassel	2. Primary branch is modified into ears
3. The leaves along the lateral branches are fully formed and composed of leaf blade and sheath	3. Leaves of the lateral branch are modified into husks which cover the ear
4. Secondary lateral branch is modified into ears	4. Secondary lateral branches are extremely rare
5. Ears are covered loosely by a single or few husks	5. Ears are covered tightly by many husks
6. Each ear possesses only two kernel rows	6. Each ear possesses about 8–22 kernel rows
7. Ear possesses about 10–12 kernels	7. Ear bear about 250–500 kernels
8. Each kernel is sealed tightly in a stony casing or fruitcase	8. Each kernel is naked and not covered by any fruitcase
9. During development, out of two spikelets one is aborted, hence each fruitcase holds a single-spikelet	9. Maize evolution involved the derepression of the second spikelet primordium, hence there are two mature spikelets
10. At maturity, fruitcase having the kernel shatter and become the dispersal units	10. At maturity, kernels do not shatter, and remain attached with ears
11. Majority of teosintes possess varying degree of seed dormancy	11. Seeds of maize do not possess dormancy

teosinte exhibit quantitative variation while others reported Mendelian pattern. Based on the conclusions from various studies, the genes responsible for transformation of teosinte into maize might be categorized into three overlapping classes: (1) major loci with simple inheritance, phenotypic effects of which are affected to lesser degree by environment; (2) QTL having complex inheritance are influenced by both genetic background and environment; and, (3) QTLs whose effects are dependent to a greater extent on epistatic interactions. This type of QTLs could exist in teosinte with little or no contribution to the phenotypic variation (Lauter and Doebley 2002). The available facts indicate that most of the morphological changes brought about during maize domestication could be accounted by QTL of measurable additive effects. Doebley and Stec (1993) and Doebley et al. (1994) mapped QTLs governing key contrasting traits using two F_2 populations generated between teosinte and maize. Key morphological traits differing between maize and teosinte have been located on chromosome 1 L (plant architecture), chromosome 2S (ear rank), chromosome 4S (cupulate fruitcase), chromosome 1 L, 1S and 3 L (pattern of spikelet: single vs. paired). Ear disarticulation showed complex inheritance as the observations were not consistent in two populations. However, a few loci having major effects, like *teosinte branched1* (*tb1*) and *teosinte glume architecture1* (*tga1*), may have played a larger role in transforming teosinte to modern maize.

4.3.1 *Teosinte Branched1* (*tb1*)

The *teosinte branched1* (*tb1*) is a mutant of maize for plant architecture or apical dominance, and maize plant homozygous for mutant *tb1* shows lateral branches like teosinte but terminating with sterile tassels (Doebley et al. 1995, 1997; Doebley 2004). This gene has been reported as class II member of TCP family of transcriptional regulators which is hypothesized to negatively control cell cycle genes and results in repressed organ growth (Cubas et al. 1999; Doebley 2004). Expression analysis showed that maize allele of *tb1* is expressed twice the level of the teosinte allele and confers repression of branching in maize (Doebley 2004). The difference between maize and teosinte architecture was not due to change in coding sequence of *tb1* gene, instead due to variations in regulatory region (Doebley et al. 1997; Wang et al. 1999). Phenotypic fine-mapping and population genetic analysis indicated that the selection force operated between −58.7 and −65.6 kb of the *tb1*-ORF, and haplotype analysis between maize and teosinte revealed that there are four fixed differences [two SNPs and two insertions, i.e. *Hopscotch* retrotransposon and *Tourist* miniature inverted-repeat transposable element (MITE)] between the sequences in this region. These four differences were grouped into two regions, i.e. proximal (*Hopscotch* and one SNP) and distal (*Tourist* and the other SNPs), based on point of recombination between them. Of these, *Hopscotch* insertion is the major causative polymorphism between maize and teosinte (Studer et al. 2011). Thus, modification of *tb1* expression was an important target of selection during maize domestication. Recently, Vann et al. (2015) reported the presence of *Hopscotch* element in *tb1* among some populations of teosinte and concluded that the role of *Hopscotch* differs between maize and teosinte. Analyses of *tb1* expression levels in teosinte with and without the *Hopscotch* element could possibly provide more lights on regulation of *tb1*.

4.3.2 *Teosinte Glume Architecture1* (*tga1*)

Another major difference between maize and teosinte is the presence of cupulate fruitcase for which *teosinte glume architecture1* (*tga1*) was identified as a major locus. The allele *tga1* has been found to have pleiotropic effect, and it affects several distinct aspects of fruitcase development. This suggested that *tga1* is a regulatory locus and it might get activated after the decision to become female is made. The *tga1* activates the programmes for invagination of the internode, internode elongation, three-dimensional growth of the glume, silica deposition, and the pattern of lignifications (Dorweiler et al. 1993; Dorweiler and Doebley 1997). Genetic fine mapping was used to map *tga1* to a ~6 kb segment between the marker *bnlg252* and *bm22.7* and further fine-mapped to a 1042 bp segment between SNP (-1024) and SNP (+18). BLAST searches revealed that *tga1* has homology to squamosa-promoter binding protein (SBP) family of transcriptional regulators. Six fixed differences between maize and teosinte were identified by DNA sequence analysis, of which five are SNPs at 5′ end of the coding sequence affecting *tga1* expression and the sixth difference is the substitution of lysine (K) in teosinte to asparagine (N) in maize. This substitution might have a role in altered protein stability, translation efficiency or protein function and is hypothesized to be the causative site of maize and teosinte polymorphism (Wang et al. 2005).

4.4 Diffusion of Maize

The centre of origin for maize is the Mesoamerican region now called Mexico and Central America (Watson and Dallwitz 1992). Most historians are of the view that maize was first domesticated in the Tehuacan Valley of Mexico (Roney and Hard 2009). Although maize pollen grains as old as 80,000 years have been identified from Mexico City, archaeological records suggest that domestication of maize is at least 6000 years old. However, a study by Matsuoka et al. (2002) has demonstrated that maize arose from a single domestication in Southern Mexico about 9000 years ago rather than the multiple independent domestication models. Though different studies have proved that maize was domesticated from *Z. mays* ssp. *parviglumis*, one major contradictory observation is that maize cultivars closely related to *parviglumis* are available in the Mexican highlands, whereas *parviglumis* is generally endemic to the mid- to lowlands. Domestication of maize in the highlands could be explained by the presence of *parviglumis* in the highlands at the time of domestication (Matsuoka et al. 2002; Piperno 2003) but unlikely because of cooler and drier climate existed at the time of domestication (Metcalfe 2006; Piperno et al. 2007, 2009). The other possibility could be that the early domesticate might have spread from lowlands to highlands and subsequent diffusion from highlands replaced lowland populations (Piperno 2003), which is also unlikely as it needs the highland maize to outcompete existing well-adapted populations in lowlands (Eagles and Lothrop 1994). Most acceptable hypothesis that explained this paradox is by the presence of a second subspecies of annual teosinte, *Z. mays* ssp. *mexicana* in the highland along with maize. Maize and *mexicana* are cross compatible (Ellstrand et al. 2007), and gene flow from *mexicana* to maize has also been reported (Wilkes 1972; Matsuoka et al. 2002). In comparison to maize, *mexicana* is more closely related to *parviglumis* (Matsuoka et al. 2002), so gene flow from *mexicana* can serve as indirect means of genetic similarity between highland maize populations and *parviglumis*.

As stated above, *parviglumis* is endemic to lowland tropical short-day condition, so adaptation of maize to longer days was prerequisite for its diffusion from tropical to temperate regions (Matsuoka et al. 2002; Piperno et al. 2009; van-Heerwaarden et al. 2011). One of the major genes reported to modulate photoperiod sensitivity is *ZmCCT* which encodes a CCT domain-containing protein. Adaptation under long day conditions was possible due to an insertion of a CACTA-like transposon into the *ZmCCT* promoter that can suppress the *ZmCCT* expression. Tropical maize plants with this transposable element insertion has been selected for by mankind and accumulated over time as maize adapted to range of long-day environments (Yanga et al. 2013). This insertion event helped maize to diffuse it from its centre of origin and adapt to different photoperiodic conditions around the world. Initial diffusion of maize across the America was a slow process which took over thousands of years, but following the discovery of the new world by Columbus, rapid diffusion took place around the world in the past 500 years (Dubreuil et al. 2006).

Mir et al. (2013) conducted the first genetic study at global level to find out the actual diffusion path of maize from America to different parts of the world by genotyping 799 different landraces collected around the globe using SSR markers. Initially to identify population structure within centre of diversity, 258 American landraces itself were classified into seven distinct clusters, viz. 'Northern US flints', 'Mexican highlands', 'Tropical lowlands', 'Middle North American', 'Northern South American', 'Middle South American' and 'Andes', and used as prior information to interpret which of these clusters served as source germplasm to different non-American landraces. Northern US flints germplasm spread out through Europe from Northern France eastwards and also contributed through hybridization to Pyrenean-Galician landraces (Mir et al. 2013). The same was also evident in admixed-ancestry of Portuguese landraces, suggesting a hybrid origin. Diffusion of maize germplasm in the Middle East and Eastern Africa took place from the Middle North-American maize germplasm, but introduction to Eastern Africa was a distinct independent event rather than diffusion from the Middle East. The ancestry of North Eastern Chinese germplasm is traced back to Middle North American maize germplasm. Throughout Eastern Asia, mainly along the coasts, maize ancestry is traced back to Mexican highlands cluster, with an initial introduction into Indonesia. Thereafter, diffusion took place northward and towards Japan. Diffusion of tropical lowlands cluster contributed to the ancestry of Southern Spanish maize and also Moroccan landraces. Maize at Western Asia from Nepal to Afghanistan also showed tropical lowland ancestry but Mexican ancestry becomes predominant in Southeast Asia. Other clusters also acted as ancestral to maize germplasm for different parts of the globe. They are Northern South American cluster for Southern European landraces, Northern South American and Middle South American cluster for Western sub-Saharan African landraces, and Middle South American for few Eastern Asian landraces. But no clear diffusion path was observed for Andean germplasm that could be due to its geographical isolation and less productive nature beyond their ecological niches (Gouesnard et al. 2002).

4.5 Wild Relatives

The wild relatives of crop plants constitute an increasingly important genetic resource for improving agricultural production and maintaining sustainable agroecosystems (FAO 1998). A crop wild relative (CWR) can be defined as 'a wild plant taxon that has an indirect use derived from its relatively close genetic relationship to a crop' (Maxted et al. 2006). The grass genus *Zea* comprises of seven genera with varying chromosome number and are classified into two groups, viz. old world group and new world group (Table 4.2). *Coix, Chionachne, Sclerachne, Trilobachne* and *Polytoca* belong to the old world group and originated from the Southeast Asia. New world group consists of *Zea* and *Tripsacum* and are native to Mexico and Central America. *Z. mays* ssp. *mays* is the only species of *Zea* with economic importance, and the other species of *Zea* are generally referred to as teosintes.

4.5.1 Teosintes

Teosintes are wild grasses mainly originated in Mexico and Central America and represent potential genetic resources for maize improvement. All species of teosinte can be crossed with maize (Wilkes 1977; Wang et al. 2008, 2014). The group of four species, viz. *Z. diploperennis*, *Z. nicaraguensis*, *Z. luxurians* and *Z. perennis*, and three subspecies of *Z. mays*, viz. *Z. mays spp. huehuetenangensis*, *Z. mays* ssp. *mexicana* and *Z. mays* ssp. *parviglumis*, are collectively called teosintes (Table 4.2) (Fukunaga et al. 2005). Among these, *Z. mays* spp. *mexicana* is annual type teosinte that spread across the central highlands of Mexico.

4.5.2 *Tripsacum*

Tripsacum commonly called as 'gamma grass' belongs to the secondary gene pool of maize as defined by Harlan and de Wet (1971). This genus consists of nearly 16 perennial and warm species which originated in Mexico and Guatemala but

Table 4.2 Wild relatives of maize and their origin

S. No.	Genus	Species	Place of origin	2n[a]
1.	*Zea*	*mays* ssp. *mays*	Mexico and Central America	20
2		*mays* ssp. *mexicana*	Mexico	20
3.		*mays* ssp. *parviglumis*	Southern and Eastern Mexico	20
4.		*mays* ssp. *huehuetenangensis*	Guatemala	20
5.		*diploperennis*	Jalisco and Mexico	20
6.		*luxurians*	South Eastern Guatemala	20
7.		*nicaraguensis*	Nicaragua	20
8.		*perennis*	Jalisco and Mexico	40
9.	*Tripsacum*	*australe*	South America	36
10.		*cundinamarce*	Colombia	36
11.		*floridanum*	South Florida and Cuba	36
12.		*latifolium*	North and Central America	36
13.		*laxum*	Central America	36?
14.		*bravum*	Mexico	36, 72
15.		*zopilotense*	Meso-America	36, 72
16.		*maizar*	Meso-America	36, 72
17.		*andersonii*	Central America	64
18.		*dactyloides*	Mexico and Guatemala	36, 72, 108
19.		*intermedium*	Guerrero, Chiapas, Guatemala, Honduras	72
20.		*manisuroides*	Tuxtla Gutierrez, Chiapas and Mexico	72
21.		*jalapense*	Oaxaca, Chiapas, Guatemala, El Salvador	72
22.		*lanceolatum*	South West USA	72
23.		*pilosum*	North and Central America	72
24.		*peruvianum*	South America	72, 90, 108
25.	*Coix*	*aquatica*	South East Asia	10
26.		*poilanei*		10
27.		*lacryma-jobi*		20
28.		*gigantea*		20, 40
29.		*ouwehandii*		–
30.		*puellarum*		–
31.		*gasteenii*		–
32.	*Chionachne*	*koenigii*	South East Asia	20, 40
33.		*semiteres*		20
34.		*hubberdiana*		–
35.		*biaurata*		–
36.	*Sclerachne*	*punctata*	South East Asia	20
37.	*Trilobachne*	*cookei*	South East Asia	20
38.	*Polytoca*	*macrophylla*	South East Asia	20, 40
39.		*massii*		–
40.		*digitata*		–
41.		*javanica*		–
42.		*wallichiona*		–
43.		*cyanthopoda*		–

[a]2n represents somatic chromosome number

are also distributed in the USA and South America, and some of these are also detected in Asian regions (Table 4.2) (Berthaud et al. 1995; Chaudhary et al. 2014). It possesses wide variation in chromosome number (2n=36, 64, 72, 90, 108) among different species. After teosinte, *Tripsacum* is used widely as a source of valuable traits in maize due to its wider adaptation and resistance to heat, drought and waterlogging stresses.

4.5.3 Asiatic Genera

Asiatic genera, namely, *Coix*, *Sclerachne*, *Polytoca*, *Chionachne* and *Trilobachne*, collectively called as 'Oriental Maydeae', are prevalent across Southern parts of Asia, viz. India, Myanmar, China and Malaysia (Bor 1960; Chaganti 1965). Based on botanical features, the Asiatic genera are classified into two major groups: first, *Coix* wherein the caryopsis is enclosed in modified spathe, and second, *Trilobachne*, *Chionachne*, *Polytoca* and *Sclerachne* with caryopsis enclosed in endurated lower glume. The members of the second group differ from each other with respect to nature of lower glume of fruit, kind of grain base, visibility of hilum and rachis over the length and bearing of inflorescence on plant (Hernard 1931; Nirodi 1955; Bor 1960; Venkateswarlu and Chaganti 1973; Arora 1977; Sachan and Sarkar 1985). The details of species occurrence and chromosome complements of respective genera are listed in Table 4.2.

4.5.3.1 *Coix*

Coix plants are erect, tall and of floating types, while the inflorescence is a fascicled axillary and terminal spaciform raceme (Venkateswarlu and Chaganti 1973; Sachan and Sarkar 1985). Considering fruitcase morphology, *Coix* separates into cultivated 'soft-shelled' types and the wild 'hard-shelled' varieties. Apart from common features of *Coix* genus, the each species of *Coix* is botanically different from each other for leaf and fruit traits. Among various species of *Coix*, *C. lacryma-jobi* is one of the 'soft-shelled' types popularly known as "job's tear" because of its grain shape resembling the drop of tear (Arora 1977). It contains highest protein content among cereals and seed extract which possesses therapeutic value (Ottoboni et al. 1990). At present, it spreads across the warmer parts of world and parts of India (Himachal Pradesh, Rajasthan, Uttar Pradesh, Madhya Pradesh, Andhra Pradesh, Tamil Nadu, Kerala, Karnataka, West Bengal, Orissa and Northeastern states). It is known by different names in local languages in different Indian states such as giral, kara, koa sangti, gurgur, kesai, kasi, golugu, gleun and many others (Venkateswarlu and Chaganti 1973). Taxonomical varieties of the 'job's tear' (*C. lacryma-jobi)* includes *typica*, *stenocarpa*, *mayuen* and *monilifer* that differ for kernel shape, size, colour and hardiness; of these, *mayuen* is of economic importance and other three are wild types (Watt 1904). The grains of *C. lacryma-jobi* var. *mayuen* is utilized by Garo, Karbi and Naga tribes as peanuts, brewing beers and making pounded flours even green leaves for silage purpose (Hore and Rathi 2007). Based on repetitive sequence analyses, *Coix* has been found to be close to sorghum genome as well (Cai et al. 2014).

4.5.3.2 Other Asiatic Genera

Trilobachne genus has fruit with trilobite lower glume, larger central spike, linear hilum, 2–3 axial female spikelets and mainly staminate terminal inflorescence, while *Polytoca* genus possesses male spikes at terminal portion and female or mixed spikes at lateral portion with invisible hilum and hollowed out base of the grain (Sachan and Sarkar 1985). *Trilobachne cookei* is an annual herb reported to be uncultivated. The overlapping margin of lower glume which enclosed rachis segment is present in *Sclerachne*, but on the contrary, *Chionachne* contains non-overlapping margin with visible rachis. The profuse tillers and hardwood stems of *C. semiteres* and *C. koenigii* are used as fodder and making huts, respectively (Janaki-Ammal 1945). Although all these four genera include several species, most of which are not cultivated (Watson and Dallwitz 1992).

4.6 Wild Relatives and Cross Compatibility Barriers

Introgressive hybridization technique has been used extensively to transfer alien genes through method of repeated backcrossing of hybrid derived either from teosinte or *Tripsacum* with maize (Mano and Omori 2013; Wang et al. 2014). While intra- and interspecific hybridization aim at widening the gene pool of crop plants vis-a-vis crop improvement, a combination of spatiotemporal and physiological factors involving the genetic complexity in resulting hybrids prevents indiscriminate hybridization. There are barriers to hybridizing different species or genotypes in maize, and the underlying phenomena are 'genetic incompatibility' and 'genomic incongruity' (Allard 1960). When normal congruous relationship between pollen and pistil is disrupted, fertilization fails to occur as pre- and postfertilization processes are not consummated due to incongruity (Hogenboom 1973). The incompatibility barrier to hybridization may be complete if no seed is set despite the use of viable pollen on potentially receptive silks. Partial compatibility is detected by differential pollen tube growth and the distorted recovery of alleles coupled with incompatibility genes. Both complete and partial incompatibilities have been reported in maize when crossed with wild relatives.

4.6.1 Cross Compatibility

In *Z. mays*, intraspecific hybridization is fully cross-fertile (Purseglove 1972), though there is great cross compatibility between maize and other *Zea* species such as annual teosinte, except for the tetraploid, *Z. perennis*, that exhibits low fitness (Wilkes 1977; Galinat 1988). Though they are cross compatible, the rate of spontaneous or natural hybridization of teosintes with maize is species-dependent. For instance, *Z. mays* ssp. *luxurians* rarely hybridizes, *Z. mays* ssp. *mexicana* naturally hybridizes at a low rate (1 %), whereas Z. *mays* ssp. *parviglumis* hybridizes at a high rate (50 %) with maize (Ellstrand et al. 2007). Intergeneric hybridization between *Zea* species and *Tripsacum* species (*T. dactyloides*, *T. floridanum*, *T. lanceolatum*, and *T. pilosum*) is difficult, but hybrids can be realized successfully through manual crossing. The pioneer publications by Mangelsdorf and Reeves (1931) demonstrated the possibility of cross between maize and *Tripsacum*. Moreover, several researchers supported the possibility of gene transfer from *Tripsacum* to maize (Galinat 1973; Bernard and Jewell 1985). However, *Tripsacum* is not easily crossable to maize; hence, few pathways have been developed to accomplish successful gene transfer from *Tripsacum* to maize background (Savidan and Berthaud 1994; Leblanc et al. 1996). Since maize-*Tripsacum* hybrid plants are sterile in nature, 'tripsacorn' from a cross of *Tripsacum* and *Z. diploperennis* has been used as bridge species to accomplish gene transfer from *Tripsacum* to maize (Kindiger and Backett 1992; Eubanks 2006). However, these hybrids exhibit high pollen sterility with limited female fertility due to the unequal number of chromosomes across genera and lack of pairing between them (Eubanks 1997). *Tripsacum* has been primarily utilized in several studies to understand its role in maize evolution and to characterize the resistance genes. *T. dactyloides* has been the focus of extensive breeding work to transfer the apomictic reproductive nature from *Tripsacum* to maize (Kindinger et al. 1996). There are no reports on compatibility between maize and the genera belonging to the old world group such as *Coix*, *Sclerachne*, *Polytoca*, *Chionachne*, and *Trilobachne*, and genetic studies do reveal that these genera are distinct from both maize and teosintes of the new world group (Katiyar and Sachan 1992).

4.6.2 Genetic and Molecular Basis

Understanding the molecular mechanisms of crossing compatibility barriers in wide hybridization is of great importance. These barriers may be classified into two types, viz. (1) pre-fertilization or pre-syngamous and (2) post-fertilization or post-syngamous barriers (Stebbins 1958; Gutierrez-Marcos et al. 2003). As they contribute to species isolation and speciation, pre-fertilization barriers relating to pollen tube germination, growth, guidance and sperm delivery are considered the major hybridization controls in nature (Dresselhaus et al. 2011). However, in incompatible crosses, stigmatic surface proteins are not involved in incompatibility responses, since germination occurs normally before pollen tube reaches, and are arrested at or near the surface of the stigma (Heslop-Harrison and Heslop-Harrison 1982). So, the initial steps in the process of hybridization such as stigmata pollen capture, adhesion, hydration, and germination do not seem to represent a pre-fertilization barrier in maize and other related grasses (Dresselhaus et al. 2011). The first pre-fertilization barrier, which involves the invasion of the stigma by pollen tubes, begins with the crossing of the cuticle of the stigma and the invasion of the intercellular space of the sporophytic tissues and pollen tube guidance towards the transmitting tract (Heslop-Harrison 1982). The second pre-fertilization barrier appears to be pollen tube growth support inside the transmitting tract in maize. Intensive cell-cell interactions between the pollen tubes and the female tissues of the pistil take place for stimulating and guiding the pollen tube growth (Dresselhaus and Franklin-Tong 2013). The growth of *Tripsacum* pollen tubes in maize pistil is arrested after about 2 cm growth, probably after pollen resources are consumed (Lausser et al. 2010).

Three genetic loci responsible for governing the compatibility barriers have been identified in North American flint and dent maize varieties. The first genetic system, *Ga1/Ga1*, is partially dominant. Genotype with *ga1/ga1* generates pollen that cannot fertilize with *Ga1-s/Ga1-s*, but is partially receptive on *Ga1-s/ga1-s* silks (Nelson 1952) and fully fertile in reciprocal crosses (House and Nelson 1958). Further, the two other genetic systems that have been described are the *Teosinte crossing barriers1* (*Tcb1*) and the *gametophytic factor2* (*ga2*) loci that restrict the crossability of maize pollen to silks of some annual maize subspecies teosinte (*Z. mays* ssp. *mexicana*) populations (Evans and Kermicle 2001; Kermicle 2006; Kermicle and Evans 2010). In vivo assays of the dynamics of incompatible pollen tube growth and morphology in silks carrying the *Tcb1-s*, *Ga1-s* or *Ga2-s* barriers showed slower pollen tube growth than in compatible crosses at early stages and growth ceased 24 h after pollination (Lu et al. 2014).

Cross-compatible barriers in most intergeneric crosses prevent the formation of hybrid zygotes, as the pollen tubes are usually arrested in the proximity of or inside the transmitting tract (Heslop-Harrison 1982; Lausser et al. 2010). There are reports that chemotactic factors or molecules are secreted by the transmitting tract cells, to guide self, but not alien pollen tubes in maize and other grass species. A large percentage of *Tripsacum* pollen tubes are mistargeted to the stigma cortex instead of reaching the transmitting tract of maize silks, whereas in the reciprocal cross maize pollen tubes occur exclusively in the transmitting tract of *Tripsacum* (Lausser and Dresselhaus 2010; Lausser et al. 2010). In the case of a cross between maize and *C. lacryma-jobi* var. *frumentacea*, the pollen tubes could not reach the base of the maize style leading to strict reproductive isolation. The ratio of abnormal pollen tubes in the cross between maize and *Coix* was higher than that between maize and *Tripsacum*, suggesting that maize is closer to *Tripsacum* than to *Coix* (Duan et al. 2008).

Post-fertilization barriers arise due to hybrid embryo abortion or failure in endosperm development. In the cross between maize and *Tripsacum*, pollens of *Tripsacum* could germinate and release androgamete into embryo sac of maize, but normal hybrids were not obtained because of embryo sac incompatibility or embryo abortion (Duan et al. 2008). Endosperm development is regulated by specific balance of gene expression and epigenetic imprinting mechanisms (characterized by their uniparental expression) from either the maternal or paternal genomes. Any alteration in this balance resulting from changes in allelic copy number, sequence or epigenetic imprints can cause endosperm failure and eventual embryo abortion (Gutierrez-Marcos et al. 2003). Histological studies on hybrids of maize × *Tripsacum* showed abnormalities of endosperm development at very early stages ranging from defects at the chalazal pole (Cooper and Brink 1942) to abnormal free nuclear division rates (Brink and Cooper 1940).

Hybrid sterility, abnormal male and female gametes in hybrid, can also be considered as a barrier in wide hybridization. Disturbances in pollen development of maize × *T. dactyloides* hybrids, derived from both diploid and tetraploid *Tripsacum* sources, are commonly observed due to anomalies in the development of the microspore. The anomalies observed are failure of the chromosomes to congregate at the metaphase plate, development of a tripolar spindle and failure of cytokinesis at the first and second meiotic divisions (Kindiger 1993). Pre-hybridization barriers are typically stronger than post-hybridization barriers in establishing discriminate hybridization in plants (Rieseberg and Willis 2007). Even though the mechanisms on how a crossing barrier arises in sexually reproducing plants are largely unknown (Lu et al. 2014), the possibilities of wide crosses and successfully introgressing desired traits from wild relatives into maize are nevertheless limited; however, different approaches can be employed to overcome these barriers.

4.7 Wild Relatives in Maize Improvement

With the advent of climate change and greater ecosystem instability, CWRs are likely to prove a critical resource in ensuring food security in the coming years. The wild maize germplasm contains significant variation for plenty of economically important traits that have remained untapped (Hoisington et al. 1999). The alien gene introgressions especially from teosinte and *Tripsacum* possess enormous potential for maize improvement to (1) increase agronomic performance, (2) transfer genes for quality traits, (3) introgress disease/pest resistance genes and (4) improve level of tolerance to abiotic stresses. Some of the traits that have been improved through wide hybridization of maize with teosinte and *Tripsacum* are presented (Table 4.3).

4.7.1 Improvement of Agronomic Performance

Cohen and Gallinat (1984) have described the importance of wild chromosome segments of teosinte and *Tripsacum* for improving maize agronomic performance. The upscaling of quantitative traits in cultivated maize has been achieved by using sexual hybridization to teosinte and *Tripsacum* (Ramirez 1997). Several experiments showed the better performance of introgressed maize lines produced from utilizing teosinte for yield, tillering and green matter production, over the maize hybrid checks (Pasztor and Borsos 1990; Savidan and Berthaud 1994). SD00100, an introgression line selected from the maize × mexicana population, possesses big ears and stay-green characters (Wang et al. 2008). Petrov (1984) suggested the development of apomictic maize genotypes by introgressing apomixis genes from *T. dactyloides* to maize. Consequently, many studies were conducted for this purpose (Table 4.3). Furthermore, bulked segregant analysis performed by Leblanc et al. (1995) identified molecular markers linked to diplospory in *T. dactyloides* and showed that RFLP marker on long arm of chromosome 6 cosegragates with diplosporous mode of apomixis. Later, efforts were made to transfer apomixis from diplosporous tetraploid *Tripsacum* into maize through conventional backcrossing (Kindiger et al. 1996; Leblanc et al. 1996).

4.7.2 Enhancement of Quality Attributes

Wild relatives have been utilized for enhancement of quality traits in maize. Introgressed maize genotype from maize × teosinte possesses better silage quality over hybrid checks (Sidorov and Shulakov 1962). Introgression of *Z. mays* ssp. *mexicana* into maize resulted in lines with higher protein content, as well as higher in amino acids such as lysine, methionine and/or phenylalanine (Table 4.3) (Pasztor and Borsos 1990; Wang et al. 2008). Angiotensin converting

Table 4.3 List of some of the economically important characters introgressed from teosinates and *Tripsacum*

S. No.	Characters	Donor genera/species	Reference(s)
Agronomic traits			
1.	Higher yield, profuse tillering, biomass production, increased height	*Z. mays* spp. *mexicana*	Pasztor and Borsos (1990) and Wang et al. (2008)
2.	Apomixis	*Tripsacum dactyloides*	Petrov (1984), Savidan and Berthaud (1994), Leblanc et al. (1995, 1996), and Kindiger et al. (1996)
Quality traits			
4.	Superior silage making ability	Teosinte	Sidorov and Shulakov (1962)
5.	Higher protein, amino acid balance	*Z. mays* spp. *mexicana*	Pasztor and Borsos (1990) and Wang et al. 2008
6.	Reduced hypertension	*Z. mays* spp. *mexicana*	Wang et al. (2014)
Biotic stress resistance: *Disease-pests*			
7.	Anthracnose, Northern corn leaf blight, Southern corn leaf blight, Stewart's bacterial wilt, Common rust	*T. dactyloides*	Bergquist (1979, 1981)
8.	Northern corn leaf blight	*T. floridanum*	Hooker and Perkins (1980)
9.	Northern corn leaf blight, Southern corn leaf blight	*Z. diploperennis*	Wei et al. (2003)
10.	Northern corn leaf blight	Teosinte	Ott (2008)
11.	Corn smut	Teosinte	Chavan and Smith (2014)
12.	Downy mildew	*Z. mays* spp. *mexicana*, *Z. diploperennis*	Ramirez (1997)
13.	Ear rot	*Z. mays* spp. *mexicana*	Pasztor and Borsos (1990)
	Maize chlorotic dwarf virus	*Z. diploperennis*	Findley et al. (1983)
Biotic stress resistance: *insect pests*			
14.	Corn borer	*Z. mays* spp. *mexicana*	Pasztor and Borsos (1990)
15.	Asiatic corn borer	*Z. mays* spp. *mexicana*, *Z. perennis*, *Z. diploperennis*	Ramirez (1997)
16.	Corn rootworm	*T. dactyloides*	Eubanks (2002, 2006), and Prischmann et al. (2009)
Biotic stress resistance: *parasitic weeds*			
17.	Striga	*Z. diploperennis*	Menkir (2006), Amusan et al. (2008), and Yallou et al. (2009)
Abiotic stress tolerance			
18.	Flooding	*Z. luxurians*, *Z. nicaraguensis*, *Z. mays* ssp. *huehuetenangensis*	Ray et al. (1999), Mano et al. (2005), and Mano and Omori (2007, 2013)
19.	Salinity	*T. dactyloides*	Pesqueira et al. (2003, 2006) and Shavrukov and Sokolov (2015)

enzyme (ACE) is one of the key regulators of blood pressure, and ACE inhibitors (ACEI) are widely used to treat hypertension in human. Drugs such as captopril and enalapril possess various side effects and naturally found ACEI can be potentially utilized as the drug (Vyssoulis et al. 2001). Introgression line SD00100 generated from a cross between maize and *Z. mays* ssp. *mexicana* possesses high ACEI peptides (Wang et al. 2014). This novel inbred line thus can be utilized as functional foods or nutraceuticals in the near future.

4.7.3 Resistance to Biotic Stresses

Scientists have introgressed genes imparting resistance to various diseases, viz. anthracnose, northern corn leaf blight, southern corn leaf blight, bacterial wilt, common rusts, common smut, downy mildew, ear rot and maize chlorotic dwarf virus, from teosinte and *Tripsacum* to cultivated maize through typical sexual hybridization (Table 4.3) (Bergquist 1979, 1981; Hooker and Perkins 1980; Findley et al. 1983; Pasztor and Borsos 1990; Ramirez 1997; Wei et al. 2003; Ott 2008; Chavan and Smith 2014) (Table 4.3). Further, resistance against corn borers and root worms have been introgressed into maize from wild relatives (Pasztor and Borsos 1990; Ramirez 1997; Eubanks 2002, 2006; Prischmann et al. 2009). Striga is an obligate root hemiparasite of cereals including maize spread in regions of Africa and parts of Asia. Striga development is dependent on exudates and solutes of host roots that affect host plant health (Rich and Ejecta 2008). For the development of host plant resistance, Lane et al. (1997) reported *Z. diploperennis* as the source of striga-resistant genes for maize improvement, while Gutierrez-Marcos et al. (2003) reported *T. dactyloides* as secondary source for striga resistance in maize. Rich and Ejecta (2008) highlighted the importance of both sources over the landraces for the development of striga-resistant maize in African regions. The scientists of International Institute of Tropical Agriculture (IITA) have developed ZD05, *Striga hermonthica*-resistant inbred from *Z. diploperennis* accession and tropical maize germplasm (Menkir 2006; Amusan et al. 2008; Yallou et al. 2009).

4.7.4 Tolerance to Abiotic Stresses

Maize succumbs to yield loss by abiotic factors such as drought, heat, flooding and salt due to limited availability of cultivated genotypes possessing tolerance to these abiotic factors (Table 4.3) (Wilkes 1967; Ramirez 1997; Chaudhary et al. 2014). Therefore, search for genes for abiotic stress tolerance has been diverted towards wild germplasm. The teosintes especially *Z. nicaraguensis*, *Z. luxurians* and *Z. mays* ssp. *huehuetenangensis* are well-characterized for its high tolerance for flooding on account of more adventitious roots above the soil surface and well-formed aerenchyma in adult plants (Ray et al. 1999; Mano et al. 2005; Mano and Omori 2007, 2013). Similarly, *T. dactyloides* was reported as tolerant species to excess salt conditions through screening of maize × *Tripsacum* hybrids for survival on exposure to NaCl concentrations (Pesqueira et al. 2003, 2006). Likewise, studies have confirmed the chromosome segment belonging to *Z. mays* ssp. *huehuetenangensis*, *Z. nicaraguensis* and *Z. luxurians* have led to flooding tolerance in maize (Mano and Omori 2007, 2013). Recently, Shavrukov and Sokolov (2015) developed lines with salinity tolerance from maize × *T. dactyloides*.

4.8 Landraces of Maize and Its Use in Crop Improvement

As defined by Camacho-Villa et al. (2005), a landrace is 'dynamic population(s) of a cultivated plant that has historical origin, distinct identity, and lacks formal crop improvement, as well as often being genetically diverse, locally adapted, and associated with traditional farming systems'. Maize possesses tremendous genetic diversity compared to other crops that offers potential opportunities for the improvement of maize for its grain yield, tolerance to biotic and abiotic stresses and specialty traits. In landraces, variation tends to be partitioned within populations, rather than between populations, and levels of variation can be very high. Landraces worldwide have been characterized using various morphological characters. With the advent of DNA-based markers, characterization using molecular markers is now the preferred choice, mainly due to their abundance and environment-neutral behaviour (Prasanna 2010, 2012). However, landraces being heterogeneous in nature, large number of individuals in each of the accessions are required to be genotyped and the use of molecular markers are very expensive and time-consuming (Prasanna 2012). A new method for SSR analysis has been formulated in which genotyping of pools of individuals from a population than genotyping of multiple individuals per population is very effective and accurate (Dubreuil et al. 2006). A minimum of 15 individuals are required to be characterized to adequately represent the allelic diversity present in a population (Dubreuil et al. 2006; Warburton et al. 2010). This bulk DNA fingerprinting technique using SSR markers developed at CIMMYT has been used to differentiate the improved open-pollinated varieties (OPVs) or synthetics (Warburton et al. 2010). Large number of studies have undertaken molecular characterization of landraces to understand the extent of genetic diversity and their relationships (Barcaccia et al. 2003; Carvalho et al. 2004; Qi-Lun et al. 2008; Sharma et al. 2010; Singode and Prasanna 2010; Warburton et al. 2011; Wasala and Prasanna 2013; Molin et al. 2013; Semagn et al. 2014; Oppong et al. 2014). The natural variability of landraces has been exploited for various biotic stress resistance, abiotic stress tolerance and quality improvement of maize (Prasanna 2010, 2012), apart from understanding the diffusion routes of maize. Some of the examples of their utilization are as follows.

4.8.1 Resistance to Biotic Stresses

Because of vast geographical distribution, maize encounters a large number of diverse pest and pathogens during its life-span. The foliar diseases like turcicum leaf blight (TLB) and maydis leaf blight (MLB) are severe constraint for maize productivity throughout the world. Tuxpeno Crema a subpopulation of landrace Tuxpeno from Mexico is well known for its resistance to tropical foliar diseases (Rodriguez et al. 1998). A popcorn landrace, Palomero Toluqueno, was found to have resistance to the maize weevil, *Sitophilus zeamais* (Arnason et al. 1994). Few Caribbean landraces were found tolerant to larger grain borer (*Prostephanus truncatus*) (Kumar 2002).

4.8.2 Tolerance to Abiotic Stresses

Some of the Mexican maize landraces with abiotic stress tolerance comprise La Posta Sequia, Conica, Conica Nortena, Bolita, Breve de Padilla, Nal Tel, Tuxpeno (drought tolerant), Oloton (acid soil tolerant) and Chalqueno × Ancho de Tehuacan cross (alkalinity tolerant) (Prasanna 2012). Drought-tolerant population-1 (DTP-1) and drought-tolerant population-2 (DTP-2) developed for drought tolerance at CIMMYT are being exploited for development of inbred lines. DTP-1 is developed by 74 half-sib crosses of 13 drought-tolerant sources, while DTP-2 was constituted by introgression of 25 new drought-tolerant landraces in DTP-1 by half-sib crossing and further recombination (Edmeades et al. 1996). Some of the derived lines from La Posta Sequia were also found to be tolerant to both drought and heat stresses (Cairns et al. 2013).

4.8.3 Improvement of Specialty Traits

There are several landraces, viz. Pepitilla, Bolita, Azul, Tlacoya and Oaxaqueno, which are very popular and preferred for their tortilla quality. One of the prolific landrace, Murlimakai, from the Northeastern states of India has been exploited for development of a baby corn composite 'VL Baby Corn' (Prasanna 2012). The sequencing of Palomero, a popcorn landrace in Mexico, would further boost the understanding of popping quality improvement as well as structure of landraces (Vielle-Calzada et al. 2009).

4.8.4 Source for Genetic Diversification

In 1987, Latin American Maize Programme (LAMP) was initiated for generating information by evaluating landraces in Latin American countries with an objective to evaluate over 14,000 accessions found in Latin America and the US germplasm banks (Salhuana et al. 1991). Along with yield evaluation, landraces were identified for lodging tolerance, earliness and plant characteristics. The subset of landraces was made available to breeders to broaden the genetic base for hybrid development. Later, it was realized that LAMP materials had to be enhanced for utilization in breeding programme, as those germplasm were not adapted to temperate corn belt of America. During the time, the US-germplasm enhancement of maize (US-GEM) was conceived with the idea of introgression of valuable genes. This project, a collaboration among public and private institutions, aims at enhancing productivity and improving diversity of maize in the USA. It was initiated with improvement of 24 tropical and 27 temperate accessions identified on the basis of yield performance in LAMP; furthermore, seven hybrids donated by DeKalb Genetics have been included (Salhuana and Pollak 2006). The recent initiative of CIMMYT, 'Seeds of Discovery' (SeeD), was launched with an aim to find the allelic variation in the genetic resources of maize and wheat. This programme utilizes high-density genotyping, phenotyping and bioinformatic tools to know the desirable alleles and haplotypes linked with priority traits to be used by breeders (Prasanna 2012).

4.8.5 Understanding Diffusion Routes

Landraces are grown by the farming community at specific places for a long period of time. Unique diversity preserved in specific landraces help in tracking the insights into domestication events in maize (Matsuoka et al. 2002) and the migration routes of maize from the centres of origin (Rebourg et al. 2003; Vigouroux et al. 2008). 258 landrace accessions from America, 237 from Africa, 13 from Middle East, 148 from Europe and 143 from Asia were characterized using SSR markers for understanding the diffusion routes of maize worldwide (Mir et al. 2013). This was carried out as a part of the Generation Challenge Programme (GCP), involving researchers from CIMMYT-Mexico, INRA-France, IITA and national programmes of China, India, Indonesia, Thailand, Vietnam and Kenya. This led to the assessment of genetic relationships among landraces/populations worldwide and understanding of global diffusion of maize from primary centre of origin (Mir et al. 2013). Besides, landraces provided a clear understanding of phylogenetic relationships and gene flow between maize landraces and the wild progenitor, teosinte (Warburton et al. 2011; van-Heerwaarden et al. 2011).

Sequencing of the most popular and widely used inbred, B73 (Schnable et al. 2009) and Palomero, a popcorn landrace in Mexico (Vielle-Calzada et al. 2009), are important landmarks in maize genome research. With the advances in high-throughput SNP-based genotyping facilities and reduced cost of genotyping, it would be possible to understand the evolution of maize at finer level. The Palomero genome is smaller (22 %) than that of B73 genome and shows a large number of unreported sequences, indicating the presence of a large number of unexplored alleles. More than 12 genes related to heavy metal detoxification and environmental stress tolerance were found to be conserved in B73 and Palomero but are absent from teosinte. This suggests that these genes were possibly involved in the domestication process during the evolution of maize (Vielle-Calzada et al. 2009).

Further, 'Sikkim Primitive', a landrace of high altitude regions of Northeastern Himalayan State of India, Sikkim, has been the centre of attraction among the maize breeders for decades (Sachan and Sarkar 1986). Locally known as 'Murli makai', the landrace is characterized by the absence of apical dominance, high prolificacy (having five to nine uniform ears per plant), erect leaves, tall with drooping tassel, popcorn type kernels and top bearing habit (Sachan and Sarkar 1982). It also has stay-green character, which make it suitable for fodder use. Due to its primitive features, the landrace was phrased as 'Sikkim Primitive' (SP1 and SP2) (Dhawan 1964). Though 'Sikkim Primitive' has been characterized using cytological, biochemical and molecular markers (Sachan and Sarkar 1982, 1986; Bhat et al. 1998; Sharma et al. 2010), its utilization in the maize breeding programme has been greatly limited due to its poor expression outside the Northeastern Himalayan State (Sharma et al. 2010). Sequencing of 'Sikkim Primitive' landrace would further provide insight into the genome architecture, the regulation of key genes responsible for its primitive nature, and would in turn help in understanding its evolution and devising strategies for better utilization of the valuable landraces in the breeding programme.

4.9 Conclusions

Wild relatives and landraces of maize possess enormous potential as donors for important alleles/genes. However, these unexplored otherwise valuable germplasm have been utilized to a less significant extent in the breeding programmes. The inbreds derived from these genetic resources may not be directly utilized in the breeding programme; however, they can serve as important donors for specific traits. Transfer of genes from wild relatives is not always straight forward, and utilization of bridging species and special techniques such as embryo rescue may provide solutions. Comprehensive evaluation of landrace accessions for biotic/abiotic stresses and nutritional quality attributes would provide opportunity to the breeders to utilize them in the specific breeding programme. Effective use of available genome sequences of 'B73' and 'Palomero' landrace, coupled with strong 'prebreeding programme', can potentially enhance the utilization of these unexplored germplasm in the maize breeding programme.

References

Allard RW (1960) Principles of plant breeding. Wiley, New York, pp 485

Amusan IO, Rich PJ, Menkir A, Housley T, Ejeta G (2008) Resistance to *Striga hermonthica* in a maize inbred line derived from *Zea diploperennis*. New Phytol 178:157–166

Arnason JT, Baum B, Gale J et al (1994) Variation in resistance of Mexican landraces of maize to maize weevil *Sitophilus zeamais*, in relation to taxonomic and biochemical parameters. Euphytica 74:227–236

Arora R (1977) Job's-tears (*Coix lacryma-jobi*) – a minor food and fodder crop of North Eastern India. Econ Bot 31:358–366

Barcaccia G, Lucchin M, Parriani P (2003) Characterization of flint maize (*Zea mays* var. *indurata*) Italian landraces, II. Genetic diversity and relatedness assessed by SSR and Inter-SSR molecular markers. Genet Resour Crop Evol 50:253–271

Beadle GW (1939) Teosinte and the origin of maize. J Hered 30:245–247

Beadle GW (1972) The mystery of maize. Chicago. Field Mus Nat Hist Bull 43:2–11

Beadle GW (1978) Teosinte and the origin of maize. In: Walden DB (ed) Maize breeding and genetics. Wiley, New York, pp 113–128

Bennetzen JE, Buckler V, Chandler J et al (2001) Genetic evidence and the origin of maize. Lat Am Antiq 12:84–86

Bergquist RR (1979) Selection for disease resistance in a maize breeding programme. II. Introgression of an alien genome from *Tripsacum dactyloides* conditioning resistance in *Zea mays*. Proceedings of the tenth meeting of the Maize and Sorghum Section of Eucarpia, Varna, Bulgaria, pp 200–206

Bergquist RR (1981) Transfer from *Tripsacum dactyloides* to corn of a major gene locus conditioning resistance to *Puccinia sorghi*. Phytopathology 71:518–520

Bernard S, Jewell DC (1985) Crossing maize with *Sorghum, Tripsacum*, and millet: the products and their level of development following pollination. Theor Appl Genet 70:474–483

Berthaud J, Savidan Y, Barre M, Leblanc O (1995) *Tripsacum*: its diversity and conservation. In: Taba S (ed) Maize genetic resources. Maize program special report, Mexico DF, CIMMYT, pp 74–85

Bhat KV, Chandel KPS, Shivakumar TM, Sachan JKS (1998) Isozyme diversity in Indian primitive maize landraces. J Plant Biochem Biotechnol 7:23–27

Bor NL (1960) The grasses of Burma, Ceylon, India and Pakistan. Pergamon Press, Oxford, pp 767

Bouis H, Hotz C, McClafferty B, Meenakshi JV, Pfeiffer W (2011) Biofortification: a new tool to reduce micronutrient malnutrition. Food Nutr Bull 32:S31–S40

Brink RA, Cooper DC (1940) Double fertilization and development of the seed in angiosperms. Bot Gaz 102:1–25

Cai Z, Liu H, He Q et al (2014) Differential genome evolution and speciation of Coix lacryma-jobi L. and Coix aquatica Roxb. hybrid guangxi revealed by repetitive sequence analysis and fine karyotyping. BMC Genom 15:1025. doi:10.1186/1471-2164-15-1025

Cairns JE, Crossa J, Zaidi PH et al (2013) Identification of drought, heat, and combined drought and heat tolerant donors in maize. Crop Sci 53:1335–1346

Camacho Villa TC, Maxted N, Scholten M, Ford-Lloyd B (2005) Defining and identifying crop landraces. Plant Genet Resour 3:373–384

Carvalho VP, Ruas CF, Ferreira JM, Moreira RMP, Ruas PM (2004) Genetic diversity among maize (*Zea mays* L.) landraces assessed by RAPD markers. Genet Mol Biol 27:228–236

Chaganti RSK (1965) Cytogenetic studies of maize-*Tripsacum* hybrids and their derivatives. The Bussey Institute, Harvard University, Cambridge, MA, pp 1–93

Chaudhary HK, Kaila V, Rather SA (2014) Maize. In: Pratap A, Kumar J (eds) Alien gene transfer in crop plants, vol 2, Achievements and impacts. Springer, New York, pp 27–50. doi:10.1007/978-1-4614-9572-7_2

Chavan S, Smith SM (2014) A rapid and efficient method for assessing pathogenicity of *Ustilago maydis* on maize and teosinte lines. J Vis Exp 83, e50712. doi:10.3791/50712

Cohen JI, Gallinat WC (1984) Potential use of alien germplasm for maize improvement. Crop Sci 24:1011–1015

Cooper DC, Brink RA (1942) The endosperm as a barrier to interspecific hybridization in flowering plants. Science 95:75–76

Cubas P, Lauter N, Doebley J, Coen E (1999) The TCP domain: a motif found in proteins regulating plant growth and development. Plant J 18:215–222

Dhawan NL (1964) Primitive maize in Sikkim. Maize Genet Coop Newsl 38:67–70

Doebley JF (2001) George Beadle's other hypothesis: one-gene, one-trait. Genetics 158:487–493

Doebley J (2004) The genetics of maize evolution. Ann Rev Genet 38:37–59

Doebley JF, Stec A (1991) Genetic analysis of the morphological differences between maize and teosinte. Genetics 129:285–295

Doebley JF, Stec A (1993) Inheritance of the morphological differences between maize and teosinte: comparison of results for two F_2 populations. Genetics 134:559–570

Doebley JF, Bacigalupo A, Stec A (1994) Inheritance of kernel weight in two maize-teosinte hybrid populations: implications for crop evolution. J Hered 85:191–195

Doebley JF, Stec A, Gustus C (1995) *Teosinte branched1* and the origin of maize: evidence for epistasis and the evolution of dominance. Genetics 141:333–346

Doebley J, Stec A, Hubbard L (1997) The evolution of apical dominance in maize. Nature 386:485–488

Dorweiler J, Doebley JF (1997) Developmental analysis of *teosinte glume architecture1*: a key locus in the evolution of maize (Poaceae). Am J Bot 87:1313–1322

Dorweiler J, Stec A, Kermicle J, Doebley J (1993) *Teosinte glume architecture1*: a genetic locus controlling a key step in maize evolution. Science 262:233–235

Dresselhaus T, Franklin-Tong N (2013) Male–female cross-talk during pollen germination, tube growth and guidance, and double fertilization. Mol Plant 6:1018–1036

Dresselhaus T, Lausser A, Marton ML (2011) Using maize as a model to study pollen tube growth and guidance, cross-incompatibility and sperm delivery in grasses. Ann Bot 108:727–737

Duan TL, Mu JY, Tong QL, Rong TZ, Wang P (2008) Sexual incompatibility between maize and its wild relatives *Tripsacum* L. and *Coix* L. Acta Agron Sin 34:1656–1661

Dubreuil P, Warburton M, Chastanet M, Hoisington D, Charcosset A (2006) More on the introduction of temperate maize into Europe: large-scale bulk SSR genotyping and new historical elements. Maydica 51:281–291

Eagles HA, Lothrop JE (1994) Highland maize from central Mexico: its origin, characteristics, and use in breeding programs. Crop Sci 34:11–19

Edmeades GO, Banziger M, Beck D, Bolanos J (1996) Development and per se performance of CIMMYT maize populations as drought tolerant sources. In: Edmeades G et al (eds) Developing drought- and low N-tolerant maize. Proceedings of a symposium, CIMMYT, Mexico, March 25–29, 1996, pp 254–262

Ellstrand NC, Garner LC, Hegde S, Guadagnuolo R, Blancas L (2007) Spontaneous hybridization between maize and teosinte. J Hered 98:183–187

Eubanks MW (1995) A cross between two maize relatives: *Tripsacum dactyloides* and *Zea diploperennis*. Econ Bot 49:172–182

Eubanks MW (1997) Molecular analysis of crosses between *Tripsacum dactyloides* and *Zea diploperennis* (Poaceae). Theor Appl Genet 94:707–712

Eubanks MW (2001) The mysterious origin of maize. Econ Bot 55:492–514

Eubanks MW (2002) Investigation of novel genetic resource for rootworm resistance in corn. In: NSF (ed) Proceedings of the NSF design, service and manufacturing conference. Iowa State University, San Juan, pp 2544–2550

Eubanks MW (2006) A genetic bridge to utilize *Tripsacum* germplasm in maize improvement. Maydica 51:315–327

Evans MMS, Kermicle JL (2001) *Teosinte crossing barrier1*, a locus governing hybridization of *teosinte* with maize. Theor Appl Genet 103:259–265

FAO (1998) The state of the world's plant genetic resources for food and agriculture. FAO, Rome

FAOSTAT (2013) http://faostat.fao.org

Findley WR, Nault LR, Styer WE, Gordon DT (1983) Inheritance of maize chlorotic dwarf virus resistance in maize×*Zea diploperennis* backcrosses. Maize Newsl 56:165–166

Fukunaga K, Hill J, Vigouroux Y et al (2005) Genetic diversity and population structure of teosinte. Genetics 169:2241–2254

Galinat WC (1973) Intergenomic mapping of maize, teosinte and *Tripsacum*. Evolution 27:644–655

Galinat WC (1988) The origin of corn. In: Sprague GF, Dudley JW (eds) Corn and corn improvement, agronomy monographs no. 18. American Society of Agronomy, Madison, WI, pp 1–31

Gouesnard B, Rebourg C, Welcker C, Charcosset A (2002) Analysis of photoperiod sensitivity within a collection of tropical maize populations. Genet Resour Crop Evol 49:471–481

Gupta HS, Hossain F, Muthusamy V (2015a) Biofortification of maize: an Indian perspective. Indian J Genet 75:1–22

Gupta HS, Hossain F, Nepolean T, Vignesh M, Mallikarjuna MG (2015b) Understanding genetic and molecular bases of Fe and Zn accumulation towards development of micronutrient-enriched maize. In: Rakshit A et al. (eds) Nutrient use efficiency: from basics to advances. Springer, New Delhi, pp 255–282. doi:10.1007/978-81-322-2169-2_17

Gupta HS, Hossain F, Muthusamy V (2015c) Development of biofortified maize through molecular breeding. ISB News Report, Virginia Polytechnic Institute and State University, Blacksburg, pp 1–5

Gutierrez-Marcos JF, Pennington PD, Costa LM, Dickinson HG (2003) Imprinting in the endosperm: a possible role in preventing wide hybridization. Philos Trans R Soc Lond 358:1105–1111

Harlan JR, de Wet JMJ (1971) Toward a rational classification of cultivated plants. Taxon 20:509–517

Hernard JT (1931) A contribution to the knowledge or Indian Maydeae. Meded Rijks Herb Leiden 67:1–17

Heslop-Harrison J (1982) Pollen–stigma interaction and cross incompatibility in the grasses. Science 215:1358–1364

Heslop-Harrison J, Heslop-Harrison Y (1982) The pollen–stigma interaction in the grasses. 3. Features of the self-incompatibility response. Acta Bot Neerl 31:307–319

Hogenboom NG (1973) A model for incongruity in intimate partner relationships. Euphytica 22:229–233

Hoisington D, Khairallah M, Reeves T, Ribaut JM, Skovmand B, Taba S, Warburton M (1999) Plant genetic resources: what can they contribute toward increased crop productivity? Proc Natl Acad Sci U S A 96:5937–5943

Hooker AL, Perkins JL (1980) *Helminthosporium* leaf blights of corn the state of the art. Proc Ann Corn Sorghum Res Conf 35:68–87

Hore DK, Rathi RS (2007) Characterization of job's tears germplasm in North-East India. Nat Prod Radiance 6:50–54

House LR, Nelson OE (1958) Tracer study of pollen-tube growth in cross-sterile maize. J Hered 49:18–21

Iltis HH (1983) From teosinte to maize: the catastrophic sexual transmutation. Science 222:886–894

Janaki-Ammal EK (1945) Chromosome Atlas of cultivated plants. In: Darlington CD, Janaki-Ammal EK (eds). Allen and Unwin Limited, London, pp 114

Katiyar S, Sachan JKS (1992) Scanning electron microscopic studies of pollen grains in the tribe Maydeae. Maize Genet Coop Newsl 66:91–92

Kermicle JL (2006) A selfish gene governing pollen-pistil compatibility confers reproductive isolation between maize relatives. Genetics 172:499–506

Kermicle JL, Evans MM (2010) The *Zea mays* sexual compatibility gene *ga2*: naturally occurring alleles, their distribution, and role in reproductive isolation. J Hered 101:737–749

Kindiger B (1993) Aberrant microspore development in hybrids of maize×*Tripsacum dactyloides*. Genome 36:987–997

Kindiger B, Beckett JB (1992) Popcorn germplasm as a parental source for maize×*Tripsacum dactyloides* hybridization. Maydica 37:245–249

Kindiger B, Sokolov V, Khatypova JV (1996) Evaluation of apomictic reproduction in a set of 39 chromosome maize-*Tripsacum* backcross hybrids. Crop Sci 36:1108–1113

Kumar H (2002) Resistance in maize to the larger grain borer, *Prostephanus truncates* (Horn) (Coleoptera: Bostrichidae). J Stored Prod Res 38:267–280

Lane JA, Child DV, Moore THM, Arnold GM, Bailey JA (1997) Phenotypic characterization of resistance in *Zea diploperennis* to Striga. Maydica 42:45–51

Lausser A, Dresselhaus T (2010) Sporophytic control of pollen tube growth and guidance in grasses. Biochem Soc Trans 38:631–634

Lausser A, Kliwer I, Srilunchang KO, Dresselhaus T (2010) Sporophytic control of pollen tube growth and guidance in maize. J Exp Bot 61:673–682

Lauter N, Doebley J (2002) Genetic variation for phenotypically invariant traits detected in teosinte: implications for the evolution of novel forms. Genetics 160:333–342

Leblanc O, Grimanelll D, Gonzalez de Leon D, Savidan Y (1995) Detection of the apomictic mode of reproduction in maize-*Tripsacum* hybrids using maize RFLP markers. Theor Appl Genet 90:1198–1203

Leblanc O, Grimanelli D, Islam-Faridi N, Berthaud J, Savidan Y (1996) Reproductive behavior in maize-*Tripsacum* polyhaploid plants: implications for the transfer of apomixis into maize. J Hered 87:108–111

Lu Y, Kermicle JL, Evans MMS (2014) Genetic and cellular analysis of cross-incompatibility in *Zea mays*. Plant Reprod 27:19–29

Mangelsdorf PC (1974) Corn: its origin, evolution and improvement. Harvard University Press, Cambridge, MA

Mangelsdorf PC, Reeves RG (1931) Hybridization of maize, *Tripsacum*, and *Euchlaena*. J Hered 22:339–343

Mangelsdorf PC, Reeves RG (1938) The origin of maize. Proc Natl Acad Sci U S A 24:303–312

Mangelsdorf PC, Reeves RG (1939) The origin of Indian corn and its relatives. Texas AES Bull 574:1–315

Mano Y, Omori F (2007) Breeding for flooding tolerant maize using "teosinte" as a germplasm resource. Plant Root 1:17–21

Mano Y, Omori F (2013) Flooding tolerance in interspecific introgression lines containing chromosome segments from teosinte (*Zea nicaraguensis*) in maize (*Zea mays* ssp. *mays*). Ann Bot 112:1125–1139

Mano Y, Muraki M, Fujimori M, Takamizo T, Kindiger B (2005) Identification of QTL controlling adventitious root formation during flooding conditions in teosinte (*Zea mays* ssp. *huehuetenangensis*) seedlings. Euphytica 142:33–42

Matsuoka Y, Vigouroux Y, Goodman MM, Sanchez GJ, Buckler E, Doebley J (2002) A single domestication for maize shown by multilocus microsatellite genotyping. Proc Natl Acad Sci U S A 99:6080–6084

Maxted N, Ford-Lloyd BV, Jury SL, Kell SP, Scholten MA (2006) Towards a definition of a crop wild relative. Bio Con 15:2673–2685

Menkir A (2006) Assessment of reactions of diverse maize inbred lines to *Striga hermonthica* (Del) Benth. Plant Breed 125:131–139

Metcalfe SE (2006) Late quaternary environments of the northern deserts and central transvolcanic belt of Mexico. Ann Mo Bot Gard 93:258–273

Mir C, Zerjal T, Combes V et al (2013) Out of America: tracing the genetic footprints of the global diffusion of maize. Theor Appl Genet 126:2671–2682

Molin D, Coelho CJ, Maximo DS, Ferreira FS, Gardingo JR, Matiello RR (2013) Genetic diversity in the germplasm of tropical maize landraces determined using molecular markers. Genet Mol Res 12:99–114

Nelson OE (1952) Non-reciprocal cross-sterility in maize. Genetics 37:101–124

Nirodi N (1955) Studies on Asiatic relatives of maize. Ann Missouri Bot Gard 42:103–130

Oppong A, Bedoya CA, Ewool MB et al (2014) Bulk genetic characterization of Ghanaian maize landraces using microsatellite markers. Maydica 59:1–8

Ott OO (2008) The search for novel resistance alleles: screening teosinte-maize introgression lines for resistance to northern leaf blight. PhD thesis submitted to the College of Agriculture and Life Sciences, Cornell University, US, pp 1–26

Ottoboni LM, Leite A, Targon MLN, Crozier A, Arruda P (1990) Characterization of the storage protein in seed of *Coix lacryma-jobi* var. Adlay. J Agric Food Chem 38:631–635

Oxford English Dictionary (2015) Online Edition © Oxford University Press

Pasztor K, Borsos O (1990) Inheritance and chemical composition in inbred maize (*Zea mays* L.) × teosinte (*Zea mays* ssp. *mexicana* (Schrader) Iltis) hybrids. Crop Prod 39:193–213

Pesqueira J, Garcia MD, Mollina MC (2003) NaCl tolerance in maize (*Zea mays* ssp. *mays*) x *Tripsacum dactyloides* L. hybrid calli and regenerated plants. Spanish J Agril Res 1:59–63

Pesqueira J, Garcia MD, Staltari S, Mollina MC (2006) NaCl effects in *Zea mays* L. × *Tripsacum dactyloides* L. hybrid calli and plants. Electron J Biotechnol 9:286–290

Petrov DF (1984) Apomixis and its role in evolution and breeding. Oxonian Press, New Delhi, pp 267

Piperno DR (2003) A few kernels short of a cob: on the Staller and Thompson late entry scenario for the introduction of maize into northern South America. J Archaeol Sci 30:831–836

Piperno DR, Moreno JE, Iriarte J et al (2007) Late Pleistocene and Holocene environmental history of the Iguala valley, central Balsas watershed of Mexico. Proc Natl Acad Sci U S A 104:11874–11881

Piperno DR, Ranere AJ, Holst I, Iriarte J, Dickau R (2009) Starch grain and phytolith evidence for early ninth millennium B.P. maize from the Central Balsas river valley, Mexico. Proc Natl Acad Sci U S A 106:5019–5024

Prasanna BM (2010) Phenotypic and molecular diversity of maize landraces: characterization and utilization. Indian J Genet 70:315–327

Prasanna BM (2012) Diversity in global maize germplasm: characterization and utilization. J Biosci 37:843–855

Prasanna BM (2014) Maize research-for-development scenario: challenges and opportunities for Asia. In: Prasanna BM et al (eds) 12th Asian maize conference and expert consultation on maize for food, feed and nutritional security, book of extended summaries, 30 October–01 November, Bangkok, Thailand, pp 2–11

Prischmann DA, Dashiell KE, Schneider DJ, Eubanks MW (2009) Evaluating *Tripsacum* introgressed maize germplasm after infestation with western corn rootworms (Coleoptera: Chrysomelidae). J Appl Entomol 133:10–20

Purseglove JW (1972) Tropical crops: monocotyledons 1. Longman Group Limited, London

Qi-Lun Y, Ping F, Kang K-C, Guang-Tang P (2008) Genetic diversity based on SSR markers in maize (*Zea mays* L.) landraces from Wuling mountain region in China. J Genet 87:287–291

Ramirez DA (1997) Gene introgression in maize (*Zea mays* ssp. *mays* L.). Philipp J Crop Sci 22:51–63

Ray JD, Kindiger B, Sinclair TR (1999) Introgressing root aerenchyma into maize. Maydica 44:113–117

Rebourg C, Chastanet M, Gouesnard B, Welcker C, Dubreuil P, Charcosset A (2003) Maize introduction into Europe: the history reviewed in the light of molecular data. Theor Appl Genet 106:895–903

Rich PJ, Ejecta G (2008) Towards effective resistance to *Striga* in African maize. Plant Signal Behav 3:618–621

Rieseberg LH, Willis JH (2007) Plant speciation. Science 317:910–914

Rodriguez MG, Miguel-Chavez RS, Larque-Saavedra A (1998) Physiological aspects in Tuxpeno maize with improved drought tolerance. Maydica 43:137–141

Roney J, Hard R (2009) The beginnings of maize agriculture. Archaeology Southwest 23:4–5

Rosegrant MR, Ringler C, Sulser TB, Ewing M, Palazzo A, Zhu T (2009) Agriculture and food security under global change: prospects for 2025/2050. IFPRI, Washington, DC

Sachan JKS, Sarkar KR (1982) Plant type of 'Sikkim Primitives'. Maize Genet Coop Newsl 56:122–124

Sachan JKS, Sarkar KR (1985) The present status of maydeae. Indian J Genet 45:480–491

Sachan JKS, Sarkar KR (1986) Discovery of Sikkim Primitive precursor in the Americas. Maize Genet Coop Newsl 60:104–106

Salhuana W, Pollak L (2006) Latin American Maize Project (LAMP) and Germplasm Enhancement of Maize (GEM) project: generating useful breeding germplasm. Maydica 51:339–356

Salhuana W, Jones Q, Sevilla R (1991) The Latin American Maize Project: model for rescue and use of irreplaceable germplasm. Diversity 7:40–42

Savidan Y, Berthaud J (1994) Maize × *Tripsacum* hybridization and the potential for apomixes transfer for maize improvement. In: Bajaj YPS (ed) Biotechnology in agriculture and forestry 25. Springer, Berlin, pp 69–83

Schnable PS, Ware D, Fulton RS et al (2009) The B73 maize genome: complexity, diversity and dynamics. Science 326:1112–1115

Semagn K, Magorokosho C, Ogugo V, Makumbi D, Warburton ML (2014) Genetic relationships and structure among open-pollinated maize varieties adapted to eastern and southern Africa using microsatellite markers. Mol Breed 34:1423–1435

Sharma L, Prasanna BM, Ramesh B (2010) Phenotypic and microsatellite-based diversity and population genetic structure of maize landraces in India, especially from the north east Himalayan region. Genetica 138:619–631

Shavrukov Y, Sokolov V (2015) Maize-gamagrass interspecific hybrid, *Zea mays* × *Tripsacum dactyloides*, shows better salinity tolerance and higher Na+exclusion than maize and sorghum. Int J Lat Res Sci Tech 4:128–133

Shiferaw B, Prasanna BM, Hellin J, Banziger M (2011) Crops that feed the world. 6. Past successes and future challenges to the role played by maize in global food security. Food Sec 3:307–327

Sidorov FF, Shulakov IK (1962) Hybrids of maize and teosinte. Bul App Bot Genet Plant Breed 347:6–85

Singode A, Prasanna BM (2010) Analysis of genetic diversity in the North Eastern Himalayan (NEH) maize landraces of India using microsatellite markers. J Plant Biochem Biotech 19:33–41

Stebbins GL (1958) The inviability, weakness and sterility of interspecific hybrids. Adv Genet 9:147–215

Studer A, Zhao Q, Ross-Ibarra J, Doebley J (2011) Identification of a functional transposon insertion in the maize domestication gene *tb1*. Nature 43:1160–1163

van-Heerwaarden JV, Doebley J, Briggs WH, Glaubitz JC, Goodman MM, de Jesus Sanchez Gonzalez J, Ross-Ibarra J (2011) Genetic signals of origin, spread and introgression in a large sample of maize landraces. Proc Natl Acad Sci U S A 108:1088–1092

Vann L, Kono T, Pyajarvi T, Hufford MB, Ross-Ibarra J (2015) Natural variation in teosinte at the domestication locus *teosinte branched1* (*tb1*). Peer J 3, e900. doi:10.7717/peerj.900

Venkateswarlu J, Chaganti RSK (1973) Job's lears (*Coix lacryma-jobi* L.). ICAR Publication, New Delhi, pp 44

Vielle-Calzada JP, de la Vega OM, Hernandez-Guzman G et al (2009) The palomero genome suggests metal effects on domestication. Science 326:1078. doi:10.1126/science.1178437

Vigouroux Y, Glaubitz JC, Matsuoka Y, Goodman MM, Sanchez J, Doebley J (2008) Population structure and genetic diversity of new world maize landraces assessed by DNA Microsatellites. Am J Bot 95:1240–1253

Vyssoulis GP, Karpanou EA, Papavassiliou MV, Belegrinos DA, Giannakopoulou AE, Toutouzas PK (2001) Side effects of antihypertensive treatment with ACE inhibitors. Am J Hypertens 14:114A–115A

Wang RL, Stec A, Hey J, Lukens L, Doebley J (1999) The limits of selection during maize domestication. Nature 398:236–239

Wang H, Nussbaum-Wagler T, Li B et al (2005) The origin of the naked grains of maize. Nature 436:714–719

Wang L, Xu C, Qu M, Zhang J (2008) Kernel amino acid composition and protein content of introgression lines from *Zea mays* ssp. *mexicana* into cultivated maize. Cereal Sci 48:387–393

Wang L, Qu M, Zhang X, Qiao Y (2014) Production of antihypertensive peptides by enzymatic zein hydrolysate from maize-*Zea mays* ssp. *mexicana* introgression line. Pak J Bot 46:1735–1740

Warburton ML, Setimela P, Franco J et al (2010) Toward a cost-effective fingerprinting methodology to distinguish maize open-pollinated varieties. Crop Sci 50:1–11

Warburton ML, Wilkes G, Taba S et al (2011) Gene flow between different teosinte species and into the domesticated maize gene pool. Genet Resour Crop Evol 58:1243–1261

Wasala SK, Prasanna BM (2013) Microsatellite marker-based diversity and population genetic analysis of selected lowland and mid-altitude maize landrace accessions of India. J Plant Biochem Biotechnol 22:392–400

Watson L, Dallwitz MJ (1992) The grass genera of the World. CAB International, Oxon, pp 1038

Watt G (1904) *Coix* spp. or Job's-tears – a review of all available information. Agric, Ledger No. 13:513–553

Wei WH, Zhao WP, Song YC, Liu LH, Guo LQ, Gu MG (2003) Genomic in situ hybridization analysis for identification of introgressed segments in alloplasmic lines from *Zea mays* × *Zea diploperennis*. Hereditas 138:21–26

Wilkes HG (1967) Teosinte: the closest relative of maize. The Bussey Institute, Harvard University, Cambridge, MA

Wilkes HG (1972) Maize and its wild relatives. Science 177:1071–1077

Wilkes HG (1977) Hybridization of maize in Mexico and Guatemala and the improvement of maize. Econ Bot 31:254–293

Yallou CG, Menkir A, Adetimirin VO, Kling JG (2009) Combining ability of maize inbred lines containing genes from *Zea diploperennis* for resistance to *Striga hermonthica* (Del.) Benth. Plant Breed 128:143–148

Yanga Q, Lia Z, Lib W et al (2013) CACTA-like transposable element in ZmCCT attenuated photoperiod sensitivity and accelerated the post domestication spread of maize. Proc Natl Acad Sci U S A 110:16969–16974

Barley

5

Naval Kishore, Vishnu Kumar, and R.P.S. Verma

Abstract

Barley being the earliest domesticated cereal is one of the major world crops today. Wide adaptation of barley to ecogeographic conditions and higher response to low input and stressful environments make it globally cultivated. The diploid and self-pollinated nature of barley renders it as an important physiological and genetic research material. The recent yield trends in barley have shown improved productivity but the area under the crop has reduced gradually. Besides other sources of genetic variation, barley landraces continue to be an important tool of genetic diversity conserved ex situ in gene banks and need suitable phenotypic and genotypic characterisation. Gene flow involving wild progenitors and domesticated cultivars has attracted breeders' interest in order to broaden the genetic base of barley. Biotic and abiotic factors affecting barley yield have to be determined specifically. Global warming generated issues such as tolerance to drought and disease management in barley have become increasingly important and need to be addressed in current times. Use of molecular markers and high-throughput techniques in barley has enhanced the precision for introgressing specific traits of importance. Consensus maps have allowed comparative mapping in barley to explore information available in other crops. Recent approaches like advanced backcross QTL (AB-QTL) analysis and association mapping (AM) studies have aided in understanding of the complex agronomic traits in barley. Continued improvements in transformation methods will increase transgenic efficiencies in different barley genotypes. Rapid selection for malt quality using molecular markers

N. Kishore (✉)
CSK Himachal Pradesh Agricultural University, HAREC, Bajaura, Kullu 175 125, Himachal Pradesh, India
e-mail: naval13@gmail.com

V. Kumar
Indian Institute of Wheat and Barley Research, Karnal 132 001, Haryana, India

R.P.S. Verma
International Centre for Agricultural Research in Dry Areas, 6299, Rabat, Morocco

M. Singh, S. Kumar (eds.), *Broadening the Genetic Base of Grain Cereals*, DOI 10.1007/978-81-322-3613-9_5

has to be achieved for developing high-yielding malt barley cultivars. Interspecific and intergeneric hybridization in barley can be beneficial for introgression of useful traits in the barley genetic background. Owing to health benefits of barley, industrial interest for introduction of different barley products in human diets has increased considerably. Further, consortia efforts are in progress to sequence the barley genome by the International Barley Genome Sequence Consortium. To sum up, the future barley breeding programme should combine the application of new tools and techniques with traditional and efficient plant breeding methods to achieve desired goals rapidly.

Keywords
Barley • Evolution • Gene pool • Genetic diversity • Wide hybridization

5.1 Introduction

Barley (*Hordeum vulgare* L. emend Bowden.) is one of the few initially domesticated crops and ancient among cereals. It has played a significant role in the development of human civilization. Cultivated barley is a self-pollinating diploid species with $2n = 14$ chromosomes and large genome size (>5.1 gigabases) consisting of highly repetitive sequences, almost 12 times the size of rice genome (Bennett and Smith 1976; IBGSC 2012). Today barley is grown over diverse ecogeographic environmental conditions as compared to other crop species. Because of its hardiness, in many countries around the world, it is often considered the only possible rainfed cereal crop under low input and stressful environments. This adaptability to extreme and marginal conditions has led to widespread cultivation of this cereal throughout the world (Bothmer et al. 1995). Barley fields can be found from the tropics to high latitudes (>60°N) in Iceland and Scandinavia as well as in high latitudes up to 4500 m above sea level in the Himalayas (Bothmer et al. 2003a; Ceccarelli et al. 2008). As per the Roman literature, owing to its rich dietary fibre, barley was awarded to the gladiators also called as barley men or 'hordearii' (Andrew 2008). Due to its higher nutritional value, barley is still consumed as a staple food in the hilly areas of Central Asia, in Southwest Asia and Northern Africa.

In terms of total food production, barley ranks fourth among cereals and fifth among various crops after wheat, maize, rice and soya bean (FAOSTAT 2012). It is grown by nearly 100 countries on about 50 million hectares. The largest barley producers in the world are France, Germany, Russia, Spain, Ukraine, Canada, Australia and Turkey (Table 5.1). Majority of the area under barley crop has been replaced by wheat cultivation. Globally, the area under barley cultivation has reduced from 80 million hectares in the 1970s to less than 50 million hectares in 2012 (FAOSTAT 2012). FAO's latest forecast for world output of barley in 2013 stands at about 142 million tonnes, 8.1% up from the 2012 level. Outputs have increased in most of the world's major barley-producing countries, most notably in North Africa and the CIS countries in Europe.

Barley has considerable economic importance both in agriculture and industry in many countries. Around 55–60 % of barley is used for feed, 30–40 % for malt, 2–3 % for food and 5 % for seed (Ullrich 2010). It is the major dietary source for ruminant and non-ruminant livestock, poultry and fish. Generally, the feed barley varieties yield more (10–20 %) than the malt barley varieties (Blake et al. 2010). Worldwide, a significant increase of about 48 % has been observed in the malt export and import values. The continuous increase in yield via adoption of high-yielding varieties in all the barley-growing areas has

Table 5.1 Area, production and productivity of major barley-growing countries of the world from 1981 to 2012

	1981–1990			1991–2000			2001–2010			2012		
Continent	Area (m ha)	Production (m tons)	Av. yield (tons/ha)	Area (m ha)	Production (m tons)	Av. yield (tons/ha)	Area (m ha)	Production (m tons)	Av. yield (tons/ha)	Area (m ha)	Production (m tons)	Av. yield (tons/ha)
World	77.65	167.03	2.16	65.83	151.65	2.31	55.12	141.98	2.58	49.53	132.89	2.68
Europe	48.26	115.40	2.40	34.25	94.09	2.78	28.08	89.26	3.18	24.38	79.10	3.24
Asia	12.05	17.95	1.50	15.86	24.23	1.53	11.63	21.11	1.81	10.56	19.25	1.82
North America	8.64	23.19	2.68	6.90	21.03	3.05	5.01	15.39	3.10	3.37	12.81	3.80
Africa	5.08	4.80	0.94	4.70	4.48	0.93	4.74	5.43	1.13	5.00	6.02	1.21
Aust.-NZ	2.76	4.33	1.56	3.14	5.98	1.89	4.34	7.79	1.76	3.78	8.66	2.29
South America	0.60	0.85	1.41	0.73	1.32	1.79	0.92	2.22	2.37	2.10	6.01	2.86
Cent. America	0.27	0.51	1.89	0.25	0.51	2.08	0.32	0.79	2.56	0.33	1.03	3.14
Country												
Russian Fed.	–	–	1.42	11.97	18.65	1.53	8.68	17.22	1.96	7.64	13.95	1.83
Germany	2.82	13.52	4.80	2.23	12.39	5.57	1.95	11.65	5.98	1.68	10.42	6.19
France	2.10	10.21	4.92	1.58	9.48	5.97	1.70	10.70	6.30	1.68	11.35	6.74
Canada	4.75	12.45	2.62	4.30	12.86	2.99	3.54	10.42	2.96	2.06	8.01	3.89
Ukraine	–	–	2.54	3.89	8.90	2.25	4.43	9.77	2.21	3.29	6.94	2.11
Spain	4.08	8.63	2.09	3.63	8.62	2.40	3.14	8.54	2.71	2.68	5.98	2.23
Turkey	3.28	6.39	1.98	3.57	7.76	2.17	3.40	7.97	2.34	2.75	7.10	2.58
Australia	2.66	3.90	1.46	3.07	5.63	1.82	4.37	7.44	1.70	3.72	8.22	2.21
UK	1.94	9.59	4.95	1.23	6.91	5.60	1.02	5.89	5.75	1.00	5.52	5.51
USA	3.88	10.74	2.75	2.59	8.18	3.16	1.47	4.97	3.43	1.31	4.80	3.65
Denmark	1.17	5.29	4.60	0.76	3.78	5.00	0.69	3.54	5.16	0.72	4.06	5.61
Poland	1.21	3.88	3.19	1.14	3.34	2.93	1.12	3.50	3.13	1.16	4.18	3.60
China	1.11	2.96	2.69	1.50	3.96	2.63	0.78	2.94	3.78	0.63	1.80	2.86
Iran	2.23	2.55	1.14	1.78	2.74	1.55	1.55	2.88	1.85	1.68	3.40	2.02
Morocco	2.30	2.22	0.95	2.21	1.87	0.81	2.14	2.03	0.95	1.89	1.20	0.63
Czech Republic	–	–	4.70	0.59	2.20	3.87	0.49	2.00	4.10	0.38	1.62	4.23
Kazakhstan	–	–	0.89	3.69	3.74	0.93	1.70	1.98	1.16	1.64	1.49	0.91
Belarus	–	–	2.49	0.97	2.20	2.23	0.65	1.89	2.92	0.56	1.92	3.44
Finland	0.57	1.58	2.78	0.53	1.71	3.22	0.54	1.86	3.45	0.45	1.58	3.50

(continued)

Table 5.1 (continued)

Continent	1981–1990			1991–2000			2001–2010			2012		
	Area (m ha)	Production (m tons)	Av. yield (tons/ha)	Area (m ha)	Production (m tons)	Av. yield (tons/ha)	Area (m ha)	Production (m tons)	Av. yield (tons/ha)	Area (m ha)	Production (m tons)	Av. yield (tons/ha)
Sweden	0.59	2.20	3.73	0.45	1.75	3.91	0.37	1.53	4.18	0.37	1.70	4.60
India	1.35	1.79	1.35	0.85	1.51	1.79	0.66	1.36	2.05	0.78	1.62	2.08
Ethiopia	–	–	1.11	0.93	0.93	1.03	1.05	1.35	1.30	1.02	1.78	1.75
Argentina	0.12	0.23	1.91	0.22	0.55	2.41	0.39	1.25	3.05	1.69	5.16	3.04
Algeria	0.99	0.72	0.70	0.81	0.78	0.88	0.78	1.08	1.32	1.03	1.59	1.54

Source: (faostat.fao.org)

stabilised the overall production. However, the prime focus of most of the breeding programmes is still on the biotic and abiotic factors which account for majority of the losses to barley production and needs to be addressed strategically to achieve stability. Barley is mainly used for feeding, malting, distilling and to less extent also as calorie food source for human consumption mainly in marginal areas with problematic soils and scanty rainfall (Grando and Macpherson 2005).

Recent research regarding dietary composition in food barley has renewed the interest confirming the health benefits of barley in human diets (Brockman et al. 2013; Sullivan et al. 2013). In comparison to other cereal crops, barley has a better fodder value including grain and straw. In most of the developed countries, barley straw is used for animal bedding, whereas it is used as animal feed in the developing countries. Even under variable climate within the growing season, barley gives comparably better yield than wheat and other small grain cereals. Being one of the most widely adapted crops, the barley germplasm pool contains enough genetic diversity to breed for adaptation to different environmental conditions. Moreover, the ample barley germplasm resources available worldwide (Bockelman and Valkoun 2010) likely contain beneficial allelic variation that new genomic and breeding technologies can exploit (Newton et al. 2011). Owing to its vast morphological and environmental adaptability, various types of barley (winter, spring, two rowed, six rowed, awned, awnless, hooded, covered, naked, malting, feed and food types) are grown throughout the world.

5.2 Evolution and Domestication

Today's cultivated barley (*Hordeum vulgare* L.) is one of the first domesticated crops and evolved from its wild progenitor *Hordeum spontaneum* (C. Koch) Thell or *H. vulgare* ssp. *spontaneum* indicating Fertile Crescent as the earliest archaeological remains of wild barley (Badr et al. 2000; Kilian et al. 2009; Salamini et al. 2002; Zohary and Hopf 2000) (Fig. 5.1). With the development and advancement of molecular markers in recent years, more precise information on origin and domestication history of barley is emerging. Wild barley (*H. spontaneum*) is widely distributed throughout the Fertile Crescent covering the geographic area in Israel, Jordan, Lebanon, Syria, southeastern Turkey, northern Iraq and western Iran (Azhaguvel and Komatsuda 2007; Zhou et al. 2014). The cultivated barley spread from the Mediterranean region to Africa and eastwards through Iran and Afghanistan into India and China (Zohary and Hopf 2000). Recently, Orabi et al. (2007) reported Eritrea and Ethiopia regions as other domestication sites of barley, whereas of late Dai et al. (2012) has also indicated Tibet Plateau and its vicinity to be one of the centres of domestication of cultivated naked barley. Hulless or naked barley (*H. vulgare* L. subsp. *vulgare*) differs from hulled barley by the loose husk cover of caryopses that is easily separable upon threshing. The hulless grain character is controlled by single recessive gene '*nud*' located on the long arm of chromosome 7H (Kikuchi et al. 2003). The cultivation of naked barley is less common worldwide than hulled barley. The domestication of naked barley is believed to occur after the hulled type around 6500 B.C. (Zohary and Hopf 2000). Taketa et al. (2004) suggested a monophyletic origin of naked barley as a single mutation event either from wild barley (*H. vulgare* subsp. *spontaneum*) or from domesticated hulled barley (*H. vulgare* subsp. *vulgare*). Its distribution is skewed towards East Asia, namely, to the Himalayas (Nepal, India, Bhutan and Tibet), China, Korea and Japan, where it accounts for up to 95 % of the domesticated barley in some regions (Sun and Wang 1999; Takahashi 1955). Besides East Asia, it is cultivated in Ethiopia at a low frequency (Assefa and Labuschagne 2004). The cultivation is rare in the western world (Europe, North America) and in Australia where hulled types are prevalent. Hulless barley is mainly used as animal

Fig. 5.1 The Fertile Crescent, domestication area of cultivated barley (*Hordeum vulgare* ssp. *vulgare*) (Feuillet et al. 2008)

feed; however, it is an important human food in the Himalayas and in Ethiopia.

5.3 Taxonomy

Barley belongs to genus *Hordeum* of Poaceae family and tribe Triticeae along with wheat and rye. The genus *Hordeum* comprises of 32 species and 45 taxa. It is classified into four sections (Bothmer 1992), although six sections have also been suggested (Reid and Wiebe 1978). Linnaeus (1753), in his book *Plantarum*, treated six species of *Hordeum*, of which four are cultivated forms, viz. *H. vulgare*, *H. distichum*, *H. hexastichum* and *H. zeocrithon*. The section *Hordeum* consists of both annual and perennial species. All annuals except *Hordeum vulgare* were placed in a separate section (Nevski 1941; Bell 1965). According to Rasmussion (1985), since the different annual species differentiated from various unrelated perennials in different groups, they should be treated together with their probable closest perennials relatives. All the species of this genus have the basic chromosome number of $x=7$ and are morphologically similar with three one flowered spikelets at each rachis mode called triplets (Bothmer et al. 1995). The two lateral florets are pedunculate, or sessile, and may be sterile as in two-rowed barley or fertile as in six-rowed barley (Bothmer et al. 2003a).

Hordeum vulgare is the most intensively studied species in barley consisting of both cultivated and wild forms, which are easily crossable and closely related (Fedak 1985). The cultivated form is referred to as subspecies *vulgare*, while the wild form is referred as subspecies *spontaneum* C. Koch. Both subspecies are diploids with $2n=2x=14$ chromosomes. The subspecies *vulgare* may be two- or six-rowed spikes, whereas *spontaneum* is of two-rowed type with brittle rachis. These cultivated forms are morphologically highly variable, under cultivation for long in wide geographical areas under different climatic conditions and subjected to intense breeding processes. Subspecies *spontaneum* has a wide distribution starting from Eastern Mediterranean in Greece to Afghanistan, north of India, western China and Near East (Harlan 1968; Witcobe 1978). High level of genetic diversity also prevails within the *Hordeum* species which are tetraploid ($2n=4x=28$) or hexaploid ($2n=6x=42$) (Komatsuda et al. 1999). These include some annuals (*H. marinum* Huds., *H. murinum* L. and *H. pusillum* Nutt.) and some perennials (*H. bulbosum* L. and *H. brevisubulatum* (Trin.) Link). *Hordeum murinum* ($2n=14$, 28, 42) is a highly variable species mainly native to Europe, Mediterranean region, Iran, Pakistan and Nepal. It has a characteristic feature of large inflated lateral spikelets. The central spikelet is self-pollinated while the lateral spikelets are mainly male sterile with open flowers. This species is further subdivided into three subspecies, viz. subsp. *murinum* (4*x*), *glaucum* (2*x*) and *leporinum* (4*x* and 6*x*). *Hordeum bulbosum* ($2n=14$ & 28) is a strictly cross-pollinated species with a self-incompatibility system with two *S* loci though most of them are inbreeding (Bothmer et al. 2003b; Lundqvist 1962). This species is mainly found in Mediterranean area to Afghanistan and Central Asia.

The two species, *H. vulgare* and *H. bulbosum*, are considered to share a common basic genome I, which is not related to any other genome in the genus. The genomes of *H. marinum* and *H. murinum* seem not closely related to the other genomes in *Hordeum* and have been designated as X and Y, respectively. The remaining diploid *Hordeum* species are closely related and share the H genome (Bothmer 1992).

5.4 Crop Gene pool

The barley gene pool can be divided into six major groups based on the practical utilisation by breeders/geneticists:

(a) Modern cultivars in current use
(b) Obsolete cultivars often the elite cultivars of the past and often found in the pedigrees of the modern cultivars
(c) Landraces
(d) Wild relatives in the genus *Hordeum*
(e) Genetic stocks
(f) Breeding lines

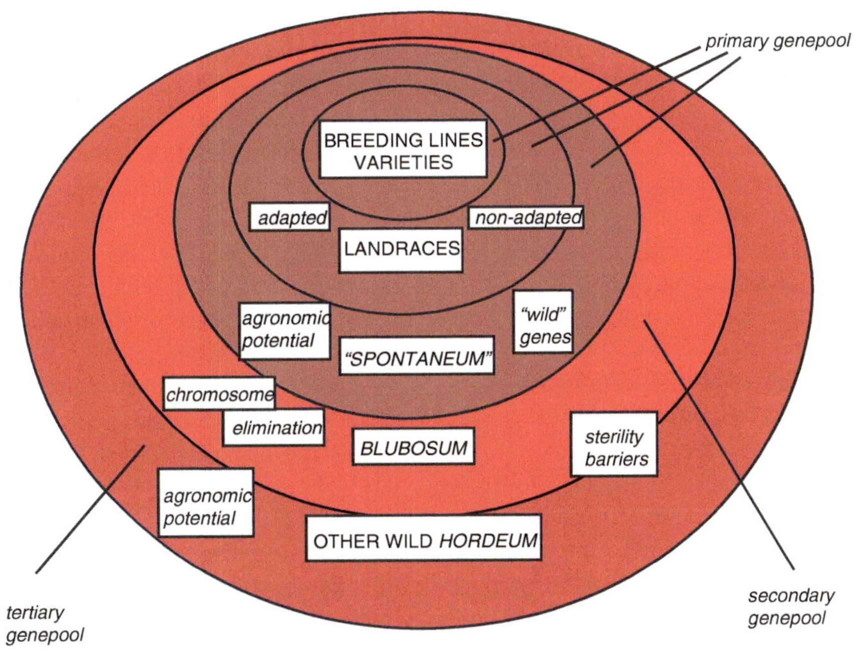

Fig. 5.2 Gene pools in barley (*Hordeum vulgare*) (Bothmer et al. 2003a)

The broad pool can also be subdivided into primary, secondary and tertiary gene pools based on their crossability and association (Harlan and deWet 1971; Bothmer et al. 2003a) (Fig. 5.2).

The primary gene pool includes the landraces, elite breeding material, cultivars and the wild ancestors of cultivated barley, *H. vulgare* ssp. *spontaneum*. Gene transfer occurs freely within the primary gene pool. Some feral hybrid types have been known to occur in the areas where they grow alongside each other. The secondary gene pool consists of only one species, *H. bulbosum* L. sharing the basic H genome with barley. Though, crossing ability between these two species is difficult but gene transfer is possible. There is elimination of the *bulbosum* chromosomes after pollination of *H. vulgare* with *H. bulbosum* and haploids are recovered. This approach further leads to the development of doubled haploids by consequent chromosome doubling. This technique is being rapidly replaced by anther and pollen culture techniques (Pickering and Johnston 2005; Sanei et al. 2011). *H. bulbosum* has agronomical advantage where in few instances some disease and insect resistant genes have been introgressed. The tertiary gene pool of barley includes the remaining wild *Hordeum* species that are mostly polyploids with tetraploid and hexaploid genome types and have a strong crossability barrier with cultivated barley. However, gene transfer between distantly related species can be accomplished through biotechnological tools (Ceccarelli et al. 2008). Autopolyploidy has been shown in *H. bulbosum* and *H. brevisubulatum*. Most of other polyploids are allopolyploids and mainly segmental allopolyploids.

5.5 Assessment of Gene Flow in Crop Improvement

Gene flow is the migration of genetic information between populations through vectors such as pollen and seeds. In a predominantly inbreeding species such as barley, the migration of pollen and natural seed dispersal are expected to be small and non-existent (Parzies et al. 2004). Gene flow between traditional variety populations and regions is a consequence of farmer seed systems (the human networks through which farmers exchange or sell the seeds of traditional varieties) and can affect metapopulation structure by creating sources and sinks of genetic variation (Alvarez et al. 2005). Consequently, gene flow found between local barley populations may be due to seed exchange between farmers or to a limited extent outcrossing with foreign pollen (Zoghlami et al. 2011). In wild plant populations, gene flow is frequently limited by the geographical isolation and the genetic structure in the pattern of isolation by distance (Wright 1943). In cultivated crop plants, gene flow by seed dispersal is usually linked with human action.

In self-fertilising crops, gene flow via pollen dispersal is limited. Considering the barley breeding system and short pollen dispersal (up to 60 m), gene flow occurs mainly due to the seeds rather than pollen dispersal in the fields. One of the examples for seed flow is use of same combine harvester among the farmers to harvest landraces and varieties. Bitocchi et al. (2009) reported low level of introgression to be population dependent, and different agronomic practices, viz. land preparation sowing and harvesting by the farmers show that gene flow from varieties to landraces is limited. Despite high levels of selfing, landraces maintained by the farmers have shown high variability, and most of the morphological, genetic and phenotypic variation has been found within fields and populations (Backes et al. 2009; Hadado et al. 2010). Partitioning of the variation between wild and cultivated barley has re-emphasised comparatively the narrow genetic base of barley, with about 50 % of the alleles being unique to wild barley and absent in the cultivated barley.

To exploit the existing genetic diversity of any species, the understanding of the genetic architecture has to be accounted for easy access and use by breeders. This requires effective breeding strategy to broaden the genetic base of modern cultivars through selection of right parents. Population structuring can happen due to various factors, viz. distance, mating preferences, barriers to dispersal or when the local populations show less dispersal rates between them which allow genetic differentiation (Slatkin 1995). One of the methods to study population structure involves use of spatial data with non-random distribution of genetically similar individuals within populations (McCauley 1997) and determined by the effects of genetic drift and natural selection. DNA markers such as simple sequence repeats (SSRs) have been widely used to study genetic make-up of populations owing to their high polymorphism, codominance and reproducibility. Bellucci et al. (2013) using SSRs in barley landraces and modern varieties have reported that structure analysis is an important approach to study gene flow for the detection of genotypes derived from hybridisation and introgression between varieties and landraces.

The future farming requires sustainable production of crops for the increasing population to be achieved with improved resistance to pests and diseases, greater fertiliser efficiency and economic water usage. Marker-assisted breeding schemes provide favourable solutions to this problem but prior to utilising these schemes, gene maps showing genes of importance are to be synthesised.

5.6 Gene Flow Constraints

Various factors tend to limit regular gene flow among different populations (Papa and Gepts 2004). Gene flow has shown to be mostly species specific and to depend on both genetic and environmental factors. Different studies have reported varying frequencies of cross-fertilisation across years, most likely due to changing environmental factors (Gatford et al. 2006; Matus Cadiz et al. 2004; Ritala et al. 2002). In self-pollinating plants, frequency of gene flow is very low. In barley, frequencies of 0–7 % gene flow have been reported (Ghani et al. 2004). Gene flow between wild and domesticated populations is also limited by their phenology, geographical distribution and spatial arrangement. The rate of gene flow decreases rapidly at distance beyond few metres (Gustafson et al. 2005). However, various studies suggest that pollen dispersal and cross-fertilisation between barley plants are possible over distance up to 50 m (Gatford et al. 2006; Ritala et al. 2002) with similar observations also being reported in wild barley (Wagner and Allard 1991). Hermannsson et al. (2010) showed that cross-pollination of closed flowering barley varieties occurs with a frequency of less than 0.002 % within a distance of 0.65 m under Icelandic environmental conditions. The reproductive and propagation systems of domesticated crops also represent crucial factors affecting the rate of gene flow between wild and domesticated populations. Though reproductive barrier is prevalent owing to the different ploidy levels between domesticated and wild progenitors of barley, hybridisation can still take place and produce fertile progeny. Barley reflect a low rate of gene flow in domesticated to wild hybridisations; however, gene flow from wild to domesticated population will only occur during sexual reproduction to obtain new cultivars.

Among the important factors limiting gene flow, farmers themselves play a greater role through seed exchange networks (Jensen et al. 2013). This activity can be controlled by the farmers by selecting the desired seeds for planting in the next generation, thus monitoring the gene flow similar to post-zygotic reproductive barrier. Needless to say, plant breeders have experienced difficulty to cross cultivated barley with wild barley species (Baum et al. 1992; Pickering and Johnston 2005). Therefore, controlled gene flow measures related to good cultivation practices such as proper isolation distances and crop rotation need to be adopted.

5.7 Level of Diversity in Crop Germplasm

Genetic biodiversity and its utilisation are topics for a wide range of research, since genetic diversity of crops is considered to be one of the main resources supporting human life. The use of high diverse germplasm in breeding increases the chances for success in developing highly productive cultivars with good quality over a long period of time (Bockelman and Valkoun 2010; Horsley et al. 1995). Barleys in the East Asian regions differ from European and North African landraces as they have frequently naked caryopsis and are mainly six-rowed types (Knupffer et al. 2003). The wild relatives of barley have brittle rachis and shattering takes place as soon seeds are ripened. Brittleness of rachis is controlled by two linked complementary genes *Btr1* and *Btr2* concentrated in all wild barley. However, the recessive allele *btr1* has been reported to be present in most occidental cultivars, whereas the *btr2* allele in most oriental cultivars (Komatsuda et al. 2004; Zohary 1999).

Various studies on diversity in barley germplasm have been attempted in order to identify the region of domestication. Badr et al. (2000) while studying 317 wild and 57 domesticated cultivars using 400 AFLP loci reported monophyletic origin of barley domestication in the Israel-Jordan area. The barley populations from this region showed more similarity to the domesticated gene pool than any other wild barley population. Based on haplotype frequencies, two origins of barley domestication were found by Morrell and Clegg (2007). One belonged to the Fertile Crescent that gave rise to majority of the European and American cultivars, and second was the western foothill region of the Zagros Mountains, about 1500–3000 km east of Fertile Crescent. However, till date the sites of domestication are subject of debate (Molina-Cano et al. 2005; Orabi et al. 2007). Over the years, barley domestication has gradually accumulated various traits important for agricultural production. Continuous selection has led to the changes in plant architecture where both unconscious (environmental) and deliberate (by choice of humans) selections have played a major role. Selection for non-brittle rachis, six-rowed spike and huskless grain types are three key traits involved in cultivated germplasm (Kilian et al. 2009). Mutation in photoperiod genes led to photoperiod insensitivity in barley and its further distribution to the diverse geographical areas across continents (Jones et al. 2008).

5.7.1 Landraces

Barley landraces are genetically heterogeneous populations comprising inbreeding lines and hybrid segregates generated by a low level of random outcrossing in each generation (Nevo 1992). These are heterogeneous plant varieties that are reproduced by farmers as populations that are subjected to both artificial and natural selection. Barley landraces (also called farmer varieties) have a reduced geographic range and are often diverse within particular types adapted to local conditions (Brush 1995; Harlan 1995). Thus, in a landrace the diversity is structured between and within populations (at the field/farmer level). The genetic structure of these landraces may be considered as an evolutionary approach to survival and performance under arid and semi-arid conditions (Schulze 1988).

The landraces of barley are principal drivers of agricultural production, representing over 10 million hectares worldwide comprising nine countries with Canada, the USA and Syria (through ICARDA) being the major contributors (Altieri 2004). In many developing countries, farmers maintain traditional varieties independently with seed often obtained from relatives, neighbours or local markets (McGuire 2008). Seed exchange networks are the most important factor in the movement and use of crop germplasm for the landraces, and they have a strong influence on the genetic diversity and structure of cultivated crops (Hodgkin et al. 2007; Jensen et al. 2013). This germplasm movement among farmers does not follow a simple pattern of isolation by distance (Pusadee et al. 2009), but is structured according to interactions among many factors including agroclimatic zones, biotic stressors, different linguistic regions, access to markets and agricultural extension services (Deletre et al. 2011; Jones et al. 2011; Samberg et al. 2013).

The diversity of regional landraces, local cultivars and related species can have a beneficial role in crop improvement. Exploiting biodiversity for genetic gain is not a new concept. Many major genes from wild relatives have been transferred into the cultivated gene pools of many crops (Hajjar and Hodgkin 2007). In the initial collections of landraces held in gene banks, at least 50–60 % of the total genetic variation captured resides within the landraces, the remainder being accounted for by differences between landraces (Brown and Munday 1982). This richness of genetic diversity in wild barley and its occurrence over a wide range of habitats in the region suggest that the genetic resources of wild barley in the Fertile Crescent and Central Asia can be exploited for improvement of cultivated barley (Jana and Pietrzak 1988). A considerable yield advantage of the landraces over modern varieties in low rainfall conditions and with little or no use of inputs has been reported by Ceccarelli and Grando (1996). Barley landraces have developed abundant patterns of variation and would represent a largely untapped reservoir of useful genes for adaptation to biotic and abiotic stresses (Brush 1995) to contribute to the improvement of modern cultivars (Hadjichristodoulou 1995; Veteläinen 1994). As an example for biotic stresses, nineteen major genes *(Rph)* for resistance against *Puccinia hordei* have been identified and mapped in barley landraces and wild barley (*H. vulgare* ssp. *spontaneum*) (Weerasena et al. 2004).

Barley landraces are valuable germplasm for sustainable agriculture in the context of future climate change, and using in situ conservation strategies based on farmer, we can preserve the genetic identity of landraces while allowing adaptation to local environments (Bellucci et al. 2013). Conventional breeding and high-yielding varieties have not been able to improve the grain and straw quality important for small ruminants, mostly sheep. Indirectly it has proved beneficial in preserving the biodiversity as in these environments all the barley grown are landraces which are popular among farmers for their good feeding quality. Generally, in both developing and developed countries, the use of exotic or primitive germplasm such as *H. spontaneum* and landraces is not common in barley breeding (Grando et al. 2001). The breeding programme involves newly released cultivars and elite germplasm. Though most of the recent breeding material is derived from the previous landraces, still the breeding efforts rely on a narrow gene pool. Many landraces are lost during the last century but still different types are available in the germplasm collections. These landrace collections are important sources of germplasm to enrich the modern barley cultivars (Tester and Langridge 2010). Exploitation of these landraces in modern crop breeding requires understanding of their phenotypic characteristics and environmental adaptations along with their evolutionary relationships.

5.7.2 Commercial Cultivars

Over the past 100 years, the pure line varieties have almost replaced the landraces resulting in reduced diversity. Numerous high-yielding barley cultivars have been evolved through extensive plant breeding efforts and strict selection approaches. Heritable variation is created mainly by controlled crosses between adapted high-yielding cultivars and breeding lines. Although variety breeding is based on elite germplasm, specific traits may be introgressed from wild barley and landraces in backcrossing programmes (Nevo 1992). Mutations in *H. spontaneum* as well as mutations induced by radiation or chemical treatments have also been used for cultivar development (Briggs 1978). A shortcut to homozygosity can be achieved in barley by producing doubled haploid lines either from immature pollen grains by anther or microspore culture or through interspecific crosses between barley and *H. bulbosum* with subsequent chromosome elimination (Mirzaei et al. 2011; Pickering and Devaux 1992; Pickering and Johnston 2005). Both methods allow faster production of new cultivars than using conventional breeding methods in commercial barley breeding programmes, and several doubled haploid varieties have been released.

The genetic uniformity of current cultivars, due to decades of breeding with elite materials, may lead to greater vulnerability to the negative effects of climate change, and it will also limit the future genetic gains (Gepts 2006). A new generation of breeding strategies has to be focused to find novel sources of genetic variation that can be incorporated into breeding and thus improving both productivity and quality while at the same time responding to climate change (Feuillet et al. 2008; Tester and Langridge 2010).

5.8 Genetic Bottleneck

Since domestication, the crop plants have witnessed change in their genetic base owing to certain selection forces. The cultivated crops possess several distinguishing traits that differentiate them from their wild progenitor and are collectively called as 'domestication syndrome' (Hammer 1984). This has resulted in the fixation and narrowing of the genetic base in cultivated barley in comparison to wild species (Tanksley and McCouch 1997). Wild gene pools and landraces can be utilised for increasing the crop productivity under changing environmental conditions (Bockelman and Valkoun 2010; Kilian et al. 2006; Xu et al. 2012). Many alleles have lost during the process of domestication and modern breeding and warrant evolvement of new genes in barley (Kilian et al. 2006). The genetic resources need to be investigated for potential of yield improvement as linkage drag often hampered their applicability in plant breeding. However, with the advent of molecular markers, usefulness of wild relatives or crop improvement has opened gates in practical terms in comparison to the conventional approaches (Xiao et al. 1996).

5.8.1 Genetic Resources

Diverse genetic resources (wild relatives, landraces, cultivars) are essentially required to ensure continued availability of genetic materials with characteristics required for development of stable productive and high quality cultivars in any breeding programme. To meet this demand, it is important to collect, evaluate and maintain genetic resources available from different geographical areas throughout the world. Conservation of the genetic diversity of landraces in barley can be done through static or dynamic methods (Brown 2000). Dynamic conservation has a specific focus, adaptive processes and this can be realized with the cultivation of plant populations of both in situ and ex situ (Frankel et al. 1995). However, due to rapid urbanization and increased human activity in situ conservation is always under threat of getting eroded. As a result ex situ conservation of genetic diversity has become paramount. The gene banks can be classified as 'base' or active or working collection. The base collection is a long term preservation (>20 years) germplasm as a safeguard to ensure availability in case it is not available in working collection. Information regarding global base collection being maintained in different countries is given in Table 5.2.

Out of more than 400000 accession collection available worldwide, 15 % are wild relatives, 44 % landraces, 17 % breeding materials, 9 % genetic stocks and 15 % cultivars. The majority of the wild relative collections are represented by barley wild progenitor, *Hordeum vulgare* ssp. *spontaneum*. The largest wild collections are maintained in ICARDA (Country), Israel, Canada, the USA and Germany. Ecologically and geographically, the most diverse collection of wild relatives are available at ICARDA with more than 2000 accessions originated from 20 countries and 730 different collection sites. Among the secondary and tertiary gene pool, the maximum collection are maintained in Canada and Sweden. Landraces constitute the largest percentage of barley germplasm conserved worldwide. Major landrace collection are held in gene bank at ICARDA, country; CAAS, China; IBC, Ethiopia; PGRC, Canada; USDA, the USA; IPK, Germany; and RIB, Japan with ICARDA having the largest collection of 15500 accessions.

Breeding material constitutes the second largest category of collection of germplasm with major barley improvement around the world including ICARDA, Syria (has global mandate for barley improvement under CGIAR system); USDA, the USA; NIAR, Japan; and INRA, France. Among genetic stocks, a large collection of 11000 accessions is being held by the Nordic Genetic Resource Centre, Sweden, followed by PGRC, Canada; USDA, the USA; and NIAR, Japan. Among the global germplasm collection, a large genetic variability is available for different traits (Table 5.3) (Gulati and Verma 1988).

Table 5.2 Major world barley collections based on FAO corporate inventory (FAO 2010)

Country	Gene bank	Accessions
Canada	Plant Gene Center of Canada, Saskatoon	40031
USA	USDA-ARS National Small Grains Collection, Aberdeen, Idaho	29874
Brazil	Recursos Geneticos e Biotecnologia, EMBRAPA/CENARGEN	29227
Syria	International Centre for Agricultural Research for Dry Areas, Aleppo, Syria	26679
Japan	Research Institute for Bioresources, Okayama University	23471
Germany	Leibniz Institute of Plant Genetics and Crop Plant Research, Gatersleben	22093
China	Institute of Crop Germplasm Resources, CAAS, Beijing	18617
Rep. of Korea	National Institute of Agricultural Biotechnology, Suwon	17660
Russia	N.I. Vavilov Institute of Plant Industry, St. Petersburg, Russia	16791
Ethiopia	Institute of Biodiversity Conservation, Addis Ababa, Ethiopia	16388
Mexico	International Maize and Wheat Improvement Centre, Mexico	15473
Sweden	Nordic Genetic Resources Centre, Alnarp	14109
UK	John Innes Centre, Norwich	10838
India	National Bureau of Plant genetic Resources, New Delhi	9161
Australia	Australian Winter Cereal Collection, Calala, New South Wales	9031
Iran	National Gene Bank of Iran, Genetic Resources Division, Karaj	7816
Israel	Institute of Cereal Crops Improvement, Tel Aviv University	6658
Poland	Plant Breeding and Acclimatization Institute, Radzikow	6184
Bulgaria	Institute of Plant Genetic Resources, Sadovo, Bulgaria	6171
	Others (180)	140259

Table 5.3 Genetic variability in barley in relation to geographical diversity

Characters	Countries
Yellow rust resistance	Anatolia, Ethiopia, Germany, Nepal, Netherlands, Tibet, Yugoslavia
Aphid resistance	Bulgaria, Crete, Greece, Mediterranean region, Taiwan
Powdery mildew resistance	Ethiopia, Europe, USA
Nematode resistance	North Africa
Dwarfness/short height	Afghanistan, China, Indonesia, Japan, Korea, Thailand
Late winter types, thermo-photosensitive	Bulgaria, India, Nepal, Netherlands, Peru, Poland, Russia, Sweden
Hull-less and hulled spring/winter types	Afghanistan, China, Hungary, Japan, Mexico, Peru, Poland, Russia, South Korea, Sweden
Hull-less grain, hooded, maturity and nodding ears	Himalayan ranges, Japan, Ladakh, Lahaul and Spiti in India, Manchuria, Tibet
Malting quality	Argentina, Australia, Canada, Denmark

5.9 Core Collection

In order to facilitate the better access and utilisation, development of the barley core collection (BCC) was been initiated by Hintum (1992) which represents minimum repetitiveness of the genetic diversity in barleys and its relatives with improved accessibility, because of its small size. The BCC have the five proposed components of germplasm (<2000 accessions) that represent the whole diversity of barley (Table 5.4). It includes 1700 accessions from different countries.

5.10 Factors Limiting Crop Production

Since early work on developing new barley cultivars, the main goal has been to increase the grain yield potential, with breeding being further expanded to include other traits, primarily quality and resistance to biotic and abiotic limiting factors (Knezevic et al. 2007; Madic et al. 2012). Among the abiotic stresses limiting barley yields, different factors such as nutrients, temperature,

Table 5.4 Composition of the barley core collection

Category	Accessions
Cultivars	500–800
Landraces	500
H. spontaneum (wild progenitor)	150
Other wild *Hordeum* spp.	50–80
Genetic stocks	150–200

water, soil toxicity, etc. have been reported. Nitrogen, potassium and phosphorus are the main nutrients and manganese, copper and zinc to a lesser extent are required for successful production of barley. In malting barley, nitrogen application must be balanced to maximise the yield without increasing the protein level in the grain (Verma et al. 2003). Deficiency of nitrogen may lead to poor tiller development and retarded protein concentration in the grain. Potassium deficiencies can lead to poor root growth and leaf development and to fewer and smaller grains. Reduced tolerance to environmental stresses and increased susceptibility to foliar diseases have also been observed. Plants deficient in phosphorus produce reduced head and grain number. Barley grows best in temperatures of 15–30 °C, but it can tolerate high temperatures under less humid conditions (Nevo 1992). High temperatures at post anthesis, however, can reduce grain weight and change malting characteristics.

Barley is considered to be the most drought and salinity tolerant among cereals (Belaid and Morris 1991). Nevertheless, it remains an important abiotic stress for barley, often grown in environments prone to drought (Stanca et al. 2003). It was shown that water stress exertion in seed filling stage decreased yield and yield component of barley (Babaeian et al. 2011). In contrast, water logging is also an important constraint to barley production, especially in high rainfall areas affecting the crop between germination and emergence stages (Zhou et al. 2012). Barley is particularly sensitive to soil acidity, the condition which encourages aluminium toxicity and can be a major constraint to growth in these areas. As barley is adaptive to wide range of soils, it is more tolerant to salinity and alkalinity conditions than other cereals (Chen et al. 2007). Co-occurrence of water logging with salinity may seriously decrease yield potential of barley.

5.11 Traits of Importance

The increasing global demand for food under climate change scenario has posed a great challenge to the breeders. Rise in global temperature has predicted increased levels of CO_2, thereby decreasing crop yields. Higher temperature, in turn, will produce significant changes in precipitation, salinity and occurrence of crop diseases and pest incidences (Ceccarelli et al. 2010). Grain yield in barley also depends upon many other factors such as length of the crop growing season, lodging resistance, frost resistance, disease and pest resistance, etc. Important agronomic traits such as heading date, spikelet number per spike, 1000 grain weight and number of effective tillers which could directly or indirectly affect the yield are governed by major and minor quantitative trait loci (QTL) but are narrow to specific populations (Fekadu et al. 2014; Ren et al. 2013). Owing to the environmental influence, studying these traits at the molecular level is difficult and needs vigorous genotyping and phenotyping. Some of the major traits influencing the factors affecting barley crop yield has been discussed below in brief.

5.11.1 Drought

Drought is a major environmental stress that limits crop production in many countries worldwide, more importantly in the rainfed areas of the world (Chaves and Oliveria 2004; Jaleel et al. 2009). Due to increase in global average temperature, increase in frequency of droughts has been predicted which are expected to cause water shortages for about two-thirds of the world population (Ceccarelli et al. 2004). Irrigation is the most comprehensive operation that might mitigate the adverse effects of drought stress. Due to earliness and its ability to escape terminal drought stress, barley would be a suitable crop in areas where irrigation is poorly available (Kilik et al. 2010). Studies in barley have shown drought to cause significant reduction in all the agronomic traits (Naghaii and Asgharipour 2011; Zare 2012). Therefore, grain yield and its components remain major selection criteria for improved adaptation to a stress environment in many breeding programmes (Kutlu and Kinaci 2010). Genetic control of drought resistance is very complex, and QTLs identified for a range of drought response traits show variation with the environment and genetic background (Diab et al. 2004; Forester et al. 2000; Tondelli et al. 2006).

5.11.2 Soil Toxicity and Deficiency

Soil salinity and sodicity are major problems in arid and semi-arid regions where irrigation is essential for crop production. Nearly 20 % of the worlds cultivated area and half of the world's irrigated lands are affected by salinity (Zhu 2001). The increased salinisation is expected to result in a 50 % loss of arable land by the middle of the twenty-first century (Wang et al. 2007). Salt tolerance in cereals is known to be associated with control of Na^+ content in the shoots. The tolerant cultivars can remove sodium from their cells (Dubcovsky et al. 1996). Combined with water logging, salinity can cause greater damage to plants (Zeng et al. 2013). Reduced grain number and straw weight in barley due to salinity results in less grain yield (Katerji et al. 2009). Application of potassium in saline soils has shown improved barley yield (Endris and Mohammad 2007). Manganese (Mn) deficiency leads to change in soil pH producing unfavourable conditions for barley growth which is more centric in alkaline soils. Pallotta et al. (2000) has revealed the presence of a Mn efficiency locus (*Mel1*) on chromosome 4HS in barley. Cultivars tolerant to manganese deficiency have shown better root and shoot growth (Tong et al. 1997). Aluminium (Al) toxicity in cereals is generally linked to acid soils, and barley is most sensitive to changes to Al levels (Gallardo et al. 1999; Minella and Sorrells 1992). A single major locus on chromosome 4H has been reported to control Al tolerance in barley (Raman et al. 2003). Similarly, boron phytotoxicity at higher levels is reported in certain areas, and a major gene conferring boron tolerance has been located on chromosome 4H (Sutton et al. 2007).

5.11.3 Heading Date

Plant growth habit and heading date are the basic traits involved in barley adaptation to a production region. Heading date is one of the important quantitative traits governed by three main factors: vernalisation response, photoperiodic response and earliness per se genes in barley (Gallagher et al. 1991; Roberts et al. 1988). Vernalisation response loci, *Vrn-H1*, *Vrn-H2* and *Vrn-H3*, are located on chromosomes 5HL, 4HL and 7HS, respectively (Laurie et al. 1995; Takahashi and Yasuda 1971), and their candidate genes and functions have been reported (Casao et al. 2010; Yan et al. 2006). Photoperiod response loci, *Ppd-H1* and *Ppd-H2*, are located on chromosomes 2HS and 1HL, respectively (Karsai et al. 1997; Laurie et al. 1995). The *earliness per se* genes *eam7*, *eam8*, *eam9* and *eam10* are located on chromosomes 6HS, 1HL, 4HL and 3HL, respectively (Gallagher et al. 1991), and the *eps* loci are located on chromosomes 2H, 3H, 4H, 5H, 6H and 7H (Laurie et al. 1995). Early maturing cultivars complete the yield process before the dry period, whereas forced maturity leads to reduced grain yield in late maturing cultivars as long season cultivars spend a large part of dry matter to form vegetative parts. Heading date is controlled by a large number of QTLs (Hayes et al. 1993; Sameri and Komatsuda 2004; Wang et al. 2010). QTL *Qhd7-5* on chromosome 7H has been detected with the effect on shortening heading date (Ren et al. 2013). Location of QTLs for important agronomic traits on specific chromosomes in barley has been summarised in Table 5.5.

Table 5.5 Location of major agronomic traits identified by QTL analysis in barley

Trait	Chromosome
Yield	1H, 2H, 3H, 4H, 5H, 6H, 7H
Grain number per ear	1H, 2H, 3H, 4, 5H, 6H
Grain weight per year	1H, 2H, 3H, 4H, 5H, 6H, 7H
Grain weight per plant	1H, 2H, 3H, 4H, 5H, 6H, 7H
Thousand grain weight	2H, 3H, 4H, 5H, 6H, 7H
Ear breaking	2H, 4H, 6H
Ear length	2H
Spike number	3H
Root length	5H
Plant height	1H, 2H, 3H, 4H, 5H, 6H, 7H
Heading date	1H, 2H, 3H, 4H, 5H, 6H, 7H
Kernel weight	1H, 2H, 4H, 5H
Kernel length	4H, 7H
Kernel plumpness	1H, 2H, 4H, 5H, 7H
Kernel shape	1H
Dwarfing	2H, 3H, 4H

Source: Hussain (2006)

5.11.4 Spike Morphology

Spike morphology is an important consideration in barley as six-rowed barley has apparent yield advantage over the two rowed due to the presence of three fertile spikelets at each rachis node (Bothmer et al. 2003b); however, recent two-rowed cultivars compensate yield with more tillers per plant and higher 1000 grains weight. In six-rowed cultivated barley, all three spikelets are fertile and seed bearing, whereas in wild and cultivated two-rowed barley, the lateral spikelets are suppressed and do not produce any seed. Five independent genes *Vrs1*(2H), *Vrs2*(5H), *Vrs3*(1H), *Vrs4*(3H) and *Vrs5*(4H) control the six-rowed spike phenotype in barley (Lundqvist et al. 1996). The major locus *Vrs1* suppresses the lateral spikelets in six-rowed genotypes resulting in two-rowed phenotype (Komatsuda et al. 2007) and was recently cloned (*HvHox1*) using map-based cloning approach. This dominant allele was found to be widespread in wild barley and thus confirmed that six-rowed barley evolved from two-rowed ancestral barley. Locus *Vrs5*(*Int-C*) was reported to control male fertility and enlarged lateral spikelets in two-rowed types (Ramsay et al. 2011).

5.11.5 Disease and Pest Incidence

Barley being adapted to varying environments is likely to encounter different pathogens and succumb to diseases and pests. The genetic variability in barley has gradually decreased with the replacement of local populations and landraces by improved cultivars, and many of these share related pedigrees (Graner et al. 2000). Barley is mostly attacked by fungi such as rusts, blights and powdery mildews. Most of the fungal pathogens of barley destroy the foliage; however, some infect roots, planted seeds and lower part of stems. Breeding for disease resistance is a major goal in any breeding programme by identifying source of disease resistance and develop resistant cultivars, in addition to a higher level of production technology and fungicide applications (Przulj and Momcilovic 1995). The major problem in barley selection is a high pathogen variability causing resistant cultivars to rapidly become susceptible. The development of resistant cultivars is a continuous long-lasting plant breeding process and needs selection for 5–7 years. There are certain examples of cultivars retaining resistance to certain pathogens for a long period of time. Changes in pathogen structure and relationship favour the multiplication of pathogens to which they are resistant and become susceptible to them after several years. Susceptibility also occurs as a result of emergence of new pathogen strains that are virulent to the cultivar. The study from vast literature references and field trials has indicated an average disease induced yield loss in barley up to 10.5 % (Oerke and Dehne 1997). Graner (1996) provided an extensive review of molecular mapping of qualitative and quantitative disease-resistant loci. Studies in breeding for disease resistance have been summarised by Kleinhofs and Han (2002), Chelkowski et al. (2003) and Weibull et al. (2003), which includes the reported QTLs for the barley genome (Hayes et al. 2003) (Table 5.6).

Table 5.6 Major resistance genes against barley diseases and pests

Gene symbol	Pathogen/pest	Number of genes
Rph	*Puccinia hordei* (leaf rust)	17
Rpg	*Puccinia graminis* (stem rust)	4
Rps	*Puccinia striiformis* f. sp. *hordei* (stripe rust)	4
Ml (Reg)	*Erysiphe graminis* f. sp. *hordei* (powdery mildew)	23
Rcs	*Cochliobolus sativus* (spot blotch)	5
Rpt	*Pyrenophora teres* (net blotch)	6
Rdg (Rhg)	*Pyrenophora graminea* (barley stripe)	3
Rrs (Rh)	*Rhynchosporium secalis* (scald)	14
Run (un)	*Ustilago nuda* (loose smut)	8
Ung	*Ustilago nigra* (semiloose smut)	1
Ruh	*Ustilago hordei* (covered smut)	4
Rsp	*Septoria passerini* (leaf blotch)	3
Rti	*Typhula incarnata* (grey snow mould)	1
Fb	*Fusarium* spp. (scab)	1
Ryd	BYDV (barley yellow dwarf virus)	2
Rym (ym)	BaYMV (barley yellow mosaic virus)	13
	BaMMV (barley mild mosaic virus)	
Rsm (sm)	BSMV (barley stripe mosaic virus)	5
Rsg	*Schizaphis graminum* (green bug aphid)	3
Rha	*Heterodera avenae* (cereal cyst nematode)	3

Source: Chelkowski et al. (2003) and Weibull et al. (2003)

5.11.6 Malting Quality

The wide range of quality requirements for malting grain makes it a complex problem for selection and complicated by various positive and negative correlations between grain and malt traits as well as between these traits and agronomic characters (Briggs 1978). The malt quality phenotype is difficult and expensive to measure because there are many interrelated component traits that lead to malt of high quality (Hayes and Jones 2000). Malting barley requires low protein and high starch content. The genetic factors and environments affect the malting quality as shown by the negative correlation between malt extract yield and protein primarily due to *hordeins* (MacLeod 2000). Among the grain physical parameters, the grain test weight (hectolitre weight) is important for achieving higher malt extract (Verma et al. 2008), followed by grain husk and protein content, and can be used as a good selection criteria for a large-scale germplasm evaluation. Cropping practices such as late sowing, low moisture sowing and higher doses of nitrogen application may also affect grain concentration. Most breeding programmes resort to making relatively few good crosses and ultimately testing a limited number of advanced lines for the target trait. As a consequence, there have been modest genetic gains for malting quality (Han et al. 1997b; Munoz-Amatriain et al. 2010b). Therefore, breeders need to develop malting barley cultivars for brewing with maximum malt extract, sufficient nutrients, yeast growth, fermentable sugars for alcohol production and desired combination of high molecular weight compounds for flavour quality. The use of molecular markers has allowed rapid selection and mapping of many agronomic and quality traits in a single generation. The first report for QTL analysis for malting quality comprised detection of five QTLs around the amylase loci on 2H, 4H and 6H (Hayes et al. 1993). A comprehensive data analysed over the years indicated that a region on chromosome 7H near the centromere, a complex QTL region, controlled many traits such as malt extract, α-amylase activity, diastatic power and β-glucanase activity, etc. (Hayes et al. 2003). Among the large number of QTLs reported for seed dormancy and germination, a major dormancy locus on chromosome 5H was confirmed in all the studies (Gao et al. 2003; Han et al. 1999; Islamovic et al. 2014). To conclude, understanding of the complex biochemical and chemical pathways occurring during the malting process has to be achieved for developing new trait-specific modified malting barley cultivars.

5.12 Interspecific Hybridisation

Interspecific breeding of barley has been attempted by a number of researchers using various approaches in order to transfer new trait-specific alleles. With a few exceptions (*Hordeum bulbosum* and *Hordeum vulgare*), there is little or no opportunity for interspecific gene transfer in nature between the other species of genus *Hordeum* as there is no chromosome pairing at meiosis. Pickering and Johnston (2005) have provided a detailed review of the interspecific crosses of *H. vulgare* with other incompatible species. As per Bothmer (1992), interspecific crosses between *H. vulgare* and *H. bulbosum* have shown chromosome pairing but have proven difficult. The *H. bulbosum* chromosomes in crosses between these two species get eliminated and hence have been used to produce haploid *H. vulgare* plants, popularly called the *bulbosum* method (Kasha and Kao 1970). The method has been widely deployed to produce genetically identical progeny from barley crosses. *H. bulbosum* has many desirable traits that could be used for barley improvement (Johnston et al. 2009). It has been used to transfer powdery mildew resistance to *H. vulgare* (Pickering et al. 1995). Several interspecific crosses between *H. vulgare* and wild *Hordeum* species have been attempted using tissue culture techniques. The viability of the hybrids is dependent on the ploidy level and taxonomic group. Most of the crosses suffer from seed lethality and sterile pollen (Ellstrand 2003). *Hordeum spontaneum* is the only wild *Hordeum* species that is totally compatible in crosses with cultivated barley and is a subspecies of *H. vulgare*. Mating between these two subspecies is not a true interspecific cross, however, used in barley improvement programmes to generate variability. The crosses are inter-fertile and show normal chromosome pairing and segregation at meiosis producing completely fertile hybrids demonstrating the genetic affinity between the two barley species (Bothmer 1992). The major barriers hindering interspecific hybridisation are reported to be chromosome elimination, infertile hybrids, reduced recombination, endosperm degeneration, linkage drag, hybrid instability, and chromosome pairing.

Intergeneric hybridisation (barley x wheat crosses) has provided additional sources of genetic material. Kruse (1973) provided the first evidence of crossability of diploid *H. vulgare* as maternal plant with diploid, tetraploid and hexaploid wheat. Significant reviews have since been published in the initial years of wheat-barley hybridisation (Fedak 1985; Shepherd and Islam 1992). Recently, Molnar-Lang et al. (2013) have provided a comprehensive review on intergeneric breeding of *Hordeum* species (cultivated and wild) with wheat. Wheat-barley hybridisation has the potential to incorporate earliness, salt and drought tolerance, desired amino acid integration and high tillering ability of barley into wheat, whereas the favourable characteristics of stem strength and winter hardiness of wheat can be introgressed into barley.

5.13 Genetic Mapping

Barley has an excellent system for genome mapping and map-based analysis (Costa et al. 2003), due to its chromosomes being homoeologous to cultivated wheat and rye (Hori et al. 2003). Several essential features of barley have established it as a model species for genetic and physiological studies (Koornneef et al. 1997). The cytogenetic studies indicate barley as a true diploid ($2n = 2x = 14$), self-fertile and having seven chromosomes identified and labelled based on their sizes and characteristics. Thus, barley chromosomes from 1 to 7 are officially redesignated as 7H, 2H, 3H, 4H, 1H, 6H and 5H, respectively (Linde-Laursen 1997). Chromosomes 1 through 5 differ in their sizes measured at mitotic metaphase with chromosome 1 being the longest and chromosome 5 being the shortest, whereas chromosomes 6 and 7 have larger and smaller satellites, respectively (Kleinhofs and Han 2002). The barley genome in the variety 'Betzes' was the reference genome where translocations and short-arm/long-arm reversals were standard in all species. New approaches especially trisomic analysis has been successfully used in the mapping of barley chromosomes (Tsuchiya 1986). Cytogenetic linkage maps were established in the early 1950s using translocation analysis and primary trisomic method. A well-developed classical genome map was constructed using isozyme and morphological markers (Nielsen and Johansen 1986).

5.14 Molecular Markers

During the last two decades, breeding programmes have been enhanced by the implementation of modern biotechnology tools, like the doubled haploid technique and marker-assisted selection (Friedt and Rasmussen 2003). With the advent of molecular markers and their use in genome analysis has provided unprecedented insight into structural features of the barley genome (Graner et al. 2004). Molecular markers can be broadly classified in the following three groups:

1. First-generation hybridisation-based DNA markers such as restriction fragment length polymorphism (RFLPs)
2. Second-generation PCR-based DNA markers such as random amplified polymorphic DNAs (RAPDs), sequence-characterised amplified regions (SCARs), inter-simple sequence repeats (ISSRs), simple sequence repeats (SSRs) and sequence-tagged sites (STS)
3. Third-generation DNA chip and sequencing-based DNA markers such as single nucleotide polymorphisms (SNPs)

In recent years, approaches like allele mining, advanced backcross QTL analysis (AB-QTL), multi-parent advanced generation intercross (MAGIC) population, nested association mapping (NAM) population and association mapping (AM) methods have been developed in crops that can be used to broaden the genetic base of the domesticated germplasm through exotic and unused alleles (Cavanagh et al. 2008; Korff et al. 2004; Pillen et al. 2003). Several genetic linkage maps have been developed for barley using different markers (Haseneyer et al. 2010a).

The first genetic map on barley was developed on chromosome 6H using RFLP by Kleinhofs et al. (1988), followed by other workers (Heun et al. 1991; Kleinhofs et al. 1993; Shin et al. 1990). The RFLP technology is a very reliable and a powerful tool for barley comparative mapping studies among species of *Triticeae* (Sherman et al. 1995), resulting in the construction of a consensus map (Langridge et al. 1995) and gene mapping (Hinze et al. 1991; Laurie et al. 1995). PCR-based arbitrarily primed genetic assay called RAPD (Williams et al. 1990), AP-PCR (Caetano-Anolles et al. 1991) or DAF (DNA amplification fingerprinting) (Welsh and McClelland et al. 1990) was widely used for the construction of genetic maps (Bowditch et al. 1993; Williams et al. 1993). In general RAPD markers generated relatively dense linkage maps in a short period of time. Compared with RFLPs, RAPD technique has the advantage of requirement of small amounts of DNA (5–20 ng), rapid screening for polymorphisms, generation of a large number of markers for genomic mapping and automation of the process. Mapping of RAPD markers in barley was used for tagging the genes resistant to barley blotch (Kutcher et al. 1996; Molnar et al. 2000) and barley yellow dwarf virus (Wang and Zhang 1996; Zhang et al. 2001). The amplified fragment length polymorphism (AFLP) technology was considerably better and several whole genome maps were produced (Qi et al. 1998; Waugh et al. 1997b). Compared with allozymes, RAPDs and RFLPs, AFLPs have superior performance in time and cost efficiency, replicability and resolution, except that the markers are mostly dominant and cannot be easily transferred to crosses with different parents. Since the first report of AFLP mapping in barley (Becker et al. 1995), several maps of AFLP markers have been constructed in barley (Mano et al. 2001; Qi and Lindhout 1997). Various studies using AFLP technique have been undertaken in barley with regard to origin of barley (Allaby and Brown 2003; Badr et al. 2000), diversity studies (Ellis et al. 1997; Fu 2012; Ozkan et al. 2005), mapping of QTLs (Hori et al. 2005; Pillen et al. 2004; Powell et al. 1997), detection of disease resistance genes (Bilgic et al. 2005; Toojinda et al. 1998) and fine mapping of disease resistance genes such as *mlo* (Simons et al. 1997), *Mla* (Schwarz et al. 1999) and *Rph15* (Weerasena et al. 2004).

Another PCR-based DNA marker sequence-tagged site (STS) (Olson et al. 1989) is a unique single copy segment of the genome whose DNA sequence is known by sequencing and can be amplified by specific PCR. STS markers have the advantage of speed with which they can be analysed once PCR primer pairs have been identified. After the report of mapping of STS markers in barley (Tragoonrung et al. 1992), a large amount of RFLP markers were converted to STSs for cultivar fingerprinting purpose (Faccioli et al. 1995), for physical mapping (Korzun and Künzel 1997) and for mapping specific genes (Blake et al. 1996; Larson et al. 1996). Later, the entire barley genome was covered with a single genetic map (Mano et al. 1999), and EST sequencing provided new sources for the development of STS markers in barley (Michalek et al. 1999). One of the popular and widely used PCR technologies based on simple sequence repeats (SSRs) (Tautz 1989) has gained much importance among the breeders and the geneticists for map building owing to their codominant nature, ease of use, high information content, abundance in barley genomes, potential for automation and readily transferrable among crosses. SSRs are also called microsatellites which are stretches of DNA consisting of tandemly repeated short units of 1–6 base pairs of length (Johansson et al. 1992). Comprehensive and integrated barley consensus maps were developed using multiple mapping populations and markers combining RFLP, SSR, SNP techniques and ESTs (Close et al. 2009; Lamara et al. 2013; Rostoks et al. 2005; Sato et al. 2009; Stein et al. 2007; Varshney et al. 2007; Wang et al. 2014; Wenzl et al. 2006).

Though SSR markers are the current choice for the breeders and geneticists, high-throughput technologies have been developed for barley genome mapping. These include diversity arrays technology (DArT) and single nucleotide polymorphisms (SNPs). DArT is a whole genome profiling technology based on DNA/DNA hybridisation using complexity-reduced representation of genomes (Jaccoud et al. 2001). DArT assays a large number of SNP and polymorphisms at many thousands (10,000–50,000) of restriction enzyme (RE) sites distributed throughout the genome. The first comprehensive genotyping DArT array was reported for barley by Wenzel et al. (2006). This was also a first report on creating a medium density genetic map on any crop species based on detecting many hundreds of markers in a single assay. Consensus maps in barley have been developed using DArT markers (Ariyadasa et al. 2014; Aslop et al. 2010; Hearnden et al. 2007; Lamara et al. 2013; Steffenson et al. 2007; Wang et al. 2014; Waugh et al. 2009). SNPs are, by far, the most abundant molecular markers identified in all genomes (Brookes 1999). Numerous SNP markers have been developed for barley (Close et al. 2009; Kota et al. 2003; Sato et al. 2009). Currently, high-throughput SNP genotyping platforms with a capacity to genotype several hundred genotypes with thousands of SNPs are available at low costs and have revolutionised the practical uses of genomics (Illumina™, Veracode, Goldengate and the iSelect chip assay, Affymetrix gene chip) (Atwell et al. 2010; Close et al. 2004, 2009; Schmalenbach et al. 2011). Due to their abundance and slow mutation rate within generation, SNPs are thought to be the next generation of genetic markers that has huge potential for candidate gene discovery using synteny-based searches in sequenced grass genomes (Druka et al. 2011; Guo et al. 2013; Mayer et al. 2011; Munoz et al. 2011; Poland et al. 2012). Introduction of targeting induced local lesions in genome (TILLING), a reverse genetic approach that represents an extension of the use of spontaneous and induced mutants in plant breeding, allows the direct identification of beneficial genetic changes in genes with known functions and their use as the genetic markers for selection (McCallum et al. 2000b). Efforts were made to generate large TILLING mutant populations in barley (Caldwell et al. 2004; Gottwald et al. 2009; Talame et al. 2008). Several genes involved in plant development were reported and cloned from barley mutants.

The advanced backcross quantitative trait loci analysis (AB-QTL) strategy was introduced by Tanksley and Nelson (1996), to transfer exotic genes into modern crop varieties. They integrated the mapping of favourable QTL alleles and their introgression into one process utilising exotic germplasm as the genetic donor for the quantitative traits and conducted the marker and phenotype analysis in advanced backcross generations like BC_2. AB-QTL-derived lines contain only one single introgression from the wild parent and are referred as introgression lines (ILs) achieved after several rounds of backcrossing to the recurrent parent and marker-assisted selection. Therefore, AB-QTL analysis can accelerate the process of marker-based breeding because the end products are close to NIL carrying favourable alleles. The first study on AB-QTL analysis in barley was reported by Pillen et al. (2003). Thereafter, similar studies have been conducted for different traits (Ibrahim et al. 2012; Honsdorf et al. 2014; Korff et al. 2005, 2006, 2008, 2010; Saal et al. 2011; Sayed et al. 2012).

The recent approach of association mapping (AM) or linkage disequilibrium (LD) mapping refers to the non-random association of alleles at different loci, and loci that are near to each other tend to have strong LD (Slatkin 2008). The extent of LD determines the resolution of marker-trait association. Therefore, high LD in a population indicates that fewer markers are required to detect the marker-trait association, but with low resolution (Myles et al. 2009), and vice versa (Waugh et al. 2009). Though association mapping at a genome-wide level is highly successful in human genetics (Syvanen 2005), however, genome-wide association studies (GWAS) have recently made inroads to study plant populations and statistical framework to investigate the population structure of natural plant populations (Flint-Garcia et al. 2003; Rafalski 2010). The use of natural populations for association mapping has added advantage over the conventional QTL mapping (biparental mapping) as there is no need to develop the mapping population. In barley, low LD has been observed in landraces and wild populations (Caldwell et al. 2006; Morrell et al. 2005). One of the initial studies in barley using SNP markers has revealed that few thousand SNP markers can cover the whole genome in elite barley as the LD is large. Since then, several association studies have been published in barley using candidate gene re-sequencing and different throughput marker sets (Cockram et al. 2008, 2010; Comadran et al. 2009, 2011b; Haseneyer et al. 2010b; Pasam et al. 2012; Munoz et al. 2014; Stracke et al. 2009). GWAS approach has also been reported in many studies for different traits (Cai et al. 2013; Cockram et al. 2010; Comadran et al. 2011a; Kraakman et al. 2004; Pasam et al. 2012; Roy et al. 2010; Visioni et al. 2013; Varshney et al. 2012). These studies indicated that despite the large barley genome, few thousand markers were used and significant marker-trait associations were observed relating to presence of larger LD (~10 cM) in cultivars. In some cases, the syntenic information from related species has further added value to GWAS (Cockram et al. 2010; Houston et al. 2012).

Traditional and conventional breeding of barley has not been replaced by the use of marker techniques and genetic maps, but these newer approaches act as tools that breeders can utilise in the identification and location of genes and their effects on trait expression (Rao et al. 2007). The information resources of latest updates in current developments in mapping strategies and associated genetic research on the barley genome have been generated (Table 5.7).

5.14.1 Molecular Cytogenetics

Advances in the molecular cytogenetic techniques of fluorescence in situ hybridisation (FISH), which enables the direct chromosomal localisation of labelled DNA probes, and genomic in situ hybridisation (GISH), which determines the interspecies distribution of repeated sequences, have enabled a resurgence of cytogenetic analysis in plant genome research and molecular breeding. Several studies of application of these tools in barley have been made to investigate chromosome mapping, genome analysis, phylogenetic relationships, chromosomal aberrations and alien chromatin in plant breeding programmes (Brown et al. 1999; Caudrado et al. 2008; Cseh et al. 2011; Dolezel et al. 2007; Juchimiuk et al. 2007; Linde-Laursen et al. 1997; Mukai and Gill 1991; Pedersen et al. 1997;

Table 5.7 Barley genomic databases/resources

Databases	Website URL	Application
Barley Genetic Stocks	http://ace.untamo.net/cgibin/ace/searches/basic	Provides information on the morphological and genetic background of barley mutants and contains information on 736 barley translocation and duplication lines
US Barley Germplasm	http://barleyworld.org/northamericanbarley/germplasm.php	Contains information on barley recombinant chromosome substitution lines and North American barley mapping populations
EBDB	http://pgrc-35.ipk-gatersleben.de/portal/page/portal/PGBICGH/PBICGH/PBICGHRESOURCES/PBICGHIRESEBDB	The European Barley Database (EBDB) contains passport and evaluation data of 155,000 barley accessions including the international barley core collection
ICARDA Barley varieties	http://www.icarda.cgiar.org/CropsVarieties.htm#Barley	Provides an index of barley variety releases from ICARDA, 1977–2005
Barley TILLING	http://www.scri.ac.uk/research/genetics/BarleyTILLING	A reverse genetics platform, which can be screened for 8600 barley (cultivar 'optic') EMS mutagenised lines
CR-EST	http://pgrc.ipk-gatersleben.de/crest/index.php	Barley EST database containing sequences, functional annotation and clustering information of more than 232,000 ESTs
HarvEST	http://harvest.ucr.edu	Barley EST database containing unigene sequences and the oligo design of Barley1 Affymetrix array. It also includes a 1000 barley SNP loci genetic map showing syntenic information with rice
HvGI	http://compbio.dfci.harvard.edu/tgi/cgi-bin/tgi/gimain.pl?gudb=barley	This *Hordeum vulgare* Gene Index provides functional annotation, information, 70-mer oligo predictions and in silico gene expression data for 50,000 unigenes
NCBI Barley genome view	http://www.ncbi.nlm.nih.gov/mapviewmapsearch.cgi?taxid=4513	Provides an overview about the available genomic and genomic survey sequences (GSS) of barley
IBSC	http://barleygenome.org	Activities of the International Barley Genome Sequencing Consortium (IBSC) are highlighted
Barley genome	http://phymap.ucdavis.edu:8080/barley	Barley physical mapping database and available BAC clones together with the accompanying ESTs
Barley physical map	http://pgrc.ipk-gatersleben.de/kuenzel/barleymap.html	Barley translocation break points integrated into the Igri/Franka-derived RFLP linkage map
Barley genomics	http://barleygenomics.wsu.edu	Contains information about barley molecular markers, genetic maps, BACs and mutants
Barley DB	http://ukcrop.net/perl/ace/search/BarleyDB	Contains information about barley germplasm, molecular markers, genetic maps and BACs
Gramene	http://www.gramene.org	Provides an overview of comparative maps of cereals including available updated molecular markers and maps of barley

Grain genes	http://wheat.pw.usda.gov/GG2/index.shtml	*Triticeae* database provides an overview about available maps, genetic markers, QTLs and gene expression data
Barley dbEST SSRs	http://www.genome.clemson.edu/projects/barley/ssr.dbest.html	15,182 barley simple sequence repeats (SSR) were predicted using the available 328,724 dbEST dataset
Barley SNP database	http://bioinf.scri.ac.uk/barley_snpdb	Barley SNP linkage map
Barley RFLP database	http://pgrc.ipk-gatersleben.de/rflp/rflp.html	Contains data of mapped barley RFLP markers from IPK
Barley DArT	http://www.triticarte.com/content/barleydiversityanalysis.html	High-density consensus map of barley DArT markers linking to existing SSR, RFLP and STS loci
BarleyBase	http://www.plexdb.org/plex.php?database=Barley	An online dataset for storing and visualising gene expression data of the Barley 1 GeneChip Affymetrix array
BDC-GED	http://pgrc.ipk-gatersleben.de/seeds	Contains barley developing caryopses gene expression data

Source: Sreenivasulu et al. (2008)

Stephens et al. 2004). Mayer et al. (2011) recently studied the barley genome by chromosomal and comparative genetics using these tools. FISH and GISH studies have enabled investigations into the homoeologous relationships and differences in pairing pattern of the barley genome. The interspecific introgressions in *H. vulgare* x *H. bulbosum* hybrids and their progenies have been studied cytogenetically in detail (Pickering and Johnston 2005; Pickering et al. 2004). The phenomenon of chromosome elimination in the process of haploid formation has been reported by Gernand et al. (2006).

5.15 Genetic Transformation

The transgenic technology has been applied in barley by using direct and indirect methods of transformation. Wan and Lemaux (1994) were the first to report fertile stably transformed barley plants. Simultaneously, transformation studies using different approaches were also attempted by Jahne et al. (1994) and Ritala et al. (1994). Publications have been made on the history, methodology and progress of transgenic research citing the difficulties encountered and successes introducing foreign genes into fertile stably transformed barley (Dahleen et al. 2001; Dahleen and Manoharan 2007; Goedeke et al. 2007; Ganeshan et al. 2008; Horvath et al. 2002; Lemaux et al. 2008; Harwood 2014; Harwood et al. 2009). Transformation tools can be used to produce better barley plants as genetic engineering provides a new way to produce nutritionally improved foods for areas of the world where barley is the staple food. There is also potential to produce many useful compounds such as enzymes, antibiotics, hormones and other biological agents. Methods of transforming plant cells include the indirect transformation which involves the use of soil phytopathogen, *Agrobacterium tumefaciens*, that induces tumours known as crown galls. *A. tumefaciens* mediated transformation of barley was first reported by Tingay et al. (1997). Several studies using this vector have been carried out since then (Matthews et al. 2001; Murray et al. 2004; Harwood 2014; Hensel et al. 2008; Holme et al. 2008; Trifonova et al. 2001). The direct transformation method does not involve any intermediary organism for the movement of genes from one organism to another. Transfer of foreign DNA into barley has been achieved to some degree of success by microinjection and electroporation procedures (Ganeshan et al. 2008). Currently, particle bombardment approach is a much preferred choice for producing transgenic barley which involves bombardment of plant tissues with DNA-coated particles and subsequent regeneration during in vitro culture. The method of transformation is rightly termed as *biolistics* and is executed with the help of gene gun (PDS/1000/He Bio-Rad, Hercules, CA) (Lemaux et al. 1996). Recent studies in barley transformation using particle bombardment have been performed for different traits (Ibl et al. 2014; Kapusi et al. 2014; Kronbak et al. 2014; Travella et al. 2005; Um et al. 2007; Zalewski et al. 2012).

5.16 Barley as Food

Barley was a staple food as far back as 18,000 years ago in many parts of the world including North Africa, the Middle East, northern and eastern Europe (Morocco, Ethiopia, Iran, Russia, Finland, England, Denmark and Poland) and Asia (Japan, India, Tibet and Korea) (Chatterjee and Abrol 1977; Newman and Newman 2006; Ryu 1979). As reported by Grando and Macpherson (2005), barley grains, mostly of landraces, have been used for making traditional dishes in developing countries. Barley grain is an excellent source of soluble and insoluble dietary fibre (DF) and other bioactive constituents, such as vitamin E, B complex vitamins, minerals and phenolic compounds. The dietary fibre (3–11 %) comprises of pentosans, β-glucan and cellulose. The nutritional benefits of barley food have attracted the attention of health-conscious people. More recently, whole-grain barley and its products have been allowed to claim that they reduce the risk of coronary heart diseases by the US Food and Drug Administration (FDA, http://www.fda.gov). According to Brady et al. (2005), nutritional composition of per 100 g edible portion of barley consists of protein (11.6 g), fat (2 g), crude fibre (6.2 g), neutral detergent fibre (16.9 g) and acid detergent fibre (6.2 g). Protein levels in barley grain are highly variable ranging from 7 to 25 % based on USDA study (Ullrich et al. 2010) and between 9 and 13 % as commonly reported in typical barley (Newman and Newman 2008).

During the last 15 years, the research efforts associated with food barley have increased significantly, and new cultivars have been generated specifically for food use, possessing increased β-glucan, desirable starch composition profiles and improved milling/processing traits. In 2006, the US Food and Drug Administration (FDA) formally recognised that the β-glucan soluble fibre in whole-grain and dry-milled barley reduces the risk of heart disease, and this health claim can now be legally displayed on food products containing sufficient levels of barley β-glucan (National Barley Foods Council, USA). Besides numerous health claims attributed to barley (Ames and Rhymer 2008), several studies have demonstrated the health-promoting properties of β-glucans (Wood and Beer 1998; Newman and Newman 1992). β-Glucans have been implicated in lowering plasma cholesterol, improving lipid metabolism and reducing glycaemic index and the risk of colon cancer (Behall et al. 2006; Brennan and Cleary 2005; Keenan et al. 2007; Li et al. 2003a). Pearled barley typically represents 60–70 % of the original barley kernel. Pearling by-products have interesting amounts of bioactive compounds (dietary fibre and β-glucans) and could therefore be proposed as potential ingredients for the manufacture of functional foods (Marconi et al. 2000). The nutritional and clinical importance of barley foods and public consciousness regarding quality of daily diet, i.e. cereal diversification, may have a positive impact on the demand of barley as a human food in the future (Table 5.8).

Table 5.8 Medicinal applications of *Hordeum vulgare*

Disease	Application
Anti-cough	Decoction of *H. vulgare* seeds with apples, dried figs and pears
Bladder inflammation	A decoction of dried seeds is used orally for bladder inflammation in Iran
Blood glucose level	Seeds of *H. vulgare* 125 g are roasted and mixed with each of 50 g of *Cicer arietinum* and *Elettaria cardamomum* and used half teaspoon with water thrice a day to control blood glucose level
Cholera	Powdered flower of *Calotropis procera*, fruits of *Piper nigrum*, seed ash of *H. vulgare* and rose water are taken orally for cholera in India
Dermatitis	Hot water extract of dried seeds is also used externally for dermatitis in Guatemala
Diabetes	This remedy is used as dietary supplement to control diabetes
Inflammations	Hot water extract of dried seeds is used externally for inflammations

Source: Mohammad et al. (2013)

5.17 Conclusions

Barley has a long history as a domesticated crop, and investigations are still in progress into the controversial origin of cultivated barley, though Fertile Crescent is the most accepted theory of origin. It remains the fifth most produced crop in the world losing the all important cultivated area and choice as food grain to wheat due to change in food habits of the farming classes. In contrast, the productivity of barley cultivars has risen by an annual rate of approximately 1–2 % owing to the improved agricultural production technology, high-yielding cultivars responsive to fertilisers and efficient disease and insect control. According to Wiebe (1978), 'progress of barley improvement depends on a supply of good genes and on breeding techniques for assembling these into superior genotypes'. Prior to the mid-1990s, genetic improvement of barley was limited to traditional plant breeding practices which included existing cultivars, breeding lines, mutations and exotic plant material as the primary sources of genotypic variation and continues so in the near future. Barley landraces are a major source of diversity for breeders and still represent an important tool for variety release and gene discovery. The genetic diversity of barley is conserved ex situ in gene banks as unique accessions which is readily available for utilisation in breeding and research. These accessions (wild relatives, landraces, breeding materials, genetic stocks and cultivars) representing the wide genetic variation are maintained under the terms of the International Treaty on Plant Genetic Resources for Food and Agriculture (ITPGRFA). However, accessing novel genetic variation in gene bank collections will require thoughtful and renewed characterisation at the genotypic and phenotypic levels.

Gene flow and introgressions between domesticated cultivars and their wild progenitors in barley vary in their intensity and effects, environment, genome location, agroecosystem and human activities. The subject has gained interest among the breeders, and knowledge of the role of genetic variability reserved in wild progenitors will be helpful in conservation and breeding. In order to broaden the genetic base of barley, plant breeders have utilised interspecific and intergeneric hybridisation with some success. Among the various plant selection schemes, modified bulk pedigree has been found the most effective in generating an allelic-rich and broad genetic-based plant material.

Tolerance to abiotic stresses such as drought, cold, salinity, heat, waterlogging, nutrient use efficiency and mineral toxicity has traditionally enabled cereal crops to cope with the prevalent local stresses with a balance of traits that were fine-tuned to optimise economic yields in their environments. Global warming has rendered drought being an increasingly important trait to be addressed in the current times. Disease prevention and control in barley is a prerequisite for competing production with the breeding of genetically disease-resistant crops being one of the most environmentally and economically desirable ways to manage plant diseases. Understanding the host pathogen biology is the first step towards minimising risks represented by plant diseases. New durable and efficient sources of resistance will have to be sought from barley landraces and wild relatives.

The progress in biotechnology has opened up enormous possibilities, both for introgression of specific traits and for genetic base broadening through pre-breeding. Even the large size of the barley genome has not restricted the progress in molecular mapping studies. The techniques rendering this development include EST resources for gene-based markers, development of microarray base of 23000 barley genes and synthesis of DArT markers for specific chromosomal regions. Comprehensive consensus maps provide means to select markers for chromosomal regions and allow comparative mapping by exploring information available for other grasses. High-throughput genotyping (SNP, DArT) has increased the data generation speed manifolds. Consortial efforts are in progress for the sequencing of barley genome (http://barleygenome.org) which will have an added advantage to rapidly clone genes and increase the number of markers for mapping QTL and marker-assisted selection. Continued improvements in transformation methods will increase transgenic efficiencies in a wide variety of genotypes and will allow further barley improvement. Marker technologies together with doubled haploid formation have almost halved the time of variety development in barley breeding programmes. Genome-wide association studies (GWAS) with information on linkage disequilibrium (LD) have revolutionised the QTL studies to unravel the genetic architecture of complex agronomic traits. Therefore, the strategy for future barley breeding should contain a broad knowledge of and access to modern technology and combining the application of new tools and techniques with traditional and efficient plant breeding methods to achieve final goals rapidly.

References

Abdel-Ghani AH, Parzies HK, Omary A, Geiger HH (2004) Estimating the outcrossing rate of barley landraces and wild barley populations collected from ecologically different regions of Jordan. Theor Appl Genet 109:588–595

Allaby RG, Brown TA (2003) AFLP data and the origins of domesticated crops. Genome 46:448–453

Altieri MA (2004) Linking ecologists and traditional farmers in the search for sustainable agriculture. Front Ecol Environ 2:35–42

Alvarez N, Garine E, Khasah C, Dounias E, Hossaert-McKey M, McKey D (2005) Farmers' practices, metapopulation dynamics, and conservation of agricultural biodiversity on-farm: a case study of sorghum among the Duupa in sub-sahelian Cameroon. Biol Conserv 121:533–543

Ames NP, Rhymer CR (2008) Issues surrounding health claims for barley. J Nutr 138:1237–1243

Andrew C (2008) The Gladiator diet, How to eat, exercise, and die a violent death. Archaeology Mag 61

Ariyadasa R, Mascher M, Nussbaumer T et al (2014) A sequence-ready physical map of barley anchored genetically by two million single-nucleotide polymorphisms. Plant Physiol 164:412–423

Aslop BP, Farre E, Wenzl P et al (2010) Development of wild barley-derived DArT markers and their integration into a barley consensus map. Mol Breed. doi:10.1007/s11032-010-9415-3

Assefa A, Labuschagne MT (2004) Phenotypic variation in barley (*Hordeum vulgare* L.) landraces from north Shewa in Ethiopia. Biodivers Conserv 13:1441–1451

Atwell S, Huang YS, Vilhjalmsson BJ et al (2010) Genome-wide association study of 107 phenotypes in Arabidopsis thaliana inbred lines. Nature 465:627–631

Azhaguvel P, Komatsuda T (2007) A phylogenetic analysis based on nucleotide sequence of a marker linked to the brittle rachis locus indicates a diphyletic origin of barley. Ann Bot 100:1009–1015

Babaeian M, Esmaeilian Y, Tavassoli A, Asgharzade A, Sadeghi M (2011) The effects of water stress, manure and chemical fertilizer on grain yield and grain nutrient content in barley. Sci Res Ess 6:3697–3701

Backes G, Orabi J, Wolday A, Yahyaoui A, Jahoor A (2009) High genetic diversity revealed in barley (*Hordeum vulgare*) collected from small-scale farmer's fields in Eritrea. Genet Resour Crop Evol 56:85–97

Badr A, Muller K, Schafer-Pregl R et al (2000) On the origin and domestication history of barley (*Hordeum vulgare*). Mol Biol Evol 17:499–510

Baum M, Lagudah ES, Appels R (1992) Wide crosses in cereals. Annu Rev Plant Physiol Plant Mol Biol 43:117–143

Becker J, Vos P, Kuiper M, Salamini F, Heun M (1995) Combined mapping of AFLP and RFLP markers in barley. Mol Gen Genet 249:65–73

Behall KM, Scholfield DJ, Hallfrisch J (2006) Barley β-glucan reduces plasma glucose and insulin responses compared with resistant starch in men. Nutr Res 26:644–650

Belaid A, Morris ML (1991) Wheat and Barley production in rainfed marginal environments of west Asia and north Africa. Problems and prospects, CIMMYT economics working paper. CIMMYT, Mexico, p 91

Bell GCH (1965) The comparative phylogeny of temperate cereals. In: Hutchinson J (ed) Essays on crop plant evolution. Cambridge University Press, Cambridge, pp 70–102

Bellucci E, Bitocchi E, Rau D et al (2013) Population structure of barley landrace populations and gene flow with modern varieties. PLoS One 8:e83891

Bennett MD, Smith JB (1976) Nuclear DNA amounts in angiosperms. Philos Trans R Soc Lond B Biol Sci 274:227–274

Bilgic H, Steffenson BJ, Hayes MP (2005) Comprehensive genetic analyses reveal differential expression of spot blotch resistance in four populations of barley. Theor Appl Genet 111:1238–1250

Bitocchi E, Nanni L, Rossi M et al (2009) Introgression from modern hybrid varieties into landrace populations of maize (*Zea mays* ssp. *mays* L.) in central Italy. Mol Ecol 18:603–621

Blake TK, Kadyrzhanova D, Shepherd KW, Islam AKMR, Langridge PL, McDonald CL, Talbert LE (1996) STS-PCR markers appropriate for wheat-barley introgression. Theor Appl Genet 93:826–832

Blake T, Blake VC, Bowman JGP, Abdel-Haleem H (2010) Barley feed uses and quality improvement. In: Barley: production, improvement and uses. Wiley-Blackwell, Ames, pp 522–531

Bockelman HE, Valkoun J (2010) Barley germplasm conservation and resources. In: Ullrich SE (ed) Barley: production, improvement, and uses. Wiley-Blackwel, Ames, pp 144–159

Bowditch BM, Albright DG, Williams JGK, Braun MJ (1993) Use of randomly amplified polymorphic DNA markers in comparative genome studies. Mol Evol: Producing the Biochem Data 224:294–309

Brady KC, O' Kiely P, Forristal PD, Fuller H (2005) Schipophyluim commune on hay-bale grass silage in Ireland. Mycologist 19:30–35

Brennan CS, Cleary LJ (2005) The potential use of cereal (1–3) (1–4)-β-D-glucans as functional food ingredients. J Cereal Sci 42:1–13

Briggs DE (1978) Barley. Chapman & Hall, London, p 612

Brockman DA, Chen X, Gallaher DD (2013) Consumption of a high β-glucan barley flour improves glucose control and fatty liver and increases muscle acylcarnitines in the Zucker diabetic fatty rat. Eur J Clin Nutr 52:1743–1753

Brookes AJ (1999) The essence of SNP's. Genes Dev 234:177–186

Brown AHD (2000) The genetic structure of crop landraces and the challenge to conserve them *in situ* on farms. In: Brush SB (ed) Genes in the field. On farm conservation of crop diversity. IPGRI/IDRC/Lewis Publishers, Boca Raton, pp 19–48

Brown AHD, Munday J (1982) Population-genetic structure and optimal sampling of landraces of barley from Iran. Genetica 58:85–96

Brown SE, Stephens JL, Lapitan NLV, Knudson DL (1999) FISH landmarks for barley chromosomes (*Hordeum vulgare* L.). Genome 42:274–281

Brush SB (1995) *In situ* conservation of landraces in centers of crop diversity. Crop Sci 35:346–354

Caetano-Anollés G, Bassam BJ, Gresshoff PM (1991) High resolution DNA amplification fingerprinting using very short arbitrary oligonucleotide primers. Bio Tech 9:553–557

Cai S, Dezhi W, Zahra J, Huang Y, Huang Y, Zhang G (2013) Genome-wide association analysis of aluminum tolerance in cultivated and Tibetan wild barley. PLoS One 8:e69776

Caldwell DG, McCallum N, Shaw P, Muehlbauer GJ, Marshall DF, Waugh R (2004) A structured mutant population for forward and reverse genetics in Barley (*Hordeum vulgare* L.). Plant J 40:143–150

Caldwell KS, Russell J, Langridge P, Powell W (2006) Extreme population-dependent linkage disequilibrium detected in an inbreeding plant species, *Hordeum vulgare*. Genetics 172:557–567

Casao MC, Igartua E, Karsai I, Lasa JM, Gracia MP, Casas AM (2010) Expression analysis of vernalization and day-length response genes in barley (*Hordeum vulgare* L.) indicates that *VRNH2* is a repressor of *PPDH2(HvFT3)* under long days. J Exp Bot 62:1939–1949

Caudrado A, Cardoso M, Jouve N (2008) Physical organisation of simple sequence repeats (SSR's) in Triticeae: structural, functional and evolutionary implications. Cytogenet Genome Res 120:210–219

Cavanagh C, Morell M, Mackay I, Powell W (2008) From mutations to MAGIC: resources for gene discovery, validation and delivery in crop plants. Curr Opi Plant Biol 11:215–221

Ceccarelli S, Grando S (1996) Drought as a challenge for the plant breeders. Plant Growth Regul 20:149–155

Ceccarelli S, Grando S, Baum M, Udupa SM (2004) Breeding for drought resistance in a changing climate. In: Rao SC, Ryan J (eds) Challenges and strategies for dryland agriculture. ASA and CSSA, Madison, pp 167–190

Ceccarelli S, Grando S, Capettini F, Baum M (2008) Barley Breeding for Sustainable Production. In: Breed Maj food staples. Blackwell Publishing, Iowa, pp 193–225

Ceccarelli S, Grando S, Maatougui M et al (2010) Plant breeding and climate changes. J Agric Sci 148:627–637

Chatterjee SR, Abrol YP (1977) Protein quality evaluation of popped barley grain (Sattu). J Food Sci Tech 14:247–250

Chaves MM, Oliveria MM (2004) Mechanisms underlying plant resilience to water deficits: prospects of water-saving agriculture. J Exp Bot 55:2365–2384

Chelkowski J, Tyrka M, Sobkiewicz A (2003) Resistance genes in barley (*Hordeum vulgare* L.). J. Appl Genet 44:291–309

Chen Z, Zhou M, Newman I, Mendham N, Zhang G, Shabala S (2007) Potassium and sodium relations in salinised barley tissues as a basis of differential salt tolerance. Funct Plant Biol 34:150–162

Close TJ, Wanamaker SI, Caldo RA et al (2004) A new resource for cereal genomics: 22K barley GeneChip comes of age. Plant Physiol 134:960–968

Close TJ, Bhat PR, Lonardi S et al (2009) Development and implementation of high-throughput SNP genotyping in barley. BMC Genomics 10:582

Cockram J, White J, Leigh FJ et al (2008) Association mapping of partitioning loci in barley. BMC Genet 9:16

Cockram J, White J, Zuluaga DL et al (2010) Genome-wide association mapping to candidate polymorphism resolution in the unsequenced barley genome. Proc Natl Acad Sci U S A 107:21611–21616

Comadran J, Thomas WT, van Eeuwijk FA et al (2009) Patterns of genetic diversity and linkage disequilibrium in a highly structured *Hordeum vulgare* association-mapping population for the Mediterranean basin. Theor Appl Genet 119:175–187

Comadran J, Ramsay L, MacKenzie K et al (2011a) Patterns of polymorphism and linkage disequilibrium in cultivated barley. Theor Appl Genet 122:523–531

Comadran J, Russell JR, Booth A et al (2011b) Mixed model association scans of multi-environmental trial data reveal major loci controlling yield and yield related traits in *Hordeum vulgare* in Mediterranean environments. Theor Appl Genet 122:1363–1373

Costa JM, Corey A, Hayes PM et al (2003) Identification of QTLs associated with Fusarium head blight resistance in Zhedar 2 barley. Theor Appl Genet 108:95–104

Cseh A, Kruppa K, Molnár I, Rakszegi M, Doležel J, Molnár-Láng M (2011) Characterization of a new 4BS.7HL wheat-barley translocation line using GISH, FISH, and SSR markers and its effect on the β-glucan content of wheat. Genome 54:795–804

Dahleen LS, Manoharan M (2007) Recent advances in barley transformation. In Vitro Cell Dev Biol Plant 43:493–506

Dahleen LS, Okubara PA, Blechl AE (2001) Transgenic approaches to combat fusarium head blight in wheat and barley. Crop Sci 41:628–637

Dai F, Nevo E, Wu D et al (2012) Tibet is one of the centres of domestication of cultivated barley. PNAS 109:16969–16973

Delêtre M, McKey DB, Hodkinson TR (2011) Marriage exchanges, seed exchanges, and the dynamics of manioc diversity. Proc Nat Acad Sci 108:18249–18254

Diab AA, Teulat B, This D, Ozturk NZ, Benscher D, Sorrells ME (2004) Identification of drought inducible genes and differentially expressed sequence tags in barley. Theor Appl Genet 109:1417–1425

Dolezel J, Kubalakova M, Paux E, Bartos J, Feuillet C (2007) Chromosome based genomics in the cereals. Chrom Res 15:51–66

Druka A, Franckowiak J, Lundqvist U et al (2011) Genetic dissection of barley morphology and development. Plant Physiol 155:617–627

Dubcovsky J, Maria GS, Epstein E, Luo MC, Dvorak J (1996) Mapping of the K+/Na+discrimination locus in wheat. Theor Appl Genet 92:448–454

Ellis R, Forester BP, Waugh R et al (1997) Mapping physiological traits in barley. New Phytol 137:149–157

Ellstrand NC (2003) Dangerous liaisons? When cultivated plants mate with their wild relatives. The John Hopkins University Press, Baltimore, pp 1–244

Endris S, Mohammad MJ (2007) Nutrient acquisition and yield of barley exposed to salt stress under different levels of potassium nutrition. Int J Env Sci Tech 4:323–330

Faccioli P, Pecchioni N, Terzi V (1995) Thionin STS marker for barley genotype identification. Barley Genet Newsl 24:54–55

FAO (2010) The second report on the state of the world's plant genetic resources for food and agriculture. Commission on Genetic Resources for Food and Agriculture. FAO, Rome, pp 245–246

FAOSTAT (2012) Food and Agriculture Organization (FAO) of the United Nations. FAOSTAT crop production and trade. http://faostat3.fao.org/home/index.html#VISUALIZE

Fedak G (1985) Wide crosses in *Hordeum*. In: Rasmusson DC (ed) Barley, Agronomy Monograph 26. American Society of Agronomy, Crop Science Society of America, and Soil Science Society of America, Madison, pp 155–186

Fekadu W, Berhane L, Zewdie W (2014) Advance in improving morpho-agronomic and grain quality traits of barley (*Hordeum vulgare* L.) in central highland of Ethiopia. Adv Sci J Agric Sci 1:11–26

Feuillet C, Langridge P, Waugh R (2008) Cereal breeding takes a walk on the wild side. Trends Genet 24:24–32

Flint-Garcia SA, Thornsberry JM, Buckler ES (2003) Structure of linkage disequilibrium in plants. Annu Rev Plant Biol 54:357–374

Forester BP, Ellis RP, Thomas WTB, Bahri DMH, Ben-Salem M (2000) The development and application of molecular markers for abiotic stress tolerance in barley. J Exp Bot 51:19–27

Frankel OH, Brown AHD, Burdon JJ (1995) The conservation of plant biodiversity. Cambridge University Press, Cambridge

Friedt W, Rasmussen M (2003) Modern European barley cultivars: genetic progress in resistance, quality and yield. Paper presented at EUCARPIA cereal section meeting 2003, Italy

Fu YB (2012) Genetic structure in a core subset of cultivated barley germplasm. Crop Sci 52:1195–1208

Gallagher LW, Soliman KM, Vivar H (1991) Interactions among loci conferring photoperiod insensitivity for heading time in spring barley. Crop Sci 31:256–261

Gallardo F, Borie F, Alvear M, von Baer E (1999) Evaluation of aluminium tolerance of three barley cultivars by two short term screening methods and field experiments. Soil Sci Plant Nutr 45:713–719

Ganeshan S, Dahleen LS, Tranberg J, Lemaux PG, Chibbar RN (2008) Barley. In: Cole C, Hall TC (eds) A compendium of transgenic crop plants. Wiley-Blackwell, Hoboken

Gao W, Clancy JA, Han F, Prada D, Kleinhofs A, Ullrich SE (2003) Molecular dissection of a dormancy QTL region near the chromosome 7 (5H) L telomere in barley. Theor Appl Genet 107:552–559

Gatford KT, Basri Z, Edlington J et al (2006) Gene flow from transgenic wheat and barley under field conditions. Euphytica 15:383–391

Gepts P (2006) Plant genetic resources conservation and utilization: the accomplishments and future of a societal insurance policy. Crop Sci 46:2278–2292

Gernand D, Rutten T, Pickering R, Houben A (2006) Elimination of chromosomes in *Hordeum vulgare* x *H. bulbosum* crosses at mitotis and interphase involves micro nucleus formation and progressive heterochromatinization. Cytogenet Genom Res 114:169–174

Goedeke S, Hensel G, Kapusi E, Gahrtz M, Kumlehn J (2007) Transgenic barley in fundamental research and biotechnology. Transgenic Plant J 1:104–117

Gottwald S, Bauer P, Komatsuda T, Lundqvist U, Stein N (2009) TILLING in the two-rowed barley cultivar 'Barke' reveals preferred sites of functional diversity in the gene *HvHox1*. BMC 2:258

Grando S, Macpherson H (2005) Food barley: importance, uses and local knowledge. ICARDA, Aleppo

Grando S, von Bothmer R, Ceccarelli S (2001) Genetic diversity of barley: use of locally adapted germplasm to enhance yield and yield stability of barley in dry areas. In: Cooper HD, Spillane C, Hodgink T (eds) Broadening the genetic base of crop production. CABI, New York, pp 351–372

Graner A (1996) Molecular mapping of genes conferring disease resistance: the present state and future aspects. In: Scoles G, Rossnagel B (eds) Proceedings of the V international oat conference and VII international barley genetics symposium, Inv papers, Saskatoon, Canada. University Extension Press, Saskatoon, pp 157–166

Graner A, Michalek W, Streng S (2000) Molecular mapping of genes conferring resistance to viral and fungal pathogens. In: Proceedings of 8th int. barley genetic symposium. University of Adelaide, Adelaide, pp 6–10

Graner A, Kota R, Perovic D et al (2004) Molecular mapping: shifting from the structural to the functional level. In: Proceedings of the 9th international barley genetics symposium, 2004, Brno, Czech Republic. Agricultural Research Institute Ltd, Kromeriz, pp 49–57

Gulati SC, Verma NS (1988) Genetic resources in barley, their diversity and utilization. In: Paroda RS, Arora RK, Chandel KPS (eds) Plant genetic resources-Indian perspective. NBPGR, New Delhi, pp 134–149

Guo P, Li R, Salvato N, Filos M, Xia Y (2013) SNP detection by using affymetrix barley array in barley. Adv Mater Res 641–642:803–807

Gustafson DI, Horak MJ, Rempel CB, Metz SG, Gigax DR, Hucl P (2005) An empirical model for pollen-mediated gene flow in wheat. Crop Sci 45:1286–1294

Hadado TT, Rau D, Bitocchi E, Papa R (2010) Adaptation and diversity along an altitudinal gradient in Ethiopian barley (*Hordeum vulgare* L.) landraces revealed by molecular analysis. BMC Plant Biol 10:121–141

Hadjichristodoulou A (1995) Evaluation of barley landraces and selections from natural outcrosses of *H. vulgare* ssp. *spontaneum* with ssp. *vulgare* for breeding in semi-arid areas. Genet Resour Crop Evol 42:83–89

Hajjar R, Hodgkin T (2007) The use of wild relatives in crop improvement: a survey of developments over the last 20 years. Euphytica 156:1–13

Hammer K (1984) Das Domestikationssyndrom. Kulturpflanze 32:11–34

Han F, Romagosa I, Ullrich SE, Jones BL, Hayes PM, Wesenberg DM (1997) Molecular marker-assisted selection for malting quality traits in barley. Mol Breed 3:427–437

Han F, Ullrich SE, Clancy JA, Romagosa I (1999) Inheritance and fine mapping of major barley seed dormancy QTL. Plant Sci 143:113–118

Harlan JR (1968) On the origin of barley. US Department of Agriculture, Washington, DC, Agriculture handbook 338:9–31

Harlan JR (1995) The living fields: our agricultural heritage. Cambridge University Press, Cambridge

Harlan JR, de Wet JMJ (1971) Toward a rational classification of cultivated plants. Taxon 20:509–517

Harwood WA (2014) A protocol for high-throughput *Agrobacterium*-mediated barley transformation. Methods Mol Biol 1099:251–260

Harwood WA, Bartlett JA, Alves SA, Perry M, Smedley MA, Leyland N, Snape JW (2009) Barley transformation using *Agrobacterium* mediated techniques. In: Jones HD, Shewry PR (eds) Mehods in molecular biology, transgenic wheat, barley and oats. Springer, New York, pp 137–147

Haseneyer G, Stracke S, Paul C et al (2010a) Population structure and phenotypic variation of a spring barley world collection set up for association studies. Plant Breed 129:271–279

Haseneyer G, Stracke S, Piepho HP, Sauer S, Geiger HH, Graner A (2010b) DNA polymorphisms and haplotype patterns of transcription factors involved in barley endosperm development are associated with key agronomic traits. BMC Plant Biol 10:5

Hayes PM, Jones BL (2000) Malting quality from a QTL perspective. In: 8th int. Barley genet. symp., Adelaide convention centre, Adelaide, south Australia. 22–27 Oct. 2000, vol 8. Glen Osmond, Waite Campus, Adelaide University, Adelaide, pp 99–105

Hayes PM, Liu BH, Knapp SJ (1993) Quantitative trait locus effects and environment interaction in a sample of North American barley germplasm. Theor Appl Genet 87:392–401

Hayes PM, Castro A, Marquez-Cedillo L et al (2003) Genetic diversity for quantitative inherited agronomic and malting quality traits. In: von Bothmer R, van Hintum T, Knüpffer H, Sato K (eds) Diversity in barley. Elsevier Science, Amsterdam, pp 201–226

Hearnden PR, Eckermann PJ, McMichael GL, Hayden MJ, Eglinton JK, Chalmers KJ (2007) A genetic map of 1000 SSR and DArT markers in a wide barley cross. Theor Appl Genet 115:383–391

Hensel G, Valkov V, Middlefell-Williams J, Kumlehn J (2008) Efficient generation of transgenic barley: the way forward to modulate plant-microbe interactions. J Plant Physiol 165:71–82

Hermannsson J, Kristjansdottir TA, Stefansson TS, Hallsson JH (2010) Measuring gene flow in barley fields under Icelandic sub-arctic conditions using closed-flowering varieties. Icel Agric Sci 23:51–59

Heun M, Kennedy AE, Anderson JA, Lapitan NLV, Sorrells ME, Tanksley SD (1991) Construction of a restriction fragment length polymorphism map for barley (*Hordeum vulgare*). Genome 34:437–447

Hintum van T (1992) Barley core collection. In: Munck L (ed) Barley genetics. Proc. 6th IBGS, (22–27 July 1991). Helsingborg, Sweden, 2:703–707

Hinze K, Thompson RD, Ritter E, Salamini F, Schulze-Lefert P (1991) Restriction fragment length polymorphism-mediated targeting of the *mlo* resistance locus in barley (*Hordeum vulgare*). Proc Nat Acad Sci USA 88:3691–3695

Hodgkin T, Rana R, Tuxill J et al (2007) Seed systems and crop genetic diversity in agro ecosystems. In: Jarvis DI, Padoch C, Cooper HD (eds) Managing biodiversity in agricultural ecosystems. Columbia University Press, New York, pp 77–116

Holme IB, Brinch-Pedersen H, Lange M, Holm PB (2008) Transformation of different barley (*Hordeum vulgare* L.) cultivars by *Agrobacterium tumefaciens* infection of *in vitro* cultured ovules. Plant Cell Rep 27:1833–1840

Honsdorf N, March J, Berger B, Tester M, Pillen K (2014) High-throughput phenotyping to detect drought tolerance QTL in wild barley introgression lines. PLoS ONE 9:e97047

Hori K, Kobayash T, Shimizu A, Sato K, Takeda K, Kawasaki S (2003) Efficient construction of high-density linkage map and its application to QTL analysis in barley. Theor Appl Genet 107:806–813

Hori K, Kobayashi T, Sato K, Takeda K (2005) QTL analysis of fusarium head blight resistance using a high density linkage map in barley. Theor Appl Genet 111:1661–1672

Horsley RD, Schwarz PB, Hammond JJ (1995) Genetic diversity in malt quality of north-american 6-rowed spring barley. Crop Sci 35:113–118

Horvath H, Huang J, Wong OT, von Westtstein D (2002) Experiences with genetic transformation of barley and characteristics of transgenic plants. In: Slafer GA, Molina-Cano JL, Savin R, Araus JL, Romagosa I (eds) Barley science: recent advances from molecular biology to agronomy of yield and quality. Haworth Press, Binghamton, pp 143–175

Houston K, Druka A, Bonar N et al (2012) Analysis of the barley bract suppression gene *Trd1*. Theor Appl Genet 125:33–45

Hussain SS (2006) Barley genetics and genomics: a review. Proc Pak Acad Sci 43:63–84

IBGSC (2012) A physical, genetic and functional sequence assembly of the barley genome. Nature 491:711–716

Ibl V, Kapusi E, Arcalis E, Kawagoe Y, Stoger E (2014) Fusion, rupture, and degeneration: the fate of in vivo-labelled PSVs in developing barley endosperm. J Exp Bot 65:3249–3261

Ibrahim SE, Schubert A, Pillen K, Léon J (2012) Comparison of QTLs for drought tolerance traits between two advanced backcross populations of spring wheat. Int J Agric Sci 2:216–227

Islamovic E, Obert DE, Budde AD (2014) Quantitative trait loci of barley malting quality trait components in the Stellar/01Ab8219 mapping population. Mol Breed 34:59–73

Jaccoud D, Peng K, Feinstein D, Kilian A (2001) Diversity arrays: a solid state technology for sequence information independent genotyping. Nucleic Acids Res 29:e25

Jahne A, Becker D, Brettschneider R, L¨orz H (1994) Regeneration of transgenic, microspore-derived, fertile barley. Theor Appl Genet 89:525–533

Jaleel CA, Manivannan P, Wahid A, Farooq M, Al-Juburi HJ, Somasundram R, Panneerselvam R (2009) Drought stress in plants: a review of morphological characteristics and pigments composition. Int J Agric Biol 11:100–105

Jana S, Pietrzak LN (1988) Comparative assessment of genetic diversity in wild and primitive cultivated barley in a centre of diversity. Genetics 119:981–990

Jensen HR, Belqadi L, Santis PD, Sadiki M, Jarvis DI, Schoen DJ (2013) A case study of seed exchange networks and gene flow for barley (*Hordeum vulgare* subsp. *vulgare*) in Morocco. Genet Resour Crop Evol 60:1119–1138

Johansson M, Ellegren H, Adersson L (1992) Cloning and characterization of highly polymorphic porcine microsatellites. Heredity 83:196–198

Johnston PA, Timmerman-Vaughan GM, Farnden KJ, Pickering R (2009) Marker development and characterization of *Hordeum bulbosum* introgression lines: a resource for barley improvement. Theor Appl Genet 118:1429–1437

Jones H, Leigh FJ, Mackay I et al (2008) Population-based resequencing reveals that the flowering time adaptation of cultivated barley originated east of the Fertile Crescent. Mol Biol Evol 25:2211–2219

Jones H, Civáň P, Cockram J et al (2011) Evolutionary history of barley cultivation in Europe revealed by genetic analysis of extant landraces. BMC Evol Biol 11:320

Juchimiuk J, Hering B, Maluszynska J (2007) Multicolour FISH in an analysis of chromosome aberrations induced by N-nitroso-N-methylurea and maleic hydrazide in barley cells. J Appl Genet 48:99–106

Kapusi E, Arcalis E, Kawagoe Y, Stoeger E (2014) Fusion, rupture, and degeneration: the fate of *in vivo*-labelled PSVs in developing barley endosperm. J Exp Bot. doi:10.1093/jxb/eru175

Karsai I, Meszaros K, Hayes PM, Bedo Z (1997) Effects of loci on chromosomes 2 (2H) and 7(5H) on developmental patterns in barley (*Hordeum vulgare* L.) under different photoperiod regimes. Theor Appl Genet 94:612–618

Kasha KJ, Kao KN (1970) High frequency haploid production in barley (*Hordeum vulgare* L.). Nature 225:874–876

Katerji N, Mastrorilli M, van Hoorn JV, Lahmer FZ, Hamdy A, Oweis T (2009) Durum wheat and barley productivity in saline-drought environments. Eur J Agron 31:1–9

Keenan JM, Goulson M, Shamliyan T, Knutson N, Kolberg L, Curry L (2007) The effects of concentrated barley β-glucan on blood lipids and other CVD risk factors in a population of hypercholesterolemic men and women. Br J Nutr 97:1162–1168

Kikuchi S, Taketa S, Ichii M, Kawasaki S (2003) Efficient fine mapping of the naked caryopsis gene (nud) by HEGS (High Efficiency Genome Scanning)/AFLP in barley. Theor Appl Genet 108:73–78

Kilian B, Ozkan H, Kohl J et al (2006) Haplotype structure at seven barley genes: relevance to gene pool bottlenecks, phylogeny of ear type and site of barley domestication. Mol Genet Genom 276:230–241

Kilian B, Özkan H, Pozzi C, Salamini F (2009) Domestication of the triticeae in the fertile crescent. In: Muehlbauer GJ, Feuillet C (eds) Genetics and genomics of the triticeae. Springer, New York, pp 81–119

Kilik H, Akar T, Kendal E, Saim I (2010) Evaluation of grain yield and quality of barley varieties under rainfed conditions. Afr J Biotechnol 9:7825–7830

Kleinhofs A, Han F (2002) Molecular mapping of the barley genome. In: Barley science: recent advances from molecular biology to agronomy of yield and quality. Theor Appl Genet 76:815–829

Kleinhofs A, Chao S, Sharp P (1988) Mapping of nitrate reductase genes in barley and wheat. In: Miller TE, Koebner RMD (eds) Proceeding of the seventh international wheat genetics symposium. Institute of Plant Science Research, Cambridge, pp 541–546

Kleinhofs A, Kilian A, Maroof MAS et al (1993) A molecular, isozyme and morphological map of the barley (*Hordeum vulgare*) genome. Theor Appl Genet 86:705–712

Knežević D, Paunović A, Madić M et al (2007) Koncept oplemenjivanja strnih žitadostignuća i perspektive. XII Savetovanje o biotehnologiji, Agronomski fakultet, Zbornik radova, Čačak, 2–3:271–283

Knupffer H, Terentyeva I, Hammer K, Sato K (2003) Ecogeographical diversity- a Vavilovian approach. In: von Bothmer R, van Hintum T, Knüpffer H, Sato K (eds) Diversity in barley (*Hordeum vulgare*). Elsevier Science BV, Amsterdam, pp 53–76

Komatsuda T, Tanno K, Salomon B, Bryngelsson T, von Bothmer R (1999) Phylogeny in the genus *Hordeum* based on nucleotide sequences closely linked to the *vrs1* locus (row number of spikelets). Genome 42:973–981

Komatsuda T, Maxim P, Senthil N, Mano Y (2004) High-density AFLP map of nonbrittle rachis 1 (*btr1*) and 2 (*btr2*) genes in barley (*Hordeum vulgare* L.). Theor Appl Genet 109:986–995

Komatsuda T, Pourkheirandish M, He C et al (2007) Six-rowed barley originated from a mutation in a homeodomain-leucine zipper I-class homeobox gene. Proc Natl Acad Sci U S A 104:1424–1429

Koornneef M, Alonso-Blanco C, Peeters AJM (1997) Genetic approaches in plant physiology. New Phytol 137:1–8

Korzun L, Künzel G (1997) Integration of 31 translocation breakpoints of chromosome 4H into the corresponding Igri/Franka-derived RFLP map. Barley Genet Newsl 28:23–25

Kota R, Rudd S, Facius A et al (2003) Snipping polymorphisms from large EST collections in barley (*Hordeum vulgare* L.). Mol Genet Genomics 270:24–33

Kraakman AT, Niks RE, Van den Berg PM, Stam P, Van Eeuwijk FA (2004) Linkage disequilibrium mapping of yield and yield stability in modern spring barley cultivars. Genetics 168:435–446

Kronbak R, Ingvardsen CR, Madsen CK, Gregersen PL (2014) A novel approach to the generation of seamless constructs for plant transformation. Plant Methods 10:10. doi:10.1186/1746-4811-10-10

Kruse A (1973) *Hordeum*×*Triticum* hybrids. Hereditas 73:157–161

Kutcher HR, Bailey KL, Rossnagel BG, Legge WG (1996) Identification of RAPD markers for common root rot and spot blotch (*Cochliobolus sativus*) resistance in barley. Genome 39:206–215

Kutlu I, Kinaci G (2010) Evaluation of drought resistance indicators for yield and its components in three triticale cultivars. J Tekirdag Agric Faculty 7:95–103

Lamara M, Zhang LY, Marchand S, Nicholas AT, Belzile F (2013) Comparative analysis of genetic diversity in Canadian barley assessed by SSR, DarT and pedigree data. Genome 56:351–358

Lang MM, Linc G, Szakacs E (2013) Wheat barley hybridization: the last 40 years. Euphytica. doi:10.1007/s10681-013-1009-9

Langridge P, Karakousis A, Collins N, Kretschmer J, Manning S (1995) A consensus linkage map of barley. Mol Breed 1:389–395

Larson SR, Kadyrzhanova D, McDonald C, Sorrells M, Blake TK (1996) Evaluation of barley chromosome-3 yield QTLs in a backcross F_2 population using STS-PCR. Theor Appl Genet 93:618–625

Laurie DA, Pratchett N, Bezant JH, Snape JW (1995) RFLP mapping of five major genes and eight quantitative trait loci controlling flowering time in a winter×spring barley (*Hordeum vulgare* L.) cross. Genome 38:575–585

Lemaux PG, Cho MJ, Louwerse J, Williams R, Wan Y (1996) Bombardment mediated transformation methods of barley. Bio-Rad US/EG Bull 2007:1–6

Lemaux PG, Cho MJ, Zhang S, Bregitzer P (2008) Transgenic cereals: *Hordeum vulgare* L. (barley). In: Vasil IK (ed) Molecular improvement of cereal crops. Kluwer Academic Publishers, Dordrecht, pp 215–316

Li J, Kaneko T, Qin LQ, Wang J, Wang Y (2003) Effects of barley intake on glucose tolerance, lipid metabolism, and bowel function in women. Nutrition 19:926–929

Linde-Laursen I (1997) Recommendations for the designation of the barley chromosomes and their arms. Barley Genet Newsl 26:1–3

Linde-Laursen I, Heslop-Harrison JS, Shepherd KW, Taketa S (1997) The barley genome and its relationship with the wheat genomes. A survey with an internationally agreed recommendation for barley chromosome nomenclature. Hereditas 126:1–16

Linnaeus C (1753) Species Plantarum, exhibentes plantas rite cognitas, ad genera relatas, cum differentiis specificis, nominibus trivialibus, synonymis selectis, locis natalibus, secundum systema sexuale digestas. Holmiae, Impensis Laurentii Salvii. [L. Salvius, Stockholm.] 1st ed. Vol. I

Lundqvist A (1962) Self incompatibility in diploid *Hordeum bulbosum* L. Heridatas 48:138–152

Lundqvist U, Franckowiak J, Konishi T (1996) New and revised descriptions of barley genes. Barley Genet Newsl 26:22–43

Macleod LC (2000) Breeding Barley for malt and beer. In: Proceedings of 8th barley genetic symposium, pp 81–86

Madic M, Paunovic A, Djurovic D, Knezevic D, Tanaskovic S (2012) Breeding barley (*Hordeum vulgare* L.) for abiotic and biotic limiting factors. Third International Scientific Symposium "Agrosym Jahorina 2012", pp 257–262

Mano Y, Sayed-Tabatabaei BE, Graner A, Blake T, Takaiwa F, Oka S, Komatsuda T (1999) Map construction of sequence-tagged sites (STSs) in barley (*Hordeum vulgare* L.). Theor Appl Genet 98:937–946

Mano Y, Kawasaki S, Takaiwa F, Komatsuda T (2001) Construction of a genetic map of barley (*Hordeum vulgare* L.) cross 'Azumamugi'×'Kanto Nakate Gold' using a simple and efficient amplified fragment-length polymorphism system. Genome 44:284–292

Marconi E, Graziano M, Cubadda R (2000) Composition and utilization of barley pearling by-products for making functional pastas rich in dietary fibre and β-glucans. Cereal Chem 77:133–139

Matthews PR, Wang MB, Waterhouse PM, Thornton S, Fieg SJ, Gubler F, Jacobsen JV (2001) Marker gene elimination from transgenic barley, using co-transformation with adjacent "twin T-DNAs" on a standard *Agrobacterium* transformation vector. Mol Breed 7:195–202

Matus-Cádiz MA, Hucl P, Horak MJ, Blomquist LK (2004) Gene flow in wheat at the field scale. Crop Sci 44:718–727

Mayer KF, Martis M, Hedley PE et al (2011) Unlocking the barley genome by chromosomal and comparative genomics. Plant Cell 23:1249–1263

McCallum CM, Comai L, Greene EA, Henikoff S (2000) Targeting induced local lesions in genomes (TILLING) for plant functional genomics. Plant Physiol 123:439–442

McCauley DE (1997) The relative contribution of seed and pollen movement to the local genetic structure of *Silene alba*. J Hered 88:257–263

McGuire SJ (2008) Securing access to seed: Sicial relations and sorghum seed exchange in eastern Ethiopia. Hum Ecol 36:217–229

Michalek W, Kunzel G, Graner A (1999) Sequence analysis and gene identification in a set of mapped RFLP markers in barley (*Hordeum vulgare*). Genome 42:849–853

Minella E, Sorrells ME (1992) Aluminium tolerance in barley: genetic relationships among genotypes of diverse origin. Crop Sci 32:593–598

Mirzaei M, Kahrizi D, Rezaeizad A (2011) Androgenesis and spontaneous chromosome doubling in *Hordeum vulgare* L. Researches of the 1st International Conference (Babylon and Razi universities), pp 248–252

Mohammad AJ, Ali A, Russel HB, Mona A, Haider KZ, Mohammed AK (2013) Characterization and antimicrobial activity of barley grain extract. Int J Curr Microbiol App Sci 2:41–48

Molina-Cano JL, Russell JR, Moralejo MA, Escacena JL, Arias G, Powell W (2005) Chloroplast DNA microsatellite analysis supports a polyphyletic origin for barley. Theor Appl Genet 110:613–619

Molnar SJ, James LE, Kasha KJ (2000) Inheritance and RAPD tagging of multiple genes for resistance to net blotch in barley. Genome 43:224–231

Morrell PL, Clegg MT (2007) Genetic evidence for a second domestication of barley (*Hordeum vulgare*) East of the Fertile Crescent. Proc Natl Acad Sci U S A 104:3289–3294

Morrell PL, Toleno DM, Lundy KE, Clegg MT (2005) Low levels of linkage disequilibrium in wild barley (*Hordeum vulgare* ssp. *spontaneum*) despite high rates of self-fertilization. Proc Natl Acad Sci U S A 102:2442–2447

Mukai Y, Gill BS (1991) Detection of barley chromatin added to wheat by genomic *in situ* hybridization. Genome 34:448–452

Munoz AM, Xiong Y, Schmitt MR et al (2010) Transcriptome analysis of a barley breeding program examines gene expression diversity and reveals target genes for malting quality improvement. BMC Genom 11:653. doi:10.1186/1471-2164-11-653

Munoz AM, Matthew JM, Prasanna RB et al (2011) An improved consensus linkage map of barley based on flow-sorted chromosomes and single nucleotide polymorphism markers. The Plant Genome 4:238–249

Munoz AM, Marcos AC, Endelman JB et al (2014) The USDA barley core collection: genetic diversity, population structure, and potential for genome-wide association studies. PLoS One 9:e94688

Murray F, Brettell R, Matthews P, Bishop D, Jacobsen J (2004) Comparison of *Agrobacterium*-mediated transformation of four barley cultivarsusing the GFP and GUS reporter genes. Plant Cell Rep 22:397–402

Myles S, Peiffer J, Brown PJ, Ersoz ES, Zhang Z, Costich DE, Buckler ES (2009) Association mapping: critical considerations shift from genotyping to experimental design. Plant Cell 21:2194–2202

Naghaii V, Asgharipour MR (2011) Difference in drought stress responses of 20 barley genotypes with contrasting drought tolerance during grain filling. Adv Env Biol 5:3042–3049

Nevo E (1992) Origin, evolution, population genetics and resources of wild barley, *Hordeum spontaneum* in the fertile crescent. In: Shewry PR (ed) Barley: genetics, biochemistry, molecular biology and biotechnology. CAB International, Wallingford, pp 19–43

Nevski SA (1941) Beitrage zur Kenntnis der wildwachsenden Gersten in Zusammenhang mit der Frage uber der Ursprung von *Hordeum vulgare* L. und *Hordeum distichon* L. (Versuch einer Monographic der Gattung *Hordeum*) Trudu Bot Inst Akad Nauk SSSR, Ser 1,5:64–255

Newman CW, Newman RK (1992) Nutritional aspects of barley seed structure and composition. In: Shewry PR (ed) Barley: genetics, biochemistry, molecular biology and biotechnology. C.A.B International, Wallingford, pp 351–368

Newman CW, Newman RK (2006) A brief history of barley foods. Cereal Foods World 51:4–7

Newman CW, Newman RK (2008) Barley for food and health: science, technology and products. Wiley Publishers, New York

Newton AC, Flavell AJ, George TS et al (2011) Crops that feed the world. Barley: a resilient crop? Strengths and weaknesses in the context of food security. Food Secur 3:141–178

Nielsen G, Johansen HB (1986) Proposal for the identification of barley varieties based on the genotypes for 2 *hordein* and 39 isoenzyme loci of 47 reference varieties. Euphytica 35:815–833

Oerke EC, Dehne HW (1997) Global crop production and the efficacy of crop protection current situation and future trends. Europ J Plant Pathol 103:203–215

Olson M, Hood L, Cantor C, Botstein D (1989) A common language for physical mapping of the human genome. Science 245:1434–1435

Orabi J, Backes G, Wolday A, Yahyaoui A, Jahoor A (2007) The Horn of Africa as a centre of barley diversification and a potential domestication site. Theor Appl Genet 114:1117–1127

Ozkan H, Kafkas S, Sertac Ozer M, Brandolini A (2005) Genetic relationships among South-East turkey wild barley populations and sampling strategies of *Hordeum spontaneum*. Theor Appl Genet 112:12–20

Pallotta MA, Graham RD, Langridge P, Sparrow DHB, Barker SJ (2000) RFLP mapping of manganese efficiency in barley. Theor Appl Genet 101:1100–1108

Papa R, Gepts P (2004) Asymmetric gene flow and introgression between domesticated and wild populations. In: den Nijs HCM, Bartsch D, Sweet J (eds) Introgression from genetically modified plants into wild relatives. CABI, Wallingford

Parzies HK, Spoor W, Ennos RA (2004) Inferring seed exchange between farmers from population genetic structure of barely landrace Arabi Aswad from Northern Syria. Genet Resour Crop Evol 51:471–478

Pasam RK, Sharma R, Malosetti M, van Eeuwijk FA, Haseneyer G, Kilian B, Graner A (2012) Genome-wide association studies for agronomical traits in a world wide spring barley collection. BMC Plant Biol 12:16

Pedersen G, Zimny J, Becker D, Jahne-Gartner A, Lorz H (1997) Localization of introduced genes on the chromosomes of transgenic barley, wheat and triticale by fluorescence *in situ* hybridization. Theor Appl Genet 94:749–757

Pickering RA, Devaux P (1992) Haploid production: approaches and use in plant breeding. In: Shewry PR (ed) Barley: genetics, biochemistry, molecular biology and biotechnology. CAB International, Oxford, pp 519–547

Pickering R, Johnston PA (2005) Recent progress in barley improvement using wild species of *Hordeum*. Cytogenet Genom Res 109:344–349

Pickering R, Hill AM, Michel M, Timmerman-Vaughan CM (1995) The transfer of a powdery mildew-resistant gene from *Hordeum bulbosum* to barley (*H. vulgare* L.) chromosome 2(2I). Theor Appl Genet 91:1288–1292

Pickering RA, Hudakova S, Houben A, Johnston PA, Butler RC (2004) Reduced metaphaseI associations between the short arms of homoeologous chromosomes in a *Hordeum vulgare* L. x *Hordeum bulbosum* L. diploid hybrid influences the frequency of recombinant progeny. Theor Appl Genet 109:911–916

Pillen K, Zacharias A, Leon J (2003) Advanced backcross QTL analysis in barley (*Hordeum vulgare* L.). Theor Appl Genet 107:340–352

Pillen K, Zacharias A, Leon J (2004) Comparative AB-QTL analysis in barley using a single exotic donor of *Hordeum vulgare* ssp. *spontaneum*. Theor Appl Genet 108:1591–1601

Poland JA, Brown PJ, Sorrells ME, Jannink JL (2012) Development of high-density genetic maps for barley and wheat using a novel two-enzyme genotyping-by-sequencing approach. PLoS ONE 7:e32253

Powell W, Thomas WTB, Baird E et al (1997) Analysis of quantitative traits in barley by the use of amplified fragment length polymorphisms. Heredity 79:48–59

Pržulj N, Momčilović V (1995) Oplemenjivanje pivarskog ječma. Pivarstvo 28:161–163

Pusadee T, Jamjod S, Chiang YC, Rerkasem B, Schaal BA (2009) Genetic structure and isolation by distance in a landrace of Thai rice. Proc Natl Acad Sci 106:13880–13885

Qi X, Lindhout P (1997) Development of AFLP markers in barley. Mol Gen Genet 254:330–336

Qi X, Stam P, Lindhout P (1998) Use of locus-specific AFLP markers to construct a high density molecular map in barley. Theor Appl Genet 96:376–384

Rafalski JA (2010) Association genetics in crop improvement. Curr Opin Plant Biol 13:174–180

Raman H, Karakousis A, Moroni JS et al (2003) Development and allelic diversity of microsatellite markers linked to the aluminium tolerance gene *Alp* in barley. Aust J Agric Res 54:1315–1321

Ramsay L, Comadran J, Druka A et al (2011) INTERMEDIUM-C, a modifier of lateral spikelet fertility in barley, is an ortholog of the maize domestication gene *TEOSINTE BRANCHED* 1. Nat Genet 43:169–172

Rao HS, Basha OP, Singh NK, Sato K, Dhaliwal HS (2007) Frequency distributions and composite interval mapping for QTL analysis in 'Steptoe' × 'Morex' barley mapping population. Barley Genet Newsl 37:520, http://www.wheat.pw.usda.gov/ggpages/bgn

Rasmussion DC (1985) Barley. American Society of Agronomy, Crop Science Society of America, Soil Science Society of America Publishers, Madison

Reid DA, Wiebe GA (1978) Taxonomy, botany, classification, and world collection. In: Barley: origin, botany, culture, winter hardiness, genetics, utilization, pests, vol 338, Agriculture handbook. US Department of Agriculture, Washington, pp 78–104

Ren X, Sun D, Sun G, Li C, Dong W (2013) Molecular detection of QTL for agronomic and quality traits in a doubled haploid barley population. Aust J Crop Sci 7:878–886

Ritala A, Apegren K, Kurt'en U et al (1994) Fertile transgenic barley by particle bombardment of immature embryos. Plant Mol Biol 24:317–325

Ritala A, Nuutila AM, Aikasalo R, Kauppinen V, Tammisola J (2002) Measuring gene flow in the cultivation of transgenic barley. Crop Sci 42:278–285

Roberts EH, Summerfield RJ, Cooper JP, Ellis RH (1988) Environmental control of flowering in barley (*Hordeum vulgare* L.). I. Photoperiod limits to long-day responses, photoperiod-insensitive phases and effects of low-temperature and short-day vernalization. Ann Bot 62:127–144

Rostoks N, Mudie S, Cardle L et al (2005) Genome-wide SNP discovery and linkage analysis in barley based on genes responsive to abiotic stress. Mol Genet Genom 274:515–527

Roy J, Smith K, Muehlbauer G, Chao S, Close T, Steffenson B (2010) Association mapping of spot blotch resistance in wild barley. Mol Breed 26:243–256

Ryu S (1979) Grain quality of barley for human diet. In: Proceedings of joint barley utilization seminar. Korean Science and Engineering Foundation, Suweon, pp 94–108

Saal B, von Korff M, Leon J, Pillen K (2011) Advanced-backcross QTL analysis in spring barley:

IV. Localization of QTL x nitrogen interaction effects for yield related traits. Euphytica 177:223–239

Salamini F, Ozkan H, Brandolini A, Schafer-Pregl R, Martin W (2002) Genetics and geography of wild cereal domestication in the near east. Nat Rev Genet 3:429–441

Samberg LH, Fishman L, Allendorf FW (2013) Population genetic structure in a social landscape: barley in a traditional Ethiopian agricultural system. Evol Appl 6:1133–1145

Sameri M, Komatsuda K (2004) Identification of quantitative trait loci (QTLs) controlling heading time in the population generated from a cross oriental and occidental barley cultivars (*Hordeum vulgare* L). Breed Sci 54:327–334

Sanei M, Pickering R, Kumke K, Nasuda S, Houben A (2011) Loss of centromeric histone H3 (*CENH3*) from centromeres precedes uniparental chromosome elimination in interspecific barley hybrids. Proc Natl Acad Sci U S A 108:E498–E505

Sato K, Nankaku N, Takeda K (2009) A high-density transcript linkage map of barley derived from a single population. Heredity 103:110–117

Sayed MA, Schumann H, Pillen K, Naz AA, Léon J (2012) AB-QTL analysis reveals new alleles associated to proline accumulation and leaf wilting under drought stress conditions in barley (*Hordeum vulgare* L.). BMC Genet 13:61

Schmalenbach I, March TJ, Bringezu T, Waugh R, Pillen K (2011) High resolution genotyping of wild barley introgression lines and fine-mapping of the threshability locus *thresh-1* using the Illumina GoldenGate Assay. G3 (Bethesda) 1:187–196

Schulze ED (1988) Adaptation mechanisms of non cultivated arid-zone plants: useful lesson for agriculture? In: Bidinger ER, Johansen C (eds) Drought research priorities for the dryland tropics. ICRISAT, Patancheru

Schwarz G, Michalek W, Mohler V, Wenzel G, Jahoor A (1999) Chromosome landing at the *Mla* locus in barley (*Hordeum vulgare* L.) by means of high-resolution mapping with AFLP markers. Theor Appl Genet 98:521–530

Shepherd KW, Islam AKMR (1992) Progress in the production of wheat–barley addition and recombination lines and their use in mapping the barley genome. In: Shewry PR (ed) Barley: genetics, biochemistry, molecular biology and biotechnology. CAB International, Wallingford, pp 99–114

Sherman JD, Fenwick AL, Namuth DM, Lapitan NLV (1995) A barley RFLP map: alignment of three barley maps and comparisons to gramineae species. Theor Appl Genet 91:681–690

Shin JS, Chao S, Corpuz L, Blake T (1990) A partial map of the barley genome incorporating restriction fragment length polymorphism, polymerase chain reaction, isozyme, and morphological marker loci. Genome 33:803–810

Simons G, van der Lee T, Diergaarde P et al (1997) AFLP-based fine mapping of the mlo gene to a 30-kb DNA segment of the barley genome. Genomics 44:61–70

Slatkin M (1995) A measure of population subdivision based on microsatellite allele frequencies. Genetics 139:457–462

Slatkin M (2008) Linkage disequilibrium-understanding the evolutionary past and mapping the medical future. Nat Rev Genet 9:477–485

Sreenivasulu N, Graner A, Ulrich W (2008) Barley genomics: an overview. Int J Plant Genom. doi:10.1155/2008/486258

Stanca AM, Romagosa I, Takeda K et al (2003) Diversity in abiotic stresses. In: Bothmer von R, Knupffer H, Hintum van T, Sato K (eds) Diversity in barley (*Hordeum vulgare* L.). Elsevier Science B.V, Amsterdam, pp 179–199

Steffenson BJ, Olivera P, Roy JA et al (2007) A walk on the wide side: mining wild wheat and barley collections for rust resistance genes. Aust J Agric Res 58:532–544

Stein N, Prasad M, Scholz U et al (2007) A 1,000-loci transcript map of the barley genome: new anchoring points for integrative grass genomics. Theor Appl Genet 114:823–839

Stephens JL, Brown SE, Lapitan NLV, Knudson DL (2004) Physical mapping of barley genes using an ultra sensitive fluorescence *in situ* hybridization technique. Genome 47:179–189

Stracke S, Haseneyer G, Veyrieras J-B et al (2009) Association mapping reveals gene action and interactions in the determination of flowering time in barley. Theor Appl Genet 118:259–273

Sullivan P, Arendt E, Gallagher E (2013) The increasing use of barley and barley by-products in the production of healthier baked goods. Trends Food Sci Technol 29:124–134

Sun L, Wang X (1999) Genetic diversity of Chinese hull-less barley germplasm and its utilization. Plant Genet Resour Newsletter 120:55–57

Sutton T, Baumann U, Hayes J et al (2007) Boron toxicity tolerance in barley arising from efflux transporter amplification. Science 318:1446–1449

Syvanen AC (2005) Toward genome-wide SNP genotyping. Natl Genet 37:5–10

Takahashi R (1955) The origin and evolution of cultivated barley. Adv Genet 7:227–266

Takahashi R, Yasuda S (1971) Genetics of earliness and growth habit in barley. In: Barley genetics. Washington State University Press, Washington

Taketa S, Kikuchi S, Awayama T et al (2004) Monophyletic origin of naked barley inferred from molecular analyses of a marker closely linked to the naked caryopsis gene (*nud*). Theor Appl Genet 108:1236–1242

Talame V, Bovina R, Sanguineti MC et al (2008) TILLMore, a resource for the discoversy of chemically induced mutants in barley. Plant Biotechnol J 6:477–485

Tanksley SD, McCouch SR (1997) Seed banks and molecular maps: unlocking genetic potential from the wild. Science 277:1063–1066

Tanksley SD, Nelson JC (1996) Advanced backcross QTL analysis: A method for the simultaneous discovery and transfer of valuable QTLs from unadapted germplasm into elite breeding lines. Theor Appl Genet 92:191–203

Tautz D (1989) Hypervariability of simple sequences as a general source for polymorphic DNA markers. Nucl Acids Res 17:6463–6471

Tester M, Langridge P (2010) Breeding technologies to increase crop production in a changing world. Science 327:818–822

Tingay S, McElroy D, Kalla R et al (1997) *Agrobacterium tumefaciens* – mediated barley transformation. Plant J 11:1369–1376

Tondelli E, Francia E, Barabaschi D et al (2006) Mapping regulatory genes as candidates for cold and drought stress tolerance in barley. Theor Appl Genet 112:445–454

Tong YP, Rengel Z, Graham RD (1997) Interactions between nitrogen and manganese nutrition of barley genotypes differening in manganese efficiency. Ann Bot 79:53–58

Toojinda T, Baird E, Booth A (1998) Introgression of quantitative trait loci (QTLs) determining stripe rust resistance in barley: an example of marker-assisted line development. Theor Appl Genet 96:123–131

Tragoonrung S, Kanizin V, Hayes PM, Blake TK (1992) Sequence tagged-site-facilitated PCR for barley genome mapping. Theor Appl Genet 84:1002–1008

Travella S, Ross SM, Harden J et al (2005) A comparison of transgenic barley lines produced by particle bombardment and *Agrobacterium*-mediated techniques. Plant Cell Rep 23:780–789

Trifonova A, Madsen S, Olesen A (2001) *Agrobacterium* mediated transgene delivery and integration into barley under a range of in vitro culture conditions. Plant Sci 161:871–880

Tsuchiya T (1986) Chromosome mapping in barley by means of trisomic analysis. In: Sddiqui KA, Faruqui AM (eds) New approaches to crop improvement. PIDC Press, Karachi

Ullrich SE (2010) Significance, adaptation, production, and trade of barley. In: Ullrich SE (ed) Barley: production, improvement and uses. Wiley-Blackwell, Ames, pp 3–13

Ullrich S, Nair S, Blake T et al (2010) Variation in kernel hardness and associated traits in US barley breeding lines. Cereal Chem 87:461–466

Um MO, Park T, Kim YJ et al (2007) Particle bombardment-mediated transformation of barley with an *Arabidopsis NDPK2* cDNA. Plant Biotechnol Rep 1:71–77

Varshney RK, Marcel TC, Ramsay L et al (2007) A high density barley microsatellite consensus map with 775 SSR loci. Theor Appl Genet 114:1091–1103

Varshney RK, Paulo MJ, Grando S et al (2012) Genome wide association analyses for drought tolerance related traits in barley (*Hordeum vulgare* L.). Field Crop Res 126:171–180

Verma RPS, Sharma RK, Nagarajan S (2003) Influence of nitrogen and irrigation on malt and wort quality in barley. Cer Res Comm 31:437–444

Verma RPS, Sarkar B, Gupta R, Varma A (2008) Breeding barley for malting quality improvement in India. Cer Res Comm 36:135–145

Veteläinen M (1994) Exotic barley germplasm: variation and effects on agronomic traits in complex crosses. Euphytica 79:127–136

Visioni A, Tondelli A, Francia E et al (2013) Genome-wide association mapping of frost tolerance in barley (*Hordeum vulgare* L.). BMC Genom 14:424

von Bothmer R (1992) The wild species of *Hordeum*: relationships and potential use for improvement of cultivated barley. In: Shewry PR (ed) Barley: genetics, biochemistry, molecular biology and biotechnology. CAB International, Wallingford, pp 3–18

von Bothmer R, Jacobsen N, Baden C, Jørgensen RB, Linde-Laursen I (1995) An ecogeographical study of the genus *Hordeum*, 2nd edn, Systematic and ecogeographical studies on crop genepools. IBPGR, Rome

von Bothmer R, Sato K, Komatsuda T, Yasuda S, Fischbeck G (2003a) The domestication of cultivated barley. In: von Bothmer R, van Hintum Th, Knüpffer H, Sato K (eds) Diversity in Barley (Hordeum vulgare). Elsevier BV, Amsterdam, pp 9–27

von Bothmer R, Sato K, Knüpffer H, van Hintum T (2003b) Barley diversity-an introduction. Dev Plant Genet Breed 7:3–8

von Korff M, Wang H, Leon J, Pillen K (2004) Development of candidate introgression lines using an exotic barley accession (*Hordeum vulgare* ssp. *spontaneum*) as donor. Theor Appl Genet 109:1736–1745

von Korff M, Wang H, Leon J, Pillen K (2005) AB-QTL analysis in spring barley. I. Detection of resistance genes against powdery mildew, leaf rust and scald introgressed from wild barley. Theor Appl Genet 111:583–590

von Korff M, Wang H, Leon J, Pillen K (2006) AB-QTL analysis in spring barley: II. Detection of favourable exotic alleles for agronomic traits introgressed from wild barley (*H. vulgare* ssp. *spontaneum*). Theor Appl Genet 112:1221–1231

von Korff M, Wang H, Leon J, Pillen K (2008) AB-QTL analysis in spring barley: III. Identification of exotic alleles for the improvement of malting quality in spring barley (*H. vulgare* ssp. *spontaneum*). Mol Breed 21:81–93

von Korff M, Leon J, Pillen K (2010) Detection of epistatic interactions between exotic alleles introgressed from wild barley (*H. vulgare* ssp. *spontaneum*). Theor Appl Genet 121:1455–1464

Wagner DB, Allard RW (1991) Pollen migration in predominantly self-fertilizing plants-barley. J Heredity 82:302–304

Wan Y, Lemaux PG (1994) Generation of large numbers of independently transformed fertile barley plants. Plant Physiol 104:37–48

Wang RR, Zhang XY (1996) Characterization of the translocated chromosome using fluorescence *in situ* hybridization and random amplified polymorphic DNA on two *Triticum aestivum-Thinopyrum intermedium* translocation lines resistant to wheat streak mosaic or barley yellow dwarf virus. Chrom Res 4:583–587

Wang YR, Kang SZ, Li FS, Zhang L, Zhanf JH (2007) Saline water irrigation through a crop-water-salinity production function and a soilwater-salinity dynamic model. Pedosphere 17:303–317

Wang JM, Yang JM, McNeil DL, Zhou MX (2010) Identification and molecular mapping of a dwarfing gene in barley (*Hordeum vulgare* L.) and its correlation with other agronomic traits. Euphytica 175:331–342

Wang J, Yang J, Jia Q et al (2014) A new QTL for plant height in barley (*Hordeum vulgare* L.) showing no negative effects on grain yield. PLoS ONE 9:e90144. doi:10.1371/journal.pone.0090144

Waugh R, Bonar N, Baird E et al (1997) Homology of AFLP products in three mapping populations of barley. Mol Gen Genet 255:311–321

Waugh R, Jannink JL, Muehlbauer GJ, Ramsay L (2009) The emergence of whole genome association scans in barley. Curr Opin Plant Biol 12:218–222

Weerasena JS, Steffenson BJ, Falk AB (2004) Conversion of an amplified fragment length polymorphism marker into a co-dominant marker in the mapping of the Rph15 gene conferring resistance to barley leaf rust, Puccinia hordei Otth. Theor Appl Genet 108:712–719

Weibull J, Walther U, Sato K, Habekus A, Kopahnke D, Proeseler G (2003) Diversity in resistance to biotic stresses. In: von Bothmer R, van Hintum T, Knupffer H, Sato K (eds) Diversity in barley (*Hordeum vulgare*). Elsevier Science BV, Amsterdam, pp 143–178

Welsh J, McClelland M (1990) Fingerprinting genomes using PCR with arbitrary primers. Nucl Acids Res 18:7213–7218

Wenzl P, Li H, Carling J et al (2006) A high-density consensus map of barley linking DArT markers to SSR, RFLP and STS loci and agricultural traits. BMC Genom 7:206

Wiebe GA (1978) Introduction of barley into the new world. In: Barley: origin, botany, culture, winter hardness, genetics, utilization, pests, vol 338, Agriculture handbook . U.S. Department of Agriculture, Washington, pp 1–9

Williams JGK, Kubelik AR, Livak KJ, Rafalski JA, Tingey SV (1990) DNA polymorphisms amplified by arbitrary primers are useful as genetic markers. Nucl Acids Res 18:6531–6535

Williams JGK, Hanafey MK, Rafalski JA, Tingey SV (1993) Genetic-analysis using Random Amplified Polymorphic DNA markers. Methods Enzymol 218:704–740

Witcobe JR (1978) Two row and six row wild barley from western Himalaya. Euphytica 27:601–604

Wood PJ, Beer MU (1998) Functional oat products. In: Mazza G (ed) Functional foods-biochemical and processing aspects. Technomic Publishing Co., Inc, Lancaster, pp 1–37

Wright S (1943) Isolation by distance. Genet 38:114–138

Xiao J, Grandillo S, Ahn SN, McCouch SR, Tanksley SD, Li J, Yuan L (1996) Genes from wild rice improve yield. Nature 384:223–224

Xu X, Liu X, Ge S et al (2012) Resequencing 50 accessions of cultivated and wild rice yields markers for identifying agronomically important genes. Nat Biotech 30:105–111

Yan L, Fu D, Li C et al (2006) The wheat and barley vernalization gene *VRN3* is an orthologue of FT. Proc Natl Acad Sci U S A 103:19581–19586

Zalewski W, Orczyk W, Gasparis S, Orczyk AN (2012) *HvCKX2* gene silencing by biolistic or *Agrobacterium*-mediated transformation in barley leads to different phenotypes. BMC Plant Biol 12:206

Zare M (2012) Evaluation of drought tolerance indices for the selection of Iranian barley (*Hordeum vulgare*) cultivars. Afr J Biotech 11:15975–15981

Zeng F, Shabala L, Zhou M, Zhang G, Shabala S (2013) Barley responses to combined waterlogging and salinity stress: separating effects of oxygen deprivation and elemental toxicity. Front Plant Sci 4:313. doi:10.3389/fpls.2013.00313

Zhang ZY, Xin ZY, Larkin PJ (2001) Molecular characterization of a *Thinopyrum intermedium* Group 2 chromosome (2Ai-2) conferring resistance to barley yellow dwarf virus. Genome 44:1129–1135

Zhou MX, Johnson P, Zhou GF, Li CD, Lance R (2012) Quantitative trait loci for water logging tolerance in a barley cross of Franklin × YuYaoXiangTian Erleng and the relationship between waterlogging and salinity tolerance. Crop Sci 52:2082–2088

Zhou F, Gonzales AM, Clegg MT et al (2014) Two genomic regions contribute disproportionately to geographic differentiation in wild barley. G3: Genes|Genomes|Genetics. doi:10.1534/g3.114.010561

Zhu JK (2001) Over expression of delta-pyrroline-5-caboxylate synthetase gene and analysis of tolerance to water and salt stress in transgenic rice. J Plant Sci 6:66–72

Zoghlami N, Bouagila A, Lamine M, Hajri H, Ghorbel A (2011) Population genetic structure analysis in endangered *Hordeum vulgare* landraces from Tunisia: conservation strategies. Afr J Biotech 10:10344–10351

Zohary D (1999) Monophyletic vs polyphyletic origin of the crops on which agriculture was founded in the Near East. Genet Resour Crop Evol 46:133–142

Zohary D, Hopf M (2000) Domestication of plants in the old world: the origin and spread of cultivated plants in West Asia, Europe and the Nile Valley. Clarendon, Oxford

Oats

6

Rahul Kapoor and Chinka Batra

Abstract

Oats rank sixth in world cereal production statistics and are being grown as multipurpose crop for grain, pasture, and forage or as a rotation crop in many parts of the world. The cultivated oat (*A. sativa* L.) has been known for its economic value in human nutrition and health care.

The collection of genotypes and the conservation of gene pools of cultivated and related wild species are essential for genetic research and plant breeding. This chapter focuses on the studies being performed at the micro- and the macrolevels in oats and explains the mechanisms underlying its genome evolvement. It also explains the application on how to support the discovery of genes and work on various methods of crop improvement, explores the opportunities that are available in our surrounding, and also focuses on future prospects of the same. The main objective of this chapter is to summarize various achievements on genetic resources and their taxonomy, collection, evaluation, conservation, and utilization in oat breeding for enhancing its economic importance.

Keywords

Oats • Evolvement • Gene pool • Genetic diversity • Wide hybridization

6.1 Introduction

Oat is a cereal grain crop with a haploid chromosome number of seven belonging to family Gramineae (Poaceae). The genus *Avena* has diploid, tetraploid, and hexaploid ploidy levels. Little history of these oats was known during the ancient times. Oats were formerly known as a weed-like plant in other cereals for a long time before it was being cultivated by itself. Many scientists believed that our present oats are developed as a result of mutation in wild oats. The ancient history of oats is somewhat unclear as there were so many varieties of different species which made it difficult to identify oats. The main

R. Kapoor (✉) • C. Batra
Department of Plant Breeding and Genetics, Punjab Agricultural University, Ludhiana, India
e-mail: rahulkapoor@pau.edu

M. Singh, S. Kumar (eds.), *Broadening the Genetic Base of Grain Cereals*,
DOI 10.1007/978-81-322-3613-9_6

epicenter of oats is in Asia Minor where most subspecies are related with one another. Currently, oat is an important grain and forage crop in various parts of the world, grown on 14.1 million hectares with a grain production of 28.3 million metric tons in 2005 (USDA, Foreign Agricultural Service, Commodity production, supply and disposition database http://fas.usda.gov/psd). Oat grain is an important source of livestock feed and also a good source of minerals, protein, and fiber, and is grown for use as grain as well as for forage and fodder, straw for bedding, hay, silage, and chaff in various parts of the world. High-quality oats (*Avena sativa* L.) are grown in Northwest Europe, Canada, and Australia and are well adapted to the cooler climates. It also has high nutrition value. Health benefits of oats for human consumption have been authenticated by the American Food and Drug Association.

In countries like India, the breeding of oats begun in the early 1980s and is termed as the most important cereal fodder crop in various parts of the country. Many research centers and agroecological zones have been developed for promoting suitable varieties for different zones and different systems of fodder production. India also has a well-defined system for evaluating oat breeding lines for further release and notification for cultivation by farmers.

It has various qualities like excellent growth habit, good quality herbage, and quick recovery after cutting. Oat requires a cool climate for its growth; therefore, it is successfully grown in the plainş and hilly areas of the country like Punjab, Haryana, Jammu and Kashmir, Uttar Pradesh, Himachal Pradesh, Rajasthan, Madhya Pradesh, Maharashtra, and Bengal. A wide variety of soils is suitable for cultivation of oats. Fertile soil is suited if temperature and moisture conditions are in favor, although maximum oat yields are usually achieved by adding sufficient lime to enhance the pH level of the soil to 5.5. The total area covered under oat cultivation in the country is about 5, 50,000 ha. Oats are considerably important in India as it serves as a dual-purpose crop, i.e., for fodder and grain for animal feed particularly calves, horses, sheep, and poultry.

A well-distributed rainfall of 410 mm and an optimum temperature range of 18–30°C from September to April are sufficient to grow as a fodder crop. Highly improved oats are especially successful for stall-feeding across a wide range of climates, ecologies, and altitudes from 400 m to higher than 2300 m.

The cultivated oat (*A. sativa* L.) has been known for its economic value in human nutrition and health care. An alcoholic extraction of the cultivated oat (*Avena* sativa L.) is deterrent for smoking. Oat also serves as antispasmodic, cyanogenetic, antitumor, diuretic, neurotonic, nervine, stimulant, and vulnerary. Prolamine (avenin) is a minor protein which is almost equivalent in quality as compared to soy protein, which has been proved by WHO (World Health Organization) being equal to milk, meat, and egg protein. Protein content of the hull-less oat kernel (groat) is the highest among cereals (ranges from 12 to 24 %). Porridge oats are a valuable food for patients. Oat is the unique cereal containing a globulin and avenalin, as the major storage protein of almost 80 %. Contents of protein, fat, β-glucan, antioxidants, minerals, and hormone analogues in oat grouts are unique in composition and are the best quality among the cereal grains (Welch et al. 2000; Peterson 2001).

The demand of oats in the recent years has increased as far as human consumption is concerned, mainly because of its various dietary benefits (Stewart and McDougall 2014). In the medical field, oat bran has received considerable attention as it helps in regulating gastrointestinal function and also reduces blood cholesterol due to presence of beta-glucans and water-soluble fibers. It also helps in preventing heart disease and is recommended by nutritionists. Many breakfast cereals and bread products are made from oat flour and rolled oat products. It serves as a raw material for fermentation to furfural (a chemical solvent used in refining minerals and for making resin). Due to more emphasized research on food oats, it has resulted in additional selection criteria as far as quality is concerned.

Furthermore, in several parts of the world, oats are grown as winter forage and multipurpose crop for grain, pasture, or as a rotation crop. Oat fodder is a must for dairy fodder as it can be fed green and the surplus can be converted into silage or hay for use during the scarcity period. In agriculture, oat is highly suitable for extensive and sustainable production systems. It is an important component as it reduces the pressure of diseases in cereal crop rotations. Cultivation of oats assures production of maximum quantities of early and nutritious forage supplies in the fodder-deficit periods during freezing temperatures in the winter season. It is also one of the major sources of green forage available in winter for the starving cattle. It also helps in the maintenance of body weights, improves the health, and also enhances milk yields of milking cattle which otherwise decline sharply during the winter forage lean periods. It is therefore safe to claim that the oat crop has brought a winter forage green revolution.

As a fodder crop, it has excellent protein quality, fat, and mineral content. It has high nutrition value and is a succulent and palatable crop. The average green fodder yield varies from 42 to 60 t per hectare. It forms an excellent combination when fed along with other winter legumes like berseem, lucerne, senji, shaftal, and wild pea or vetch. When mixed with berseem, oat provides balanced feed to milch animals. It has softer straw and high superiority over wheat and barley.

6.2 Crop Gene Pool, Evolutionary Relationships and Systematic

6.2.1 Gene Pools

The collection of genotypes and the conservation of gene pools of cultivated and related wild species are essential for genetic research and plant breeding. The largest world collection of cultivars and genetic stocks of cultivated oats is maintained by the US Agricultural Department. Various small collections of seeds were received from several workers through seed exchanges. The largest world collection of wild species of the genus *Avena* (CAV collection) was maintained in 1972 by the Department of Agriculture of Canada and published the first edition of wild oat gene pool to describe the collection. Since then, the number of samples in the CAV collection has doubled.

The second edition of wild oat gene pool has been published to update and enlarge the listing first published in 1972. The diploid entry with a question mark instead of a specific epithet in the first edition has been described as a new species, *A. damascena* (Rajhathy and Baum 1972), in this second edition. The accession of *A. macrostachya*, a perennial species, which is considered as *Avena*, also appears here as part of the *Avena* gene pool.

Mainly there is a group of three gene pools of wild species of *Avena* on the basis of their inter-fertility with cultivated hexaploid oat where the primary and tertiary gene pools are of great diversity and are extensive too. In *Avena*, the transfer of traits has been mostly from within the primary gene pool. On the other hand the secondary gene pool is small and least represented in ex situ collections. The primary gene pool includes all hexaploid *Avena* species with minor restrictions on gene flow, whereas the secondary gene pool consists of the tetraploids *A. murphyi, A. maroccana*, and *A. insularis*. Various desirable traits represent the secondary gene pool, and various ways in respect to collection as well as evaluation are to be explored. Backcrossing increases the hybrids of *A. sativa* which are partially female fertile. Recombination occurs best with *A. insularis*.

For the purpose of domestication, the introgression from wild into cultivated and vice versa has been considered. In tertiary gene pool, the tetraploid and all diploid species with a gene flow restriction to the hexaploid oat is considered. Also it is considered as a rich reservoir of diversity for oat breeding.

6.2.1.1 The Primary Gene Pool

The gene pool involved in production of fertile hybrids and in which intermating is easy is known as primary gene pool. All crop species belong to a primary gene pool together with such material with which they produce completely fertile crosses through hybridization. Plants of the same species or of closely related species are included thereby, producing completely fertile offsprings on intermating. In such gene pool, genes can be exchanged between lines simply by making normal crosses. This is also known as gene pool one (GP1). This material is of prime breeding importance.

The hexaploids form the primary gene pool in order to transfer desired traits to the cultivated hexaploid oat. Due to high interfertility, conventional crossing and backcrossing procedures are used for the accomplishment of introgression of traits from wild hexaploid. But, greater frequencies of meiotic abnormalities like univalents and micronuclei were also observed in wild species hybrids of *A. sterilis* and *A. fatua* than in intraspecific hybrids.

A. sterilis, also termed as the progenitor species of cultivated oat, is quite a rich source of diverse traits, preferably resistant to diseases, used in oat breeding programs. In the release of several oat cultivars, mostly crown rust resistance *(Pc)* genes from *A. swills* are used in several oat breeding programs. These *Pc* genes from *A. refills* have been backcrossed into susceptible oat lines forming sets of differentials being used to monitor changes in rust virulence patterns providing rust race-specific resistance (Carson 2008). For the effective use of these genes, many schemes have been developed over the years.

6.2.1.2 The Secondary Gene Pool

The genetic material leading to partially fertility on crossing with primary gene pool is referred to as secondary gene pool. Plant groups containing certain barriers against crossing and that are related species belong to the secondary gene pool. Usually the hybrids are sterile and to some extent some progenies can be crossed with primary gene pool. Gene transfer from such material to primary gene pool is different but shows various possibilities. This type of gene pool is also known as gene pool two (GP2). There is partly restriction in the gene flow using the secondary gene pool as the backcrossing to the cultivated parent is required and also the F1 hybrids are pentaploids.

The high groat protein content of more than 25 % and large groat size in certain accessions provoked the initial interest in genes transfer to cultivated oats. AACC tetraploid species such as *A. maroccana* are included in the secondary gene pool as the F_1 plants are highly self-sterile in hybridizations with *A. sativa*. Fertile F_1 female enables crossing to hexaploid, thereby producing low seed set. Chromosome pairing of the tetraploid and hexaploid species allows sufficient recombination occurrence. The recovery of backcrosses' derivatives of the pentaploid F_1 containing 42 chromosomes involves the transfer of genes from the AACC tetraploid to hexaploid cultivated oat.

With the use of synthetic hexaploid being derived from colchicines treated F_1 hybrid between diploid *A. longiglumis* accession CW57 and *A. magna* accession CI 8330, successful introgression for resistance from crown rust was made into hexaploid cultivated oat. The successful incorporation of these polygenically inherited traits through selection efforts and backcrossing into high-yielding cultivars has not been proved possible. Ladizinsky and Feinstein (1977) proposed its alternative by transferring more simply inherited domestication traits from *A. sativa* into *A. murphyi* and *A. magna*, thereby producing domesticated tetraploid oats.

6.2.1.3 The Tertiary Gene Pool

The genetic material leading to the production of sterile hybrids on crossing with primary gene pool is termed as tertiary gene pool or gene pool three (GP3). Those groups that can only be crossed with the help of radical new techniques are involved in the tertiary gene pool. In tertiary gene pool, the hybrids are sterile and include material which can be crossed with primary gene pool. With the help of special techniques, transfer of gene from such material to primary gene pool is possible. Tertiary gene pool consists of all the diploid and tetraploid *Avena* species.

It is difficult to characterize this pool as its members mainly do not readily hybridize with *A. sativa* to produce F_1 plants. Highly sterile F_1 plants are produced in self-fertilization and in backcrosses to the F_1; there is difficulty in introgressing the desired traits into stable cultivated breeding lines free to accompany the deleterious genes. The use of the lower ploidy female parents in crosses helps in enhancing the frequency and seed set quality. To recover F_1 plants, embryo rescue is usually required in many diploid x hexaploid crosses. Treatment of plants with colchicine doubles the chromosome number, thereby producing plants with the full sets of chromosomes from both parents and thus helps in attainment of fertility in these highly sterile interspecific F_1 plants. For the improvement of oats, knowledge about the *Avena* gene pools has therefore critically become more important to search for new genes sources (Fu et al. 2007).

The dimensions of the gene pool have been increased by various gene technologies being available for cultivar improvement as well as have increased our capacities to detect appropriate genetic variation. These new breeding tools are increasing the rate of development of new and higher-performing cultivars.

6.2.2 Evolutionary Pathways and Genomic Relatedness

The genus *Avena* includes diploid, tetraploid, and hexaploid belonging to the Gramineae family. There is a wide variability in the taxonomic classification of genus *Avena*. With the idea of primary, secondary, and tertiary gene pools, a rational taxonomic classification system for crops was being suggested by Harlan and de Wet. According to them, in classification system cultivated and wild hexaploid *Avena* species come in the primary gene pool category, as they are easy to cross with each other.

Higher conservation priority should be given to closest wild relatives in each crop gene pool than distantly related species as these taxa are more easy to use for the improvement of crop by the use of conventional breeding methods.

The morphological or the biological species concept forms the basis for the taxonomy of *Avena*. By the use of morphological concept, 33 species are described and just 14 species on using the biological concept. Several gene banks represent most of those species which have also been used for the breeding purposes. With common oats, the AC genome tetraploid, *Avena murphyi*, shows weaker genetic ties as compared to *Avena insularis* though it is still included in the common oat secondary gene pool. The recent discovered species *Avena insularis* has appeared to be the tetraploid progenitor of common oat and all the hexaploid oat types.

The Cv genome diploid, *Avena ventricosa*, could be the C genome donor of the hexaploids as being revealed by molecular data. *Avena hirtula*, As genome diploid, is considered the wild progenitor of the slim oat. Al genome diploid, *Avena longiglumis* and *A. strigosa*, is found to be the possible donor of the A genome in evaluating the hexaploids.

Numerous genome combinations are found in the nature, and identification of four *Avena* genomes (A, B, C, and D) is done. The all three ploidy groups involve A and C genomes, while the B genome is found only in tetraploids and D genome in hexaploids. Higher similarity is found in A and D genome (Chen and Armstrong 1994) whereas the genome C is related more distantly (Leggett and Markhand 1997; Drossou et al. 2004). Similarity is also observed in the genome B and genome A. There is still no clarity in the origination of the cultivated hexaploid *Avena sativa* with the three A, C, and D genomes. Higher probability is that diploid members of the genome A and genome C were involved in the process regardless of information on the progenitor for the genome D. The hexaploid oats both cultivated and wild occasionally cross in nature and are also interfertile (Loskutov 2001).

The genetic diversity of these *Avena* species are rarely addressed by previous molecular studies and hence offered little resolution in understanding the exotic oat gene pools. This quietly clears that the wild hexaploid oats are the direct progenitor to *A. sativa* but clouded its occurrence. Several molecular markers such as microsatellite (Li et al. 2000), amplified fragment length polymorphism (AFLP) (Drossou et al. 2004), isozymes, and restriction fragment length polymorphism (RFLP) (Nocelli et al. 1999) confirmed various species relationships. However, different molecular markers show limitations to a small set of various *Avena* species, thereby making lesser compatible genetic relationships.

The two different A and C genomes are involved in the evolution of the genus and few other distant derivatives of genomes (A′=B, A″=D). In diploid species different variants of the genome A have been described that includes *A. prostrate*, *A. longiglumis*, *A. hirtula*, *A. strigosa*, and *A. wiestii*. *A. prostrata* is considered to show compatibility with all other species, while *A. longiglumis* is found to be crossable with *A. strigosa*, and incompatible species are *A. damascena* and *A. canariensis*. Several incompatible variants Cp and Cv represented genome C. In the section (Eu) *Avena*, *A. ventricosa* from the genome A is considered as the most primitive species along with *A. bruhnsiana*. Under section Denticulatae, along with the other cultivated forms, these species are classified. In cultivated oats, *A. magna* is considered as the progenitor of the genome AC, and wild hexaploids showed great morphologic similarity, evidence in meiosis, and reciprocal crosses along with its diploid ancestors *A. ventricosa* and *A. canariensis*. As related to *A. pilosa* and *A. barbata*, the outstanding, primitive, and autotetraploid *A. macrostachya* is considered (Loskutov 2001).

A single biological species complex was formed with all these species including *A. nuda* and *A. brevis*. *A. longiglumis* and *A. prostrata* showed more relatedness with each other as compared to *A. strigosa*, from which there is different rearrangement of at least five chromosomes. Difference of more than three translocations is observed in *A. strigosa* from *A. damascena* and *A. canariensis* and a greater one in *A. canariensis*, assuming *A. damascene* as a common ancestor, and distinct phylogenetic relationship is assumed between *A. canariensis* and *A. prostrata* (Rodionov et al. 2005).

Multilocus genetic structure showed large differences. In early hypothesis, Rajhathy and Sadasivaih (1969) described *A. ventricosa* as the donor of genome C and *A. magna* as the donor of both the genomes. *A. barbata* showed morphological similarity with *A. murphyi* and *A. magna* and other hexaploids. Loskutov (2001) considers it to be crossed easily with the latter and probably to be involved in their evolution. A satellite repeat sequence probe used in hybridization considers *A. damascena* and *A. canariensis* as species belonging to genome D and for the Pachycarpa section (*A. insularis* and *A. maroccana*) suggested CCDD (Linares et al. 1998). In Sicily and Tunisia, recent analysis of intraspecies hybridization and C-banding differentiated populations of *A. insularis* suggested it as the donor of CD genome and tetraploid species (Ladizinsky and Jellen 2003). On the basis of chromosome numbers, *A. sterilis* shows difference from *A. insularis* and *A. magna* (Rajhaty and Sadasivaih 1969). Moreover, *A. sterilis* is also considered as the progenitor of the hexaploid cultivated species and *A. byzantina* being the eldest hexaploid cultivated form. Both cultivated and wild species were coevolved and distributed in all the directions (Loskutov 2005).

The inheritance of genome A from *A. longiglumis* and *A. strigosa* is considered as a result of interspecific crosses between species with different ploidy levels and diploid species. The A genome diploid chromosomes and cultivated oat revealed partial homeology. On the basis of morphological traits, *A. canariensis* is considered to be the ancestor of tetraploid and hexaploid species while between their chromosomes; incomplete homeology is revealed in cytological studies. It was revealed that the diploid donor of the genome C could be *A. ventricosa* for tetraploid and hexaploid species on basis of ample evidence.

Based on morphology, chromosome conjugation pattern and karyotypes, all tetraploid species can be classified into four groups. *A. vaviloviana*, *A barbata*, and *A. abyssinica* are included in Group 1. They possess genome AB and are genetically uniform. According to Maltsev, they belong to the series *Eubarbatae* Malz., together with *A. strigosa* and other diploid species bearing the genome A. The tetraploid species of this group are autotetraploids being evolved from the diploid species and later confirmed their relationships. The original diploid genome A showed divergence resulting in AB genome. Recently, it is designated as AA′. These species are easily crossable with the genus *Avena* L. species, excluding the genome C diploids.

A. murphyi and *A. magna* are included in Group 2. It includes crossing between diploid species and *A. magna*, thereby yielding partially sterile hybrids. On the cytological basis of F_1 hybrids (*A. barbata* and *A. magna*), these species are not related with each other. Furthermore, sterile F_2 hybrids are yielded from cross between *A. sativa* x *A. magna*. The important role in evolvement of hexaploid species is demonstrated by the reciprocal crosses of *A. magna* with cultivated oat and other hexaploid species. Some scientists concluded that the tetraploids with the AC genome originated from diploid species with its components as a result of interspecific crosses. Another alternate hypothesis was that the genomes A were identical and the genomes C were related which was discarded later on. The diploid and tetraploid species included in Group 2 are either not crossed directly or yield sterile hybrids in coming generations or even require tissue-culture methods for hybridization. With the use of wild species along with the collection of new specimens, the genetic pool for the selection of oats is, however, extended.

Group 3 includes perennial autotetraploid species, *A. macrostachya* (C), with the genomic formula AA. According to Maltsev, the morphology of *Avena* species allows it to be assigned to the section Avenastrum Koch, the most primitive section. Its crosses with diploids including various derivatives of the A genome describe negligible chromosome homeology. Several studies revealed that *A. macrostachya* and *A. pilosa* were the most viable F_1 hybrids forming trivalents in metaphase.

The morphological differences of *A. macrostachya* and a lower chromosome conjugation percentage in its hybrids with *A. sativa* and *A. murphyi* revealed that *A. macrostachya* showed no involvement in the tetraploid or hexaploid species evolution. Furthermore, *A. macrostachya* genome is closer to the genome C (*A. pilosa*) than to the genome A (*A. strigosa*) being demonstrated in recent studies describing it to be evolutionarily primitive.

Group 4, including the species *A. agadiriana* Baum et Fed., can be considered dependent as this species is rarely studied. The discovery of *A. agadiriana* was in the collections of the diploid species of tetraploid forms in *A. canariensis* rather than to be discovered naturally. It is considered to be having the AB genome being related more closely to *A. barbata* than other tetraploid species. Moreover, *A. agadiriana* shows similarity with *A. magna* and other hexaploids in the form of lodicules structure, upper part of the lemma, etc. Accordingly, *A. agadiriana* is involved in the evolution of hexaploids and by the good crossability with the hexaploids.

The genomic constitution of the most important group of hexaploids is determined by interspecific crosses. Crosses of diploid species with the As genome (*A. strigosa* group) and tetraploid species with AB genome with *A. sativa* demonstrated the presence of A genome in hexaploid species and lacking the B genome. Chromosome morphology in the hybrids was analyzed in cultivated oat having the genomic formula ACD. The genomes A and C showed similarity in the corresponding diploid and tetraploid genomes, whereas the origination of genome D is still unknown according to recent studies. *A. canariensis*>*A. magna*>*A. sterilis* evolutionary pathway was being proposed. But later on, the A genomes of *A. sterilis* and *A. magna* proved to be closer to each other as compared to the A genomes of *A. sativa* and *A. abyssinica*. *A. magna* was considered as the donor of two genomes (AC) of the hexaploid *Avena* L. species. Leggett proposed the similarity in genomes A and D differentiating them from genome C. The genome A of the diploid progenitor is observed to be the precursor of the A and D genomes of hexaploid species. Like the B genome, genome D is likely to be a variant of the A″ genome but differs from B genome being demonstrated in recent studies.

Fertile hybrids are partially obtained with hexaploid species. As compared to *A. fatua*, the *A. sterilis* species is more likely to be the hexaploid ancestor of cultivated oat. The evolution of the resulting allopolyploids is assumed to be involved in structural chromosome rearrangements making them partially homologous. Thus, the modern oat species are complex differentiated ecologic systems, whose evolution was affected by certain environmental conditions and selection.

In the transitional ancestral forms of hexaploid oat evolution, species with the genomes C and AC are considered with no other (direct) cultivated analogues. Under subsection Aristulatae, species with the A, AB (AA') genome has few cultivated analogues along with widely distributed and weedy species. Hence, this group is not considered as the part in hexaploid oat development as it has reached its evolutionary climax.

A. barbata is distributed widely all over the world. It belongs to autotetraploid AA' genome thereby constituting an interfertile group with little structural differentiation. Consideration of different genomes in the world collection has to be taken. Involvement of the *A. barbata* group in tetraploids is ruled out. Based on cytological evidence, a tetraploid, *A. agadiriana*, is considered to currently undergo evolutionary differentiation. Currently there are no cytological evidences that could shed light on the involvement of diploid oats species in the evolution of hexaploid oats.

6.3 Assessment of Gene Flow in Crop Improvement

6.3.1 Gene Flow Occurrence in Nature

In nature, the transfer of genetic information occurs between different generations and individuals and across spatial dimensions. The phenomena of gene flow helps in maintaining the biological diversity, ensuring long-term survival of species and populations in different environments. Thus, gene flow is the movement of genes among sympatric or neighboring populations of crops and related taxa or the introduction of genes from related taxa into the recipient gene pool of a crop, i.e., it describes the processes also known as introgression which is also possible if plant populations with introduced genes persist at low frequency for several generations (a bottleneck) followed by conditions more favorable to their reproduction and competition.

We must differ between artificial or induced and spontaneously occurring gene flow. Artificial or human-caused gene flow has a rather long history in plant breeding and applied genetics and is documented by the efforts to use wild or weedy relatives of crops as donors of economically important characters. The successful breeding, e.g., of disease-resistant cultivars in many crop species by the transfer of resistance genes from wild taxa are well-known examples. The phenomenon of spontaneous occurrence of gene flow within relatives and crops however is poorly understood. The gene flow processes between wild taxa and crops in the past led to obvious evolutionary changes, e.g., the formation of new species. Actual gene flow between crops and relatives can be detected only by careful field studies. In general, such gene flow is a rare event and can happen only in crops if their primary gene pools include related taxa and if crop and relatives are more or less sympatric.

In plants, such exchange of genetic information typically occurs via pollen dispersal. This ordinary phenomenon occurs naturally in conventional (non-biotech) and biotech crops, but interest in understanding and managing gene flow has increased with the development of biotech crops. Various human beings have involved in selection, adaptation, and improvement of crops from diverse species for number of purposes. Gene flow is not an inherently adverse phenomenon. Gene flow is nearly ubiquitous in the biological world, thereby playing a major role in the development of several species of plants as they exist today. In fact, gene flow among sexually compatible plants is thought to be so widespread that the concepts of species based on the absence of gene flow are rendered useless in application to higher plants (Raven 1980).

The main result of gene flow from transgenic crops is dependent on frequency of species containing the transgenes of interest, the future recruitment of their seeds from the soil bank, and the relative reproductive success of these plants. When plants with these genes are at a low frequency, they may disappear from the population if their recruitment from the soil bank is low and if they have no selective advantage (e.g., an herbicide-resistant plant is not sprayed with that particular herbicide) or reproductive advantage over other plants. This disappearance from the population during random fluctuations in gene frequency is called genetic drift. If recruitment from the soil seed bank is high and there is a selective or reproductive advantage for this population, the genes are more likely to be retained and spread and may be inculcated in the gene pool of the recipient population. In the context of this discussion, the transport of genetic traits via seed movement and/or dissemination due to mechanical means (e.g., shattering), wind, water, or distribution by birds or animals is considered as a separate mechanism, not gene flow. In general, however, seed traits that encourage dispersal—shattering from the mother plant, structures that help seeds stay airborne or catch the fur of animals, or hard coatings to survive digestion by animals—are absent from most domesticated crops. Seeds of certain agricultural weeds mimic crop seeds to aid dispersal by humans. Seeds also can confer the ability to disperse genes by remaining dormant in soil for extended periods before germinating.

Transgenes in crops have the potential to move within and among populations like any other gene. As in non-biotech crops, pollen from a biotech crop can and does fertilize other biotech or non-biotech varieties of the same crop, biotech pollen can and does fertilize sexually compatible wild plants, and seed from a biotech crop can and does become mixed with seed of different varieties or even a different crop species. Gene flow is likely to contribute proportionately more to the total adventitious presence in several crop species having high natural levels of cross-pollination.

Numerous genetic improvement programs that confer a variety of useful traits to crops have been established in the USA and many parts of world. Well-established approaches of classical breeding, including induced mutagenesis or wide crossing, as well as direct gene transfer via transformation, have been used in crop improvement. All these approaches are applied to a highly diverse group of traits, including resistance to insects, herbicides, diseases, and viruses, male sterility, delayed ripening, improved nutrition, and freezing tolerance.

6.4 Gene Flow Constraints

6.4.1 Artificial Gene Flow

This introduction by humans of foreign genes into the gene pool of a crop is a process which has been applied by plant breeders and applied geneticists for almost 100 years. By this method desirable characters not present in the crop itself are transferred to the crop's gene pool. Such breeding programs focus mainly on traits like resistance against biotic and abiotic environmental factors and on improvement of quality of the crop in a broad sense. Many resistant or improved cultivars had been selected by these means, and in plant breeding the use of wild relatives is well defined (Harlan 1976; Watson 1970; Ladizinsky 1989). Two almost classical examples may illustrate this approach: *Aegilops* species had been successfully used as donors of disease resistance characters. Leaf rust resistance from *A. umbellulata* and yellow rust resistance from *A. comosa* were transferred to *Triticum aestivum* with the help of crossbreeding and complicated chromosome manipulations (Sears 1956; Riley et al. 1968). Artificial gene flow will become much more important in the future, because the new molecular techniques will allow the use of a much broader range of gene donors than formerly possible.

6.4.2 Spontaneous Gene Flow

Gene flow can happen spontaneously only when crop cultivation and the distribution of its relatives are more or less sympatric or at least overlapping locally and if they are cross compatible and produce at least somewhat fertile offspring. This means that in the primary gene pool of crop species, gene flow occurs mainly according to the classification of Harlan and de Wet (1971). Crops in which domestication resulted in true speciation and strong reproductive isolation (like *Vicia faba, Allium cepa*) cannot be involved in the gene flow processes. There are prehistorical gene flow events which led after subsequent chromosome doubling to the evolution of new amphiploid crop species.

Actual gene flow—although expected to be a rather widespread phenomenon—is still a poorly understood and badly documented process. It can be detected only by careful field studies and genetical and biochemical analysis.

Gene flow can occur at different taxonomic levels. For the sake of convenience, the following types can be distinguished: intraspecific, interspecific, and intergeneric. We must be aware however that this grouping may be very much influenced by the taxonomic concept applied by the researchers. The extent of gene flow is governed by the breeding systems of the partners involved, the availability of pollinating agents, the planting design of the crop, the distribution pattern of the partners, the distance between them, and their population sizes (Levin and Kerster 1974; Klinger et al. 1992). Even for the same partners, it depends strongly on local conditions. The same is true for the direction of gene flow, which can go to and from the crop or in both ways.

6.4.3 Gene Flow Potential with Wild/Weedy Relatives

Weediness is a gene flow concern to the extent that there is gamete transfer among populations. If hybridization of a biotech crop being compared to a non-biotech crop or wild or weedy species occurs, gene flow is an issue. If a biotech crop escapes to the wild because of seed transfer and becomes a weedy feral population, this is not true gene flow because no genetic information is exchanged between populations, even though the economic or environmental consequences of such seed transfers could be important in some instances.

Thus, key considerations for using biotech crops are the history and impact of gene flow from non-biotech food and feed crops to the similar crop or to wild and weedy crop relatives, and the likelihood that flow of genes to weedy relatives will result in aggressive weedy populations of weed-crop crosses.

The potential for gene flow often does not or cannot translate into actual gene flow for a variety of reasons, including separation of species due to differing habitats or geographic distribution. In many instances, the geographical distributions of major production areas of crops and their compatible weedy relatives do not overlap appreciably. Genetic barriers to wide crossing and non-synchronized flowering periods may further limit gene flow. These facts mainly emphasize on numerous crosses of weed crop mainly leading to small problems that are not manageable and very troublesome.

The nature of biotech-specific traits introduced using biotechnology methods is central in evaluating the possible outcome of gene flow to weeds and wild relatives. Crossing of weed with biotech crops that are tolerant to herbicides are more likely to be involved in few herbicides using agricultural fields. The crossed weed-biotech crops will not be favored in areas where very less or none of the herbicide is used such as in native lands. More risks are involved in developing weed-resistant crops by the use of selection pressure from applying herbicides repeatedly in crops that are herbicide resistant, that is, in the absence of gene flow, than that from gene flow to the related weed species (Beckie et al. 2001; Warwick et al. 2004).

Stress-tolerant biotech crops, for example, to salt stress, insects, diseases, water deficits, or deficiency in nutrients, are in more need of scrutiny as their crosses with weedy relatives in both agricultural and nonagricultural areas may be selectively advantageous (Andow and Zwahlen 2006). Hence, few traits evolved from biotech crops could facilitate in the development of problem-causing wild or weedy species. But highly domesticated crops such as maize or wheat (whether as volunteers or as crosses with wild relatives) still will be relatively less invasive than less domesticated crops such as rapeseed, cultivated oats, sugarcane, and rice (Warwick and Stewart 2005).

Seeds or grain of virtually all crop plants have been distributed globally for the past two centuries, and associated ecological problems in virtually all instances came from weeds present in the seed lots or from wild plants that were thought to have some useful or aesthetic purpose and were, therefore, purposely introduced. Various examples in which several agronomic or vegetable crops have become problematic to the environment are rarely found, even though those crops have been bred to exhibit disease and insect resistance and other traits similar to the traits currently being developed through genetic engineering.

6.5 Level of Diversity in Germplasm

6.5.1 Genetic Diversity Among Cultivars

During the twentieth century, landraces have increasingly been replaced by modern cultivars, resulting in significant reduction in genetic diversity. Thus, an important problem both in natural plant populations and in important crop species is the loss of genetic diversity. More emphasis is laid on cultivated gene pools than on exotic gene pools representing wild relative species in current molecular characterization of ex situ plant germplasm (Karp 2002). Although these modern cultivars may be higher yielding under high-input systems, landraces have considerable potential for use in improving disease and abiotic stress tolerance. Transfer of beneficial traits from landraces is relatively straightforward in that there is no barrier to crossing, as there can be with the use of crop wild relatives. Thus, several studies suggest that for several breeding programs, landraces might be a good example of new allelic diversity. However, better characterization of exotic germplasm is needed to facilitate its use in research and in plant breeding (Jellen and Leggett 2006), so as to involve the exotic germplasm for changing quality traits into various plant breeding programs. To date, identification of oat cultivars has relied on morphological and phenological characteristics that may be influenced by environmental factors. In addition, some cultivars are morphologically similar, making it difficult to distinguish them visually.

The precincts associated with morphological and biochemical markers for diversity analysis can be amended by molecular markers techniques. In different individuals within species, a region of DNA showing sequence polymorphism is represented by molecular markers. Various types of DNA markers include amplified fragment length polymorphism (AFLP), simple sequence repeats (SSR), restriction fragment length polymorphism (RFLP), random amplified polymorphic DNA (RAPD), and many more presently available to evaluate polymorphism in sample genomes. Selection of a correct marker system depends upon the type of study to be commenced and whether that marker system would fulfill at least a few of the required characteristics such as easy availability, higher polymorphism, Mendelian inheritance, high reproducibility, neutral behavior, frequent occurrence in genome, easy and fast assay, free of pleiotropy and epistasis, etc. (Weising et al. 1995). DNA markers are environmental friendly and show segregation as a single gene unlike protein markers. The analysis of DNA is labor and cost competitive and is easily extracted from plant materials.

Variations in small DNA regions are detected by the use of two approaches basically. In nucleic acid hybridization-based markers, the visualization of DNA profiles by hybridization of the restriction enzyme digesting the DNA to a labeled probe sharing considerable homology with interested fragment, as identified by complementary base pairing. This is the foundation of the RFLP marker system. Another approach is based on the specific sequences amplification by the use of PCR (polymerase chain reaction). Gel electrophoresis technique is used to amplify a target segment and the primers of known length and sequence are used to amplify genomic and cDNA sequences. The invention of PCR, a very versatile and extremely sensitive technique (Saiki et al.

1995) based on a thermostable DNA polymerase (Saiki et al. 1988), totally changed the scenario of molecular biology. Many possibilities in the molecular marker research have been opened up. Some of the molecular markers can be detected using hybridization and amplification approach. High-density oligonucleotide arrays (Gupta et al. 1999) and the matrix-assisted laser desorption ionization time of flight (MALDI-TOF) mass spectrometry are opened recently, allowing parallel genotyping of biallelic loci in large numbers. These new methods have made possible the genotyping of either one or least individuals for thousands of markers or that of thousands of individuals for one or less markers.

Almost all the studies on genetic diversity involves molecular methods which has now become an integral part in the analysis of breeding systems (Ramanatha Rao and Hodgkin 2002). Marker analysis also gives information on their phylogenetic relationship and various chances to new and useful genes as the accession with distinct DNA profiling and likely to be containing the numerous novel alleles. Marker-based genetic analysis to make such sampling decision is now underway in most crop plants. DNA-based profiling is also necessary in understanding the genetic variations between the exiting cultivars and comparing them with wild relatives.

For characterization of germplasm, specific as well as nonspecific DNA markers such as SSR, AFLP, ISSR, RAPD, and RFLP are being used in different crops. Recently microarray-based DNA marker technique (Gupta et al. 2008) has also been developed for genetic analysis. Few polymerase chain reaction-based markers and SSR types available currently in oat (*Avena sativa* L.) are available due to limitations in the information of genomic sequence. Li et al. 2000, constructed oat microsatellite-enriched libraries for isolation of microsatellite sequences and evaluating polymorphism level in oat cultivars and *Avena* species. From three oat microsatellite libraries, they isolated and sequenced 100 clones. Thus, an average of four alleles was detected among the *Avena* species and 3.8 alleles among the cultivars. To date, genetic studies in hexaploid oat have been more difficult than in other species, mainly due to large genome size (Bennett and Smith 1976) and polyploidy resulting in inherent complexities for mapping, and also include numerous linkage groups, detecting multiple loci by use of single probe and co-migration of fragments from different loci involved in interpreting genetic analyses and allelic relationships (Iannucci et al. 2011). Random amplified polymorphic DNA (RAPDs) (Baohong et al. 2003; Paczos-Grzeda 2004), amplified fragment length polymorphisms (AFLPs) (Achleitner et al. 2008; Fu and Williams 2008), and microsatellites have been used earlier in oat for assessment of genetic diversity. Particularly, microsatellites have been used to dissect genetic diversity in several *Avena* spp. (Li et al. 2000, 2007) and to examine allelic diversity changes over 100 years of oat breeding in both Nordic countries (Nersting et al. 2006), Canada (Fu et al. 2007), and North Europe (He and Bjornstad 2012). Association of genetic markers with regions of the genome controlling different traits would enable efficient and precise transfer of useful alleles from landraces to modern cultivars while minimizing linkage drag of non-beneficial alleles.

6.6 Limiting Factors Related to Production

Each of the breeding and selection procedures has some advantage and disadvantage, and often the choice of a procedure depends on the preference of the breeder, the objectives of the breeding program, and available land, greenhouse, and labor resources. Superior cultivars are developed from any one or from a combination of the procedures. The identification of superior genotypes requires careful observation and testing at number of locations for several years. These testing phases are absolutely essential and require a major amount of time and other resources. Because of this, only lines that are clearly superior should be continued, otherwise the breeding program will become cluttered with mediocre material that has little chance of producing superior cultivars.

Advantages of breeding and selection procedures include important rotational crop, development of extensive root system, low-input efficient crop varieties (including pesticides and fertilizers), high genetic diversity (oil content), helps in establishment of crop and straw and seed cake as coproducts. These procedures also have some disadvantages such as lack of cold tolerance, limitations in production of both food and feed and invasive weeds as relatives.

The breeding procedures include various opportunities such as development of robust and high-yielding oil crop and breeding for high oil content. But it involves few threats such as public perception against use of oats for feedstock production and competition from oil crops such as rape.

6.7 Research and Development Needs

Focus on breeding programs is required in increasing the oil content of the oat grain, which necessarily involves the understanding of metabolic pathways and flux control between oil and starch.

There is need to develop winter oat varieties which will further add in increasing the yield of oat oil as in general, winter crop varieties yields up to 30 % more oil as compared spring varieties. Fast-track breeding programs that do not involve GM should be encouraged as the major risk of oat is that it can cross-fertilize with other oat species known as invasive weeds. In particular, higher recommendation is the use of tilling for development of new varieties. However, requirement of GM technology is mainly for the development of the high oleic acid cultivars. The uses of marker-free technologies should be included. Thus, in short, the research and development is needed in order to develop oat into a high-yielding oil crop such as breeding (conventional and tilling) for increased oil content, breeding for high oleic and very high oil oat varieties, breeding (conventional, tilling, and GM) for winter oat varieties, increasing knowledge about metabolism (flux between oil and starch), and developing efficient seed and or endosperm-specific promoters for oat.

6.8 Traits of Importance

Some traits are of primary importance irrespective of the particular use of the crop. These traits of importance in oat cultivar are determined by whether to be used on farm or marketed off the farm and by other purposes. Virtually all oat breeding programs are aimed to improve grain yield, lodging resistance, disease resistance, test weight, and grain protein yield; however, the value of other traits is determined by the end use of the oats, and some traits valuable for oat millers may be less desirable for on-farm use. For example, on-farm oats might prefer a cultivar with high groat oil percentage, low β-glucan content, high protein content, hull-less kernels, high straw yield, high forage yield and quality, grazing tolerance, and good companion cropping ability with forage legumes. An ideal oat cultivar for oat millers might possess high groat percentage, low groat oil content, high groat protein content, and high groat β-glucan content.

6.8.1 Crown Rust Resistance

Puccinia coronata, crown rust, is considered to be the most damaging oat disease on a worldwide basis. Crown rust can reduce grain yields and quality, forage yields and quality, and promote lodging (Ohm and Shaner 1992). Seed weight, groat percentage, and protein yield are also reduced by crown rust (Simons et al. 1979).

Durability of genetic resistance to crown rust is through "gene deployment," wherein the release of cultivars is required with different sets of resistance genes by the breeders in different regions, which should put disruptive, selection pressure on the pathogen population (Frey et al. 1973), and "gene pyramiding" is to combine multiple race-specific resistance genes into a single line.

6.8.2 Barley Yellow Dwarf Virus Tolerance

The second most damaging disease of oats worldwide is barley yellow dwarf virus (BYDV). The most important vectors of the disease for oats are the aphid species *Rhopalosiphum padi* L., *Macrosiphum avenae* Fab., *Schizaphis graminum* R, and *Metopolophium dirhodum.* Jedlinski et al. (1977) found that field tolerance (reduced symptoms) was associated with lower virus titer levels, suggesting that mechanisms of true resistance exist in oats.

Tolerance is normally assessed by visual scoring of germplasm under field infection. Disease expression in the field could be improved by space planting, late planting, planting aphid trap crops early in disease nursery, rearing infective aphids in the greenhouse to use for artificial infestations, and replicating entries, and a number of cultivars were cited with good tolerance to the virus. Comeau (1982) screened extensive collections of many species of *Avena* for tolerance to BYDV and identified many useful sources of tolerance, particularly in *A. sterilis* and *A. occidentalis.*

6.8.3 Resistance to Other Diseases

Other diseases, such as stem rust, powdery mildew, nematodes, and smuts, are important in particular regions. Breeding for resistance to such diseases has been reviewed by Ohm and Shaner (1992).

6.8.4 Lodging Resistance

The increased use of intensive management practices, such as fertilization and irrigation, on oats has increased opportunities for lodging along with serious losses in grain yield and test weights. Difficulties of breeding for improved lodging resistance include low heritability, difficulty in measuring the trait because of high environmental influences, tendency to underestimate lodging severity in small plots, and the difficulty of combining lodging resistance with high yield (Marshall 1992). The "snap test" described by Marshall (1992) and the "cLr" method by Grafius and Brown (1954) are the two indirect methods of estimating lodging resistance.

6.8.5 Protein Content

The oat groat contains higher levels of all of the "essential" amino acids and, therefore, has a nutritionally superior amino acid profile compared with wheat, rye (*Secale cereale* L.), or barley. The superior amino acid profile of oats seems to be constant over varying levels of total protein content. Therefore, an important breeding objective has been for increased protein content in improving the feeding value of oat grain for animals and also the nutritional quality of the grain for humans. McFerson and Frey (1990) demonstrated that selection on the basis of groat protein yield was successful in improving yield without a reduction in protein content.

6.8.6 Milling Quality

The value of the oat grain to millers producing food for human consumption is referred to as milling quality. These include physical factors, such as kernel weight, size, shape, and groat percentage, and also biochemical factors, such as oil, protein, and β-glucan contents of the groats. Milling yield is the mass of complete grain required in producing a particular mass of rolled oats, such that low values of milling yield are desired by millers.

6.8.7 Test Weight

Due to the ease and speed of grading oats in the marketplace, test weight is the standard indicator of grain quality. Test weight has broad-sense heritability in the range of 60–90 % across many different populations. Test weight seems to be controlled primarily by additive gene action with lesser contributions from epistatic gene action

(Pixley and Frey 1991). The following traits influenced test weight, kernel and groat size and shape, groat percentage, groat density, the presence of awns, disease resistance and hull length, and thickness. As differences in some of these traits, such as kernel shape and size, disease resistance, and the presence of awns, can be easily scored visually thus can be used as indirect selection criteria for test weight in early generations (Forsberg and Reeves 1992).

6.8.8 Groat Percentage

Groat percentage is an important indicator of grain quality but has been used less extensively by breeders and grain marketers mainly because rapid, accurate measures of groat percentage are not available. Reasonably quick and reliable estimates of groat percentage, however, can be made using small samples (as few as 15–25) of primary kernels. Estimates of broad-sense heritability of groat percentage have been reported to range from 20 to 92 % in different populations (Wesenberg and Shands 1971).

6.8.9 β-Glucan Content

Because of the desirable effects of β-glucan on human health, cultivars with high levels of β-glucan in the groat are of interest to millers. Furthermore, the oat soluble-fiber fraction, which consists primarily of β-glucan, seemed to be responsible for the effect. Rather complex, expensive, and time-consuming chemical methods have traditionally been used to estimate β-glucan content (Lim et al. 1992). Although reliable, such methods preclude the routine screening of large numbers of breeding lines, especially in early generations.

6.8.10 *On-Farm* Uses

The utility of oat cultivars for on-farm use depends on their ability to fulfill some needs created by farm animals. Oats have many on-farm uses, such as a source of animal feed, fodder, and bedding. Breeders should attempt to determine which roles oats could serve on-farm that might provide growers with some benefit not supplied by other crop species. Special-purpose cultivars might be developed or more multipurpose cultivars might be envisaged to provide growers flexibility.

6.8.11 Forage Yield and Quality

In colder climates, spring oats can serve as a source of forage for grazing cattle and also be used in a variety of ways to produce forage. In the spring, they can continue to be grazed or they can be allowed to mature and produce grain. Harvesting of green oats for hay or silage, while the grain is in milk or early dough stage, can be done. This can be followed by planting a second crop, such as soybeans. Oats can also be sown as a "companion crop" with forage legumes to help establish pastures and harvested either for grain or for forage.

Oat cultivars that produce both high forage and grain yields may be preferred by growers because they could choose to use the oats for either type of feed, depending upon the changing needs of their animals during the growing season. Forage quality can be measured via acid detergent fiber, in vitro digestible dry matter, neutral detergent fiber, and crude protein. High-quality forage should have high digestibility and protein with low fiber components.

6.8.12 Oil Content

According to Welch, edible oil provides more than twice the dietary energy per unit of weight as carbohydrates or protein. Oat grains have the highest oil contents and, consequently, the highest gross energy value of any temperate small grain crop species; however, the presence of the fibrous hull of the grain reduces the metabolizing energy of oat grain being less than that of barley, wheat, or maize. The improvements in the feeding value of oats could be accomplished by either reducing the hull percentage of the grain or by increasing the oil content of the grain.

6.8.13 Hull-less Oats

The major factor contributing to the low feeding value of oat grain is the non-nutritious fibrous hull. This limits the use of whole oats primarily to horse, cow, and sheep diets. Whole oats are unpalatable to pigs. Feed value of the grains can be improved by dehulling the oats, but this is not always an economically feasible procedure. Thus, breeders can attempt to improve feeding value of oats by reducing the hull percentage of the grain or, more drastically, by breeding oats that have groats that thresh free of the hulls, as occurs with naked-seeded or hull-less oats. Cudderford (1995) concluded that naked oats are superior to other cereals (wheat, barley, or maize) as an animal feed. Hull-less oats could be equally useful on-farm and off-farm, and this is an example of new markets being developed because of progress from plant breeding.

6.9 Interspecific Hybridization

Crosses between species of the same or different genera with equal or different ploidy levels have immensely contributed in crop improvement, understanding of chromosome behavior gene, and genome mapping and evolution in oats. The ultimate motive of hybridization is to transfer useful genes from alien species into cultivated species. Various environmental factors also influence the development of embryo of interspecific hybrids and hence the crossability potential. Therefore, an understanding of the extent of crossability is essential for successful production of hybrids and their derivatives.

The combining properties of different species that are drifted apart in the process of evolution involve the practical importance of interspecies hybridization. The formation of extensive intraspecific diversity of characteristics includes larger area of distribution of wild relatives of crop species. Numerous traits are being lost in the cultivated species in most cases inherent in their wild ancestors. Resistance to pathogens, resistance to unfavorable environmental factors, a number of characters framing increased productivity and quality, and wide range of adaptation to different edaphic and climate conditions are present unique sources of variation for crop improvement. In the past years, the goal of breeding is the fusion of high yield and different valuable parameters in one cultivar. According to many scientists, the utilization of inter- and intraspecific hybridization in conventional breeding along with various modern breeding techniques may increase plant adaptation and the percentage of qualitative kernel components in the crop to a very high level.

The genus *Avena* L. has three ploidy levels which includes many species representing di-, tetra-, and hexaploids. Most of these are wild species. Cultivated crops are present in each ploidy group: *A. strigosa*, *A. abyssinica*, *A. byzantine*, and *A. sativa* L. Many wild *Avena* species have commercially important traits. The diploid species *A. pilosa*, *A. ventricosa*, *A. hirtula*, and *A. Prostrate* are found to be resistant to powdery mildew; *A. wiestii* Steud is highly resistant to *Septoria* leaf rust (*Septoria avenae* Frank.); *A. longiglumis* Dur. is a genetic intermediate in crossing otherwise uncrossable tetraploids with cultivated oat (Thomas 1989). Tetraploid species such as *A. Magna* and *A. murphyi* possess high contents of protein (to 30 %), lysine, and oil in grains, are also resistant to crown rust and powdery mildew, have large grains, and demonstrate high productive tillers. The species *A. barbata* Pott is resistive to stem and crown rusts and powdery mildew and sensitive to barley yellow dwarf virus (BYDV).

The wild hexaploid oat species play an important role in breeding. Virtually, *A. sterilis* L includes all commercially important traits such as large grains, high contents of oil, high protein content and its balanced amino acid composition, and β-glucans in the grain. The schematic representation of its crossing with *A. sativa* has been presented in Fig. 6.1. Another hexaploid species, *A. fatua*, is also widely used in oat breeding programs owing to its short culm, early ripeness,

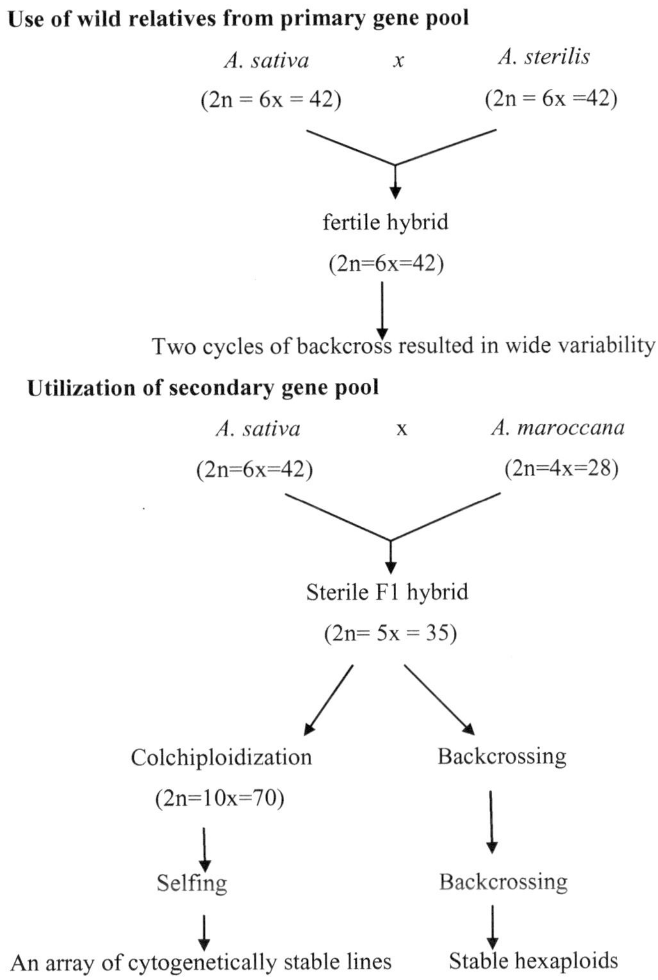

Fig. 6.1 Scheme of introgression of trait of interest through interspecific hybridization 2n=42, 44, 46, 50, 52, 58, 66, 70 (predominantly hexaploids by A11) *lines with introgression of different levels of agronomic traits*

high contents of protein and oil in grain, cold resistance, little grain shedding, resistance to stem and smuts, crown rusts, and tolerance to BYDV.

The studies carried out on oats involved crosses of the cultivated species *A. sativa* with *A. fatua*, *A. sterilis*, *A. ludoviciana*, and *A. strigosa* and, later, with *A. barbata* and other wild tetraploid species. By analyzing a great body of information on the morphology, ecology, and interspecific crosses of oat species, A.I. Mal'tsev developed the best classification of the genus *Avena* L. type section (*Euavena* Griseb). This classification was then confirmed by karyological and other studies. In experiments on crossing of diploid and tetraploid species with hexaploid ones, the parents were incompatible or F_1 plants were sterile. Species with the same ploidy levels were intercrossed successfully. Fertile interspecific oat hybrids were obtained in Japan with the use of the tetraploid species *A. barbata* and, later, with the use of cultivated and sand oats. In 1930, an attempt was made in transferring powdery mildew resistance from the wild diploid species *A. hirtula* collected in Spain to cultivated species. Most attempts to cross species of different ploidy levels made at that time were not successful.

Researchers observed in the regions of the origin and maximum diversity of species *Avena* L. related to the karyology and cytogenetics of cultivated oat species and their wild relatives. This also added in understanding the mechanisms responsible for reproductive isolation of the species. The research on wild and cultivated oat species has involved interspecific crosses conducted for several purposes. On the one hand, the main objective of these studies was in establishing the diploid ancestor of cultivated hexaploid oat, which would help in understanding the cytogenetics of this species and clarifying its genomic formula. On the other hand, the practical task of the studies involved search for commercially important traits in wild species and their introduction into cultivated species to extend their genetic potential.

Interspecific crosses allow the transfer of agriculturally important traits from wild and weed species to cultivated ones. On the basis of crossability, all *Avena* L. species are subdivided into two groups. Group 1 includes all weed species which are easy to cross with cultivated oat.

Many species were involved in successful hybridization in the 1960s, thereby confirming the possibility of transfer of whole chromosomes or its part from diploid species to the genomes of cultivated hexaploid oat. The gene for crown rust resistance was transferred from the diploid species, *A. strigosa*, into a tetraploid line, which was later used for developing multi-linear oat varieties. This line also provided the basis for developing the phenoperiod-nonsensitive variety Donald. In 1968, a fertile hexaploid bearing a part of an *A. hirtula* chromosome with an allele of gene for mildew resistance was obtained in Great Britain. In 1970s, Forsberg transferred the gene *Pc-15* for crown rust resistance from *A. strigosa* to a hexaploid species. In 1986, a hexaploid line resistant to stem rust was obtained by transferring the *Pg-16* gene from *A. barbata*. Tetraploid species are used for solving the problem of powdery mildew resistance. This trait was transferred from *A. barbata* to number of cultivars.

Incompatibility in crosses causes difficulty in transfer of genes from diploids and tetraploids to hexaploids. This can be overcome by the use of backcrosses, mutants, and genetic intermediates. Sterile pentaploid hybrids can be obtained simply by crossing the tetraploid species *A. magna* and *A. murphyi* with cultivated oat, and their fertility can be restored by backcrossing with *A. sativa* pollen. This approach for application of tetraploid species used in breeding of oats for grain quality and weight was developed in Sweden. Natural backcross by sowing pentaploid F_1 hybrids in cultivated oat plots can be used in crosses involving *A. barbata* and *A. macrostachya*. Cross incompatibility between diploids and cultivated oat is often overcome by colchicine-mediated chromosome duplication. Irradiation with thermal neutrons enables transfer of genes for crown rust resistance from *A. abyssinica* to *A. sativa*.

The most successful method for overcoming chromosome incompatibility involved the CW-57 specimen of *A. longiglumis*, a diploid species, which favored conjugation of homologous chromosomes during transfer to mildew-resistance genes from the tetraploid *A. barbata* to cultivated oat. The main example of such intermediation is the development of the hexaploid line Amagalon, bearing the allele of the gene *Pc-91* transferred from *A. magna* through *A. longiglumis*. The same approach can be used in other crosses. At present, interspecific crosses among hexaploid oat species have gained the most widespread acceptance because new alleles can be transferred by routine programs involving cyclic crosses. Many agriculturally important traits are being transferred into cultivated oat by use of these breeding methods, thereby extending its gene pool along with the gene pool of the entire genus.

6.9.1 The Concept of Pre-breeding

Plant species are separated by barriers to intercrossing. This prevents them from being submerged in one large gene pool. A breeder has to break or circumvent these barriers in order to make the genes accessible. Wild species and cultivated crops may be remotely related. In such cases, the breeder may have to use bridging species when direct crosses with cultivars do not succeed. Even when the direct crosses are successful, several backcrosses may be needed to nobilize the hybrids. Both the use of bridging species and backcrossing are laborious and usually lead to a decreased level of desirable polygenic traits. Species which are in close relationship with the cultivated crop are to be preferred.

Furthermore, the nobilization process is speeded up at the diploid level, because at the polyploid level, undesirable "wild genes" will disappear more slowly from the population. Pre-breeding is a procedure which may promote the efficiency of interspecific hybridization. It is called pre-breeding because it implies breeding within each species before crosses with cultivated material are made. Pre-breeding, in addition to selection for adaptation to the prevailing environmental conditions, may include: (1) a thorough evaluation of desired characters within the species before crossing with cultivars, (2) a study on the genetics of such characters within the wild species in order to avoid erratic genetic ratios due to unbalance in interspecific hybrids, (3) concentrating genes controlling polygenic traits, and (4) combining within the species different valuable characters, which usually are scattered over different accessions of that species.

The results of a pre-breeding program may be highly valuable genotypes that are environmentally adapted and that comprise concentrated genotypes for the required characters. A clarification of the genetics of these characters may have occurred. Especially, crosses between a crop and remotely related species should not be carried out before the unique genes of that species are combined and the characters concerned are upgraded to the highest possible level.

6.10 Barriers to Interspecific Hybridization

Interspecific barriers in the broadest sense include all isolation mechanisms between species. When discussing barriers to interbreeding, spatial isolation, nonoverlapping flowering time, and cleistogamy are left out of consideration because they act as real barriers only in nature, not in a breeding program. There are various kinds of barriers to interbreeding, and they may become manifest at different stages or even in different generations. Following Stebbins (1971), a distinction can be made between prezygotic and postzygotic barriers to interbreeding.

Prezygotic barriers in a breeding program are:

- Inhibition of germination pollen or penetration of pollen tube of the stigma
- Arrested pollen tube growth at different sites in style, ovary, or ovulum
- Single or otherwise abnormal fertilization which may lead to parthenogenesis or defective endosperm

Postzygotic barriers are:

- Disturbance of early embryo and endosperm development due to somatoplastic sterility, elimination of chromosomes, and other expressions of disharmony
- Failure of shriveled or even normal seeds to germinate
- (Sub) lethality and other abnormalities during F_1 plant growth
- Male sterility or poor flowering of F_1 plants
- Disharmonious genotypes in segregating generations causing hybrid breakdown

6.10.1 Evolutionary Aspects of Interspecific Barriers

Gene exchange between species in nature is restricted or completely absent. Without interspecific barriers, species would become submerged in one gene pool. Ecological or geographic isolation of subpopulations of a species may initiate the evolution of new species due to natural selection in different environments. An isolated population may become genetically divergent from the original species to such an extent that upon artificial intercrossing with the original species, postzygotic barriers become manifest. Prezygotic barriers may evolve between sympatric forms due to natural selection against fertilization by alien pollen. They may also evolve in the presence of previously developed postzygotic barriers.

6.10.2 Genetic Aspects of Interspecific Barriers: A Genetic Model

A cross between two species is successful if there is a normal relationship between pollen and pistil parent. A normal relationship implies that, after pollination, a chain of processes can take place unhampered. This requires perfect coordination and interaction between gene (complexes) in the pollen (parent) and matching gene (complexes) in the pistil parent. According to Hogenboom (1975), each of the matching parental genes or gene complexes controls a link in the chain of processes and interactions which are needed for successful hybridization. Incomplete matching may cause one or more missing links in the chain and may thus inhibit interbreeding. In this case, genetic information is lacking in one species about few characters in another one. For this phenomenon, Hogenboom (1973) introduced the term "incongruity," which is genetically different from incompatibility based on oppositional action of equal S-alleles in pollen (or pollen plant) and pistil. In relation to incongruity as far as it is expressed between pollination and fertilization, Hogenboom proposed the terms "penetration capacity" and "barrier capacity." Penetration capacity is determined by all genes or gene complexes in the pollen which control its capacity to overcome or circumvent barriers against hybridizing alien females. Barrier capacity comprises the genes or gene complexes controlling all barriers on the female side against being hybridized by alien pollen.

6.10.3 Ways to Overcome Interspecific Barriers

Prezygotic barriers have been overcome or circumvented more or less successfully in different ways. Using large numbers of female plants and a pollen mixture from many males may be advocated in view of genotypic differences in crossability. Cutting of styles and direct pollination of ovules has been reviewed by Rangaswamy (1977). The results of using "mentor pollen" are different. Little success has been achieved by the use of immunosuppressants (Bates 1975) in cereals. Embryo culture is an indispensable procedure when the endosperm is defective (Raghavan 1977; Jensen 1976). The embryos should be sufficiently large. F_1 sterility may be overcome by chromosome doubling if it is caused by lack of homologous pairing. This is not the case when sterility is due to sterility genes, cytoplasmic-genic interactions, or disharmonious gene combinations, but then sterility is often restricted to male gametes. However, female fertility may also be low in hybrids and may cause difficulties when backcrossing is needed.

Environmental factors are important in interspecific hybridization, because unstable genotypes are more sensitive to unfavorable conditions than stable ones. When crosses between wild species and cultivars do not succeed at all, the breeder has to resort to so-called bridging species, which are crossable with both cultivar and wild species. Problems related to this procedure have been discussed earlier. A few examples of the use of double bridges are given below.

Reciprocal differences are very common in interspecific hybridization. In most cases the success is not predicted. Therefore, it is advisable to make routinely interspecific crosses reciprocally. An important exception is the rather general phenomenon that when self-incompatible and self-compatible species are intercrossed, a cross between a self-compatible female x self-incompatible male is most successful. This "unilateral incompatibility," first described by Lewis and Crowe (1958), has been observed in nearly all plant families which comprise both self-compatible and incompatible species (Abdalla and Hermsen 1972; Nettancourt D de 1977).

6.10.4 Interspecific Crossability

Crossability between species is determined by the genotypes of both parental species according to Hogenboom's model for incongruity. Hermsen et al. (1977) reported results from interspecific crosses that suggested a gene-for-gene relationship between male and female parent; crossability spectrum found resembled reaction spectra displayed by many host-parasite systems. There are several reports on genetic determination of crossability between species (Hermsen 1966; Hermsen et al. 1974; Hogenboom 1972). In outbreeding species, only certain genotypes of one species may be crossable with certain genotypes of the other. If these crossability genotypes are scant in either species, the probability of getting a combination that fits is small. By the use of broad genetic variation of the parental species and carrying out the cross combinations on a large scale under a wide range of environmental conditions, the statement represents controversy that two species are not crossable (Hermsen 1979).

6.11 Conventional and Contemporary Approaches of Gene Transfer

Oat breeding procedures used worldwide can be grouped into three categories: selection, introduction, and hybridization followed by selection. In the USA before about 1930, all oat cultivars originated from direct introductions or by selections from those introductions. In early oat breeding programs, the first step was to introduce and evaluate available cultivars from other countries, identify the highest yielding and best adapted, and increase and make them available to farmers. Many of early introductions were mixtures of numerous genotypes from among the mixtures.

Although a number of different breeding procedures currently are used by individual breeders, a feature common to almost all of them is hybridization. Although successful hybridization of oats was reported as early as 1870, it was not until the 1930s that hybridization was widely used by oat breeders to develop segregating populations, followed by some type of selection to produce improved cultivars. So, most important activity is to select superior parents to be used in hybridization program. Parental selection will depend upon the program objectives and on germplasm availability for meeting the objectives. Because of limitation of time and other resources, the oat breeder can use a very small part of total available plant material. Ultimate success of breeding program depends on large part on ability of the breeder in selecting those parents which complement each other and whose progenies likely will be superior. Following hybridization, the most commonly used breeding procedures include backcrossing, pedigree selection, single-seed descent, bulk methods, multiline breeding, and recurrent selection.

6.11.1 Backcrossing

Backcrossing is a form of hybridization used to transfer desirable genes controlling a simple inherited character to desirable cultivar or genotype. The desirable cultivar, which should contain a preponderance of genes for desirable traits, is used as recurrent parent and the parent contributing the simply inherited characteristic is the nonrecurrent or donor parent.

In backcrossing, the F_1 of the hybrid between the two parents is crossed back to the recurrent parent. After each backcross, only those plants containing the desired characteristics from the nonrecurrent parent are crossed back to the recurrent parent. The genes from the nonrecurrent parent are reduced by one half with each successive backcross; thus most of the genetic contribution of the recurrent parent is rapidly recovered. Usually 4–6 backcrosses are considered sufficient in recovering the prototype of the recurrent parent. Transfers of single genes for disease resistance from unadapted to adapted cultivars are performed by the backcross method. For example, at the Purdue Agricultural Experiment Station, backcrossing was used to transfer specific genes for crown rust resistance that led to the development of several important cultivars including "Clintland," "Clintland 60," and "Clintland 64."

While backcrossing is an efficient, predictable, and rapid method of dealing with simply inherited characters, it does not provide an efficient means of improving characteristics that are under polygenic control. Unfortunately, many of the most important oat traits including yield, lodging, resistance, and grain quality are controlled polygenically and, therefore, are not very well adapted to the backcross procedure of breeding. Another drawback is that the upper limits of progeny performance are, for the most part, restricted to those of recurrent parent. Advances due to genetic recombination are minimal or absent.

6.11.2 Pedigree Selection

Pedigree selection is the most commonly used oat breeding procedure. In this method, selection begins in the F_2 generation. Selected F_2 plant progenies and later selfing generations are grown and selected until homozygosity is reached. This usually requires 5–7 generations before individual lines are increased for yield testing. In the pedigree method, the breeder makes the appropriate cross and grows the harvests seed from the F_2 population, which in turn is used to plant the F_2 population. Usually the individual seeds from F_1 plants are planted 5–7.5 cm apart in rows 30 cm wide to facilitate selection. The size of F_2 population will vary depending upon the objectives of program, genetic differences between parents, availability of seed, and resources available for evaluating and making selections of individual plants. Usually several hundred individual plants or individual panicles are selected from each cross. The seeds from each of the plants are visually examined and usually 25–50 are selected. F_3 progeny of each selected F_2 plant are grown in a single row approximately 90 cm long, commonly referred to as a panicle or head row. A few breeders prefer larger F_3 plots such as a single 3 m row or two 90 cm rows, to aid visual selection. Some breeders use hills instead of rows. Individual hills are usually planted at the rate of 25–30 seeds per hill with hills spaced 30–60 cm apart. In the F_3, several panicles are selected from progeny rows or hills with the best appearance. In the F_4 and later generations, similar procedure is being followed until the desired level of uniformity is reached. At this point, all plants from a selected progeny row are harvested in bulk and used for increasing seed and yield testing. The pedigree method of breeding is most effective when the characters of interest to be combined through hybridization are easy in identifying and can be readily selected in the earlier generations.

Modifications of pedigree procedure are also used sometimes. The oat breeder may begin yield testing in early generations to determine the overall potential of the cross and concentrate on selection within the better-performing crosses in later generations. Several breeders have successfully used replicated hill plots for early generation yield evaluations. Hill plot evaluation can be made with a minimal amount of seed and land area.

Pedigree method can be systematized so that large number of crosses and progeny lines can be efficiently planted, evaluated, and harvested each year as it is labor intensive and considerable effort is required. The method also requires the recording of detailed records in each generation. Only superior progenies for one or more of the desired traits need to be carried forward to next generation.

6.11.3 Single-Seed Descent

The single-seed descent method is well suited for oat breeding, and this method and its modification have been popular with oat breeders. With the single-seed descent method, the F_1 population is grown in same manner as for pedigree and bulk method. To provide the population for next generation, the F_2 and their progenies are advanced through succeeding generations by harvesting one seed from each plant in each generation. This procedure undergoes repetitions for several generations until a population of homozygous plants is available for selection. Seed from homozygous plants is harvested and planted in panicle rows or hills, and the lines with desired traits are harvested for replicated testing.

The single-seed descent method is especially well suited when generations are advanced in the green house. Optimum plant growth and development are not necessary because only one seed is needed from each plant; thus, the segregating populations can be planted in thick stands which permit maximum use of limited greenhouse space. All plants from cross are harvested in bulk and threshed to provide seed for next generation. Thus, the population can be advanced through two or three generations during winter. With this procedure, some of plants in each generation are lost; therefore, the breeder needs to maintain rather large populations throughout the segregating generations to ensure the continuation of an adequate number of different genotypes.

The drawback of the single-seed descent method is that plants grown in the green house are not subjected to adaptive stresses which have to be ultimately faced in the field. This disadvantage will be less important if both parents are elite lines with minimum agronomic deficiencies. However if one parent is lacking in one or more critical traits, the use of single-seed descent delays selection for response to an adequate sample of environmental conditions until advanced generation lines are evaluated in yield trials.

6.11.4 Bulk Breeding

In bulk breeding method, the F_2 population is planted in rows 30 cm apart at the rate of approx. 100 seeds per meter row, the plants are harvested and the sample of the bulked seed is used randomly, to plant the F_3 generation. The cycle is repeated several times, usually until the F_5 to F_7 generation, at which time the bulk population is a mixture of homozygous plants. At this time, individual panicles or plants are selected from bulk population, and each plant is grown in an individual panicle row or hill. Individual lines with the desired traits are harvested for replicated testing.

The bulk population method is inexpensive and simple and requires less labor and individual attention from oat breeder, especially during the early segregating generations. It provides an excellent opportunity for artificial and natural selection to eliminate undesirable genotypes, particularly when the bulk populations are subjected to disease epidemics and other adversities during the segregating generations. Seed of selected plants can be bulked and used to plant population for subsequent generations. Certain desirable genotypes which do not compete well in a bulk population may be eliminated through natural selection. Some genotypes that do not survive in the bulk may be superior when grown in a uniform homozygous stand. For example, short, early, lodging-resistant genotypes may be gradually eliminated, while tall, late, lodging-susceptible plants may be favored.

Some oat breeders use a modification of bulk breeding method or combine the bulk method with pedigree or other methods. One modification involves growing the F_2 and F_3 generations by bulk method, then utilizing the pedigree method for remainder of the generations.

6.11.5 Multiline Breeding

Scientists involved in research on oats at the Iowa Agricultural Experiment Station, USA, pioneered the use of multiline breeding to decrease crown rust in oats. Different individual genes for rust resistance are transferred by backcrossing into the same adapted, recurrent parent to form isolines, each with a different gene for resistance. Seed of the isolines is composited to form the multiline cultivar. The isolines making up the cultivar can be changed year by year, depending upon the prevalence of specific races of the rust pathogen. The Iowa experiment station has released several multiline cultivars, including multilines M68, M69 and M70, and "E68," "E69,"and "E70." These cultivars have performed quite well and have provided excellent protection from crown rust damage.

While the multiline breeding approach has provided an effective method for dealing with crown rust of oats, it has several disadvantages that are similar to those of the backcross method. Generally, the system provides no genetic improvement in yield or other important characteristics, except that provided by disease resistance. The labor is intensive because considerable effort is required to develop and maintain the isolines which are composited each year to form the multiline cultivar. The nature of multiline breeding dictates that releases of new and improved recurrent parents will be delayed while isolines are produced and increased. This delay often makes it complicated for the multiline cultivars to compete with other pure line cultivars, especially when yield improvements continue to occur with the more conventional procedures.

6.11.6 Recurrent Selection

Generally, oat breeders have not used the designed recurrent selection schemes to improve germplasm populations of oats. Their failure to use such schemes is partly due to the difficulties associated with making a sufficient number of intermatings which, in oats, must be made by hand crossing. Oat breeders seem to have concluded that the expected results from a carefully designed and executed recurrent selection program would not justify the time and efforts required.

Researchers of the University of Minnesota, USA, have conducted a recurrent selection program for grain yield in oats. Their program was initiated in 1968 when they intermated 12 oat genotypes. In every cycle, progeny was advanced from F_2 to F_4 by single-seed descent. Afterwards, lines from each cross were evaluated in hill plots. Following three cycles of recurrent selection, they found a total yield improvement of 10 % or 3.3 % per cycle. Each cycle required 3 years; therefore, increase in grain per year was 1.1 %.

6.11.7 Additional Considerations

Regardless of breeding method, selection for only one specific trait at a time may result in correlated responses in other important traits. For example, some researchers found that selection only for winter hardiness in winter oats resulted in a shift from erect to decumbent juvenile growth and in taller, later plants.

In advanced breeding nurseries, it is common for many of the top-yielding entries to have average-or-below protein concentrations and for lowest-yielding entries to have the highest protein concentrations. The oat breeder searches for those unique genotypes that have both high grain yields and high or above-average protein percentages. From both the practical and the genetic points of view, both protein concentration and grain yield must be considered before determining the breeding and agronomic value of a genotype. Many oat breeders use protein yield, being expressed as the produced protein weight per unit area, as a single trait reflecting both grain yield and protein percentage.

6.11.8 Concurrent Selection

The breeder usually is able to prevent or minimize undesirable correlated responses by selecting for acceptable or desired levels of several key traits concurrently. However, rate of gain for any one trait will likely be less than when only one criterion for selection is used. Two key challenges faced by oat breeder are combining (a) high grain yield and stiff straw and (b) high grain yield and high protein percentage. Both combinations are strongly influenced by physiological process during growth and development.

Followed by hybridization, tandem selection for individual traits, an alternate to concurrent selection, is considered among concurrent selection and superior types. While this procedure has worked well for combining certain agronomic and disease traits, generally it has not been successful for combination of high groat protein percentage and grain yield. For groat protein, concurrent selection for grain yield, stiff straw, disease resistance, and protein % among progenies from crosses between elite breeding stocks may be the efficient short-range procedure to follow.

An option to be considered for further improvement in percentage of protein is concurrent selection in two separate programs. One such program includes straw strength, selection for yield and resistance for disease receive foremost emphasis, and percentage of proteins is slightly less emphasized on yield, stand ability, and resistance to disease. On the basis of additive gene action, superior gene combination is favored being generated by intercrossing those superior lines from these two programs. The procedure avoids the dilution of desirable traits in one parent because of lower level in the other parent, such as when the cultivated oat is mated with the wild species *A. sterilis* that has a high protein percentage but is inferior for many agronomic traits.

6.12 Molecular Markers, Genome Mapping, and Genomics as an Adjunct to Breeding

6.12.1 Molecular Markers

Many advantages of molecular markers such as numerous markers can be identified in breeding material showing codominant mode of inheritance. Molecular marker technology allows a targeted approach in introgressing and selecting important genes from wide range of genetic resources while retaining the integrity using forward and background selection.

Specific molecular markers are established as a reliable and fast method to identify several forms constituting specific genes which considerably speed up in the genotype selection with improved agronomic traits. Until now, only several molecular markers providing a stable selection of forms consisting specific genes have been identified in oat. The number of analyses of genetic resemblance between different species belonging to the genus *Avena* which are currently carried out may explain the evolution process within this genus and may also elaborate the development of particular oat species.

In oats diploid and hexaploid species, DNA markers were used in constructing and increasing the genetic maps density. Genetic markers are transmitted from one generation to another. Genetic loci or allelic forms of genes determine these biological features and, hence, can be used as experimental tags or probes for keeping the individual track, a nucleus, a cell, a tissue, a gene, or a chromosome. In plant breeding and genetics, Xu (2010) classified genetic markers into two categories. These are DNA markers and classical markers. On the basis of different polymorphism detecting methods or techniques, such as PCR—polymerase chain reaction, southern blotting—nucleic acid hybridization, and DNA sequencing (Collard et al. 2005) including RFLP, AFLP, RAPD, SSR, SNP, etc., DNA markers have developed into many systems. RFLP technique is expressed codominantly and reflects deviation in sequences of homologous DNA that modifies the restriction fragments' length acquired by digestion with restriction enzymes' types. Such divergence is because of alteration of base pairs or other rearrangements such as inversions or translocations at the recognition site of restriction enzyme or due to inner insertion or deletion events. RFLP method is advantageous in detecting the patterns of hybridization of all the obtainable probes.

Biochemical markers, cytological markers, and morphological markers are included in classical markers. On the basis of causal similarity of SNPs with few marker systems and differences with many others, DNA markers have developed into many systems. Due to variation in sequences and length, the molecular markers can also be classified into SNPs such as RFLP and non-SNPs such as SSR, respectively (Gupta et al. 2001).

By the use of F_2 and backcross populations, recombinant inbred lines, and doubled haploids, marker techniques help in selecting multiple desired characters. Further, another popular tool for the dissection of complex traits in plants is Genome-Wide Association Study (GWAS) approach proposed by Hall et al. (2010). The main logic for using GWAS is to genotype numerous markers that are distributed across the genome so that functional alleles or phenotype will be in LD with either of the single or few markers for their use in the breeding program. But, due to limitation of LD, requirement of more number of markers is necessary for sufficient power in detecting linkage between the variation in phenotype and the marker. For detecting QTL, implementation of GWAS in oat could be of great importance to the oat community. Rafalski (2010) revealed many publications on association mapping in plants in early years.

On comparison with other cereal crops such as rice, wheat, and barley, the development and use of molecular markers in oat breeding is relatively slow. Consequently, numerous maps have been published for diploid and hexaploid oats including *A. sativa* L. and *A. byzantina* C. According to Tinker et al. (2009), in recent years, the improvement in the accuracy and coverage density of reference map of Kanota × Ogle helps in representation of linkage groups by adding markers based on Diversity Array Technology (DArT). Furthermore, the maps based on the populations provided significant base for genomic studies in the localization of quantitative trait loci (QTL) and identification of genes for important traits (De Koeyer et al. 2001, 2004; Zhu et al. 2003). For instance, a QTL was found in having a severe effect on the content of oil in groat. Later on, an enzyme catalyst, acetyl-CoA carboxylase, involved in first step of de novo fatty acid synthesis, was connected to the similar QTL.

RAPD and RFLP are main marker types available and used for marker-assisted breeding and genomic mapping. AFLP is concluded to be more efficient and reproducible method which helps in combining the PCR analysis with the reliable RFLP for identifying noteful markers in oats and also in analysis of relationship among oat cultivars and *Avena* spp. (Pal et al. 2002; Fu et al. 2004, 2005).

Simple sequence repeats (SSR) or microsatellites have been produced recently for oat (Jannink and Gardner 2005; Yu and Herrmann 2006) and more initiative is being managed by several research institutes in expanding numerous more microsatellites by focusing on collecting the molecular tools that includes mapping for marker-assisted selective breeding or MAS and molecular markers that can be desegregated into programs to produce new varieties for the organic and animal sector. Until now, a number of polymorphic microsatellites have been established and are more closely observed. Recently, Diversity Array Technology (DArT) is used as a new approach in molecular technique to develop a considerable set of frequently used markers in oat. Currently, hundreds of polymorphic markers have been generated.

6.12.2 Genetic Diversity and Marker Associations

Plant traits, pedigree information, and RFLP markers are helpful in studying genetic diversity in oat. Recently, molecular markers have become very beneficial in evaluation of genetic diversity in plant populations. Li et al. (2000), created and utilized SSR markers in analyzing genetic relationships among various *Avena* species. Comparison between AFLP and RAPD markers has been analyzed for assessment of genetic diversity in different oat varieties. Conduct of genomic studies in the experimental populations produced from a biparental cross is disadvantageous in oat genomic analysis.

Along with genetic complexity, many useful traits in oats are quantitative traits being controlled by multiple genes. Also, quantity trait loci (QTL) analysis and other techniques enable division of this genetic complexity into many simpler portions. Another alternative approach is the utilization of double haploids, i.e., plants that are homozygous at each locus.

Though numerous QTL have been observed, these are effective to be characteristic to a distinctive genetic background and finite success was noticed in result application. From mixed genetic background, the association analysis which involves detection of random set of genotypes, genes, and QTL is a viable solution to such problem. Although for several years, association analysis concept has been known as an effective tool for gene discovery, but its application showed limited success in oat. However, with large accessibility of molecular markers and the statistical tools refinement, this approach has regenerated further interest. Failure in implementation can lead to the false detection of associations between unlinked markers and traits.

6.12.3 Oat Genomics

Avena sativa L. is cultivated hexaploid oat belonging to group of complementary species of the genus *Avena* belonging to Gramineae family. Species are presumed to be evolved from ancestors in this group sharing their genome(s) with each other. An important prerequisite for analysis of genome in this genus was the invention of the appropriate basic chromosome number of seven with different number of chromosome sets and includes four identified A, B, C, and D genomes with three ploidy levels (diploidy, tetraploidy, and hexaploidy).

The cultivated oat *Avena sativa* (2n=6x=42) is a natural allohexaploid constituting three genomes (A, C, D). Polyploidization and interspecific hybridization cycles are involved in oat nuclear genome evolution. Though, accurate diploid progenitors in the hexaploid oat evolution are still unclear. The higher probability of A and C genome diploid species contributed these two genomes of hexaploid oat. The origin of the D genome is cloudy, but several studies suggested close relation between the A and D genomes (Linares et al.1998). The possibility of A genome diploid species is considered to be the donor of both A and D genomes of hexaploid oat due to absence of D genome diploid species.

AA or CC genomes are present in all diploid species that are considered to be the putative donors of the genome C of hexaploids. Morphological studies based on the number of satellite chromosomes were pointed to *Avena ventricosa* or association of *Avena canariensis* and *Avena eriantha* (Rajhathy and Thomas 1974). Oat nuclear genomes' evolution was a complicated process involving differences at diploid level from ancestral common diploid and then at the polyploid level undergoes convergence, followed by divergence (Thomas 1992). Isozyme variations, karyotype studies, and chromosome pairing in interspecific hybrids revealed divergence among the diploid species of A genomes. One karyotype comparison between *A. sativa* and diploid *Avena strigosa* revealed that the *A. strigosa* genome was equivalent to the supposed A genome of hexaploid oat.

Both A and D genomes of *A. sativa* were referred to be highly homologous to *A. strigosa* on the basis of several molecular investigations involving genomic in situ hybridization. Various attempts were found to be unsuccessful to differentiate between A and D genomes by genomic in situ hybridization by use of DNA as probe from various A genome species and distinct combinations of probe of diploid A genome species. This observance in hexaploid oat lacking DD genome diploid species pointed out the assumption of A genome diploid species to be the donor of both A and D genomes. Genome C is distinguished largely from genomes A and D.

Chromosomes of genome C possess large heterochromatic regions, while chromosomes of genomes A and D are mostly euchromatic as indicated by C-banding studies (Jellen et al. 1993). The use of *A. eriantha* in process of in situ hybridization revealed the presence of this chromatin in genome C of hexaploid species. Moreover, evolution of the hexaploids from diploid species is still contentious; observations are in favor of involvement of tetraploid oats *Avena murphyi* and *Avena maroccana* as the purported AC donors. Such evidences are based on crosses between tetraploid species and cultivated oats along with chromosome pairing (Ladizinsky 1995). Also, chromosome pairing between AACC tetraploid species and AA diploid was not found sufficient in supporting the assertion that the genome of *Avena strigosa* was present in tetraploid species. Such different points of views might be suggesting the involvement of several diploid species in the evolution of A and C genomes each.

Intergenomic translocations in polyploids are regarded as an important factor in evolution of oats. On the basis of molecular and cytological evidences, incorporation of A or D and C genomes into polyploid species have undergone various structural rearrangements (Fominaya et al. 1995). Moreover, in such translocations, it is impossible to identify the A and D genomes because of inability to differentiate between these two genomes. In hexaploid oats, establishment of detailed restriction fragment length polymorphism linkage map consisting of 38 linkage groups has been done (O'Donoughue et al. 1995), but no achievement is observed in assigning the linkage groups to the specific chromosomes. The excellent approach for the development of an oat cytogenetic map and allotting linkage groups is merger of molecular cytogenetics and genetic chromosome analysis.

The genus *Avena* shows large diversification containing both cultivated and wild species of polyploid series with a simple chromosome number seven. Within the genus, three ploidy levels are known to be present, diploids, tetraploids, and hexaploids (Table 6.1). Cultivated oat is believed to have originated from *Avena byzantina* or *A. sterilis.*

6.12.4 Marker-Assisted Breeding

Plant breeders have relied heavily on generating new gene combinations and selecting these empirically. The response in the improvement had been tremendous so far, and its efficiency can further be improved by marker-assisted selection. Important prerequisites for marker-assisted selection in a plant breeding program are:

1. Marker(s) must be closely linked or co-segregate (less than 1 cM) with the desirable character.
2. An effective means of filtering vast populations for molecular marker(s) must be procurable. PCR-based techniques fulfill this criteria to some extent.
3. In several laboratories, the filtering technique must have large reproducibility, must be frugal to use, and must be ergonomic.

Marker-assisted selection can be practiced more efficiently for characters whose phenotypic selection is difficult. For example, transferring a fertility restorer gene from one line to another line through backcrossing needs test crossing before subsequent backcrossing. If such genes are tagged with molecular markers, desirable plants with fertility restorer gene (in heterozygous condition) can be identified and backcrossed. Similarly, screening for abiotic stresses is very difficult. If desirable genes conferring tolerance to abiotic stresses are tagged, these then can be selected easily in segregating generations. The quality traits or disease resistance that is hard to manage or is uneconomical can be easily handled by the molecular markers being closely attached to quantitative trait loci or key genetic loci.

In oats, molecular marker-assisted selection (MAS) breeding is not firmly developed. However, it plays a significant role in rationalizing several researches on oats genomics. The easy accessibility of molecular markers forms an indispensable approach to MAS in any crop, which widely means vigorous PCR-based markers. In an ironic manner, RFLPs were established to be the first advanced markers in oats and evinces strength in comparative and recombination mapping. These were also linked to several genes and quantitative trait loci and are work intensive and expensive to be utilized for the large throughput genotyping requisite for marker-assisted selection. In oats, PCR-based RAPDs, although being irreproducible, are more conveniently used for tagging several genes as demonstrated in various studies (Penner et al. 1993) to be utilized for such selection. AFLPs sample numerous loci occurring at the same time because of its complicated multiband pattern which is abundant in genetic information. Such characteristic is advantageous for swift gene tagging or recombination mapping in completely managed research circumstances, but in the case of uniform loci, it is difficult to tackle in marker-assisted selection. Simple sequence repeats or microsatellite markers show various limitations till date for use in oats because finite number and unpredicted low polymorphism though have alluring distinguished feature for selection (Zhu and Kaeppler 2003). Single nucleotide polymorphisms (SNPs) are becoming prominent markers of preference in several crops but need an important straightforward establishment endeavor, and hence, some are yet to be accomplished in oat. The conversion of early developed AFLP, RAPD, or RFLP markers into more active PCR-based sequence to depict amplified region markers, SCAR, has been accomplished by scientists researching on oats for solving few issues being discussed above.

To unveil polymorphism, SCAR markers need post amplification restriction enzyme digestion which is also considered as cleaved amplified polymorphic sequences or CAPS. The procedure for conversion of markers can be cumbersome. In

Table 6.1 Avena species—genome constitution and source of desirable traits

Classification	Chromosome no.	Genome constitution	Desirable traits
Section Avenotrichon			
A. macrostachya	4x = 28		
Section Ventricosa			
A. clauda	2x = 14	CpCp	
A. eriantha	2x = 14	CpCp	
A. ventricosa	2x = 14	CvCv	
Section Agraria			
A. brevis	2x = 14	AA	
A. hispanica	2x = 14	AA	
A. nuda	2x = 14	AA	Leaf size
A. strigosa	2x = 14	AsAs	Plant height
Section Tenaicarpa			
A. agadiriana			
A. atlantica	2x = 14	AsAs	
A. barbata	4x = 28	AABB	Longer growth duration
A. canariensis	2x = 14	AcAc	Earliness
A. damascena	2x = 14	AdAd	
A. hirtula	2x = 14	AsAs	Tillering
A. longiglumis	2x = 14	AiAi	
A. lusitanica	2x = 14	AA	
A. matritensis	2x = 14		
A. prostrata	2x = 14	ApAp	Tillering
A. wiestii	2x = 14	AsAs	
Section Ethiopica			
A. abyssinica	4x = 28	AABB	Earliness, foliage protein
A. vaviloviana	4x = 28	AABB	
Section Pachycarpa			
A. maroccana	4x = 28	AACC	Foliage protein, longer growth duration
A. murphyi	4x = 28	AACC	
Section *Avena*			
A. atherantha	6x = 42	AACCDD	
A. fatua	6x = 42	AACCDD	
A. hybrida	6x = 42	AACCDD	
A. occidentalis	6x = 42	AACCDD	
A. sativa	6x = 42	AACCDD	Plant height, leaf size, longer growth duration, forage yield
A. sterilis	6x = 42	AACCDD	Plant height, tillering, foliage protein, forage yield
A. trichophylla	6x = 42	AACCDD	

numerous cases, conversion of markers is not necessary as PCR-based markers that are suitable for marker-assisted selection have been blended into the hexaploid oat map already and are attached sufficiently closer to the interested quantitative trait loci or genetic locus being utilized directly for marker-assisted selection.

For different oat populations, linkage maps based on molecular markers have been established. For various oat genes, several marker-trait connections have been predicted, and marker-assisted selection or QTL has been executed or assessed for few economic characters in oat, although not in similar extent as in several important crops. Due to advancement in technology, research on oats will be resumed by researchers as an advantage of being quicker and inexpensive and larger production rate of marker establishment, bioinformatics analyses, and DNA sequencing.

None of the gene-rich regions and complete oat genome sequencing along with establishment of large databases of genes will be available as an intellectual property in coming years. However, scientists working on oats will be able to arrange information on the structural and functional genomics originated in another species for designing markers and identifying connections between gene and trait for molecular applications in oat. To recognize genes showing characters of economic importance that are common to number of plant species enables the use of "perfect markers" that includes SNPs within the key genes, thereby allowing thorough and high-throughput selection at the genotypic level and not phenotypic level. It even helps in designing and substituting enhanced alleles for these genes by few forms of transformations that are specific to gene. Identification and mapping of gene and analyses are required to be prolonged in oat, although several biological features and traits of importance are particular to oat. For instance, oat is unparalleled with the cereals in production of active agents known as triterpenoids or avenacins for fighting certain microbial infections affecting roots. Cloning of genes for enzymes involved in avenacin biosynthesis pathway is done in diploid oat *A. strigosa* and discovered to map as a group, although the deviation in structure of enzymes suggests necessary consequences of genes arising as another alternative through a process of duplication from a single gene, thereby raising a necessary difference of opinion related to gene regulation and its origination (Qi et al. 2004).

Another important characteristic particular to oat among other cereals is the presence of a high proportional quantity of the grain oil in the endosperm, thereby providing an activity in response to stimulus for genetical analyzation and, thereby, not only contributing to the practical value of this character but also in shedding knowledge on regulating the frequency of occurrence of tissue related to gene expression and distribution of the product. In biosynthesis of oil, the relationship of high oil and a gene for ACCase key enzyme, in specific oat crosses, demonstrated information about genes contributing to a trait, thus providing certain favorable circumstances for modification and manipulation of the trait. Also, oat constitutes different characteristics or increased levels of secondary metabolites that are beneficial for health. These include tocopherols and avenanthramides, thereby providing economical approach toward new opportunities on the basis of their genetics (Peterson 2004).

In oat, cDNA sequencing basically from tissue-specific libraries and subtraction helps in identifying the expressed genes which may be linked with specific traits. The identification of genes may be done by being their homologues to ones identified in other species and ones original to oat. The accessibility of a large fragment DNA library of various oat genome equivalents grants the utilization of cDNA sequences of oat, or even ones from other species, in selection of clones with particular oat genomic regions that involves flanking sequences with elements included in regulation of interested or specific gene expression. These cDNA sequences of oat could also be involved in constructing gene chips for studying particular expression of gene as was performed similarly in barley (Close et al. 2004). Recent knowledge of oat biology obtained from functional and structural analysis of oat genome provides greater understanding of oat and its genetic manipulation for improving oat.

The evaluation of similarities and differences between repetitive element families and several DNA sequences in addition to AFLPs, RFLPs, RAPDs, and SSRs is continuously contributing in reinvigorating genomic relationships within the cultivated hexaploid species and also phylogenetic relationships among *Avena* species (Ananiev et al. 2002; Drossou et al. 2004). Such unambiguous equipments and knowledge facilitates in identification, arrangement, and manipulation of homologous alleles or genes in comparative wild species or homologous loci within the hexaploid to ameliorate the characters or trait. Even the general conserving nature of constituting the genome in segments, among other species, permits the comparison of mapping between oat that frequently recognizes the linked marker loci and homologous genes. Illustration of distinct mapping comparison with other crops and also among oat genomes depicted several limitations, however, by the complicated segmental polymorphism among the genomes of oat. Repetitive struggle in assigning linkage groups and markers to chromosome for the development of coordinated chromosomal mapping and in lowering the quantity of genetic linkage groups is required to facilitate such comparable mapping efforts.

Among the small-grained cereals, oat represents the unique biological feature of being capable in possessing whole or, with certain arrangements, segments of individual maize chromosomes in wide hybrid crosses (Kynast et al. 2004). Each oat line emerges economically as of potential and biological interest, thereby comprising a segment of the maize genome and providing various new and favorable circumstances in considering the genetic basis of such traits as resistant to particular disease. Establishment of methods in genetic engineering technologies for transferring genes from an organism into oat has been done; however, none of the oat cultivars consisting of such introduced genes have been released at the commercial level. Partly due to insufficient funds, research in oat and the genetic transformation technology has become readily slower in comparison with various major crops. Also careful attention is required in oat for introduction of genes for specific traits which includes resistance to herbicides as the cultivated oat can be easily outcrossed with its weedy wild oat relatives. Moreover, research on transformation of oats is very limited (Perrer et al. 2003; Nuutila et al. 2004) due to continuous improvement in technologies that contributes in making such gene transfer more accurate and presumed to receive the public consent. These gene introductions and transfers are considered to be the important component in recognizing or strengthening the particular gene function for their contribution in research on more typical genetic arrangements. To summarize research in oat, such as molecular breeding, mapping of genomes, and association of genomic research in oat lag, is behind to some extent in comparison with that of various major crops. In the near future, oat's characteristic commercial and biological properties, along with the ability of oat researchers to form the functional and structural genomics and awareness and divulgence in different cereals, make it probable for the establishment and relevant application of molecular breeding to play a significant role in oat genetic enhancement.

6.13 Conclusions

With ever rising demand of oats as grain crop and to fit well into the human platter even in developing world, oat breeders are facing challenge to develop varieties/genotypes most suited for easy milling. Thus, there is an immediate requirement to widen the cultivated oat gene pool by incorporating genes for economically important traits from different sources. Also, there is a need of identifying high-crossability genes. Such identification of genes can bring noncrossable species within the ambit of alien gene transfer technology. There are major gaps in gathering germplasm of wild species and their accession in oats that are required to be filled, in order to develop further inroads in alien gene introgression. Various advancements in wide-crossing techniques, such as establishment of novel crossing strategies and embryo culture, are creating larger approachability in wild oat gene pools. The rate of success in gene transfer within such wide crosses can be increased by awareness of several chromosome pairing mechanisms and their genetic control. The advanced techniques of molecular biology that include in situ hybridization and monoclonal antibodies using several DNA probes may soon make it possible in studying switching on and off of several genes in distinct tissues of the fertilized ovule and managing the levels and motion of both endogenous and exogenous growth materials inside the developing seed.

Oats, a variant of cultivated hexaploid *A. sativa* used both as a traditional grain and for fodder purposes, but without a hull also known are naked oats or hull-less grains, where the caryopsis (groat) gets separated from peripheral plant tissue during the process of threshing. These naked oat varieties offer farmers a dual-purpose cropping option, that is, as a fodder and an alternative source of food for human consumption. Hull-less grains can also be milled into flour for several purposes. Moreover, oat grains constitute higher oil content in comparison with other cereal grains such as wheat. Consequently, the shelf life of oat flour is shorter than wheat flour. Damaged oat grains or milled flour can become rancid, thereby affecting taste, but is not regarded of any distinction for animal feeding. Many countries are producing and progressing in the development of naked oats for its use as a food crop.

In regions having low temperature, diploid oat cultivars are used as an alternate fodder crop. Oat grain constitutes a husk, but there is availability of naked varieties. Recently, several oat cultivars, grown under greenhouse, showed the observations that the diploid cultivar produced a much larger root mass as compared to the traditional oat cultivars. Though, it was not a controlled experiment for root mass measurement, but there were drastic deviations in the diploid and a traditional hexaploid oat root mass. But, under field conditions if diploid cultivars generate larger root mass, then the diploids may be found to be beneficial in controlling erosion, additionally for hay or fodder uses. Moreover, the emphasis of authors shows none of the identification of research references to substantiate this observation. Wild species have demonstrated to be a significant reservoir of useful genes, thus offering great potential for the incorporation of such genes into commercial cultivars. Many useful alien genes are expected to differ from those of the cultivated species and are, therefore, useful in broadening the base of resistance to various stresses.

Nowadays, molecular markers are very well established as equipments or tools in plant breeding and genetics. Also, they have introduced a significant new approach toward plant breeding programs, thereby offering noticeable advancement in the effectiveness and enlightenment of breeding. The major growth challenges for the use of molecular markers lie in establishing the advanced innovative breeding strategies where there will be an increase in the objectives to enhance the germplasm base and increment of numerous traits that can be intrinsically selected simultaneously. The main focus of these developments will be on offering the new marker technologies with great reduction in costs of marker screening and high multiplexing capabilities. Eventually, the main movement will be toward the whole genome-based selection techniques or strategies where specific recombination events are acquired and alterations will be assessed on modifications of a genome-wide scale. In this way, researchers can look for better management of chromosome regions that may be evolved from landraces or wild relatives and trace various characteristics or traits at once and also keep the size of the population as small as possible. Hence, the availability of genetic linkage maps based on molecular markers has led to the opening of advanced opportunities in successful achievement of utilization of germplasm. Genome analysis based on molecular markers has created a huge amount of information, and numerous databases are being produced to preserve and popularize it.

References

Abdalla MMF, Hermsen JGT (1972) Unilateral incompatibility: hypotheses, debate and its implications for plant breeding. Euphytica 21:32–47

Achleitner A, Tinker NA, Zechner E, Buerstmayr H (2008) Genetic diversity among oat varieties of worldwide origin and associations of AFLP markers with quantitative traits. Theor Appl Genet 117(7):1041–1053

Ananiev EV, Vales MI, Phillips RL, Rines HW (2002) Isolation of A/D and C genome specific dispersed and clustered repetitive DNA sequences from *Avena sativa*. Genome 45:431–441

Andow DA, Zwahlen C (2006) Assessing environmental risks of transgenic plants. Ecol Lett 9:196–214

Baohong G, Zhou X, Murphy JP (2003) Genetic variation within Chinese and western cultivated oat accessions. Cereal Res Commun 31(3–4):339–346

Bates LS (1975) Chemical manipulation of crossability barriers. In: Gaul H (ed) Proceedings of 3rd international barley genetics symposium. Verlag Karl Thiemig, Munich, pp 271–273

Beckie HJ, Hall LM, Warwick SI (2001) Impact of herbicide-resistant crops as weeds in Canada. In: Proc Brighton crop protection conference–weeds. British Crop Protection Council, Farnham, pp 135–142

Bennett MD, Smith JB (1976) Nuclear-DNA amounts in angiosperms. Philos Trans R Soc B 274(933):227–274

Carson M (2008) Virulence frequencies in oat crown rust in the United States from 2001 through 2005. Plant Dis 92:379–384

Chen Q, Armstrong K (1994) Genomic *in situ* hybridization in Avena Sativa. Genome 37:607–612

Close TJ, Wanamaker SI, Caldo RA et al (2004) A new resource for cereal genomics: 22K barley GeneChip comes of age. Plant Physiol 134:960–968

Collard BCY, Jahufer MZZ, Brouwer JB, Pang ECK (2005) An introduction to markers, quantitative trait loci (QTL) mapping and marker assisted selection for crop improvement: the basic concepts. Euphytica 142:169–196

Comeau A (1982) Geographic distribution of resistance to barley yellow dwarf virus in *Avena sterilis*. Can J Plant Pathol 4:147–151

Cudderford D (1995) Oats for animal feed. In: Welch RW (ed) The oat crop: production and utilization. Chapman and Hall, London, pp 321–368

De Koeyer DL, Phillips RL, Stuthman DD (2001) Allelic shifts and quantitative trait loci in a recurrent selection population of oat. Crop Sci 41:1228–1234

De Koeyer DL, Tinker NA, Wight CP et al (2004) A molecular linkage map with associated QTLs from a hulless x covered spring oat population. Theor Appl Genet 108:1285–1298

Drossou A, Katsiotis A, Leggett JM, Loukas M, Tsakas S (2004) Genome and species relationships in genus Avena based on RAPD and AFLP molecular markers. Theor Appl Genet 109:48–54

Fominaya A, Hueros G, Loarce Y, Ferrer E (1995) Genome 38:548–557

Forsberg RA, Reeves DL (1992) Breeding oat cultivars for improved grain quality. In: Marshalls HG, Sorrels ME (eds) Oat science and technology. American Society of Agronomy and Crop Science Society of America, Madison, pp 751–775

Frey KJ, Browning JA, Grindeland RL (1973) Registration of X117 oats germplasm. Crop Sci 13:290

Fu YB, Williams DJ (2008) AFLP variation in 25 Avena species. Theor Appl Genet 117(3):333–342

Fu YB, Kibite S, Richards KW (2004) Amplified fragment length polymorphism analysis of 96 Canadian oat cultivars released between 1886 and 2001. Can J Plant Sci 84:23–30

Fu YB, Peterson GW, Williams D, Richards KW, Fetch JM (2005) Patterns of AFLP variation in a core subset of cultivated hexaploid oat germplasm. Theor Appl Genet 530:530–539

Fu YB, Chong J, Fetch T, Wang ML (2007) Microsatellite variation in germplasm accessions of the wild oat *Avena sterilis* L. Theor Appl Genet 114:1029–1038

Grafius JE, Brown HM (1954) Lodging resistance in oats. Agron J 46:414–418

Gupta PK, Varshney RK, Sharma PC, Ramesh B (1999) Molecular markers and their applications in wheat breeding. Plant Breed 118:369–390

Gupta PK, Roy JK, Prasad M (2001) Single nucleotide polymorphisms (SNPs): a new paradigm in molecular marker technology and DNA polymorphism detection with emphasis on their use in plants. Curr Sci 80:524–536

Gupta PK, Rustgi S, Mir RR (2008) Array-based high-throughput DNA markers for crop Improvement. Heredity 101:5–18

Hall D, Tegstrom C, Ingvarsson PK (2010) Using association mapping to dissect the genetic basis of complex traits in plants. Brief Funct Genomics 9(2):157–165

Harlan JR (1976) Genetic resources in wild relatives of crops. Crop Sci 16:329–333

Harlan JR, de Wet JMJ (1971) Toward a rational classification or cultivated plants. Taxon 20:509–517

He X, Bjornstad A (2012) Diversity of North European oat analyzed by SSR, AFLP and DArT markers. Theor Appl Genet 125(1):57–70. doi:10.1007/s00122-012-1816-8

Hermsen JG (1966) Crossability, fertility and cytogenetic studies in *Solanum acaule* x *S. bulbocastanum*. Euphytica 15:149–155

Hermsen JG (1979) Factors controlling interspecific crossability and their bearing on the strategy for breaking barriers to intercrossing of Solanum species. In: Proceedings of the conference on broadening the genetic base of crops, Wageningen, pp 311–318

Hermsen JG, Olsder J, Jansen P, Hoving E (1974) Acceptance of self-compatible pollen from Solanum verrucosum in dihaploids from *S. tuberosum*. In: Linskens HF (ed) Fertilization in higher plants. American Elsevier, New York, pp 37–40

Hermsen W, Bennett K, Bignami GF (1977) Nature 269:494

Hogenboom NG (1972) Breaking breeding barriers in Lycopersicon. The inheritance of the unilateral incompatibility between L. peruvianum L. and L. esculantum Mill. and the genetics of its breakdown. Euphytica 21:405–414

Hogenboom NG (1973) A model for incongruity in intimate partner relationships. Euphytica 22:219–233

Hogenboom NG (1975) Incompatibility and incongruity: two different mechanisms for the non-functioning of intimate partner relationships. Proc R Soc Lond B 188:361–375

Iannucci A, Codianni P, Cattivelli L (2011) Evaluation of genotype diversity in oat germplasm and definition of ideotypes adapted to the Mediterranean environment. Int J Agron 2011:8. doi:10.1155/2011/870925

Jannink JL, Gardner SW (2005) Expanding the pool of PCR-based markers for oat. Crop Sci 45(6):2383–2387

Jedlinski H, Rochow WF, Brown CM (1977) Tolerance to barley yellow dwarf virus in oats. Phytopathology 67:1408–1411

Jellen E, Leggett JM (2006) Cytogenetic manipulation in oat improvement. In: Singh RJ, Jensen CF (1976) Embryo culture as an implement in hybridization, haploid production and in genetical and developmental studies. News Int Ass Pl Tissue Cult 18:10

Jellen EN, Phillips RL, Rines HW (1993) Genome 36:1129–1137

Jensen CF (1976) Embryo culture as an implement in hybridization, haploid production and in genetical and developmental studies. News Int Ass Pl Tissue Cult 18:10

Karp A (2002) The new genetic era: will it help us in managing genetic diversity? In: Engels JMM, Ramanatha Rao V, Brown AHD, Jackson MT (eds) Managing plant genetic diversity. CAB International/IPGRI, Wallingford/Rome, pp 43–56

Klinger T, Arriola PE, Elistrand NC (1992) Crop-weed hybridization in radish *(Raphanus sativus)*: effects of distance and population size. Am J Bot 79(12):1431–1435

Kynast RG, Okagaki RJ, Galatowitsch MW et al (2004) Dissecting the maize genome by using chromosome addition and radiation hybrid lines. Proc Natl Acad Sci U S A 101:9921–9926

Ladizinsky G (1989) Ecological and genetic considerations in collecting and using wild relatives. In: Brown AHD, Frankel OH, Marshall DR, Williams JT (eds) The use of plant genetic resources. Cambridge University Press, Cambridge, pp 297–305

Ladizinsky G (1995) Domestication via hybridization of the wild tetraploid oats *Avena magna* and *Avena murphyi*. Theor Appl Genet 91:639–646

Ladizinsky G, Feinstein R (1977) Introgression between the cultivated hexaploid oat *Avena sativa* and the tetraploid wild *A. magna* and *A. murphyi*. Can J Genet Cytol 19:59–66

Ladizinsky G, Jellen EN (2003) Cytogenetic affinities between populations of *Avena Insularis* Ladizinsky from Sicily and Tunisia. Genet Resour Crop Evol 50:11–15

Leggett JM, Markhand GS (1997) In Aberystwyth cell genetics group 7th annual conference. Institute of Grassland and Environmental Research, Aberystwyth, p 45

Levin D, Kerster HW (1974) Gene flow in seed plants. Evol Biol 7:139–220

Lewis D, Crowe LK (1958) Unilateral interspecific incompatibility in flowering plants. Heredity 12:233–256

Li CD, Rossnagel BG, Scoles GJ (2000) The development of oat microsatellite markers and their use in identifying relationships among *Avena* species and oat cultivars. Theor Appl Genet 101:1259–1268

Li RZ, Wang SW, Duan LS, Li ZH, Christoffers MJ, Mengistu LW (2007) Genetic diversity of wild oat (Avena fatua) populations from China and the United States. Weed Sci 55(2):95–101. doi:10.1614/ws-06-108.1

Lim HS, White PJ, Frey KJ (1992) Genotypic effects of β-glucan content of oat lines grown in two consecutive years. Cereal Chem 69:262–265

Linares C, Ferrer E, Fominaya A (1998) Discrimination of the closely related A and D genomes of the hexaploid oat *Avena sativa* L. Proc Natl Acad Sci U S A 95:12450–12455

Loskutov IG (2001) Interspecific crosses in the genus *Avena*. Russ J Genet 37:581–590

Loskutov IG (2005) Classification and diversity of the genus *Avena* L. In: Lipman E, Maggioni L, Knüpffer H et al (eds) Cereal genetic resources in Europe. Report of a cereals network/report of a working group on wheat, first meeting, Yerevan, Armenia, 3–5 July 2003, Second Meeting, 22–24 September 2005, La Rochelle, France, International Plant Genetic Resources Institute (IPGRI), Rome, Italy, pp 84–90

Marshall HG (1992) Breeding oats for resistance to environmental stress. In: Marshalls HG, Sorrels ME (eds) Oat science and technology. American Society of Agronomy and Crop Science Society of America, Madison, pp 699–749

McFerson JK, Frey KJ (1990) Three selection strategies to increase protein yield in oats. J Genet Breed 44:39–48

Nersting LG, Andersen SB, von Bothmer R, Gullord M, Jorgensen RB (2006) Morphological and molecular diversity of Nordic oats through one hundred years of breeding. Euphytica 150(3):327–337

Nettancourt D de (1977) Incompatibility in angiosperms. Mono Theoretical and Appllied Genetics 3. Springer, Berlin, p 230

Nocelli E, Giovannini T, Bioni M, Alicchio R (1999) RFLP- and RAPD based genetic relationships of several diploid species of *Avena* with the A genome. Genome 42:950–959

Nuutila AM, Lehto K, Oksman-Caldentey KM (2004) Transgenic oat for improved BYDV resistance. In: Peltonen-Saino P, Topi-Hulmi M (eds) Proceedings of the 7th international oat conference, Helsinki, Finland, p 88. www.mtt.fi/met/pdf/met51.pdf

O'Donoughue LS, Kianian SF, Rayapati PJ et al (1995) A molecular linkage map of cultivated oat. Genome 38:368–380

Ohm HW, Shaner HG (1992) Breeding oats for resistance to diseases. In: Marshalls HG, Sorrels ME (eds) Oat science and technology. American Society of Agronomy and Crop Science Society of America, Madison, pp 657–698

Paczos-Grzeda E (2004) Pedigree, RAPD and simplified AFLP based assessment of genetic relationships among *Avena sativa* L. Cultivars. Euphytica 138(1):13–22

Pal N, Sandhu JS, Domier LL, Kolb FL (2002) Development and characterization of microsatellite and RFLP-derived PCR markers in oat. Crop Sci 42:912–918

Penner GA, Bush A, Wise R et al (1993) Reproducibility of random amplified polymorphic DNA (RAPD) analysis among laboratories. PCR Methods Appl 2:341–345

Perrer SP, Valentine J, Leggett MJ, Morris P (2003) Integration, expression and inheritance of transgenes in hexaploid oat (*Avena sativa* L.). J Plant Physiol 160:931–943

Peterson DM (2001) Oat antioxidants. J Cereal Sci 33:115–129

Peterson DM (2004) Oat – a multifunctional grain. In: Peltonen-Saino P, Topi Hulmi M (eds) Proceedings of the 7th international oat conference, Helsinki, Finland, pp 21–26. www.mtt.fi/met/pdf/met51.pdf

Pixley KV, Frey KJ (1991) Combining ability for test weight and agronomic traits of oat. Crop Sci 31:1448–1451

Qi X, Bakht S, Leggett M, Maxwell C, Melton R, Osbourn A (2004) A gene cluster for secondary metabolism in oat: implications for the evolution of metabolic diversity in plants. Proc Natl Acad Sci U S A 101:8233–8238

Rafalski JA (2010) Association genetics in crop improvement. Curr Opin Plant Biol 13(2):174–180

Raghavan V (1977) Applied aspects of embryo culture. In: Reinert J, Bajaj YPS (eds) Applied and fundamental aspects of plant cell, tissue and organ culture. Springer, Berlin, pp 325–397

Rajhathy T, Sadasivaiah RS (1969) The cytogenetic status of Avena magna. Can J Genet Cytol 11:77–85

Rajhathy T, Baum BR (1972) *Avena damascena*: a new diploid oat species. Can J Genet Cytol 14:645–654

Rajhathy T, Thomas H (1974) Cytogenetics of Oats (Avena L.). Miscellaneous Publications of the Genetics Society of Canada 2. Genetics Society of Canada, Ottawa, pp 1–90

Ramanatha Rao V, Hodgkin T (2002) Genetic diversity and conservation and utilization of plant genetic resources. Plant Cell Tiss Org Cult 68:1–19

Rangaswamy NS (1977) Application of *in vitro* pollination and *in vitro* fertilization. In: Reinert J, Bajaj YPS (eds) Applied and fundamental aspects of plant cell, tissue and organ culture. Springer, Berlin, pp 413–415

Raven P (1980) Hybridization and the nature of species in higher plants. Can Bot Assoc Bull Suppl 13:3–10

Riley R, Chapman V, Johnson R (1968) Introduction of yellow rust resistance of *Aegilops comosa* into wheat by genetically induced homoeologous recombination. Nature 216:383–384

Rodionov AV, Tyupa NB, Kim ES (2005) Genomic structure of the autotetraploid oat *Avena macrostachya* inferred from comparative analysis of ITS1 and ITS2 sequences: on the oat karyotype evolution during the early events of the Avena species divergence. Russ J Genet 41:1428–1433

Saiki RK, Gelfand DH, Stoffei S et al (1988) Primer-directed enzymatic amplification of DNA with a thermostable DNA polymerase. Nature 239:487–497

Saiki RK, Scharf S, Faloona F, Mullis KB, Horn GT, Erlich HA, Arnheim N (1995) Enzymatic amplification of beta-globin genomic sequences and restriction site analysis for diagnosis of sickle cell anemia. Science 230:1350–1354

Sears ER (1956) The transfer of leaf-rust resistance from *Aegilops umbellulata* to wheat. Brookhaven Symp Biol 9:1–21

Simons MD, Rothmann PG, Michel LJ (1979) Pathogenicity of *Puccinia coronata* from buckthorn and from oats adjacent to and distant from buckthorn. Phytopathology 69:156–168

Stebbins GL (1971) Processes of organic evolution, 2nd edn, Chapter 5. Prentice Hall, Inc., Englewood Cliffs, pp 88–115

Stewart D, McDougall G (2014) Oat agriculture, cultivation and breeding targets: implications for human nutrition and health. Br J Nutr 112:S50–S57

Thomas H (1989) New Species of *Avena*, Proceedings of the 3rd international oat conference, Lund, Sweden, pp 18–23

Thomas H (1992) Cytogenetic of *Avena*. In: Marshall HG, Sorrels ME (eds) Oat science and technology, agronomy monograph number 33. American Society of Agronomy and Crop Science Society of America, Madison, WI, pp 473–508

Tinker NA, Kilian A, Wight CP et al (2009) New DArT markers for oat provide enhanced map coverage and global germplasm characterization. BMC Genomics 10:39. doi:10.1186/1471-2164-10-39

Warwick SI, Stewart CN Jr (2005) Crops come from wild plants—how domestication, transgenes, and linkage together shape ferality. In: Gressel J (ed) Crop ferality and volunteerism. CRC Press, Boca Raton, pp 9–30

Warwick SI, Beckie HJ, Simard JJ, Legere A, Nair H, Seguin-Swartz G (2004) Environmental and agronomic consequences of herbicide-resistant (HR) canola in Canada. In: den Nifs HCM, Bartsch D, Sweet J (eds) Introgression from genetically modified plants (GMP) into wild relatives. CABI Publ, Wallingford, pp 323–337

Watson IA (1970) The utilisation or wild species in the breeding of cultivated crops resistant to plant pathogenes. In: Frankel OH, Bennett E (eds) Genetic resources in plants – their exploration and conservation. Blackwell Science Publication, Oxford, pp 441–457

Weising K, Nybom H, Wolff K, Meyer W (1995) DNA fingerprinting in plants and fungi. In: Arbor A (ed) CRC Press, Boca Raton, pp 1–3

Welch RW, Brown JCW, Leggett JM (2000) Interspecific and intraspecific variation in grain and groat characteristics of wild oat (*Avena*) species: very groat (1–3), (1–4)-β glucan in an *Avena atlantica* genotype. J Cereal Sci 31:273–279

Wesenberg DM, Shands HL (1971) Caryopsis percentage and related characters in early generations of *Avena sativa* L. Crop Sci 11:586–588

Xu Y (2010) Molecular plant breeding. CAB International, Wallingford

Yu J, Herrmann M (2006) Inheritance and mapping of a powdery mildew resistance gene introgressed from *Avena macrostachya* in cultivated oat. Theor Appl Genet 113(3):429–437

Zhu S, Kaeppler HF (2003) A genetic linkage map for hexaploid, cultivated oat (*Avena sativa* L.) based on an intraspecific cross 'Ogle/MAM17-5'. Theor Appl Genet 107:26–35

Zhu S, Kolb FL, Kaeppler HF (2003) Molecular Mapping of genomic regions underlying barley yellow dwarf tolerance in cultivated oat (Avena sativa L.). Theor Appl Genet 106:1300–1306

Sorghum

7

Pummy Kumari, S.K. Pahuja, Satyawan Arya, and J.V. Patil

Abstract

Sorghum [*Sorghum bicolor* (L.) Moench] is called camel's crop and it is a major dryland cereal crop which has multipurpose uses like food, feed, fodder, and bioethanol production. It is nutritionally superior to other fine cereals such as rice and wheat with high fiber and mineral content. As a model for the tropical grasses, sorghum is a logical complement to rice. Sorghum is a representative of tropical grasses in that it has C_4 photosynthesis with biochemical and morphological specializations that improve net carbon assimilation at high temperatures. Sorghum and maize share a common ancestor due to polyploidization and repetitive DNA propagation; however, the sorghum genome is much smaller (736 Mbp) and was completely sequenced in 2009. Sorghum is an even closer relative of sugarcane, arguably the most important biomass/biofuel crop worldwide. Gene flow between cultivated plants and their wild/weedy relatives play an important role in structuring the genetic variability within and among populations. The consequences of gene flow can contribute to the scientific basis (risk assessment) for managing agricultural systems, understanding evolutionary processes, and designing in situ conservation measures for genetic resources and using these resources to secure current and future plant breeding programs. Gene flow is of practical concern in crop breeding, weed evolution, or transgene movement from crops to weeds. While gene flow might pertain to seed movement with regard to volunteer plants or type "contamination," most research has been per-

P. Kumari • S.K. Pahuja (✉) • S. Arya
Department of Genetics and Plant Breeding, CCS Haryana Agricultural University, Hisar 125004, India
e-mail: pahujask66@gmail.com

J.V. Patil
Indian Institute of Millet Research, Hyderabad 500 030, India

M. Singh, S. Kumar (eds.), *Broadening the Genetic Base of Grain Cereals*, DOI 10.1007/978-81-322-3613-9_7

formed on assessing gene flow via pollen movement, with transgenes being important and convenient markers to assay. Sorghum germplasm screening for various antioxidants and nutritional qualities which might be helpful in control of various diseases is now-a-days very easy with the help of next-generation sequencing (NGS) techniques and other new sequencing strategies like genotyping by sequencing (GBS), digital genotyping (DG), etc. Keeping in view its genetic variability, a known genomic sequence, and a good seed industry, utilization of sorghum within the health food market and the bioenergy arena will make sorghum a promising renewable resource for generations to come. The present chapter reviews the evolutionary aspects of sorghum with special emphasis on the gene pool and flow for crop improvement, considering levels of diversity, traits of importance, and interspecific hybridization and concludes with molecular aspects and future prospects of sorghum breeding.

Keywords
Sorghum • Evolution • Gene pool • Genetic diversity • Wide hybridization

7.1 Introduction

Sorghum [*Sorghum bicolor* (L.) Moench] is the world's fifth most important cereal crop, valued for its grain, stalks, and leaves. Its small diploid genome ($2n = 20$) and phenotypic diversity makes it an ideal C_4 grass model as a complement to C_3 rice (Mace et al. 2013). It is known by many names: *milo*, *guinea corn* in West Africa, *kafir corn* in South Africa, *durra* in Sudan, *mtama* in eastern Africa, *jowar* in India, and *kaoliang* in China. It belongs to the family *Poaceae* and first described by Linnaeus in 1753 under the name *Holcus*. Moench later separated the genus *Sorghum* from the genus *Holcus* (Clayton 1961), being major cereal crops of arid and semiarid tropics having Africa as its center of diversity, The major sorghum-growing countries are Nigeria, India, Mexico, Sudan, China, and Argentina. Rice, wheat, maize, and potatoes are consumed more than sorghum and these main crops supply more than 80 % world's food energy. Sorghum is called "camel among crops" because of its ability to withstand prolonged droughts that is why it is primarily grown in arid and semiarid regions of the word. It is a major dietary staple for more than 500 million people in over 30 countries of the SAT regions (Cordeiro et al. 2006). The crop plays a major role in the food security of millions of people in marginal agricultural areas. Sorghum is used not only for human food but also for feed and fodder for livestock and as fuel for bioethanol production, making it a multipurpose crop.

Sorghum is unique in the sense that it is naturally drought, heat, and insect resistant. It thrives in arid areas, which makes sorghum increasingly important globally. The International Water Management Institute (IWMI) warns that, by the year 2025, 25 % of the world's population will experience severe water scarcity, and drought-tolerant crops such as sorghum will be important in meeting the food demands for those people. Thus, keeping all the points in view, this chapter reviews and discusses gene pool, evolutionary relationship and systematics, assessment of gene flow for crop improvement, levels of diversity, interspecific hybridization, molecular markers, genome mapping, and genomics.

7.2 Crop Gene Pool, Evolutionary Relationships, and Systematics

7.2.1 Origin and Distribution

Sorghum is an ancient crop and it is difficult to determine when and where domestication took place (de Wet et al. 1970). Murdock (1959) has suggested that the Mande people around the headwaters of the Niger River may have domesticated sorghum. The origin and early domestication of sorghum is hypothesized to have take place in Northeastern Africa, a part of Abyssinian Center – Ethiopia present in north of the Equators and 10°E latitude at east, approximately 5000 years ago (Mann et al. 1983). In an archeological survey, sorghum's carbonized seeds with radiocarbon treatment date back 8000 years BC at Nabta Playa near the Egyptian-Sudanese border presently (Dahlberg and Wasylikowa 1996 and Wendorf et al. 1992). In Africa at the Egyptian-Sudanese border, the largest diversity was found among wild and cultivated sorghum cultivars. Major migratory and trading paths of early Africans and Asians had followed the early domestication events. As this early domesticated sorghum spread throughout Africa and Asia, *Sorghum* plants were selected and dispersed throughout a broad range of environments via disruptive selection and their multipurpose use has widened their genetic base that has been further exploited throughout the agricultural process to develop the presently cultivated sorghum. It is supposed that African slaves brought sorghum seeds with them to the USA; thus, the crop got introduced to the largest sorghum-growing country at present. Vavilov (1992) and Damania et al. (1998) suggested that the Indian subcontinent is the secondary center of origin, with evidence for early cultivation in western parts of Rojdi (Saurashtra) dating back to about 4500 years.

De Wet et al. (1970) suggested that sorghum has diverse origin and might have arisen from *S. verticilliflorum* because of its large occurrence in areas of sorghum cultivation. There is tremendous variation in *S. verticilliflorum*. The races *durra*, *guinea*, and *kafir* are closely allied and might have arisen from *S. aethiopicum*, S. *arundinaceum*, and S. *verticilliflorum*, respectively, but morphological differences among races may have arisen because of ethnic isolation (Snowden 1936; Proteres 1951). Various studies including introgression also have indicated that present-day cultivated sorghums might have probably developed through disruptive selection (Doggett 1965).

Outside Africa, a specialized type of sorghum called broomcorn is mainly grown which was first cultivated and utilized in Europe and America, whereas later on, other cultivated sorghum types were introduced directly or indirectly from Africa (Franklin 1757; Doggett 1988). The Mediterranean region is believed to be the primarily origin of broomcorn in Europe, more precisely Italy. Becker-Dillingen (1927) reported that the broomcorn type was grown in the Piedmont region of Italy as early as 1204. Broomcorn had widespread uses in Italy up to the end of the seventeenth century when Bauhini (1658) reported that the rigid and slender dried sorghum heads were made into brooms by the Italians and used for brushing clothing in France, Germany, and Italy. Ray in 1688 (Washburn and Martin 1933) presented a full discussion on sorghum and recorded the use of its plant as corn brooms on sale in Venice and Italy. In the sixteenth century, it was moved from Italy to the Danube river basin region, spread widely into the Austro-Hungarian Empire, and finally reached to the Pannonian Plain, most probably in the eighteenth century (Sávoly 1921). Körnicke (1885) mentioned that in Italy, Croatia, and Hungary "*sorghum technicus*" was produced widely. There are documents bearing witness that broomcorn had been grown in Northern Serbia in the 1860s. In the last decades of 1700s, its uses were widespread in Europe and had attracted the interest of Benjamin Franklin, who first documented its utilization in the USA in a letter to Samuel Ward on March 24, 1757 (Martin and Leonard 1949).

7.2.2 Crop Domestication

Early domestication of sorghum commenced around 4000–3000 BC in Ethiopia and surrounding countries, but later on, various varieties of sorghum were developed through hybridization, disruptive selection, and trait-specific breeding for crop improvement (Doggett 1970). Farmers' selection had also played a major role in the development of improved cultivars of sorghum via natural selection for wild characteristics along with cultivated traits (Doggett 1970). Improved sorghum cultivars were spread via trade routes and movement of people into other regions of Africa, the Middle East (approx. 900–700 BC), and India (approx. 1500–1000 BC). Disruptive selection, geographic isolation, hybridization, and recombination of these types in different environments have led to the generation of a broad range of diversity among new sorghum types, varieties, and races (Wright 1931; Doggett 1970).

Initial focus of sorghum domestication was primarily on converting wild types with small, shattering (dehiscent) seed to improved types with larger, non-shattering seed. Disruptive selection resulted in sorghum types with vastly different characteristics in height, inflorescence type, and, of course, end uses (food, fodder, fiber, building materials, etc). Various taxonomists have described sorghum which is now placed under the family Poaceae, tribe Andropogoneae, subtribe Sorghinae, and genus *Sorghum* Moench (Clayton and Renvoize 1986).

7.2.3 Gene Pool

Domesticated *Sorghum* and its close relatives are classified/divided into three gene pools:

7.2.3.1 Primary Gene Pool

Nowadays, sorghum is a staple cereal in sub-Saharan Africa, a region of its domestication. The primary gene pool of *Sorghum* includes only two diploid species, *S. bicolor* and *S. propinquum* (Harlan and deWet 1971). The species *Sorghum bicolor* has been subdivided into three subspecies based on spikelet morphology: *Sorghum bicolor* ssp. *drumondii* (the weedy forms, which are stabilized hybrids between wild relatives and cultivated forms), *Sorghum bicolor* ssp. *verticilliflorum* (the closest wild relatives), and *Sorghum bicolor* ssp. *bicolor* (the domesticated forms). These wild sorghum types which are fully interfertile with cultivated forms grow sympatrically with landraces in various sorghum-growing regions of sub-Saharan Africa (de Wet et al. 1970; de Wet 1978; Tesso et al. 2008).

The *Sorghum* genus as currently prescribed consists of 31 cultivated species and 17 related wild species (Snowden 1936, 1955). But as per USDA ARS 2007, *Sorghum* genus comprises 25 species, but this varies in different scientific publications confirming the dynamic nature of the classification of wild relatives and cultivated sorghum. The genus is divided into five taxonomic subgenera or sections, i.e., *Eu-Sorghum*, *Chaetosorghum*, *Heterosorghum*, *Para-sorghum*, and *Stiposorghum* (Garber 1950). The four sections, *Chaetosorghum*, *Heterosorghum*, *Para-sorghum*, and *Stiposorghum*, comprise 19 wild species native to Africa, Asia, and Australia (Garber 1950; Lazarides et al. 1991). These species contain novel sources of genetic diversity for agronomic traits affecting yield, survivability, and important traits that may create new markets for sorghum products which are briefly outlined below: *Chaetosorghum* and *Heterosorghum* are found as single species primarily in Australia and the South Pacific and include the octoploid ($2n=40$) Australian species *S. macrospermum* E.D. Garber and *S. laxiflorum* F.M. Bailey, respectively. *Para-sorghum* includes seven species found in the eastern hemisphere and Central America, with five Australian species *S. grande* Lazarides, *S. leiocladum* (Hack.) C.E. Hubb., *S. matarankense* E.D. Garber and Snyder, *S. nitidum* (Vahl) Pers., and *S. timorense* (Kunth) Buse and the two African/Asian species *S. purpureo-sericeum* (Hochst. ex A. Rich.) Asch. and Schweinf. and *S. versicolor* Andersson. These species range in ploidy from $2n=10$ to $2n=40$, with *S. grande*, *S. nitidum*, and *S. timorense* showing different ploidy levels within species. *Stiposorghum* includes 10 species endemic to

Australia (Dahlberg 2000). Section *Parasorghum* also comprise of Australian endemic species from section *Stiposorghum*: *Sorghum amplum* Lazarides, *S. angustum* S.T. Blake, *S. brachypodum* Lazarides, *S. bulbosum* Lazarides, *S. ecarinatum* Lazarides, *S. exstans* Lazarides, *S. interjectum* Lazarides, *S. intrans* F. Muell. ex Benth., *S. plumosum* (R. Br.) P. Beauv., and *S. stipoideum* (Ewart & Jean White) C.A. Gardner & C.E. Hubb. Most of these species are diploid except *S. interjectum* and *S. plumosum* which are polyploid (Garber 1950; Lazarides et al. 1991).

Eu-Sorghum contains all cultivated, domesticated races and varieties such as *Sorghum bicolor* subsp. *bicolor*, as well as all weedy and wild species *S. propinquum* (Kunth) Hitchc, *S. halepense* (L.) Pers. (Johnson grass), *S. arundinaceum* (Desv.) Stapf. (progenitor of *S. bicolor*), *S. almum* Parodi, and *S. drummondii* (Steud.) Millsp. & Chase (Doggett 1988). *S. bicolor* subsp. *bicolor* ($2n=20$) described as annual, having thick culms up to 5 m, often branched with many tillers. They are classified into five basic races: *bicolor*, *guinea*, *caudatum*, *kafir*, and *durra*, with ten intermediate races, namely, guinea-bicolor, caudatum-bicolor, kafir-bicolor, guinea-caudatum, durra-bicolor, guinea-kafir, guinea-durra, kafir-caudatum, durra-caudatum, and kafir-durra (Harlan and deWet 1972). Intermediates that resulted from hybridization of races exhibit characters of both the parents, are recognized on the basis of spikelet or panicle morphology alone, and either linked back to its specific environments of first cultivation (Smith and Frederiksen 2000).

Experiments carried out by Snowden (1936) and Harlan (1995) confirmed that "all of the races belong to the same biological species and are fully fertile when hybridized." For the cereal sorghums, they identified four wild races and five cultivated races (Harlan and Stemler 1976). The wild races of *S. bicolor* are *virgatum*, *arundinaceum*, *aethiopicum*, *and verticilliflorum*. They are now placed in *S. bicolor* subspecies *verticilliflorum*, instead of formerly subspecies *arundinaceum*. The presently conceived cultivated races placed in *S. bicolor* subsp. *bicolor* are (1) *bicolor*, the primitive type, (2) *guinea*, (3) *kafir*, (4) *caudatum*, and (5) *durra*. Snowden (1936) described that sorghum had separate centers of origins for different types. According to Snowden, wild race *aethiopicum* gave rise to races *durra* and *bicolor*, *arundinaceum* to *guinea*, and *verticilliflorum* to *kafir*. de Wet and Huckabay (1967) also had the same opinion except that *durras* came out of *kafirs*. A huge diversity present in the wild relative forms reflects intervention associated with human manipulation along with the selection of domesticated types Doggett (1965).

Today, *bicolor* is distributed widely not only in African regions but also in Asia, coastwise from India to Indonesia to China (de Wet and Price 1976). How *bicolor* migrated out of Africa is unknown, but its diffusion might have taken place with migration of the people to different regions. Cultivated sorghum may have reached China from Indochina by way of the Mekong River or other river valleys. Hawkes (1973) asserted that Semitic speakers from Africa carried their culture to India before 3000 BC. This may have been one avenue on which sorghum moved off the coast of Africa and to India.

Undomesticated *Sorghum* species were adapted across a wide range of microenvironments such as ability to survive in very hot, dry, nutrient-limited environments. Due to their wide range of adaptability, various undomesticated sorghum species had developed pest and diseases resistance that affect sorghum grain production globally. Various Australian undomesticated species also had resistance to major pests/diseases of Africa and America, which are not yet preserved within their native place (Bapat and Mote 1982; Karunakar et al. 1994; Franzmann and Hardy 1996; Sharma and Franzmann 2001; Komolong et al. 2002; Kamala et al. 2002).

Harlan and Stemler (1976) considered the *guineas* to be the oldest of the specialized races because of its relatively wide distribution and diversity. These are better adapted to atmospheric conditions that are more wet and humid than other domesticated races. Their relatively lax panicle provides space for the movement of air among the seeds on individual panicle branches. Other scientists have concluded that, as *bicolor* moved west, these came into contact with wild *S.*

arundinaceum and are now found with it in mixed populations, thus picking up some adaptive genetic material through introgression. These mixed populations further provide the base for evolution of race *guinea*. All three biotypes of *guinea* can be found in feral populations today, especially across the African savannas. Plants of this race can tolerate up to 5000 mm of rainfall. *Guineas* first moved into East Africa and humid, foggy Southeastern Africa and then were transported, probably from East African ports to the Malabar Coast of India (Vishnu-Mittre 1974; Harlan and Stemler 1976; Kimber 2000).

According to Stemler et al. (1975), race *caudatum* is domesticated later than *bicolor* and *guinea*, having been segregated out of *bicolor* in the ancestral home territory of race. *Caudatum* is associated with the group of people speaking Chari-Nile language. Stemler and her associates concluded that *caudatums* of the Ethiopian highlands are intrusive, probably acquired by trade. These plants, adapted to harsh conditions, are found most commonly in areas receiving 250–1300 mm of rain annually (Stemler et al. 1975, 1977).

De Wet (1978) and Harlan et al. (1976) concluded that race *kafir* was derived from an early *bicolor* race, which had been introduced to east from the Savanna belt. Electrophoresis data collected by Schecter and de Wet (1975) suggested that the *kafirs* are more closely associated with wild species *verticilliflorum*. The *kafir* sorghums are very much associated with the Bantu-speaking people of Eastern and Southeastern Africa. It is known that the Bantu entered this part of Africa from the western forested regions to the north and may have migrated before the *guinea* race was segregated from the race *bicolor*. *Kafir* may have been taken to the Indian plateau edge ports after discovery of the monsoon wind systems by traders in the western Indian Ocean. This route is different from the route via the northern ports, so it may be a later development than the movement of *bicolors* to the Indus Valley and Punt, or it may be a second migration into the southern plateau area, preceded by *bicolor*.

Durra race, the most important grain in Ethiopia, is also found strictly in north of the equator in Africa. It is an important sorghum type in India and may have been domesticated (Harlan and Stemler 1976). Its compact panicle and white seeds predominantly adapted to low-rainfall environments with a low risk of grain mold (Mann et al. 1983). Its name derived from an Arabic root. The *durras* were almost entirely cultivated by Muslim Africans and Arabic people in Ethiopia. Harlan et al. (1973) reported that the main growers of *durra* sorghum in Ethiopia are the Muslim Oromo (Gallo), who settled the fertile warm highland almost 500 years ago and used race *durra* sorghum as the foundation of their agricultural system. *Durras* are presently distributed in the mid-altitude highlands of Ethiopia, the Nile Valley of Sudan and Egypt, and in belt 10–15° north latitude from Ethiopia to Mauritania. They are also grown in the Islamic and Hindu areas of India and Pakistan. The kaoliangs (galiangs) of China are also thought to be derived from races of *bicolor* introduced from India to China (Harlan 1995). Alternatively, they may be derived from wild diploid sorghums with which they were compatible (Harlan 1995). At one time, they were considered native to Manchuria (Quinby 1974). According to Harlan (1995), the Chinese kaoliangs are more fibrous and have been selected for non-grain uses such as basketry, fencing materials, and house construction materials. Broomcorn sorghum is thought to be part of the story, as are the sorgos such as amber cane. In some regions, the Chinese use their grains in the preparation of a fiery whiskey called Mai Tai. In Southeast Asia and Indonesia, the sorghum types are different as well. *S. propinquum* is found in southern China through Thailand, Cambodia, Malaya, and Burma to the Philippines (Burkill 1966). These sorghum types are characterized by very large, loose, open panicles and may also have a history different from those of the African-based races (Harlan 1975; Doggett 1988; Bhatti et al. 1960). In fact, recent molecular studies showed that *S. propinquum* exhibits approximately an 1.2% nucleotide difference in coding regions of the genome from *S. bicolor*, suggesting a divergence of 1–2 million years between the two sorghums (Feltus et al. 2004).

7.2.3.2 Secondary Gene Pool

Secondary gene pool of domesticated sorghum includes Columbus grass (*Sorghum* x almum Parodi) and, one of the world's worst weeds, Johnson grass *(S. halepense*). Inspite of ploidy level differences, cultivated ($2n=2x=20$) and wild sorghum ($2n=4x=40$) outcrosses easily under controlled and natural conditions (Morrell et al. 2005). F_1 hybrids are completely sterile triploids or fertile tetraploids (Hadley 1953; Godwin 2005). In triploid plants, fertility can be restored quite easily within two generations via backcrossing to the diploid parent (Arriola and Ellstrand 1996). Tetraploid hybrids are usually annual weeds, whereas sterile triploid hybrids develop extremely vigorous rhizomes that enable them to propagate vegetatively and persist over several years, even in cooler climates (Hadley 1953; Yim and Bayer 1997). Molecular marker analysis showed that introgression from domesticated sorghum might involve in the evolution of very aggressive weedy *S. halepense* biotypes (Celarier 1958; Morrell et al. 2005). Genetic evidences support the hybrid origin of weedy Columbus grass (*Sorghum* x almum Parodi), which seems to have arisen from hybridization between domesticated sorghum and Johnson grass (*S. halepense*) in America (Holm et al. 1977; Paterson et al. 1995).

7.2.3.3 Tertiary Gene Pool

The tertiary gene pool includes *Sorghum sp.* from other sections. Nowadays, sorghum breeding programs have not used species outside the section *Eu-Sorghum* as potential sources of genetic diversity due to lack of information regarding the genetic relationships between the species and their relatives. Based on the three gene sequences *ITS1*, *ndhF*, and *Adh1*, phylogenetic analysis of sorghum species had identified *S. macrospermum* and *S. laxiflorum* as the undomesticated species outside of *Eu-Sorghum* section, most closely related to cultivated sorghum varieties (Dillon et al. 2007). Thus this report helps plant breeders in suitable genes identification for resistance against biotic and abiotic stresses which can be transferred to sorghum varieties. Mostly undomesticated sorghum species present within tertiary gene pool had made gene transfer to domesticated species very difficult due to strong pre- and postzygotic reproductive barriers that exist between cultivated and wild sorghum relatives due to the differences in chromosome morphology, genome size, pollen–pistil interactions, and embryo abortions (Garber 1950; Price et al. 2005). Strong pollen–pistil incompatibilities where the pollen of undomesticated species behaves abnormally in the pistils of *S. bicolor*, resulting in no hybrid embryo formation (Hodnett et al. 2005), thus crossing only possible via artificial hybridization techniques, such as embryo rescue (Price et al. 2005). Some intergeneric crosses of sorghum have been obtained with sugarcane and maize but hybridization requires considerable human interventions which are otherwise highly unlikely to occur under natural conditions (Thomas and Venkatraman 1930; Nair 1999).

7.3 Assessment of Gene Flow in Crop Improvement

Gene flow is a continuous and major evolutionary process among cultivated ones and their wild relatives (Jarvis and Hodgkin 1999; Arnold 2004). In most of the crops, wild progenitors carry more diversity than their domestic counterparts, due to domestication processes that have induced tight bottlenecks (Papa et al. 2005). Wild relatives constitute a reservoir of useful genes for plant breeders which could be utilized for crop improvement like specific pest and disease resistance, abiotic stress tolerance, and increase in nutritional qualities (Jarvis et al. 2008). Nowadays, due to human population growth, wild relatives are threatened by the loss of natural habitats. The Convention on Biological Diversity emphasizes that conservation of wild species in nature and on farm of domesticated varieties has two major issues (CBD 1992). For efficient conservation and their use in food and agriculture growth, it is first necessary to understand genetic relationships existing between a crop and its wild relatives and to set priorities, especially when conservation resources are limited (Maxted 2006). However, gene flow from domesticated to wild relatives had increased weediness in seven major crop species (Ellstrand et al. 1999) or massive gene flow from among crop to wild and wild to crop can also lead to the displacement of genetic diversity in wild populations (Papa and Gepts 2003) and, in extreme cases, to the extinction of wild populations (Ellstrand et al. 1999). With the advancements in GM crops, gene flow from cultivated ones to their wild and weedy relatives associated with scientific interest is a possible threat to wild populations (Ellstrand 2003). To access possibilities of transgene transfers via gene flow and for proposing natural population conservation strategies, knowledge of the geographical distribution of wild relatives and landrace diversity, particularly in centers of domestication, is a prerequisite. Recently, molecular markers are successfully implemented in various species as supporting evidence of secondary centers of domestication by analyzing the structure and diversity within the wild/weedy-domesticated complex and the extent and direction of gene flow. Earlier workers had worked on these major issues for common bean in Mesoamerica (Zizumbo-Villarreal et al. 2005), pearl millet in Niger (Mariac et al. 2006a, b), and maize in North and South America (Matsuoka et al. 2002; Vigouroux et al. 2005). Knowledge about potential for crop–wild gene flow is still lacking for sorghum, especially in traditional sorghum agroecosystems in Africa. Researchers are more concerned about ecological and geographical distribution of wild and weedy sorghum in Kenya (Mutegi et al. 2010), Ethiopia, and Niger (Tesso et al. 2008). In both cases, they reported that gene flow from crop to wild was likely to occur in many agroecosystems. In a traditional farming system in a village of northern Cameroon, the dynamics of genetic diversity within the weedy–crop complex was investigated by Barnaud et al. (2009), and later on, morphological and molecular markers had also confirmed that the weedy forms were introgressed in that particular place.

Moreover, farmers' practices have enhanced the post-domestication crop-to-wild and wild-to-crop gene flow. Some farmers' practices also contribute to gene flow in the other direction also via integration of hybrid seeds resulting from weed-to-crop gene flow in the traditional seed system. Gene flow to some extent between the *guinea margaritiferum* wild type and probably a more limited contribution of *bicolor*, *guinea gambicum–guineense*, and *durra* and *caudatum* types to crop-to-wild gene flow is reported by Sagnard et al. (2011). The genetic relatedness among wild types and *guinea margaritiferum* accessions could be the result of both sorghum domestication history and preferential post-domestication crop-to-wild gene flow. Chloroplastic- or mitochondrial-inherited molecular markers should help in the estimation of the relative contributions of these two evolutionary factors. Some stable weedy hybrids between cultivated sorghum and wild types are known as *S. bicolor* subsp. *drummondii*. Bidirectional gene flow is not unique among major crops plants (Jarvis and Hodgkin 1999), and sorghum is the only major cereal crop where the wild/weedy relatives followed by cultivated ones had followed inadvertent introduction from Africa to America, Asia, and Australia; hence, crop–wild–crop gene flow occurs almost everywhere in areas where sorghum is cultivated. But further research refinements are also required to evaluate precisely the extent and direction of gene flow in different agroecosystems. But there are always some risks associated with gene flow in every crop.

7.3.1 Risks Associated with Gene Flow

Johnson grass (*Sorghum halepense* L. Pers.), being the most noxious weed affecting the world's agriculture (Holm et al. 1977), possesses a major contaminant in sorghum seed production, an alternate host and means of overwintering for pests and pathogens of both monocotyledonous and dicotyledonous crops, and a major competitor for sunlight and other growth-limiting resources. Taxonomic status of Johnson grass with cultivated sorghum (*Sorghum bicolor*, $2n=2x=20$) is clear. Its well-developed rhizomes are the only morphological feature which consistently distinguishes *S. halepense* from *S. bicolor* (Celarier 1958). Its geographical distribution and rhizomatous habits had suggested that *S. halepense* is also a descendant of *S. propinquum* Kunth. ($2n=2x=20$), a rhizomatous species found in Southeast Asia, Indonesia, and the Philippines (Doggett 1976). *S. propinquum* may account for reports of putative "*S. halepense*" with ploidy of $2n=20$ (Celarier 1958). The natural distribution of *S. halepense* overlaps the distributions of both *S. bicolor* and *S. propinquum* (Doggett 1976). *S. bicolor* has a center of diversity in Africa; however, the greatest morphological diversity of $2n=40$ sorghums is in Southeast Asia, supporting the hypothesis that *S. propinquum* contributed to *S. halepense* (Celarier 1958). Another similar weedy sorghum, *S. almum* ("Columbus grass", $2n=4x=40$), is also widespread (Doggett 1976).

In commercial use of bioengineered sorghum cultivars, gene flow from sorghum to Johnson grass would represent a serious potential risk. The prospect of sorghum's resistance against major diseases and insects, as well as tolerant to herbicides, can be utilized for improving productivity and profitability of this important dryland food and feed crop, which may particularly beneficial to dry areas. However, the sorghum biotechnology can act as a two-edged sword; exposure of Johnson grass to sorghum cultivated on more than eight million acres in the USA alone presents an enormous opportunity for gene exchange, even if such events are relatively rare. Because of the close relationship between Johnson grass and sorghum, no herbicide is known which can kill Johnson grass but not harm sorghum; thus, the 8–14 million acres of sorghum grown each year provides a "heaven" of sorts, for Johnson grass. Gene flow from sorghum to Johnson grass would not only affect weed control in sorghum but also in maize, soybean, cotton, and other crops. Major constraint associated with high gene flow rate in domesticated sorghum to its weedy and wild relatives is that it constrains the use of transgenic approaches for sorghum improvement program.

7.4 Level of Diversity in Crop Germplasm

7.4.1 Crop Wild Relatives

The genus *Sorghum* contains species belonging to two separate lineages based on chromosome morphology and genome size ($x=5$ and $x=10$). The $x=5$ lineage is typified by a large genome size and large chromosomes, while the $x=10$ lineage has smaller chromosomes and genome size. The evolutionary relationships between these two lineages and between species within the lineages have not been resolved. The *Eu-Sorghum* section has received the most attention in elucidating the genomic relationships between species. To attempt introgression breeding, the genomic relationships between parental species must be understood, and without homology, there will be little to no genetic recombination from which to recover introgression.

Sorghum bicolor ($2n=20$) belongs to the $x=10$ chromosome lineage within the genus, as do all species within *Eu-Sorghum*, *Chaetosorghum*, and *Heterosorghum*; however, the name of the lineage does not accurately describe the genomic makeup of the species. *Sorghum bicolor* is generally regarded as a diploid and regularly displays normal meiosis with the chromosomes forming 10 bivalents at metaphase (Durra and Stebbins 1952; de Wet 1978; Sangduen and Hanna 1984; Doggett 1988). Multivalent formation in diploid plants and bivalent formation in haploid plants indicate that some homology exists between the chromosomes of the haploid genome. So while *S. bicolor* functions as a diploid, it is most likely tetraploid ($2n=4x=20$) in origin (Garber 1950; Endrizzi and Morgan 1955; Celarier 1958; Doggett 1988).

Recently, cytological evidence supported the conclusion that *S. bicolor* is a disomic tetraploid. Fluorescence in situ hybridization (FISH) of a large-insert genomic clone, BAC22B2, and later a 1047-bp subclone, pCEN38, showed differential hybridization near the centromere. Ten homologous chromosomes had a strong FISH signal, while the ten remaining homologous chromosomes had a weak but detectable signal; based on that, it is concluded that *S. bicolor* is composed of two subgenomes, $A_bA_{bb}B_b$ (Gomez et al. 1998; Zwick et al. 2000). *S. halepense* ($2n=40$) is a rhizomatous species and is considered to be one of the most noxious weeds found in USA and world agriculture (Holm et al. 1977; McWhorter 1989). It has long been proposed to be a polyploidy, although the exact nature of its genomic constitution is still not resolved. Many authors have proposed genomic formulas for *S. halepense*: an autotetraploid (Casady and Anderson 1952; Durra and Stebbins 1952), an autooctoploid (Bennett and Merwine 1966), and a segmental auto-allo-octoploid (Hadley 1953; Tang and Liang 1988). The difficulty in assigning genomic constitution to *S. halepense* results from the lack of sharp differentiations between its polyploid genomes, i.e., significant homology exists between its subgenomes. Its meiotic behavior is not normal and chromosomes associate primarily as bivalents and quadrivalents, which indicates *S. halepense* is at minimum an allopolyploid (Hadley 1953; Endrizzi 1957; Tang and Liang 1988). In crosses with *S. bicolor*, the hybrid ($2n=30$) showed approximately equal numbers of univalents, bivalents, and trivalents (~4 I+4 II+6 III), which led Hadley (1953) to propose a genomic formula of AABB for *S. bicolor* and AAAABBCC for *S. halepense*. This formula predicts the interspecific hybrid (AAABBC) would form chromosome associations as 5 III within the A subgenome, 5 II within the B subgenome, and 5 I with the remaining C subgenome. The hypothetical pairing is roughly close to the empirical results obtained by multiple researchers (Hadley 1953; Endrizzi 1957; Tang and Liang 1988). Tang and Liang (1988) modified Hadley's model slightly by replacing genomes B and C with B_1 and B_2, respectively. Accordingly, the B_1 and B_2 genomes share high amounts of homology but are still distinct subgenomes; they also share some homology, though much less, with subgenome A. This modification allows the chromosomal associations to vary in that it is allowable for B_1 and B_2 chromosomes to interact and pair more frequently as well as rarely to pair with A_9 chromosomes. An explanation for the 40 chromosome hybrid example that produces a deficiency of trivalents could be that the affinity to form B-B bivalents is stronger than the affinity to form both B trivalents and exclude B chromosomes as univalents. This seems to be the most plausible explanation for the chromosome pairing observed in all ploidies of interspecific hybrids. Accordingly, *S. halepense* is considered ($2n=8x=40$) $AAAABBB_{12\,1\,2\,112}B_2$ and *S. bicolor* is considered ($2n=4x=20$) AAB_1B_1 (Tang and Liang 1988). *Sorghum propinquum* ($2n=20$) is a rhizomatous species within *Eu-Sorghum* that is native to Asia. It hybridizes readily with *S. bicolor*, produces fertile hybrids and F_2 progeny, and shows normal chromosome pairing (Celarier 1958; Doggett 1988). There is a cytogenetic difference between the two species. The smallest chromosome in *S. propinquum* contains the nucleolar organizing region; whereas, it is present in the largest chromosome of *S. bicolor* (Magoon and Shambulingappa 1961; Doggett 1988). It is considered a separate species from *S. bicolor* because of its markedly different rhizomatous habit and distinct geographic distribution (de Wet 1978). It had been thought to be a progenitor of *S. halepense*, possibly hybridizing with *S. bicolor*, followed by a doubling of chromosomes (Celarier 1958; Magoon and Shambulingappa 1961; Doggett 1970), but this hypothesis is currently not resolved. Attempts to recreate *S. halepense* by doubling the chromosomes of *S. bicolor* x *S. propinquum* hybrids have not created plants that show similar chromosome associations. Molecular evidence tentatively suggests that *S. propinquum* shares little homology with *S. halepense* as very few RFLP fragments are shared between the two species (Chittenden et al. 1994). Research on its distribution within the genus *Sorghum* revealed that the sequence is present in *S. halepense* and *S. propinquum*, although at slightly lower copy number in the latter, lowly present in other species including *S. macrospermum* and *S. laxiflorum*, and not detectably present in the remaining species (Anderson 2005). This agrees with the close evolutionary relationships in the *Eu-Sorghum* section and the considerable divergence of the other species.

7.5 Limiting Factors Related to Crop Production

Sorghum [*Sorghum bicolor* (L.) Moench] is a C_4 plant and is well adapted to arid and semiarid tropics and genetically constitutes a diverse group. There are some major factors which delimit the crop production like non-availability of good quality seed of improved cultivars/hybrids, generally commercial cultivation is done on marginal lands, non-implementation of modern INM practices, drought occurrence during crop growth and development resulting in moisture deficiency, heat stress at maturity reduces the grain size and weight and insect pest and disease attack also causes yield losses. Some sorghum varieties are more resistant to stresses than others due to genetically diverse genome (Tao et al. 1993, 2003).

7.5.1 Biotic Stresses

Among various biotic stresses, insect pest is the most important one which affect production and productivity of sorghum. About 150 insect species have been reported as pests on sorghum (Sharma 1993), of which stem borers (*Chilo partellus)*, sorghum shoot fly (*Atherigona soccata*), *Busseola fusca*, *Eldana saccharina* and *Diatraea* spp.), armyworms (*Mythimna separata*, *Spodoptera frugiperda*, and *S. exempta*), shoot bug (*Peregrinus maidis*), aphids (*Schizaphis graminum* and *Melanaphis sacchari*), spider mites (*Oligonychus* spp.), grasshoppers and locusts (*Hieroglyphus*, *Oedaleus*, *Aliopus*, *Schistocerca*, and *Locusta*), sorghum midge (*Stenodiplosis sorghicola*), mirid head bugs (*Calocoris angustatus* and *Eurystylus oldi*), and head caterpillars (*Helicoverpa*, *Eublemma*, *Cryptoblabes*, *Pyroderces*, and *Nola)* are the major pests worldwide. Annual losses due to insect pests differ in magnitude on a regional basis and climatic conditions. They have been estimated to be $1089 million in the semiarid tropics (SAT), $250 million in the USA, and $80 million in Australia (ICRISAT 1992). In India, nearly 32 % of sorghum crop is lost due to insect pests (Borad and Mittal 1983). Bent the most damaging pest of West Africa (Tenkouano 1995; Neya and Le Normand 1998; Trouche et al. 2001). In addition, sorghum midge caused by insects (*Contarinia sorghicola*, *Sitodiplosis moselana*, *Stenodiplosis sorghicola*, etc.) attack developing sorghum grains (Sharma and Hariprasad 2002; Tao et al. 2003). Sorghum production is constrained by grain mold because of sorghum caryopsis damage (Melake-Berhan et al. 1996; Waniska et al. 2001). In Africa, the absence of grain mold resistance has been cited as a constraint to adoption of improved cultivars (Audilakshmi et al. 1999). Enhanced activity of phenolic-related enzymes and the accumulation of phenolic compounds have been directly correlated with resistance of cereals to biotic stresses (Mohammadi and Kazemi 2002). Plant resistance to biotic and abiotic stresses is often regulated by the metabolism of phenolic compounds. In sorghum, various phenolic compounds, e.g., phytoalexins (3-deoxyanthocyanidins) or allelochemicals (p-hydroxybenzoates, p-coumarates, and flavanols), are involved in plant resistance to different kind of biotic stresses (Lo et al. 1999; Weston et al. 1999; Weir et al. 2004). Both biotic (fungi, insects, viruses, etc.) and abiotic (drought, temperature, photoperiod, nutrient deficiencies, etc.) stresses induce phenylalanine ammonia lyase (PAL; EC 4.3.1.5) synthesis (Chalker-Scott and Fuchigami 1989; Tovar et al. 2002). PAL activity has been detected in green shoots and leaves of sorghum, and there is PAL activity which increases with the infection of the plant with pathogens and involves a rapid accumulation of PAL mRNA (Stafford 1969; Mohan et al. 1988; Cui et al. 1996) in sorghum. Peroxidases also play a major role in stress-related resistance. POXs a polymer of cell wall lignin and suberin constitute physical barriers for both biotic and abiotic stresses (Cosgrove 1997). In sorghum, POXs are involved in thermal tolerance and resistance against fungal infection (Luthra et al. 1988; Choudhary et al. 1993). Polyphenol oxidases (PPOs, EC1.14.18.1; monophenol, 3, 4-L-dihydroxyphenylalanine: oxygen oxidoreductase) play an important role in plant defense which produces o-quinones from the oxidation of endogenous phenolic compounds, a toxic constituent to invading pathogens and pests, and its activity increases under abiotic stress conditions (Mayer and Harel 1991; Mohammadi and Kazemi 2002).) and upon fungal infections (Luthra et al. 1988). Several studies in other plant species have shown that the levels of phenolic compounds and the activities of PAL, POX, and PPO are different between resistant and susceptible plants under stresses (Lo et al. 1999; Mohammadi and Kazemi 2002).

Marker-assisted selection and transgenics use for insect-resistant cultivar development depends on the precision of resistance-screening techniques. Infester row, cage, and leaf disk screening techniques have been standardized to evaluate sorghum germplasm, breeding material development, and mapping populations for resistance to insect pests under field and greenhouse conditions (Sharma et al. 1992; Sharma and Hariprasad 2002). In screening for stem borer resistance, there is the relative importance of other traits like foliar injury, dead hearts, stem tunneling, exit holes, and tiller production (Singh et al. 2011). The effects of different damage parameters on grain yield loss are not fully understood like whether the material should be screened in each generation, alternate generations, or only after the material has become homozygous in F_5–F_6 generations. Extensive studies at ICRISAT have indicated that the material subjected to borer infestation in F_2 to F_5 generations had greater frequency of resistant progenies than the material exposed to borer infestation in the F_5 generation only. Such information needs to be generated for different insect pests screening.

7.5.2 Abiotic Stresses

Various abiotic factors affect the sorghum productivity directly or indirectly like salinity, drought, heat, etc. Sodacity is a major abiotic stress which reduces agricultural productivity (Boyer 1982). Salinity affects the plant growth in various terrestrial areas of the world. It is estimated that more than one third of the total irrigated land in the world is presently affected by salinity. This is exclusive of the regions classified as arid and desert lands (which comprise 25 % of the total land of our planet). Salinization is directly in conflict with the needs of the increasing world population which is projected to increase by 1.5 billion in the next 20 years due to loss of farmable land. The damaging effects of salt accumulation in agricultural soils have influenced ancient and modern civilizations. Food grain production boom provided by "Green Revolution" but now it has reached a plateau. Thus there is an urgent need to enhance yield potential of crop plants in available resources like optimal soils and in less productive lands (Jacobsen and Adams 1958).

Azhar and McNeilly 1988 reported both additive and dominance effects of NaCl via diallel analysis, based on assessing root tolerance to NaCl in salt-treated plants as compared with control plants. Thus it indicates that salinity tolerance is a complex trait as explained in terms of growth and yield. There are various constraints which affect the growth and yield of crop plants under stress conditions including technical and biological constraints, and to understand the biological mechanism of stress is difficult (Epstein et al. 1980). To unreveal biological mechanism of salt tolerance identification of the genetic basis of stress tolerance or the identification of genes and QTLs related to salt stress is essential to develop varieties with enhanced salinity tolerance are a prerequisite. Two approaches currently being utilized to develop stress tolerance includes (1) genetic variations exploitation, either via direct selection in stressful environments or mapping of

QTLs and subsequent use of marker-assisted selection (MAS), and (2) use of transgene technology to introduce novel genes or to alter expression levels of the existing genes that affect the degree of salt stress tolerance.

Drought is a major constrains limiting crop productivity worldwide. In sorghum, drought affect plant growth during the early vegetative seedling stage, panicle development stage, and in post-flowering, i.e., the period between grain filling and physiological maturity (Rosenow and Clark 1995; Rosenow et al. 1996). Post-flowering drought stress can result in reductions in crop yield significantly (Rosenow and Clark 1995; Rosenow et al. 1996). Sorghum is one of the major drought-tolerant crops and is an ideal model system to study physiological and molecular mechanisms underlying drought tolerance (Doggett 1988; Ludlow and Muchow 1990; Mullet et al. 2001; Sanchez et al. 2002). Stay-green phenotype after post-flowering in sorghum is related to drought stress, which is characterized by the maintenance of green stems and upper leaves under water scarcity after flowering (Subudhi et al. 2000).

Various workers had mapped many stay-green QTLs associated with post-flowering drought tolerance in sorghum (Tuinstra et al. 1997, 1998; Crasta et al. 1999; Tao et al. 2000; Xu et al. 2000; Subudhi et al. 2000; Kebede et al. 2001; Haussmann et al. 2002; Sanchez et al. 2002), and molecular markers linked to these QTLs are also available (Harris et al. 2007; Kassahun et al. 2009). Commonly used source of stay green has historically been the sorghum line, BTx642 (formerly called B35), a member in the *durra* race. Major stay-green QTLs identified by using molecular approaches were designated as Stg1, Stg2, Stg3, and Stg4 as well as many additional minor QTLs. In addition, role of transgressive segregation (Haussmann et al. 2007) and epistatic interactions involved in stay-green QTL identification (Subudhi et al. 2000) have been reported. Jordan et al. (2003) and Kassahun et al. (2009) reported that under terminal drought conditions stay green that had direct positive impact on grain yield has been observed. Tuinstra et al. (1997, 1998) reported co-localization of grain yield and stay-green QTLs under drought result in enhanced yield performance under drought stress and also reported positive associations among these traits via indirect selection based on stay-green performance of sorghum under drought stress.

Being a crop of semiarid tropics, it faces failure of seedling establishment due to heat stress which is one of the key factors that limit yield and affect stability of production (Peacock 1982). A wide range of genetic attributes for thermal tolerance has been reported in many crop species (Levitt 1980). Heat tolerance is measured as heat tolerance index (HTI). Jagtap et al. (1998) reported that in sorghum E 36–1 a heat-tolerant cultivar possessed the highly thermostable photosynthetic enzymes RuBisCo (ribulose-bisphosphate carboxylase), phosphoenolpyruvate carboxylase, components of photosystem II, and higher levels of the chloroplast chaperonin 60 compared to thermosensitive cultivars.

All these factors directly or indirectly affect the sorghum production, creating an imbalance between its demand and supply. So, some dedicated efforts are required to streamline the sorghum production to meet the growing demand because it is key dryland crop for farmers having limited resources. Therefore, there is an urgent need of capacity building to develop strategies for resolving the constraints, which limits the use of improved technologies in a cost-effective manner to reassure high production and profitability to the resource poor farmer in drylands across the world.

7.6 Traits of Importance

Mainly three factors affect the base broadening in sorghum as are discussed below in details:

7.6.1 Cytoplasmic Male Sterility

The CMS system is a valuable tool for hybrid seed production in self-pollinating crop species, like rice, rye, sorghum, sunflower, wheat, and a number of vegetable crops. Hybrids often have heterosis, more commonly known as hybrid vigor, whereas hybrid progeny exhibits superior growth characteristics as compared to either of the parental lines. CMS systems can be of considerable value in facilitating efficient hybrid seed production. Understanding regarding molecular basis of CMS, as well as other hybrid production methods involved in self-incompatibility and apomixis, is critical for continued improvements in hybrid technology.

7.6.1.1 Origin of CMS

The production of degenerated or nonviable pollen grains or non-dehiscent anthers with or without functional pollen grains resulting from a disharmonious interaction between the cytoplasmic factors (now widely identified as mitochondrial genetic factors) and nuclear genetic factors lead to CMS, a physiological abnormality, and this type interactions are common in populations incorporating divergent sources of nuclear and cytoplasmic genes (Reddy et al. 2003). Being a maternally inherited trait, CMS is often associated with unusual open reading frames (ORFs) found particularly in mitochondrial genomes (Chase and Babay-Laughnan 2004; Hanson and Bentolila 2004). CMS/*Rf* systems are also of value in the study of interactions between nuclear and mitochondrial genomes because, in many cases, it has been found that male fertility can be restored by nuclear-encoded fertility restorer (*Rf*) gene(s). On the one hand, sterility is induced by mitochondrial genes causing cytoplasmic dysfunction, and on the other hand, fertility restoration relies on nuclear genes that suppress cytoplasmic dysfunction and spontaneous induction of breeding lines, as a result of wide crosses or the interspecific exchange of nuclear and cytoplasmic genomes, or following mutagenesis (Hanson and Bentolila 2004). Sorghum is no exception to this, e.g., the A_1 CMS source in sorghum was identified in the F_2 population of cross Double Dwarf Yellow sooner milo × Texas Blackhull *kafir* by Stephens and Holland (1954), in which the milo inbred belongs to *durra* race from Sudan and the Ethiopian border (Duncan et al. 1991), and the *kafir* inbred from Eastern Africa (House and Nelson 1985). But later on number of male sterile cytoplasms had been identified and studied (Pring et al. 1999). These cytoplasms are known to have originated in geographically diverse regions in India, America, and Africa. They are broadly classified into groups designated as A_1, A_2, A_3, A_4, etc., depending upon their maintainer and restorer crosses, of which mainly the A_1 cytoplasm (milo) has so far been used for commercial hybrid seed production. However, reliance on only one type of cytoplasm, such as milo, can be disastrous if the cytoplasm happens to be susceptible to attack by a particular pathogen. It is therefore necessary to look for alternative cytoplasms for hybrid seed production. In last few years, Indian agriculturists and plant breeders under the All India Coordinated Sorghum Improvement Project have developed several male-sterile lines based on three Indian male-sterile cytoplasms designated as Maldandi, Guntur, and Vizianagaram. These cytoplasms are of Indian origin (race *durra*) and have been identified separately in the regions of Maldandi, Guntur, and Vizianagaram. They have been tentatively grouped as Indian A_4 types. Restoration to fertility of these cytoplasms has been difficult and, so far, no restorer lines are known, which has prevented use of these cytoplasms for commercial hybrid seed production.

7.6.2 Restoration of Pollen Fertility

There is a lot of diversity among genetics of fertility restoration systems like restoration of pollen fertility within and among species as being either sporophytic or gametophytic. Sporophytic restorers act prior to meiosis or in sporophytic tissues; gametophytic restorers act after meiosis in microspores or pollen grains leading to very different transmission patterns. Male sterile cytoplasm of diploid plants that is heterozygous for a restorer will produce two classes of pollen grains: those that carry the restorer and those that do not. Both genotypic classes of gametes will be functional in sporophytic restorer. But in heterozygous plants for gametophytic restorer, only those gametes that carry the restorer will be functional. There is huge diversity in restoration systems regarding number of restorer genes, but there is one or two major restorer loci that confer complete restoration but full restoration requires the concerted action of a number of genes, many of which provide only small incremental effects. For practical reasons, most research has been conducted on systems in which restoration is conditioned by relatively few genes. In most such systems, a single restorer gene is sufficient. Duplicate restorer loci that exist in a number of systems indicate that duplication of gene functions or multiple mechanisms can induce restoration. Nuclear restorers often alter the expression of CMS-associated genes and thereby presumably lessen or eliminate the deleterious effects associated with their gene products (Schnable and Wise 1996). In sorghum A_1, cytoplasmic-nuclear male sterility system is used almost exclusively for the production of commercial hybrid seed, and thus, the dominant genes that restore male fertility in F_1 hybrids are of critical importance to commercial seed production. The genetics of fertility restoration in sorghum appear complex, being controlled by at least two major genes with additional modifiers and additional gene–environment interactions. To elucidate the molecular processes controlling fertility restoration and to develop a marker screening system for this important trait, Jordan et al. (2011) have developed two sorghum recombinant inbred line populations by crossing a restorer and a non-restoring inbred line, with fertility phenotypes evaluated in hybrid combination with three unique cytoplasmic male-sterile lines. In both populations, approximately 85 and 87 % of the phenotypic variation in seed set was associated with the major Rf gene on SBI-02. Jordan et al. (2011) identified a second locus capable of partial restoration of pollen fertility along with major fertility restoration gene capable of restoring pollen fertility in both A_1 and A_2 male-sterile cytoplasms, which was located on sorghum chromosome SBI-05. Depending on the restorer parent, mapping populations exhibited fertility restoration phenotypes that ranged from nearly bimodal distribution due to the action of Rf_5 to a more normalized distribution reflecting the action of Rf_5 and additional modifier/partial restoration genes. A second fertility restoration locus capable of partially restoring pollen fertility in A_1 cytoplasm was localized to chromosome SBI-04.

7.6.3 Effect of Flowering Time

The success of sorghum production is determined to a considerable extent by the appropriateness of the flowering time for the specific production environment. The highly variable sowing dates, due in part to erratic onset of the rainy season, present an important challenge since grain maturity needs to occur at a more fixed calendar date to coincide with the end of the rainy period for successful grain filling and pest avoidance (Niangado 2001). Thus, photoperiod-sensitive flowering responses of sorghum enhance adaptation by enabling more fixed maturity dates despite variable sowing dates (Vaksman et al. 1996; Clerget et al. 2008; Haussmann et al. 2007). The transition of plant growth from vegetative to generative stage is the primary determinant of flowering time in crops of determinant growth type such as sorghum. The degree to which varieties can adjust this onset of panicle initiation with differing sowing dates, and photoperiod conditions, is called photoperiodic flowering response (Garner and Allard 1920). Photoperiod sensitivity triggers panicle initiation in short-day (SD) plants such as sorghum when they sense an appropriate decrease in day length (Thomas and Prue 1997). A series of six maturity genes have been recognized to affect flowering time and photoperiodic flowering response in sorghum: Ma_1, Ma_2, Ma_3, Ma_4, Ma_5, and Ma_6 (Rooney and Aydin 1999; Kouressy et al. 1998). The first four maturity genes inhibit flowering under LD conditions but allow early flowering under short-day conditions. Of these first four genes, Ma_1 causes the greatest sensitivity to linkage disequilibrium (LD) conditions. In contrast, Ma_2, Ma_3, and Ma_4 generally have more modest effects on sensitivity to LD conditions (Rooney and Aydin 1999). Kouressy et al. (1998) showed that photoperiod sensitivity was affected by dominant alleles of one major gene, equivalent to the Ma_5 or Ma_6 maturity loci identified by Aydin et al. (1997).

Several other studies reported on sorghum photoperiodic flowering are by Childs et al. (1992) and Paterson et al. (1995). These studies highlighted the role of phytochromes as an important gene family. MAS being an advanced plant breeding technique has the potential to accelerate the selection process substantially (Kumar 1999; Collard et al. 2005). Functional markers are the state-of-the-art molecular tools that minimize the risk of recombination between marker and QTL alleles (Andersen and Luebberstedt 2003). Association studies based on linkage disequilibrium offer a new possibility to identify marker–trait associations (Thornsberry et al. 2001). But it will be necessary to characterize the entire flowering time gene network in sorghum to reveal how selection has shaped the photoperiod pathway of flowering time and thus helped sorghum to adapt to climatic zones with different day-length conditions (Bhosale et al. 2012). Identification of molecular markers associated with photoperiodic flowering serve as powerful tools for MAS in sorghum to identify segregant genotype having specific sensitivities to photoperiods.

7.7 Interspecific Hybridization

Interspecific hybridization has proved to be a valuable tool in capturing genetic variation for improving traits in many important species. Plant breeders normally utilize elite and exotic germplasm within the cultivated species for trait improvement. These accessions form the primary gene pool, those that cross readily and produce fertile hybrids and progeny with the cultivated type (Harlan and deWet 1971). A secondary gene pool normally encompasses other species that hybridize with some difficulty; however, hybrids exhibit deleterious effects including partial sterility and low genetic recombination between genomes making gene transfer difficult. The tertiary gene pool encompasses other related species that exhibit significant barriers to recovering introgression due to difficulty in hybridization, hybrid lethality or sterility, and lack of genetic recombination. Thus, gene transfer is very difficult without extreme techniques. Plant breeders routinely utilize germplasm in the closest gene pool that contains the necessary variation. Sometimes the primary gene pool does not contain variation for the desired traits; therefore, use of more distant gene pools becomes a necessity.

Significant interspecific hybridization barriers may exist between the cultivated species and members of the secondary and tertiary gene pools. If these sexual barriers can be overcome, some cytological barriers also affect recovering of genetic variation from the wild species in the cultivated type. Introgression of genetic variation from wild species into cultigens through interspecific hybridization can take place in three ways: genetic recombination, alien translocations, and alien chromosome additions. Alien chromosome addition and substitution lines are formed when the recurrent parent genome is recovered with either an addition or substitution of alien chromosomes for recurrent parent chromosomes (Singh 2003). Alien translocation lines are formed when segments of alien chromosomes break and fuse to the recurrent parent's chromosomes. This process can occur naturally or be synthesized using radiation to induce chromosomal breaks (Sears 1993; Singh 2003). Genetic recombination occurs when the parental genomes share enough homology for homoeologous chromosomes to pair and recombine during meiosis. Chromosome pairing in the interspecific hybrid may occur in two forms: (1) autosyndesis, pairing occurs between members of the same parental genome, and (2) allosyndesis, pairing occurs between members of different parental genomes (Burnham 1962). The interspecific hybrid must also display some level of gamete fertility if introgression is to be recovered. Interspecific hybrid fertility is not solely dependent on chromosome behavior at meiosis but may be influenced by cryptic structural differences, complementary lethal genes, or differentiation in genes and chromosome structures (Stebbins 1950).

Major concern for the future of wide crossing is that many potentially beneficial donor species or local populations of wild plants are being destroyed every year by habitat degradation, industrialization, and agricultural expansion. This illustrates the need for an inventory and/or the improved conservation of wild plants and wild relatives of the major Asian crops that could possibly contribute useful genes to major crops that influence resistance against biotic as well as abiotic stresses. Gurdev Khush, former principal breeder at the International Rice Research Institute (IRRI), developer of wide crosses of rice, and 1996 World Food Prize laureate, has described wild relatives as "truly priceless seeds" (Barclay 2004). Interspecific hybridization has been very successful in transferring useful traits from wild species in many crops including rice (*O. sativa* L.), wheat (*T. aestivum* L.), oat (*Avena sativa* L.), cotton (*Gossypium hirsutum* L.), maize (*Zea mays* L.), tomato (*Solanum lycopersicum* L.), and soybean (*Glycine max* (L.) Merr.) (Brar and Khush 1997; Shi et al. 1998; Sharma and Forsberg 1977; Meredith 1991; Stalker et al. 1977; Stamova and Chetelat 2000; Riggs et al. 1998). Interspecific hybridization with *S. bicolor* is not a new concept. Within the *Eu-Sorghum* section, *S. bicolor* will readily hybridize with *S. propinquum*. Meiosis is normal in the hybrid ($2n=20$) and fertile backcrosses are readily produced (Celarier 1958; Chittenden et al. 1994). Wooten (2001) analyzed BC_3 lines and hybrids and determined that *S. propinquum* does contain useful alleles that could be used in the improvement of cultivated grain sorghum. Germplasm containing *S. propinquum* introgression is still present in the TAES sorghum improvement program, but it has not yet been used in released and improved germplasms, lines, or hybrids. Successful hybridization of *S. bicolor* with the other species of *Eu-Sorghum*, i.e., *S. halepense*, has been reported numerous times; interspecific hybrids can be $2n=30$ ($n+n$) or $2n=40$ ($2n+n$) (Endrizzi 1957; Tang and Liang 1988). Chromosome pairing in the hybrids is irregular but gene transfer into $2n=20$ backcross progeny is possible (Hadley 1953; Hadley and Mahan 1956). More recently, an interspecific hybrid with $2n=20$ chromosomes was produced between a genetic male-sterile *S. bicolor* and *S. halepense*. The chromosome number was likely generated from fertilization by a rare monohaploid *S. halepense* pollen grain (Dweikat 2005). This plant was fertile and showed normal segregation of polymorphic SSR markers in the F_2 generation. Interspecific hybrids between tetraploid sorghum ($2n=40$) and *S. halepense* have been developed easily and used in attempts to create perennial grain sorghums (Piper and Kulakow 1994; Cox et al. 2002). None of the efforts to date have succeeded in producing agronomically useful germplasm.

Until recently, hybridization between *S. bicolor* and species outside the section *Eu-Sorghum* have been unsuccessful (Karper and Chisholm 1936; Ayyanger and Ponnaiya 1941; Garber 1950; Endrizzi 1957; Tang and Liang 1988; Wu 1990; Sun et al. 1991; Huelgas et al. 1996). This can easily be explained based on the recent results identifying pollen–pistil incompatibilities as the major cause of reproductive isolation in the genus. A single hybrid was recovered between *S. bicolor* and *S. macrospermum* using standard cytoplasmic male-sterile (CMS) germplasm after much efforts (Price et al. 2005). The interspecific hybrid was confirmed cytologically to be $2n=30$ and was intermediate in phenotype. Unfortunately this hybrid was male-sterile and meiotic recombination could not be studied. The recovery of this hybrid indicates that reproductive isolation barriers are strong but not absolute within the genus. Price et al. (2005) later used a CMS *S. bicolor* accession with genotype *iap iap*, which allows the growth of pollen tubes from foreign species into its pistils (Laurie and Bennett 1989). This gene dramatically increased the ease and frequency of recovering interspecific hybrids. They supported with cytological evidence the production of three different interspecific hybrids between *S. bicolor* and *S. macrospermum*, *S. nitidum*, and *S. angustum*. Interspecific hybrids between *S. bicolor* and both *S. nitidum* and *S. angustum* did not survive to maturity, but the hybrids with *S. macrospermum* were vigorous. It was not possible to determine the meiotic chromosome pairing behavior of the interspecific hybrids and the genomic relationships between *S. bicolor* and *S. macrospermum* because of male sterility in the hybrids.

An attempt has been made by Bernard and Jewell (1985) to make intergeneric hybrids of maize, sorghum, and millet with the aim of introgressing desirable alien characteristics into maize. The products of crosses were analyzed up to their level of differentiation following pollination, and their further development on artificial culture medium was also compared. Except for one haploid maize plant, all the plants recovered proved to be classical hybrids, most of them showing the expected complement of chromosomes from each parent (10 + 36 chromosomes), a few others being slightly hyperploid ($2n=47–50$ chromosomes). No nonclassical hybrids constituted by an unreduced female gamete and a reduced male gamete were obtained.

Intergeneric hybridization specifically between *S. bicolor* and sugarcane has been successful. Most crosses used sugarcane as the female parent and sorghum as the pollen parent (Thomas and Venkatraman 1930; Janakiammal and Singh 1936); however, reciprocal hybrids have more recently been recovered (Nair 1999). The compatibility of sorghum and sugarcane is less surprising since results involving the *Eu-Sorghum*-specific repeat CEN38 implicate related genomes in the polyploidy of sugarcane (Zwick et al. 2000). Compared to other species such as wheat, rice, oat, barley, cotton, and soybean where interspecific hybridization has produced tangible results, in sorghum, it is in its infancy stage. Interspecific hybrids have just become a reality in species beyond *Eu-Sorghum* and much research remains to prove its usefulness.

7.8 Barriers to Interspecific Hybridization

Hybridization barriers are present in the genus *Sorghum* as evidenced by the multiple unsuccessful attempts in obtaining hybrids between sorghum and wild species beyond the section *Eu-Sorghum* (Karper and Chisholm 1936; Ayyanger and Ponnaiya 1941; Garber 1950; Endrizzi 1957; Tang and Liang 1988; Wu 1990; Sun et al. 1991; Huelgas et al. 1996). Gene transfer from undomesticated to domesticated species is very difficult due to strong sterility barriers because most of the undomesticated sorghum species fall within the tertiary gene pool (Harlan and deWet 1971). The nature of the sterility barriers in Sorghum have recently been identified as pollen–pistil incompatibilities whereby the pollen of undomesticated species behaves abnormally in the pistils of *S. bicolor* which hinders the hybrid embryo formation (Hodnett et al. 2005). As a result, pollen rarely grew beyond the stigma of *S. bicolor*; however, a single embryo was formed using *S. macrospermum* pollen. The embryo of this *S. bicolor* × *S. macrospermum* cross was rescued and raised through tissue culture, with the seedling verified as a hybrid based upon cytological and morphological characteristics (Price et al. 2005). Mostly barriers to interspecific hybridization are common in crop species and overcoming these is a prerequisite for utilizing wild species germplasm. Dillon et al. (2007) had recently done the phylogenetic-based analysis of all 25 *Sorghum* species based on the three gene sequences ITS1, *ndh*F, and *Adh1* and identified *S. macrospermum* and *S. laxiflorum* as the undomesticated species outside of *Eu-Sorghum* most closely related to cultivated sorghum varieties, and these results can now act as a guide for plant breeders. In addition to that, there are some other ways to overcome barriers to interspecific hybridization like use of bridge spp., doubling of chromosome number, and irradiation of pollen grains.

7.9 Conventional and Contemporary Approaches of Interspecific Gene Transfer

There are three major conventional methods of interspecific gene transfer:

7.9.1 Pedigree Method

It is the most extensively used method in handling segregating generations from crosses in self-pollinated crops and practice selection among them. It is highly suitable method for improving characters having high heritability that are quick and easy to measure such as disease resistance, plant height, and maturity and also used to improve an established variety for some specific character of interest.

7.9.2 Bulk Method

The Bulk method is a method in which F_2 and subsequent generations are harvested in bulk to grow the next generation. At the end of bulking period, individual plant selection and evaluation is carried out in the similar fashion as in the pedigree method. Homozygous lines can be isolated in less time, automatic increase in homozygosity up to F_6 or F_8 generation, after this individual plant selection can be carried out. It is used to improve considered character (yield), with minimum expenses, and is suitable for crops which are generally planted at high planting densities, e.g., small-grain crops.

7.9.3 Backcross Method

Backcross method also called "backcross breeding" is in use from decades to transfer specific character into elite lines. The variety which receives the gene is called the *recipient parent* and the variety which is the source of the gene of interest is called as the "*donor parent*." Repeated backcrossing leads to homozygosity at the same rate as in case of selfing. It is the only method for interspecific gene transfer and is independent of environment.

Crop improvement led by traditional method alone is very time-consuming and laborious, but improved breeding practices had contributed toward crop improvements over the past 50 years and consistently providing future benefits. Still there is intense pressure for crop quality and quantity improvement as a result of population growth, social demands, health requirements, environmental stresses, and ecological considerations. Major limitations to conventional plant breeding are either due to the limited gene pool or to the restricted range of organisms between which genes can be transferred due to species barriers. Improved biotechniques, in addition to traditional breeding tools, are needed to boost yield potential of the crops that feed the world (Borlaug 2002; Ortiz et al. 1998). The introduction of new techniques which help in gene transfer between organisms without crossing provides breeders with new opportunities to improve the efficiency of crop production and to enhance the utility of agricultural crops. Plants with improved traits, such as resistance to herbicides, insect pests, and viruses, have been genetically engineered using genes from unrelated organisms. However, it should be emphasized that biotechnology is not a substitute for conventional breeding methods but a means of improving on them. The major differences between traditional breeding and biotechnological tools lie in speed, precision, reliability, and scope.

In addition to conventional, contemporary approaches also play a very important role in agriculture sustainability. Some contemporary approaches used for sorghum improvement are discussed below in details:

7.9.4 Embryo Rescue

Embryo abortion occurs frequently in wide crosses, and thus embryo rescue is required for survival of the next generation. Rescues are performed either by directly transferring the excised embryo to an artificial medium or indirectly through flower (ovary), immature seed (fertilized ovule), or pod (siliqua) culture. Various techniques are available for oilseeds, cereals, legumes, and horticultural crops which alter various components of base medium like sucrose conc., vitamins, and growth regulator contents for new protocol developments. Monocot culture tended to be more direct than dicot culture, where many protocols required a multistep approach from pod to ovule culture to embryo rescue, shoot regeneration, and root induction. Each step required a specific medium and growth conditions. Hybrid embryos as young as 2 days after pollination have been recovered. However, many species such as soybean and chickpea still need procedures for rescue of very young embryos. In other species, hurdles such as poor rooting have been overcome by using grafting techniques. Embryo rescue remains a useful component in any breeding program where wide or interspecific crosses are performed, where rapid cycling through generations is used, and where germplasm preservation is required. *Sorghum* species outside *Eu-Sorghum* have not been utilized as germplasm because of strong reproductive barriers due to pollen–pistil compatibilities. *S. macrospermum's* close phylogenetic relationship and cytogenetic similarities to *S. bicolor* is of main interest to sorghum breeders because of its resistance to wide range of pests and pathogens, such as sorghum midge and sorghum downy mildew. Highly vigorous interspecific hybrid was obtained via cross between *S. bicolor* and *S. macrospermum* by using embryo rescue and in vitro culture techniques. The hybrid male sterile is like its ATx623 parent but morphologically intermediate to *S. bicolor* and *S. macrospermum* in leaf width, leaf pubescence, plant height, inflorescence morphology, chromosome number, and nuclear DNA content and is the first hybrid reported between *S. bicolor* and any *Sorghum* species outside the *Eu-Sorghum* section (Price et al. 2005).

7.9.5 Somatic Hybridization

Genetic improvement of cultivated species has been extensively investigated by breeders in their efforts toward a successful crop production. However, in conventional breeding programs, the transfer of desirable traits is restricted to sexually compatible species. A somatic hybridization protoplast-based approach may be an alternative for sexual interspecific hybridization, which is often confronted with prezygotic or postzygotic barriers (Eeckhaut et al. 2006). Plant cells from which the cell wall has been enzymatically or mechanically removed are called protoplasts. Theoretically, protoplasts are totipotent, i.e., after their isolation and subsequent culture, they have the capability to dedifferentiate, reenter the cell cycle, go through repeated mitotic divisions, and then proliferate or regenerate into various organs. In other words, applying the correct physical and chemical stimuli would be sufficient to regenerate fertile plants through tissue culture practices. This provides a multitude of opportunities for crop improvement, including a system for protoplast fusion (somatic hybridization), somaclonal variation, and plant transformation. For foreign gene introduction, somatic fusion is superior to plant gene transformation in some aspects as antibiotic resistance markers can be avoided and complex agronomic traits controlled by polygenes can be introgressed (Xia 2009). Although both mechanical isolation (Klercker 1892) and fusion (Küster 1909) were originally described more than a century ago, protoplast-related research was only well initiated after the first enzymatic digestion (Cocking 1960; Cocking et al. 1977). Some years later, availability of commercial macerating enzymes enabled worldwide protoplast research on a wide array of plant species. The first report on the full plantlet regeneration from protoplasts was described in tobacco mesophyll cells (Takebe et al. 1971). Tobacco was also the first crop in which successful interspecific somatic hybridization was reported through symmetric protoplast fusion (Carlson et al. 1972). Protoplast culture at low densities became possible for more species after the publication of the Kao and Michayluk medium composition (Kao and Michayluk 1975). Since then, the 1970s and 1980s onward, much progress has been reported on protoplast regeneration and fusion. Those were compiled in reviews by Melchers and Labib (1974) and Gleba and Sytnik (1984). In spite of regeneration problems, protoplast fusion became a common technique for the introduction of novelties in commercial crops (Brown and Thorpe 1986). However, during the 1990s, protoplast-based technologies for gene transfer were overshadowed by recombinant DNA technologies. Partly due to public antagonism toward these technologies, interest in protoplast research was renewed, although at that time only a limited number of plant families produced regenerative protoplasts (Waara and Glimelius 1995). A somatic breeding protocol can typically be subdivided into the following steps: isolation, fragmentation (in case of asymmetric hybridization), fusion, regeneration, and selection (Liu et al. 2005). Protoplast isolation is typically performed through a one-step procedure in which pectinase- and cellulase-type enzymes are jointly added to an osmotically corrected solution to separate the cells from their pectin-rich matrix and to dissolve the cellulose-rich cell wall, respectively. Several agents can be used to create an osmotic equilibrium between the protoplasts and their environment to prevent protoplast bursting: metabolically inactive sugar alcohols such as mannitol and sorbitol are most frequently applied, but glucose, sucrose, or salts can also be used (Razdan 2003). Bekele (2004) developed an extraction solution that is optimal for protoplast release in *Eragrostis tef* (Zucc.) Trotter and *Sorghum bicolor* (L.) Moench that consists of 1 % cellulysin, 0.5 % macerozyme and 2 % cellulase Onozuka, 1 mM $CaCl_2$, 5 mM 2 (n-morpholino) ethane sulfonic acid, and 0.4 M mannitol. Electroporation and heat shock treatments stimulated cell wall formation and cell division in both sorghum and tef. Electrical parameters and conditions for fusing tef and sorghum were determined. Field pulse conditions for eventual electrogene transfer in tef and sorghum consisted of 700 and 400 V/cm, respectively, at 20 °C and 40 μs field pulse duration. Carlson et al. (1972)) described the first protoplast fusion. However, this fusion between tobacco species was spontaneous. Fusions can also be induced through mechanical pushing, $NaNO_3$ treatment, or high pH/Ca^{2+} treatment (Razdan 2003). Currently, fusion is nearly exclusively performed through polyethylene glycol (PEG) (Kao et al. 1974) or electrofusion (Zimmerman and Scheurich 1981). Chemical fusogens cause the isolated protoplast to adhere to each other and lead to tight agglutination like high pH/Ca^{2+} solution is used for membrane disruption. Chemofusion is nonspecific and inexpensive but can be cytotoxic. For electrofusion, protoplasts are first aligned between electrodes in a low-strength electric field generated by alternating current, upon which fusion is induced by application of one or a few high-voltage direct current pulses. The electroporation induced by these pulses enables cell fusion. Electrofusion is less cytotoxic than chemical fusion, but more expensive. After fusion, different types of homokaryons or heterokaryons can be created along with alloplasmic hybrids (cybrids) (Liu et al. 2005). Fusion of divergent parental protoplasts leads to the formation of symmetric hybrids, combining both nuclear genomes. Symmetric fusion events, however, lead to the incorporation of total genomes, which can disturb the regeneration capacity, the development, or the fertility of the somatic hybrid. By reducing the amount of transnuclear information, these problems can be overcome. Asymmetric fusions enable us only to transfer partial genomes. Several techniques can be used for fragmentation of the so-called "donor" genome, such as X or gamma rays, ultraviolet (UV) irradiation, and microprotoplasts (Yemets and Blume 2003). Moreover, after symmetric somatic fusion, gene introgression leads to the reduction in successive no. of backcrosses. Also, cytoplasmic genomes can be recombined with nuclear genomes for applications like CMS introduction (Liu et al. 2005). Although much investigation has already been performed on fragmentation, chromosome elimination is random and unpredictable. Generally, irradiation causes the reduction of the amount of donor DNA, but this may vary from a few traits, one or a few chromosomes, to a big part of the donor genome (Waara and Glimelius 1995). Protoplast regeneration is often the bottleneck in somatic hybridization breeding programs. Evidently, regeneration within a single species is strongly genotype related. In a given crop, however, lots of parameters are usually optimized to achieve an optimal efficiency. Regeneration problems have forced researchers to come up with more innovative approaches, such as electrical stimulation, nonionic surfactants, and artificial gas carriers. A complete overview of regeneration-related parameters is presented by Davey et al. (2005). All types of fusion products formed after protoplast fusion, viz., heterokaryotes, homokaryotes, and unfused parental protoplasts. Identifying the hybrids in a population requires a stringent selection system. Usually, screening is performed during or after in vitro regeneration. Apart from morphological markers, many tools were developed (Liu et al. 2005): flow cytometry, in situ hybridization, and isoenzyme and molecular markers.

7.9.6 Genetic Engineering

In the last two decades, biotechnology has provided a powerful means of genetically enhancing various previously recalcitrant monocotyledonous cereal crops. For sorghum improvement, the following programs can be used: (1) in vitro plant regeneration, (2) molecular markers, (3) gene identification and cloning, (4) genetic engineering and gene transfer technology to integrate desirable traits into the sorghum genomes, and (5) genomic and germplasm databases. The first report of the successful transformation of sorghum appeared as early as in 1990s. Sorghum is the most recalcitrant crop for tissue culture and plant regeneration and for genetic transformation. Recalcitrance in sorghum tissue culture is largely due to the release of phenolics, lack of regeneration in long-term in vitro cultures, and a high degree of genotype dependence. Sorghum is mainly grown in low-input conditions; therefore, development of host plant resistance to biotic and abiotic stresses is a viable option. Demand for sorghum as a health food is gaining importance, and thus, value addition in sorghum is advantageous to the food industry. Worldwide, sorghum producers are facing a major threat to their crops from insect pests, the lepidopteran stem borer (*Chilo partellus*) which is the most destructive pests of sorghum, midge (*Stenodiplosis sorghicola*), and shoot fly (*Atherigona soccata*). Resistance building via conventional breeding is limited due to lack of reliable resistance sources. The concern is that transfer of the herbicide tolerance gene to *S. halepense* through gene flow would make control of the weed unattainable. The hybridization of cultivated sorghum, *S. bicolor* ($2n=20$) and *S. halepense* ($2n=40$), would produce unviable triploids Godwin (2005). Transgenic technology in sorghum is at a juvenile stage. Recently, Gao et al. (2005) established transgenics using the positive selectable marker gene *pmi* (phosphate mannose isomerase), which is biosafe and found widely in other crops.

7.10 Molecular Markers, Genome Mapping, and Omics

The genome sequence of the first African origin crop by Paterson et al. (2009) had provided vast knowledge about the evolution of grass genomes, and the hidden diversity present within *S. bicolor* taxa was also explored. Molecular analysis has also revealed that close relatives of *Sorghum* genus share many novel traits, endosperm structure, and composition with *S. bicolor*, the cultivated one to expand gene pool of cultivated sorghum. *Sorghum bicolor* is a diploid ($2n=20$) having second smallest genome (735 Mbp). In *S. bicolor* genome, various evolutionary events like genome duplication occurs much earlier than the divergence of the major cereal crops from a common ancestor (Paterson et al. 2004). Completion of sorghum genome sequencing had increased availability of the sequence data exponentially which has provided valuable information on cereal domestication in Africa. Sorghum genome sequencing has helped a lot to understand the biographic triangle of the divergence of these important grasses and ultimately in the understanding of the evolution in cereal crops between Africa, America, and Asia (Kresovich et al. 2005). The tenets of colinearity and microlinearity of grass genomes mean that our knowledge of other cereals and their evolutionary ties will also greatly improve. In the last one decade, highly advanced technologies were used to study the cereal genome due to their high economic and scientific value, and the similarity observed at the DNA level among these cereals helps in comparative genetic analysis to look for particular genes of unknown sequence between the genomes with the aim of using that information to develop new varieties or discovering new genes that could have a potential impact on traits that are of global importance in present scenario (e.g., food quality, drought resistance). With recent advancements in molecular biology, various new tools were employed to identify genes and QTLs of interest and use these genomic variations in cereal crop improvement. Once DNA markers tightly linked to the trait of interest are identified, they can be used for MAS for that particular trait. Due to availability of a large repertoire of SSR markers in most cereals, MAS is now effectively supplementing the conventional phenotypic selection. Further, simultaneous low-cost discovery and genotyping of SNP and indel makers has been accelerated by application of sample bar coding and next-generation sequencing (NGS) technologies. For the improvement of complex polygenic traits, plant breeders are using various genomic scale analysis tools like NGS (next-generation sequencing) to next-NGS (NGS of single DNA molecules), high-throughput genotyping using new marker systems (e.g., DArT and SNPs), and new selection strategies such as MARS (marker-assisted recurrent selection) and GWS (genome-wide selection) along with new technologies like genotyping by sequencing and whole-genome sequencing assisted by high-throughput phenotyping techniques. Whole-genome GBSs are very efficient and cost-effective new techniques. These techniques play a major role in genetic and molecular dissection of complex traits by integrating with diverse "omics" efforts which empowers high-resolution linkage mapping, genome-wide association studies, and training genomic selection models having broad applications in genomic-assisted plant breeding programs.

7.10.1 Molecular Markers

During the nineteenth century, genetic diversity assessment and crop improvement programs were based primarily on morphological markers. Use of few discrete morphological and isozyme markers does not provide an accurate knowledge about genetic divergence among the cultivated genotypes of sorghum. The introductions of DNA-based markers had created a boom in genetic diversity assessment and trait improvement in various crop improvement programs. DNA markers act by number of ways, like enhancing the efficiency of selection, helping in genetic diversity assessment, fingerprinting of elite genetic stocks, helping in environment–neutral selection for various traits, increasing the rate of desirable genotypes gain from selection, and helping in correct mapping of various interacting genes that condition complex agronomic traits. Genetic mapping is essential for effective manipulation of important genes. The complete scenario of breeding and agriculture sustainability was changed by the introduction of MAS because molecular marker-based selection or marker-assisted introgression helps in genetic recombination beyond the range of conventional plant breeding.

Molecular markers like AFLPs, RFLP, SSRs, SNPs, etc. have been used from the last two decades for identification and characterization of QTLs associated with several different traits in sorghum including plant height, maturity and yield (Pereira and Lee 1995), characters concerned with plant domestication (Paterson et al. 1995), disease resistance (Gowda et al. 1995), and drought tolerance (Tuinstra et al. 1996, 1997, 1998). In addition several sorghum linkage maps have been generated by various workers (Hulbert et al. 1990; Melake-berhan et al. 1993; Xu et al. 1994; Chittenden et al. 1994; Pereira et al. 1994; Dufour et al. 1997; Boivin et al. 1990). Haussmann et al. (2002) sampled 190 randomly selected sorghum accessions maintained at the ICRISAT (International Crop Institute of the Semi-Arid Tropics) from the five major cultivated races, namely, *bicolor*, *guinea*, *caudatum*, *kafir*, and *durra* using 82 RAPD primers. A wide range of genetic variation were detected among sorghum accessions, and genetic diversity within a race that was high for *bicolor* and *guinea* and low for *kafir* was reported. Partitioning the genetic variation further revealed that 86 % of the total genetic variation occurred among accessions and 14 % among races. Examination of the degree of association of accessions with their geographic areas of origin indicated that only 13 % of the total genetic variation among accessions was responsible for divergence among regions. Southern African germplasm exhibited the least amount of genetic variability suggesting a narrow genetic base of accessions from this region and West African germplasm had exhibited a high level of genetic diversity with a least number of accessions. Genetic diversity in Central and Eastern Africa as well as accessions from the Middle East was as high as that observed in accessions from West Africa. Molecular markers are valuable tools to assess genetic diversity and to identify diverse sources in crop germplasm collections. Genetic distances estimated by molecular marker analysis help in suitable germplasm identification for introgression into breeding stocks. Use of highly divergent accessions for introgression also increases the chances for extracting suitable inbred lines from backcross populations and these inbred lines are in turn used as useful sources of favorable alleles to improve the varieties/hybrids productivity. Dillon et al. (2004) evaluated Australian *Sorghum* species using simple sequence repeat markers and reported considerable amount of genetic diversity within and between species. In sorghum SSRs markers were developed by Brown et al. (1996) and Kong et al. (2000) for diversity assessment in these closely related taxa. This method has successfully evaluated diversity in the related species of many crop groups (Chen et al. 2002; González-Martínez et al. 2004; Quintero et al. 2012). Australian species evaluation using SSRs has shown significantly higher levels of genetic diversity both between (inter-) and within (intra-) species as compared to the intraspecific diversity of *S. bicolor* varieties. Thus it is concluded that simple sequence repeats are highly efficient and informative type of molecular markers which helps in genetic studies of undomesticated *Sorghum* species.

In addition to SSRs, nowadays, single nucleotide polymorphism (SNP) markers are used due to their high density which increases the rate of probability of polymorphism identification in a target gene which provides a huge advantage over previous markers that are closely linked to a locus of interest and not within. They are derived from restriction fragment length polymorphisms (RFLPs) that are linked to a gene of interest and are also extremely useful for marker-assisted selection (Jordan and Humphries 1994). The main advantage of SNPs over SSRs is that they are gel technology independent and require only single-base pair position differences in genomes. In the case of linkages, it can also happen that a linkage is lost when a marker is applied to other populations with different recombination patterns (Xu and Crouch 2008). Ganal et al. (2009) observed that SNP markers are expected to become the marker system of choice in the near future, especially as the full sequences of more plant genomes will become available and this is now reality. Majority of SNP genotyping technologies were originally developed for polymorphism detection in human genetic research. But the rapid growth and expansion of agribusiness and challenges to "increase the slope" and "stay competitive" forced major seed companies to adapt these technologies for plant genome discovery and also use in high-throughput SNP genotyping for MAS projects. Although MAS projects are diverse, in terms of sample and marker throughput, they can be divided into two major groups with opposite tasks: those which deal with >10,000 plants (large sample volume) to be genotyped with a few markers (1–96 SNPs) and second that require genotyping a fewer no. of samples, i.e., 1–300 with large number of SNPs (384 to several millions of SNPs). The projects of first category are related to marker-assisted germplasm evaluation, marker-assisted trait introgression, and marker-assisted gene pyramiding and second one was used for GWS (genome-wide selection) of complex traits that will require genotyping of several millions of SNPs in fairly small subset of samples. SNP genotyping platforms used for type one projects include competitive allele-specific PCR (KASPar) complemented with the SNP Line platform (SNP Line XL, KBiosciences, Hoddesdon, England)

and Open Array platform coupled with TaqMan chemistry and KBiosciences, but the first one is more flexible and cost-effective (Chen et al. 2010). The second one can be implemented using Illumina's Bead Array technology coupled with Golden Gate and Infinium assays. After SNP identification, there is SNP validation step, and after thatthey are considered as viable molecular markers and ready to use in various molecular genetics and plant breeding-related experiments, including gene/QTL mapping, linkage disequilibrium-based association mapping, map-based gene/QTL cloning, germplasm characterization, genetic diagnostics, event characterization, marker-assisted trait introgression, and finally marker-assisted selection (MAS) (Varshney et al. 2009).) *S. propinquum* genome was resequenced by Mace et al. (2013) and identified 8 M high-quality SNPs, 1.9 M indels, and specific gene loss and gain events in *S. bicolor*. They have observed strong racial structure and a complex domestication history involving at least two distinct domestication events. These assembled genomes enable the leveraging of existing cereal functional genomic data against the novel diversity available in sorghum, providing an unmatched resource for the genetic improvement of sorghum and other grass species. They also observed that in wild and weedy sorghum genotypes had higher number of SNPs as compared to landraces and improved inbreds. Wild-specific alleles (34 %) were more abundant than improved inbred-specific alleles (8 %) and landrace-specific alleles (18 %). Similarly, the average total number of SNPs and indels per genotype was highest in the wild species and lowest in the improved inbreds and equates to an average of 1 SNP per 1543 bp, 1282 bp, and 763 bp for the improved inbreds, landraces, and wild and weedy groups, respectively.

7.10.2 QTL Identification and Genetic Analysis

QTL analysis is increasingly becoming a common approach for evaluating the inheritance and feasibility of accelerating gains from selection for complex quantitative traits in crop plants. Application of QTL mapping for a marker-assisted selection depends on the type of breeding method, i.e., a recurrent selection to increase the population mean applied mainly in allogamous populations intolerant to inbreeding and selection force act only upon individual plants that are continually recombined (Yabe et al. 2013) the pedigree method and SSD (single-seed descent) methods that are used for self-pollinated crops.

Genetic linkage analysis helps in gene/QTL detection based on the genetic recombination event during meiosis (Tanksley 1993), and it also helps in the linkage map construction composed of genetic markers for a specific population. Association mapping helps in QTL identification by identifying particular allelic variants that are associated with the phenotypic trait of interest in a population, but the pedigree structure is created from families that are created by crossing the number of parents (in three-way or four-way crosses). Both phenotyping and genotyping help in the mapping of the possible location of QTL of interest. After marker identification with their favorable alleles, the frequency of such alleles will be increased, response to marker-assisted selection will be estimated, and markers with desirable effects will be further used in the next year of selection or in other experiments. Identification of QTLs using DNA markers was a major breakthrough in the characterization of quantitative traits (Paterson et al. 1988). Various researchers have reported a large number of DNA markers linked to genes or QTLs (Francia et al. 2005; Singh et al. 2012; Miedaner and Korzun 2012). However, there are many factors that influence the accuracy of QTL mapping such as population size and type, level of replication of phenotypic data, environmental effects, and genotyping errors. These factors are particularly important for complex quantitative traits with many QTLs, each with relatively small effects (e.g., drought tolerance, yield) (Collard et al. 2005). But now with the advancement in field of genetics and molecular biology, scientists became aware that QTL mapping alone is not directly useful in MAS. However, QTL confirmation,

QTL validation, and/or fine (or high-resolution) mapping are inevitable (Collard and Mackill 2008). Although, there are examples of highly accurate preliminary QTL mapping data as determined by subsequent QTL mapping research. Ideally, a confirmation step is preferable because QTL positions and effects can be inaccurate due to factors such as sampling biasness (Price 2006) and QTL validation generally refers to the QTL verification that it is highly effective in different genetic backgrounds (Langridge et al. 2001).

QTL analysis holds a great promise for traits of importance like drought tolerance. The genetic and physiological mechanisms underlying the expression of drought tolerance in crops are still not well understood. It is polygenetically controlled and dependent on the timing and severity of moisture stress; drought is one of the more difficult traits to study and characterize (Hillel and Rosenzweig 2002). Sorghum, being a drought-tolerant crop, exhibits rich genetic diversity for stress tolerance, making it an excellent crop model to study the genetic and physiologic mechanisms of drought tolerance. Nonetheless, even in sorghum, direct selection for drought tolerance using conventional approaches has been slow and difficult. Large number of physiological and biochemical traits have been implicated which enhance drought tolerance. Yet, only a few of these mechanisms has been demonstrated to be causally related to the expression of tolerance to drought under field conditions. Many QTLs for drought tolerance-related traits in sorghum was identified using populations such as recombinant inbred lines (RILs) (Crasta et al. 1999; Kebede et al. 2001; Subudhi et al. 2000; Tuinstra et al. 1996, 1997; Tao et al. 2000; Xu et al. 2000), and F_2 and F_3 populations (Habyarimana et al. 2010). Under terminal (late-season) drought, the stay-green trait has a positive impact on grain yield (Borrell et al. 2000; Jordan et al. 2003; Kassahun et al. 2009). Several stay-green QTLs are colocalized with QTLs for grain yield, flowering time, and plant height (PH) (Sabadin et al. 2012). Recently, several QTLs for nodal root angle were found to be colocated with QTLs associated with drought response in sorghum (Mace et al. 2012).

Drought stress at developmental stage affects panicle size, grain number, and grain yield. A total of 57 major QTLs were detected for eight agronomically important traits under two contrasting photoperiods, like heading date, plant height, numbers of nodes, stem diameter, panicle neck length, flag leaf width, etc. for understanding the genetic mechanisms of the agronomically important traits. The most common source of stay-green used in sorghum is BTx642 (formerly called B35), a member in the durra race. Sabadin et al. (2012) identified four major stay-green QTLs designated as Stg1, Stg2, Stg3, and Stg4, and many additional minor QTLs were also reported. Stg1 and Stg2 were mapped to sorghum chromosome 3, explaining approximately 20 and 30 % of the phenotypic variance, respectively (Xu et al. 2000; Sanchez et al. 2002; Harris et al. 2007). Stg3 is located on chromosome 2 and Stg4 on 5, accounting for 16 and 10 % of the phenotypic variance, respectively (Sanchez et al. 2002; Harris et al. 2007). In sorghum, drought stress at developmental stage is common, but not enough information is available regarding the genetic control of drought stress tolerance till date.

7.10.3 Mapping for *Striga* Resistance Genes

The highly feasible method for *Striga* resistance in sorghum is breeding for resistance. However, developing crop cultivars with durable *Striga* resistance is lacking behind due to non-availability of reliable screening method. Field screening is also affected by many confounding environmental factors. In this era of biological sciences, efforts have been made to track important gene(s) that governs resistance against important traits. Genetic linkage maps were constructed using molecular markers and QTLs for most crops. In sorghum, great efforts were made to construct linkage maps to identify molecular markers and QTLs in close association to *Striga* resistance. These markers and QTLs will not only ease the screening for *Striga* resistance but also help tracking important *Striga* resistance genes. The parasite control is difficult, but it is not impossible. In the last few decades, various efforts have been devoted to develop resistance to *Striga* control, but breeding efforts were hampered due to lack of knowledge for the genetics of resistance and lack of reliable and rapid screening methods (Ejeta et al. 1991). In fields also screening for resistance had failed and slowed due to the quantitative nature of the trait and G × E interactions. Genotype screening in *Striga*-infested fields is often inefficient due to the host complexity, and parasite and environment interactions (Ejeta et al. 1992). Integrated *Striga* management approach suggested by Ramiah (1987) is not so successful, i.e., the use of resistant lines, in combination with cultural practices and chemical inputs. The integrated management will not only provide effective *Striga* control but also suppress development of virulent strain of the parasite and reduce the buildup of seeds in the soil seed banks.

Molecular markers due to their abundance are considered as powerful tools that speed up and increase the precision and the effectiveness of plant breeding programs (Paterson 1996). They are also helpful in selection and manipulation of genotypes for desirable quantitative traits loci (QTLs) (Ejeta et al. 2000). In the last two decades, enormous progress has been made in molecular mapping techniques and great amount of polymorphic markers are generated, sufficient enough to build linkage maps (Korzun 2005) for abiotic and biotic resistance. Three major QTLs that are associated with *Striga* resistance were identified in sorghum cultivar N13 by Gamar and Mohamed (2013). Pyramiding of *Striga* resistance genes into elite backgrounds is becoming possible through foreground and background selection in MAS breeding.

7.10.4 Marker-Assisted Selection

Marker-assisted selection is the use of molecular markers to track the genetic makeup of plants during the varietal development program. Marker-assisted breeding (MAB) combines classical plant breeding with the tools and discoveries of molecular biology and genetics, most specifically the use of molecular markers. It is now practiced worldwide by farmers, plant breeders, and seed companies. Molecular markers act as indicators for a specific genetic difference, which could result in an improved form of trait. Marker-assisted selection greatly increase the efficiency and effectiveness especially for quantitative traits breeding that are associated with low heritability and inconsistent environmental effects as compared to conventional breeding (Kuchel et al. 2005). The fundamental advantages of MAS compared to conventional plant breeding are simpler compared to phenotypic screening and time-saving due to substitution of complex field trials with molecular tests; selection may be carried out at seedling stage, and single plants may be selected with high reliability and may also avoid the transfer of undesirable or deleterious genes ("linkage drag"; this is of particular relevance when the introgression of genes from wild species is involved) and help in selection of traits with low heritability. These advantages may translate into (1) greater efficiency or (2) accelerated line development in breeding programs. Thus, it is more relevant technique for crop improvement for ensuring food security by developing new varieties that are higher-yielding, resistant to pests and diseases, drought-resistant, or regionally adapted to different environments and growing conditions (Collard and Mackill 2008).

Marker-assisted selection (MAS) offers effective introgression of simple traits or a few genes in many crop species due to efficient selection, such as rice (*O. sativa*) (Joseph et al. 2004; Liu and Burke 2006; Steele et al. 2006; Jin et al. 2010), wheat (*T. aestivum*) (Vida et al. 2009; Kuchel et al. 2007; Wilde et al. 2007; Kumar et al. 2010), maize (*Z. mays*: Flint-Garcia et al. 2003; Abalo et al. 2009; Asea et al. 2012), potato (*Solanum tuberosum*) (Colton et al. 2006, papaya (*Carica papaya*) (de Oliveira et al. 2010), soybean (*Glycine max*) (Maroof et al. 2008), and common bean (*Phaseolus vulgaris*) Ender et al. 2008).

It is applied widely for targeted disease resistance traits. The nuclear restorer genes of CMS in sorghum are studied by Sattari et al. (2007).

Pollen sterility in sorghum A_1 cytoplasm is controlled by two major genes, rf1 and rf2, as well as several modifying genes. Both rf1 and rf2 are genetically dominant: the fertile Rf is dominant against the sterile rf and the interaction between these two loci shows a duplicate effect (Maunder and Pickett 1959; Erichsen and Ross 1963; Miller and Pickett 1964; Schertz et al. 1989). Restoration is also influenced by environmental conditions, like low temperature at flowering time favors sterility, whereas high temperature favors fertility (Downs and Marshall 1971; Brooking 1971, 1979). Simple sequence repeat (SSR) Xtxp18 marker exhibited a high accuracy (95.098 %) for selecting recessive homozygotes for the rf1 gene which was identified by Klein et al. (2005), thus promoting the possibility of using MAS for developing new male-sterile lines. Recently, Jordan et al. (2010) identified the SSR makers Xtxp406 and Xtxp408 tightly linked with rf2. Thus, MAS could be used to construct an integrated MAS system for sorghum A_1 CMS lines and to speed up the hybrid breeding programs at a low cost in the future (Gao et al. 2013).

7.10.5 Next-Generation Sequencing

The next-generation sequencing (NGS) technology is a golden approach of molecular biology which helps in the deep understanding of biological systems among crops. As compared to the traditional sequencing methods like Sanger method, the cost of NGS is dramatically decreased, and employing advanced NGS technology is more easy and feasible to sequence crop genomes. Some cash crops were considered to be less-studied/orphan crops due to a lack of sequence and marker information available. But now, by employing de novo assembly strategies, whole-genome sequences of less-studied/orphan crops are becoming feasible for crop improvement. Also, more molecular markers like SNPs and indels have been rapidly developed at lower cost, and these markers are easily applicable to MAS in crop breeding programs. Next-generation sequencing (NGS) technology, the most advanced method of genome sequencing, has become the main tool for developing novel molecular markers and identifying genes of agronomic importance (Edwards and Batley 2010). Thus, NGS becomes a "disruptive technology," radically reducing limitations in sequence information and consequently altering the types of questions and problems that can be addressed (Mardis 2010). Prior NGS, the time-consuming clone-by-clone method was used in genome sequencing (Ariyadasa and Stein 2012). Thus, NGS platforms like GS-FLX and Illumina HiSeq are the best choice for present sequencing era for employing the whole-genome shotgun (WGS) strategy for sequencing projects of various crops because huge amounts of data are produced in very short period of time. Several companies have brought different technology platforms to the market for third-generation sequencing. Egan et al. (2012) reviewed these NGS technologies, which employ three different methods: sequencing by synthesis, sequencing by ligation, and single-molecule sequencing; Roche 454 pyrosequencing, Illumina, and Ion Torrent are the sequencing platforms that are based on sequencing by the synthesis method; SOLiD and Polonator are based on sequencing by the ligation; and Helicos and Pacific Biosciences use the single-molecule sequencing method. Sequencing using single-molecule sequencing method is considered to be next-NGS (Barabaschi et al. 2012). Although the development of bioinformatic tools and storage space for huge sequence data are still a challenge for NGS, the speed of crop improvement will be much faster than before because the third generation of sequencing platforms, such as HeliScope, Ion Torrent, single molecular real-time sequencing and Oxford Nanopore, is in reality now and NGS technology also provides us with better opportunities for studying crop genomics and other post-genomics (transcriptomics, proteomics, metabolomics) more closely (Van et al. 2013).

7.10.6 NGS: Marker Development and Association Studies

As compared to the traditional Sanger method, NGS is helpful for discovering and developing SSR or microsatellite loci efficiently which are used commonly nowadays also for the construction of linkage maps, QTL mapping, MAS, cultivar fingerprinting, and studying gene flow. Using NGS, from the raw data directly, SSRs were identified without contig assembly, using bioinformatics methods to efficiently design primers. Due to difficulties of DNA sequence assembly with repeats, the use of GS-FLX 454 technology, rather than Illumina technology, is preferred due to its longer read length of SSR loci in plants. Due to rapidly developing NGS, large no. of molecular markers like SNPs have been identified, and SNP-based resources are publically available at various online portals for crop improvement and breeding programs (Kilian and Graner 2012). Using new methods, such as reduced-representation libraries (Hyten et al. 2010), complexity reduction of polymorphic sequences (van Orsouw et al. 2007), restriction site-associated DNA sequencing (Baxter et al. 2011), and low-coverage sequencing for genotyping (Huang et al. 2009; Elshire et al. 2011), genome-wide marker discovery by NGS become more feasible. Since genome-wide markers were quickly developed in large quantities using NGS technologies, association mapping, patterns of natural population structure, and the decay of linkage disequilibrium (LD) can be studied more easily by whole-genome scanning using NGS (Varshney et al. 2009; Kilian and Graner 2012). Also, whole-genome scanning using specially designed mapping populations is also feasible. Thus, different MAS strategies are used depending on the specific types of traits and breeding programs for crop improvement (Xu et al. 2012). Hamblin et al. (2005) reported that sorghum is ideal for association mapping studies due to moderate level of LD and its self-pollinating mating system. Morris et al. (2013) identified additional loci underlying variation in major agroclimatic traits of sorghum. They have performed genome-wide association studies (GWAS) on plant height components and inflorescence architecture. GWAS maps several classical loci for plant height and candidate genes for inflorescence architecture and also traced the independent spread of multiple haplotypes carrying alleles for short stature or long inflorescence branches. This genome-wide map of SNP variation in sorghum provides a basis for crop improvement through marker-assisted breeding and genomic selection.

7.10.7 NGS for Hybrid Production

A hybrid is a key contributor to sustainable food production with the heterosis of wide crosses delivering to greater production and productivity potential. Nowadays, NGS techniques are applied to the analysis and prediction of heterosis in well-studied systems like maize and sorghum (Martienssen et al. 2004). Gene expression patterns are also evaluated using NGS in hybrids (Guo et al. 2008). NGS could be extended to species in which the technology has not been well developed (Henry 2012). Omics and GWAS along with sequencing can be used to compare the genomes and transcriptomes of parental lines and to direct the crossing of individuals with greater potential for superior hybrid performance. This new technique had potential to advance the development of new crop species/varieties as a good option for production of new hybrid.

7.10.8 Digital Genotyping

Development of chip-based methods for genotyping has been hindered by the high start-up costs for this technology in sorghum. Thus, RE-based genotyping-by-sequencing method termed digital genotyping (DG), using Illumina GAIIx, 454 genome sequencing platform, and HiSeq2000, can be used due to their increased sequencing capacity. DG analyzes sorghum at different levels of complexity (number of sites per genome), using a set of methylation-sensitive restriction enzymes (REs) having nested cut sites, so that information from all assays can be easily cross-referenced. Morishige et al. (2013) reported some of the reasons why variation in sequencing depth per site occurs within the same genome and how to minimize this source of inefficiency. Digital genotyping was validated through genetic map reconstruction, QTL analysis, and haplotype/pedigree analysis.

7.10.9 Genotyping by Sequencing, SNP Identification, and MAS

In molecular biology labs, plant genetic and genomic studies are very easy with the help of available molecular markers which help in variability identification, mapping and tagging of genes/quantitative trait loci (QTLs), and segregation analysis. But because of its high cost of analysis and limited availability, most of these marker systems are constrained in their use conducted on a large scale. Among various molecular markers in use (Aggarwal 2008; Sonah et al. 2010), single nucleotide polymorphisms (SNPs) are the most abundant in a genome and suitable for analysis on a wide range of scales (Rafalski 2002; Zhu et al. 2003). But, the development of high-throughput genotyping platforms for large numbers of SNPs studies (thousands to millions) is relatively lengthy and costly.

High-throughput next-generation sequencing (NGS) technologies coupled with the growing number of genome sequences open the opportunity to redesign genotyping strategies for more effective genetic mapping and genome analysis compared to Sanger sequencing (Pareek et al. 2011). Beissinger et al. (2013) also reported that, even at relatively low coverage, GBS can produce enough information for powerful QTL mapping in biparental populations, the large number of multiplexing can be supported without any reduction in the number or depth of coverage of SNPs, and the use of two selective bases (AC) primer enhances twofold level of multiplexing results in an increased number of SNPs (+9.5 %) and depth of coverage (+37 %) by reducing the per-sample cost effectively. The use of modified protocol of GBS library preparation in using primers with selective bases resulted in an increased SNPs coverage, but not always increases SNPs number. This will likely vary from species to species and need to be discovered empirically but is very simple to optimize.

Being a straight forward sequencing approach for small genomes, target enrichment and genome complexity reduction can also be used to ensure sufficient overlap in sequence coverage for plant species having large genomes. The use of restriction enzymes is comparatively very easy and reproducible compared to other target enrichment methods such as the use of long-range PCR, molecular inversion, capture probes, etc. used for genome complexity reduction. These studies illustrate the application of GBS (Elshire et al. 2011) in introgression programs for traits of interest on the basis of its role in linkage mapping and QTL's identification studies and expansion of its use from this single locus selection approach to genomic selection (Hayes et al. 2010), applied directly in breeding populations for significantly enhanced efficiencies and genetic gains which in turn speed up the process of MAS for crop improvement. Genotyping by sequencing using NGS technologies (Illumina, 454 and others) successfully generated over 53,000 SNPs per population for use in multiple linkage and association mapping studies Thurber et al. (2013). They have also mapped several QTLs across a variety of traits (plant height, maturity, biomass accumulation, lodging, leaf length and width, growth rate, moisture content, stalk circumference, sugar content, juice volume, and nitrogen stress) with significant resolution in sorghum. Several candidate genes for sugar production, plant height, and maturity variation could be exploited to improve sorghum and related grasses for biofuel uses in the coming decades. Though, adaptation of GBS approaches has enabled us to cost-effective genotyping of genetically heterogeneous and diverse landraces and also helpful in analyzing population level distance and diversity parameters. This move from individual SNP-based genotyping to more comprehensive accession-based SNP fingerprints enables enquiry and use by those interested in germplasm development and germplasm conservationists and evolutionary biologists alike. Thus, GBS being a cost-effective option for mapping large populations, its application is very challenging for association mapping studies in plants as well as animal species due to huge variations existing in natural populations. GBS is very easy and less time-consuming because variations are typically detected by reference genome alignments, but in natural populations, the variations are not limited to SNPs but also PAVs (presence/absence variants), and PAV detection is exceptionally difficult unless comparative genome hybridization (CGH) approaches are applied along with NGS. In addition, these novel computational tools along with omic studies and deep data sequencing would also help to overcome these problems.

7.11 Conclusions

Increasing livestock population is creating a pressure on supply of forage crop which is lacking behind the demand of high-quality forage requirements for good health of livestock due to continuously degrading natural resources, water scarcity, and global warming. Thus, sorghum provides useful crop options due to its multipurpose uses as food, feed, fuel, and fiber. It is an important crop for diverse agroecological conditions having wide adaptation, abiotic stress tolerance, and ability to produce high forage yield with better quality. But sorghum yield and quality improvement can also be accelerated by the use of available diverse germplasm resources and novel screening/breeding technologies to further improve their through crop breeding. Sorghum hybrids possess 20–40 % yield enhancement over varieties. Sorghum availability of stable cytoplasmic-nuclear male sterility (CMS) system in the germplasm with high maintainer frequency (>90 %) had increased the opportunity of genetic diversification of male-sterile lines, leading to high probability for high-yielding hybrid development. Various simply inherited traits identified in sorghum can also be used to improved forage quality cultivar development.

Aviablity of detected for several that show Significant and positive association of highly heritable traits (e.g., plant height, tiller/branch number, leaf number and leaf length, photosensitivity) having high genetic variability with forage yield provides opportunities to develop forage sorghum hybrid with high seed and forage yield potential. Similarly, genetic variability for the anti-nutritional traits (HCN in sorghum) helps in good quality lines/varieties development using cost-effective screening procedures available nowadays to improve forage quality. Large variability identified for salinity, heat, and sodacity tolerance provides opportunities to develop cultivars/varieties for cultivation in the unusable lands with the help of molecular tools along with NGS, GBS, and digital genotyping which has opened the new avenues for the breeders and agriculturists to copup various constrains which has slowed down the pace of crop improvement in last decades.

References

Abalo G, Tongoona P, Derera J, Edema R (2009) A comparative analysis of conventional and marker assisted selection methods in breeding for maize streak virus (MSV) resistance in maize. Crop Sci 49:509–520

Aggarwal PK (2008) Global climate change and Indian agriculture: impacts, adaptation and mitigation. Indian J Agri Sci 78:911–919

Andersen JR, Luebberstedt T (2003) Functional markers in plants. Trends Plant Sci 8:554–560

Anderson JC (2005) The use of CEN38 in assessing evolutionary relationships in the genus *Sorghum*. M.S. Thesis. Texas A&M University, College Station

Ariyadasa R, Stein N (2012) Advances in BAC-based physical mapping and map integration strategies in plants. J Biomed Biotechnol. doi:10.1155/2012/184854

Arnold MJ (2004) Natural hybridization and the evolution of domesticated, pest and disease organisms. Mol Ecol 13:997–1007

Arriola PE, Ellstrand NC (1996) Crop-to-weed gene flow in the genus Sorghum (Poaceae): spontaneous interspecific hybridization between johnson grass, *Sorghum halepense*, and crop sorghum *S. bicolor*. Amer J Botany 83(9):1153–1160

Asea G, Vivek BS, Lipps PE, Pratt RC (2012) Genetic gain and cost efficiency of marker-assisted selection of maize for improved resistance to multiple foliar pathogens. Mol Breed 29:515–527

Audilakshmi S, Stenhouse JW, Reddy TP, Prasad MVR (1999) Grain mould resistance and associated characters of sorghum genotypes. Euphytica 107:91–103

Aydin S, Rooney WL, Miller FR (1997) Identification and characterization of the Ma5 and Ma6 maturity loci in sorghum. In: Proceedings of the international conference on genetic improvement of sorghum and pearl millet, INTSORMIL and ICRISAT, 22–27 September 1996. Lubbock, USA, pp 641–642

Ayyanger GNR, Ponnaiya BWX (1941) Studies in *Para-Sorghum*-the group with bearded nodes. Proc Ind Acad Sci 14:17–24

Azhar FM, McNeilly T (1988) The genetic basis for salt tolerance in *Sorghum bicolor* (L) Moench seedlings. Plant Breed 101:114–121

Bapat DR, Mote UN (1982) Sources of shoot fly resistance in Sorghum. J Maharashtra Agri Uni 7:238–240

Barabaschi D, Guerra D, Lacrima K et al (2012) Emerging knowledge from genome sequencing of crop species. Mol Biotechnol 50:250–266

Barclay A (2004) Ferel Play. Rice Today, pp 15–19

Barnaud A, Deu M, Garine E et al (2009) A weed–crop complex in sorghum: the dynamics of genetic diversity in a traditional farming system. Amer J Bot 96(10):1869–1879

Bauhini C (1658) Theatri botanici sive historia plantarum. Basel

Baxter SW, Davey JW, Johnston JS, Shelton AM, Heckel DG, Jiggins CD, Blaxter ML (2011) Linkage mapping and comparative genomics using next-generation RAD sequencing of a non-model organism. PLoS One 6, e19315

Becker-Dillingen J (1927) Handbuch des Getreidebaues einschließlich Mais, Hirse und Buchweizen. Paul Parey, Berlin

Beissinger TM, Hirsch CN, Sekhon RS et al (2013) Marker density and read-depth for genotyping populations using genotyping-by-sequencing. Genetics Online, published 10.1534/genetics.112.147710

Bekele E (2004) Electric field mediated fusion of *Erugrostis Tef* and *Sorghum Bicolor* protoplasts and their electroporation conditions. Hereditas 123(3):199–203

Bennett HW, Merwine NC (1966) Meiotic behavior of a Hodo Sorgo x Johnsongrass hybrid. Crop Sci 6:127–131

Bernard S, Jewell DC (1985) Crossing maize with sorghum, *Tripsacum* and millet: the products and their level of development following pollination. Theor Appl Genet 70(5):474–483

Bhatti AG, Endrizzi JE, Reeves RG (1960) Origin of Johnson grass. J Hered 51:107–110

Bhosale SU, Benjamin S, Frederick H et al (2012) Association analysis of photoperiodic flowering time genes in west and central African sorghum [*Sorghum bicolor* (L.) Moench]. BMC Plant Biol 12:32–42

Boivin K, Deu M, Rami JF, Trouche G, Hamon P (1990) Towards a saturated sorghum map using RFLP and AFLP markers. Theor Appl Genet 98:320–328

Borad PK, Mittal VP (1983) Assessment of losses caused by pest complex to sorghum hybrid CSH 5. In: Krishnamurthy Rao BH, Murthy KSRK (eds) Crop losses due to insect pests. Entomological Society of India, New Delhi, pp 271–278

Borlaug NE (2002) Feeding a world of 10 billion people: the miracle ahead. Vitro Cellular and Developmental Biology-Plant 38(2):221–228

Borrell AK, Hammer GL, Henzel RG (2000) Does maintaining green leaf area in sorghum improve yield under drought? II. Dry matter production and yield. Crop Sci 40:1037–1048

Boyer JS (1982) Plant productivity and environment. Science 218:443–448

Brar DS, Khush GS (1997) Alien introgression in rice. Plant Mol Biol 35:35–47

Brooking IR (1971) Male sterility in *Sorghum bicolor* (L.) Moench induced by low night temperature. I. Timing of the stage of sensitivity. Aust J Plant Physiol 3:589–596

Brooking IR (1979) Male sterility in *Sorghum bicolor* (L.) Moench induced by low night temperature. II. Genotypicdifferences in sensitivity. Aust J Plant Physiol 6:143–147

Brown DCW, Thorpe TA (1986) Plant regeneration by organogenesis. In: Vasil IK (ed) Cell culture and somatic cell genetics of plants, vol 3. Acedemic Press, New York, pp 49–65

Brown SM, Hopkins MS, Mitchell SE et al (1996) Multiple methods for the identification of polymorphic simple sequence repeats (SSRs) in sorghum [*Sorghum bicolor* (L.) Moench]. Theor Appl Genet 93:190–198

Burkill IH (1966) A dictionary of the economic products of the Malay Peninsula, vol 1, 2nd edn. A.H. Government of Malaysia and Singapore, Kuala Lumpur

Burnham C (1962) Discussions in cytogenetics. Burgress, Minneapolis

Carlson PS, Smith HH, Dearing RD (1972) Parasexual interspecific plant hybridization. PNAS 69:2292–2294

Casady AJ, Anderson KL (1952) Hybridization, cytological and inheritance studies of a sorghum cross – autotetraploid sudangrass x (Johnsongrass x 4*n* sudangrass). Agronomy 44:189–194

CBD (1992) Convention on biological diversity. Text and Annexes Secretariat of Convention on Biological Diversity , Monetreal

Celarier RP (1958) Cytotaxonomy of the Andropogoneae. HI. Subtribe Sorgheae, Genus Sorghum. Cytologia [Tokyo] 23:395–417

Chalker-Scott L, Fuchigami LH (1989) The role of phenolic compounds in plant stress responses. In: Paul HL (ed) Low-temperature stress physiology in crops. CRC Press, Boca Raton, pp 27–40

Chase C, Babay-Laughnan S (2004) Cytoplasmic male sterility and fertility restoration by nuclear genes. In: Daniell H, Chase C (eds) Molecular biology and biotechnology of plant organelles. Kluwer Academic Publishers, Dordrecht, pp 593–622

Chen X, Cho YG, McCouch SR (2002) Sequence divergence of rice microsatellites in *Oryza* and other plant species. Mol Genet Genom 268:331–343

Chen R, Davydov EV, Sirota M, Butte AJ (2010) Non-synonymous and synonymous coding SNPs show similar likelihood and effect size of human disease association. PLoS One 5(10):e13574

Childs KL, Cordonnier-Pratt MM, Pratt LH, Morgan PW (1992) Genetic regulation of development in Sorghum bicolo. VII. m3 R flowering mutant lacks a phytochrome that predominates in green tissue. Plant Physiol 99:765–770

Chittenden ML, Schertz KF, Lin YR, Wing RA, Paterson AH (1994) A detailed RFLP map of *Sorghum bicolor and S. Propinquum*, suitable for high density mapping, suggests ancestral duplication of sorghum chromosomes or chromosomal segments. Theor Appl Genet 87:925–933

Choudhary S, Bohra SP, Swarnkar PL (1993) Heat shock responses in sorghum. J Phytol Res 6:39–42

Clayton WD (1961) Proposal to conserve the generic name Sorghum Moench (*Gramineae*) versus Sorgum Adans. (*Gramineae*). Taxon 10(8):242

Clayton WD, Renvoize SA (1986) Genera graminum grasses of the world, Kew Bulletin Addition Series XIII. Royal Botanic Gardens, Kew, pp 338–345

Clerget B, Dingkuhn M, Gozé E, Rattunde HFW, Ney B (2008) Variability of phyllochron, plastochron and rate of increase in height in photoperiod sensitive Sorghum varieties. Ann Bot 101:579–594

Cocking EC (1960) A method for isolation of plant protoplast and vacuoles. Nature 187:962–963

Cocking EC, George D, Price-Jones JJ, Power JB (1977) Selection procedures for the production of interspecific somatic hybrids of *Petunia hybridae* and *P. parodii*. II Albino Complextion selection. Plant Sci Letter 10:2–12

Collard BCY, Mackill DJ (2008) Marker-assisted selection: an approach for precision plant breeding in the twenty-first century. Philos Trans R Soc 363:557–572

Collard BCY, Jahufer MZZ, Brouwer JB, Pang ECK (2005) An introduction to markers, quantitative trait loci (QTL) mapping and marker-assisted selection for crop improvement: the basic concepts. Euphytica 142:169–196

Colton LM, Groza HI, Wielgus SM, Jiang S (2006) Marker-assisted selection for the broad-spectrum potato late blight resistance conferred by gene RB derived from a wild potato species. Crop Sci 46:589–594

Cordeiro GM, Pan YB, Henry RJ (2006) Sugarcane microsatellites for the assessment of genetic diversity in sugarcane germplasm. Plant Sci 165:181–189

Cosgrove DJ (1997) Assembly and enlargement of the primary cell wall in plants. Ann Rev Cell Dev Biol 13:171–201

Cox TS, Bender M, Picone C et al (2002) Breeding perennial grain crops. Crit Rev Plant Sci 21:59–91

Crasta RR, Xu W, Rosenow DT, Mullet JE, Nguyen HT (1999) Mapping of post-flowering drought resistance traits in grain sorghum: association of QTLs influencing premature senescence and maturity. Mol Genet Genom 262:579–588

Cui XQ, Wise RP, Schnable PS (1996) The *rf2* nuclear restorer gene of male-sterile, T-cytoplasm maize. Science 272:1334–1336

Dahlberg JA (2000) Classification and characterization of sorghum. In: Smith CW, Frederiksen RA (eds) Sorghum: origin, history, technology, and production. Wiley, New York, pp 99–130

Dahlberg JA, Wasylikowa K (1996) Image and statistical analyses of early sorghum remains (8000 B. C.) from the Nabta Playa archaeological site in the Western Desert, southern Egypt. Veg Hist Archaeobot 5:293–299

Damania A, Valkoun J, Willcox G, Qualset C (1998) The origins of agriculture and crop domestication, Ist ed. International center fro agricultural research in Dry Areas, Aleppo, Syria

Davey MR, Anthony P, Power JB, Lowe CL (2005) Plant protoplasts: status and biological perspectives. Biotechnol Adv 23:131–171

de Oliveira EJ, dos Silva AS, de Carvalho FM, dos Santos LF, Costa JL, deAmorim VBO, Dantas JLL (2010) Polymorphicmicrosatellite marker set for *Carica papaya* L. and its use in molecular-assisted selection. Euphytica 173:279–287

de Wet JMJ (1978) Systematics and evolution of Sorghum sect. Sorghum (Gramineae). Amer J Bot 65(4):477–484

de Wet JMJ, Huckabay JP (1967) The origin of sorghum bicolor II. Distribution and domestication. Evolution 21:787–802

de Wet JMJ, Price EG (1976) Plant domestication and indigenous African agriculture. In: Harlan JR, de Wet JMJ, Stemler A (eds) Origins of African plant domestication. Mouton, The Hague, pp 453–464

de Wet JMJ, Harlan JR, Price EG (1970) Origin of variability in the spontanea complex of Sorghum bicolor. Amer J Bot 57(6):704–707

Dillon SL, Lawrence PK, Henry RJ, Ross L, Price HJ, Johnston JS (2004) *Sorghum laxiflorum* and *S-macrospermum*, the Australian native species most closely related to the cultivated *S-bicolor* based on ITS1 and ndhF sequence analysis of 25 Sorghum species. Plant Syst Evol 249:233–246

Dillon SL, Shapter FM, Henry RJ, Cordeiro G, Izquierdo L, Lee LS (2007) Domestication to crop improvement: genetic resources for Sorghum and Saccharum (Andropogoneae). Ann Bot 100(5):975–989

Doggett H (1965) Disruptive section in crop development. Nature 4981:279–280

Doggett J (1970) Sorghum (Tropical Agriculture Series) Longmans. Green and Co., London

Doggett H (1976) Sorghum. In: Simmonds NW (ed) Evolution of crop plants. Longman Press, Essex, pp 112–117

Doggett H (1988) Sorghum, 2nd edn. Wiley, New York

Downs RW, Marshall DR (1971) Low temperature induced male sterility in grain sorghum. Aust J Exp Agric Ani Husb 11:352–356

Dufour P, Deu M, Grivet L et al (1997) Construction of a composite sorghum genome and comparasion with sugarcane a related complex polyploid. Theor Appl Genet 94:409–418

Duncan RR, Bramel-Cox PJ, Miller FR (1991) Contributions of introduced sorghum germplasm to hybrid development in the USA. In: Shands HL, Wiesner LE (eds) Use of plant introductions in cultivar development, part 1. CSSA Special Publications 17

Durra BN, Stebbins GL (1952) A polyhaploid obtained from a hybrid derivative of *Sorghum halepense* x *S. vulgare* var. *sudanense*. Genetics 37:369–374

Dweikat I (2005) A diploid, interspecific, fertile hybrid from cultivated sorghum, *Sorghum bicolor*, and the common Johnson grass weed *Sorghum halepense*. Mol Breed 16:93–101

Edwards D, Batley J (2010) Plant genome sequencing: applications for crop improvement. Plant Biotechnol 8:2–9

Eeckhaut T, Huylenbroeck JV, Van Laere K, Gent EV (2006) Proceedings of the Vth international symposium on in vitro culture and horticulture breeding, 1–2:117–126

Egan AN, Schlueter J, Spooner DM (2012) Applications of next-generation sequencing in plant biology. Amer J Bot 99:175–185

Ejeta G, Butler L, Hess D, Vogler R (1991) Genetic and breeding strategies for striga resistance in sorghum. In: Ransom J, Musselman LJ, Worshamand A, Parker C, (eds) Proceedings of the 5th international symposium on parasitic weeds, CIMMYT Nairobi, p 539

Ejeta G, Butler L, Babiker A (1992) New approaches to the control of striga. In: Ejeta G, Butler L, Babiker (eds) Striga research at Purdue Bulletin, pp 11–13

Ejeta GP, Tuinstra GM, Grote E et al (2000) Application of molecular markers in plant breeding. In: Haussmann B, Geiger H, Hess D, Hash C, Bramel-Cox P (eds) Training manual for a seminar held at IITA, Ibadan, Nigeria, from 16–17 August 1999. International Crops Research Institute for the semi-Arid Tropics (ICRISAT), Patancheru 502324, Andhra Pradesh, India

Ellstrand NC (2003) Current knowledge on gene flow in plants: implications for transgene flow. Phil Trans R Soc Lond B 358:1163–1170

Ellstrand NC, Prentice HC, Hancock JF (1999) Gene flow and introgression from domesticated plants into their wild relatives. Ann Rev Ecol Syst 30:539–563

Elshire RJ, Glaubitz JC, Sun Q, Poland JA, Kawamoto K, Buckler ES, Mitchell SE (2011) A robust, simple genotyping-by sequencing (GBS) approach for high diversity species. PLoS One 6, e19379

Ender M, Terpstra K, Kelly JD (2008) Marker-assisted selection for white mold resistance in common bean. Mol Breed 21:149–157

Endrizzi JE (1957) Cytological studies of some species and hybrids in the *Eusorghums*. Bot Gaz 119:1–10

Endrizzi JE, Morgan DT (1955) Chromosomal interchanges and evidence for duplication in haploid *Sorghum vulgare*. J Hered 46:201–208

Epstein E, Norlyn JD, Rush DW, Kingsbury R, Kelley DB, Wrana AF (1980) Saline culture of crops: a genetic approach. Science 210:399–404

Erichsen AW, Ross JG (1963) Irregularities at microsporogenesis in colchicine-induced male-sterile mutants in Sorghum vulgare Pers. Crop Sci 3:481–483

Feltus FA, Wan J, Schulze SR, Estill JC, Jiang N, Paterson AH (2004) A SNP resource for rice genetics and breeding based on subspecies *indica* and *japonica* genome alignments. Genome 110:453–464

Flint-Garcia SA, Thornsberry JM, Buckler ES (2003) Structure of linkage disequilibrium in plants. Annu Rev Plant Physiol Plant Mol Biol 54:357–374

Francia E, Tacconi G, Crosatti C, Barabaschi D, Bulgarelli D, Dall'Aglio E, Valè G (2005) Marker assisted selection in crop plants. Plant Cell Tissue Org 82:317–342

Franklin B (1757) Letter to Mr. Ward from Benjamin Franklin on March 24, 1757. Original letter held in trust with the Nantucket alheneum, Nantucket 02554

Franzmann BA, Hardy AT (1996) Testing the host status of Australian indigenous sorghums for the sorghum midge. In: Foale MA, Henzell RG, Kneip JF (eds) Proceedings of the third Australian sorghum conference. Tamworth, NSW, Australia, pp 365–367

Gamar YA, Mohamed AH (2013) Introgression of Striga resistance genes into a Sudanese Sorghum cultivar, Tabat, using marker assisted selection (MAS). Greener J Agri Sci 3(7):550–556

Ganal MW, Altmann T, Röder MS (2009) SNP identification in crop plants. Curr Opin Plant Bio 12:211–217

Gao Z, Xie X, Ling Y, Muthukrishnan S, Liang GH (2005) *Agrobacterium tumefaciens*-mediated sorghum transformation using a mannose selection system. Plant Biotechnol 3:591–599

Gao J, Xia B, Luo F et al (2013) Marker-assisted breeding for rf1, a nuclear gene controlling A1 CMS in sorghum (*Sorghum bicolor* L. Moench). Euphytica 193(3):383–390

Garber ED (1950) Cytotaxonomic studies in the genus *Sorghum*. Univ Calif Publ Bot 23:283–361

Garner WW, Allard HA (1920) Effect of the relative length of day and night and other factors of the environment on growth and reproduction in plants. J Agric Res 18:553–606

Gleba Y, Sytnik K (1984) Protoplast fusion: genetic engineering in higher plants. In: Shoeman R (ed) Monographs on theoretical and applied genetics. Springer, Berlin, pp 115–161

Godwin ID (2005) Sorghum genetic engineering: Current status and prospectus. In: Seetharama N, Godwin I (eds) Sorghum tissue culture and transformation. Oxford & IBH Publishing Co. Pvt. Ltd, New Delhi, pp 1–8

Gomez M, Islam-Faridi MN, Zwick M et al (1998) Tetraploid nature of *Sorghum bicolor* (L.) Moench. J Hered 89:188–190

Gonzalez-Martınez SC, Robledo-Arnuncio JJ, Collada C, Dıaz A, Williams CG, Alıa R (2004) Cross-amplification and sequence variation of microsatellite loci in Eurasian hard pines. Theor Appl Genet 109:103–111

Gowda PSB, Xu GW, Frederiksen RA, Magill CW (1995) DNA markers for downey mildew resistance genes in sorghum. Genome 38:823–826

Guo M, Yang S, Rupe M et al (2008) Genome-wide allelespecific expression analysis using massively parallel signature sequencing (MPSS (TM)) reveals cis- and trans-effects on gene expression in maize hybrid meristem issue. Plant Mol Biol 66:551–566
Habyarimana E, Lorenzoni C, Busconi M (2010) Search for new stay- green sources in *Sorghum bicolor* (L.) Moench. Maydica 55:187–194
Hadley HH (1953) Cytological relationships between *Sorghum vulgare* and *S. halepense*. Agron J 45:139–143
Hadley HH, Mahan JL (1956) The cytogenetic behavior of the progeny from a backcross (*Sorghum vulgare* x *S. halepense* x *S. vulgare*). Agron J 48:102–106
Hamblin MT, Fernandez MGS, Casa AM, Mitchell SE, Paterson AH, Kresovich S (2005) Equilibrium processes cannot explain high levels of short- and medium-range linkage disequilibrium in the domesticated grass *Sorghum bicolor*. Genetics 171:1247–1256
Hanson MR, Bentolila S (2004) Interactions of mitochondrial and nuclear genes that affect male gametophyte development. Plant Cell 16:S154–S169
Harlan JR (1975) Crops and man. American Society of Agronomy, Madison
Harlan JR (1995) The living fields: our agricultural heritage. Cambridge University Press, Cambridge
Harlan JR, deWet JMJ (1971) Towards a rational classification of cultivated plants. Taxon 20:509–517
Harlan JR, deWet JMJ (1972) A simplified classification of cultivated sorghums. Crop Sci 12:172–176
Harlan JR, Stemler A (1976) The races of Sorghum in Africa. In: Harlan JR, de Wet JMJ, Stemler A (eds) Origins of African plant domestication. Mouton, The Hague, pp 465–478
Harlan JR, de Wet JMJ, Price EG (1973) Comparative evolution of cereals. Evolution 27:311–351
Harlan JR, de Wet JMJ, Stemler ABL (1976) Origins of African plant domestication. Mouton, The Hague
Harris K, Subudhi PK, Borrell A et al (2007) Sorghum stay-green QTL individually reduce post-flowering drought-induced leaf senescence. J Exp Bot 58:327–338
Haussmann BIG, Mahalakshmi V, Reddy BVS, Seetharama N, Hash CT, Geiger HH (2002) QTL mapping of stay-green in two sorghum recombinant inbred populations. Theor Appl Genet 106:133–142
Haussmann BIG, Boureima SS, Kassari IA, Moumouni KH, Boubacar A (2007) Two mechanisms of adaptation to climate variability in West African pearl millet landraces – a preliminary assessment. SAT eJournal 3(1). http://ejournal.icrisat.org/
Hawkes J (1973) The first great civilizations. Penguin, Hammondsworth
Hayes TB, Falso P, Gallipeau S, Stice M (2010) The cause of global amphibian declines: a developmental endocrinologist's perspective. J Experiment Bio 213:921–933
Henry RJ (2012) Next generation sequencing for understanding and accelerating crop domestication. Brief Funct Genomics 11:51–56
Hillel D, Rosenzweig C (2002) Desertification in relation to climate variability and change. Adv Agron 77:1–38
Hodnett GL, Burson BL, Rooney WL, Dillon SL, Price HJ (2005) Pollen-pistil interactions result in reproductive isolation between *Sorghum bicolor* and divergent Sorghum species. Crop Sci 45:1403–1409
Holm LG, Plucknett DL, Pancho JV, Herberger JP (1977) The world's worst weeds: distribution and biology. University Press of Hawaii, Honolulu, pp 54–61
House LR, Nelson OS (1985) Tracer study of pollen-tube growth in cross-sterile maize. J Hered 49:18–21
Huang X, Feng Q, Qian Q et al (2009) High-throughput genotyping by whole-genome resequencing. Genome Res 19:1068–1076
Huelgas VC, Lawrence P, Adkins SW, Mufti MU, Goodwin ID (1996) Utilisation of the Australian native species for sorghum improvement. In: Foale MA, Henzell RG (eds) Proceedings of the Australian Sorghum Conference. 3rd, Tamsworth, NSW. 20–22 February 1996, Occasional Publ. 93. Aust Inst Agric Sci, Melbourne, VIC, pp 369–375
Hulbert SH, Richter TE, Axtell JD, Bennetzen JL (1990) Genetic mapping and characterization of sorghum and related crops by means of maize DNA probes. Proc Natl Acad Sci U S A 87:4251–4255
Hyten D, Cannon S, Song Q et al (2010) High-throughput SNP discovery through deep resequencing of a reduced representation library to anchor and orient scaffolds in the soybean whole genome sequence. BMC Genomics 11:38
ICRISAT (1992) Medium term plan. International Crops Research Institute for the Semi-Arid Tropics, Patancheru
Jacobsen T, Adams RM (1958) Salt and silt in ancient Mesopotamian agriculture. Science 128(3334):1251–1258
Jagtap V, Bhargava S, Streb P, Feierabend J (1998) Comparative effect of water, heat and light stresses on photosynthetic reactions in *Sorghum bicolor* (L.) Moench. J Exp Bot 49:1715–1721
Janakiammal EK, Singh TSN (1936) A prelimainary note on a sew *Saccharum-Sorghum* hybrid. Indian J Agric Sci 4:1050
Jarvis DI, Hodgkin T (1999) Wild relatives and crop cultivars: detecting natural introgression and famer selection of new genetic combinations in agro ecosystems. Mol Ecol 8:159–173
Jarvis A, Lane A, Hijmans RJ (2008) The effect of climate change on crop wild relatives. Agric Ecosyst Environ 126:13–23
Jin J, Liu K, Wang G, Mi L, Shen Z, Chen X, Herbert SJ (2010) Agronomic and physiological contributions to the yield improvement of soybean cultivars released from 1950 to 2006 in Northeast China. Field Crops Res 115:116–123
Jordan SA, Humphries P (1994) Single nucleotide polymorphism in exon 2 of the BCP gene on 7q31-q35. Human Mol Genet 3:1909–1915

Jordan DR, Tao Y, Godwin ID, Henzell RG, Cooper M, McIntyre CL (2003) Prediction of hybrid performance in grain sorghum using RFLP markers. Theor Appl Genet 106:559–567

Jordan DR, Mace ES, Henzell RG, Klein PE, Klein RR (2010) Molecular mapping and candidate gene identification of the Rf2 gene for pollen fertility restoration in sorghum [Sorghum bicolor (L.) Moench]. Theor Appl Genet 120(7):1279–1287

Jordan DR, Klein RR, Sakrewski KG, Henzell RG, Klein PE, Mace ES (2011) Mapping and characterization of Rf 5: a new gene conditioning pollen fertility restoration in A1 and A2 cytoplasm in sorghum (Sorghum bicolor (L.) Moench). Theor Appl Genet 123(3):383–396

Joseph M, Gopalakrishnan S, Sharma RK, Singh VP, Singh AK, Singh NK, Mohapatra T (2004) Combining bacterial blightresistance and basmati quality characteristics by phenotypicand molecular marker-assisted selection in rice. Mol Breed 13:377–387

Kamala V, Singh SD, Bramel PJ, Rao DM (2002) Sources of resistance to downy mildew in wild and weedy Sorghums. Crop Sci 42:1357–1360

Kao KN, Michayluk MR (1975) A method for high frequency intergeneric fusion of plant protoplast. Planta 115:355–367

Kao KN, Constabel F, Michayluk MR, Gamborg OL (1974) Plant protoplast fusion and growth of intergeneric hybrid cells. Planta (Berl) 120:215–227

Karper RE, Chisholm AT (1936) Chromosome numbers in sorghum. Amer J Bot 23:369–374

Karunakar RI, Narayana YD, Pande S, Mughogho LK, Singh SD (1994) Evaluation of wild and weedy sorghums for downy mildew resistance. International Sorghum and Millets Newsletter 35:2–11

Kassahun B, Bidinger FR, Hash CT, Kuruvinashetti MS (2009) Stay green expression in early generation sorghum (*Sorghum bicolor* L. Moench) QTL introgression lines. Euphytica 172:351–362

Kebede H, Subudhi PK, Rosenow DT, Nguyen H (2001) Quantitative trait loci influencing drought tolerance in grain sorghum (*Sorghum bicolor* (L.) Moench). Theor Appl Genet 103:266–276

Kilian B, Graner A (2012) NGS technologies for analyzing germplasm diversity in genebanks. Brief Func Genomics 2:38–50

Kimber C (2000) Origins of domesticated sorghum and its early diffusion to India and China. In: Smith CW, Frederickson RA (eds) Sorghum: origin, history, technology, and production. Wiley, New York, pp 3–98

Klein PE, Klein RR, Cartinhour SW et al (2005) A high-throughput AFLP-based method for constructing integrated genetic and physical maps: progress toward a sorghum genome map. Genome Res 10:789–807

Komolong B, Chakraborty S, Ryley M, Yates D (2002) Identity and genetic diversity of the sorghum ergot pathogen in Australia. Aust J Agril Res 53:621–628

Kong L, Dong J, Hart GE (2000) Characteristics, linkage-map positions, and allelic differentiation of *Sorghum bicolor* (L.) Moench DNA simple-sequence repeats (SSRs). Theor Appl Genet 101:438–448

Körnicke F (1885) Die Arten und Varietäten des Getreides. In: Handbuch des Getreidebaues. Verlag von Emil Strauss, Band 1, Bonn

Korzun V (2005) Molecular markers and their application. In: Cereals breeding. www.fao.org/BIOTECH/docs/korzun.pdf

Kouressy M, Niangado O, Dembélé T, Vaksmann M, Trouche G, Reyniers FN (1998) La sélection de sorghos photopériodiques In: Bacci L, Reyniers FN (eds) Le Futur des Céréales Photopériodiques pour une production durable en Afrique Tropicale semi-Aride. CeSIA/CIRAD, pp 247–262

Kresovich S, Barbazuk B, Bedell JA et al (2005) Toward Sequencing the Sorghum Genome. A US National Science Foundation-Sponsored Workshop Report. Plant Physiol 138:1898–1902

Kuchel H, Guoyou Y, Fox R, Jefferies S (2005) Genetic and economic analysis of a targeted marker assisted wheat breeding strategy. Mol Breed 16:67–78

Kuchel H, Williams KJ, Langridge P, Eagles HA, Jefferies S (2007) Genetic dissection of grain yield in bread wheat. I. QTL analysis. Theor Appl Genet 115:1029–1041

Kumar LS (1999) DNA markers in plant improvement: an overview. Biotechnol Adv 17:143–182

Kumar AA, Reddy BVS, Reddy CR, Blümmel M, Rao PS, Ramaiah B, Reddy PS (2010) Enhancing the harvest window for supply chain management of sweet sorghum for ethanol production, E-Journal of SAT Agricultural Research, p 8

Küster E (1909) Uber die Verschmelzung nackter Protoplasten. Ber Dtsch Bot Gesell 27:589–598

Langridge P, Lagudah ES, Holton TA, Appels R, Sharp PJ, Chalmers KJ (2001) Trends in genetic and genome analyses in wheat: a review. Aust J Agri Res 52(12):1043–1077

Laurie D, Bennett MD (1989) Genetic variation in Sorghum for the inhibition of maize pollen tube growth. Ann Bot 64:675–681

Lazarides M, Hacker JB, Andrew MH (1991) Taxonomy, cytology and ecology of indigenous Australian sorghums (Sorghum Moench: Andropogoneae: Poaceae). Aust Syst Bot 4:591–635

Levitt J (1980) Responses of plants to environmental stresses. Vol. I. Chilling, freezing, and high temperature stresses. Academic, New York

Liu A, Burke JM (2006) Patterns of nucleotide diversity in wild and cultivated sunflower. Genetics 173:321–330

Liu J, Xu X, Deng X (2005) Intergeneric somatic hybridization and its application to crop genetic improvement. Plant Cell Tissue Org Cult 82:19–44

Lo SCC, De Verdier K, Nicholson RL (1999) Accumulation of 3-deoxyanthocyanidin phytoalexins and resistance to *Colletotrichum sublineolum* in sorghum. Physiol Mol Plant Pathol 55:263–273

Ludlow MM, Muchow RC (1990) A critical evaluation of traits for improved crop yields in water-limited environments. Adv Agron 43:107–153

Luthra YP, Ghandi SK, Joshi UN, Arora SK (1988) Total phenols and their oxidative enzymes in sorghum leaves resistant and susceptible to *Ramulispora sorghicola* Harris. Acta Phytopathol Entomol 23:393–400

Mace ES, Sing V, Van Oosteron EJ, Hammer GL, Hunt CH, Jordan DR (2012) QTL for nodal root angle in sorghum (*Sorghum bicolor* L. Moench) co-locate with QTL for traits associated with drought adaptation. Theor Appl Genet 124:97–109

Mace ES, Tai S, Gilding EK et al (2013) Whole-genome sequencing reveals untapped genetic potential in Africa's indigenous cereal crop sorghum. Nature Comm 4:2320. doi:10.1038/ncomms3320

Magoon ML, Shambulingappa KG (1961) Karyomorphology of sorghum propinquum and its bearing on the origin of 40-chromosome sorghum. Chromosoma (Berl) 12:460–465

Mann JA, Kimber CT, Miller FR (1983) The origin and early cultivation of sorghums in Africa. Texas Agric Exp Stn Bull, p 1454

Mardis ER (2010) The $1000 genome, the $100,000 analysis? Genome Medicine 2:84

Mariac C, Robert T, Allinne C et al (2006a) Genetic diversity and gene flow among pearl millet crop/weed complex: a case study. Theor Appl Genet 113:1003–1004

Mariac C, Luong V, Kapran I et al (2006b) Diversity of wild and cultivated pearl millet accessions (*Pennisetum glaucum* [L.] R. Br.) in Niger assessed by microsatellite markers. Theor Appl Genet 114:49–58

Maroof MAS, Jeong SC, Gunduz I, Tucker DM, Buss GR, Tolin SA (2008) Pyramiding of soybean mosaic virus resistance genes by marker-assisted selection. Crop Sci 48:517–526

Martienssen TA, Rabinowicz PD, O'Shaughnesssy A (2004) Sequencing the maize genome. Curr Opin Plant Biol 7:102–107

Martin JH, Leonard WH (1949) Principles of field crop production. McMillan, New York

Matsuoka Y, Vigouroux Y, Goodman MM, Jesus Sanchez GS, Buckler E, Doebley J (2002) A single domestication for maize shown by multilocus microsatellite genotyping. PNAS 99(9):6080–6084

Maunder AB, Pickett RC (1959) The genetic inheritance of cytoplasmic-genetic male sterility in grain sorghum. Agron J 51:47–49

Maxted N (2006) UK landraces – a hidden resource? Plant Talk 44:8

Mayer AM, Harel E (1991) Phenoloxidases and their significance in fruit and vegetables. In: Fox PF (ed) Food enzymology, vol 1. Elsevier, London, pp 373–398

McWhorter CG (1989) History, biology, and control of Johnsongrass. Rev Weed Sci 4:85–121

Melake-Berhan A, Hulbert SH, Butler LG, Bennetzen JL (1993) Structure and evolution of the genomes of Sorghum bicolor and Zea mays. Theor Appl Genet 86:598–604

Melake-Berhan A, Ejeta LG, Menkir G (1996) Grain mold resistance, polyphenol accumulation in sorghum. J Agric Food Chem 44:2428–2434

Melchers G, Labib G (1974) Somatic hybridization of plants by fusion of protoplasts. Selection of light resistant hybrids of "haploid" light sensitive varieties of tobacco. Mol Gen Genet 135:277–294

Meredith Jr. WR (1991) Contributions of introductions to cotton improvement. In: Shands HL , Wiesner LE (eds.) Use of plant introductions in cultivar development, part 1. CSSA Special Publications 17, pp 127–146

Miedaner T, Korzun V (2012) Marker-assisted selection for disease resistance in wheat and barley breeding. Phytopathol 102(6):560–566

Miller DA, Pickett RC (1964) Inheritance of partial male-fertilityin Sorghum vulgare Pers. Crop Sci 4:1–4

Mohammadi M, Kazemi H (2002) Changes in peroxidase and polyphenol oxidase activities in susceptible and resistant wheat heads inoculated with *Fusarium graminearum* and induced resistance. Plant Sci 162:491–498

Mohan S, Purushothaman D, Jayaraj S, Rangarajan AV (1988) Phenylalanine ammonia lyase activity in the roots of Sorghum bicolor (L.) inoculated with Azospirillum. Curr Sci 57:492–493

Morishige DT, Klein PE, Hilley JL, Sahraeian SME, Sharma A, Mullet JE (2013) Digital genotyping of sorghum – a diverse plant species with a large repeat-rich genome. BMC Genom 14:448–467

Morrell PL, Williams-Coplin TD, Lattu AL, Bowers JE, Chandler JM, Patterson AH (2005) Crop-to-weed introgression has impacted allelic composition of Johnsongrass populations with and without recent exposure to cultivated sorghum. Mol Ecol 14:2143–2154

Morris GP, Ramu P, Deshpande SP et al (2013) Population genomic and genome-wide association studies of agroclimatic traits in sorghum. PNAS 110(2):453–458

Mullet JE, Klein RR, Klein PE (2001) *Sorghum bicolor*-an important species for comparative grass genomics and a source of beneficial genes for agriculture. Curr Opin Plant Biol 5:118–121

Murdock GF (1959) Africa: its peoples and their cultural history. McGraw-Hill, London

Mutegi E, Sagnard F, Muraya M et al (2010) Ecogeographical distribution of wild, weedy and cultivated *Sorghum bicolor* (L.) Moench in Kenya: implications for conservation and crop-to-wild gene flow. Genet Resour Crop Evol 57:243–253

Nair NV (1999) Production and cyto-morphological analysis of intergeneric hybrids of *Sorghum* x *Saccharum*. Euphytica 108:187–191

Neya A, Le Normand M (1998) Responses of sorghum genotypes to leaf anthracnose (Collelotrichum graminicola) under field conditions in Burkina Faso. Crop Prot 17:47–53

Niangado O (2001) The state of millet diversity and its use in West Africa. In: Cooper HD, Spillane C, Hodgin T (eds) Broadening the genetic base of crop production. IPGRI/FAO, Rome, pp 147–157

Ortiz R, Ruia-Tapia EN, Mijica-Sanchez A (1998) Sampling strategy of a core collection of Peruvian quinoa germplasm. Theor Appl Genet 96:475–483

Papa R, Gepts P (2003) Asymmetry of gene flow and differential geographical structure of molecular diversity in wild and domesticated common bean (*Phaseolus vulgaris* L.) from Mesoamerica. Theor Appl Genet 106:239–250

Papa R, Acosta J, Delgado-Salinas A, Gepts P (2005) A genome-wide analysis of differentiation between wild and domesticated Phaseolus vulgaris from Mesoamerica. Theor Appl Genet 111:1147–1158

Pareek CS, Smoczynski R, Tretyn A (2011) Sequencing technologies and genome sequencing. J Appl Genet 52(4):413–435

Paterson AH (1996) An historical perseptive. In: Paterson AH (ed) Genome mapping in plants. Academic, Austin, pp 1–5

Paterson AH, Lander ES, Hewitt JD, Peterson S, Lincoln SE, Tanksley SD (1988) Resolution of quantitative traits into Mendelian factors by using a complete linkage map of restriction fragment length polymorphisms. Nature 335:721–726

Paterson AH, Lin YRLZ, Schertz KF et al (1995) Convergent domestication of cereal crops by independent mutations at corresponding genetic loci. Science 269:1714–1718

Paterson AH, Bowers JE, Chapman BA (2004) Ancient polyploidization predating divergence of the cereals, and its consequences for comparative genomics. Proc Natl Acad Sci 101:9903–9908

Paterson AH, Bowers JE, Bruggmann R (2009) The *Sorghum bicolor* genome and the diversification of grasses. Nature 457:551–556

Peacock JM (1982) Response and tolerance of sorghum to temperature stress. In: House LR et al. (eds) Sorghum in the eighties. Proceedings of the International Symposium on Sorghum, 2–7 November 1981. ICRISAT, Patancheru, India, pp 143–160

Pereira MG, Lee M (1995) Identification of genomic regions affecting plant height in sorghum and maize. Theor Appl Genet 90:380–388

Pereira MG, Lee M, Bramel-Cox P, Woodman W, Doebly J, Whitkus R (1994) Construction of a RFLP map in sorghum and comparative mapping in maize. Genome 37:236–243

Piper JK, Kulakow PA (1994) Seed yield and biomass allocation in *Sorghum bicolor* and F_1 backcross generations of *S. bicolor* x *S. halepense* hybrids. Can J Bot 72:468–474

Price AH (2006) Believe it or not, QTLs are accurate! Trends Plant Sci 11:213–216

Price HJ, Hodnett GL, Burson BL, Dillon SL, Rooney WL (2005) A *Sorghum bicolor* x *S. macrospermum* hybrid recovered by embryo rescue and culture. Aust J Bot 53:579–582

Pring DR, Tang HV, Howad W, Kempken F (1999) A unique two-gene gametophytic male sterility system in Sorghum involving a possible role of RNA editing in fertility restoration. J Hered 90:386–393

Proteres R (1951) Une cereal mineure cultivee dans l'Quest-Africain (*Brachlaria deflexa* C.E. Hubbard. Var.sativa nov. var.). L'Agronomique Tropicale 6:39–42

Quinby JR (1974) Sorghum improvement and the genetics of growth. Texas A&M University Press, College Station

Quintero PV, Anaya-Lόpez JL, Zamarripa-Colmenero A (2012) Genetic diversity of sweet sorghum germplasm in Mexico using AFLP and SSR markers. Pesq agropec bras, Brasflia 47(8):1095–1102

Rafalski A (2002) Applications of single nucleotide polymorphisms in crop genetics. Curr Opin Plant Biol 5(2):94–100

Ramiah KV (1987) Breeding cereal grains to with-weed. In: Musslemann LJ (ed) Parasitic weeds in agriculture, vol 1. CRC Press, Boca Raton, pp 227–242

Razdan MK (2003) Introduction to plant tissue culture. Science Publishers, Enfield

Reddy BVS, Rai KN, Sarma NP, Kumar ISH, Saxena KB (2003) Cytoplasmic-nuclear male sterility: origin, evaluation, and utilization in hybrid development. In: Jain HK, Kharkwal MC (eds) Plant breeding: Mendelian to molecular approaches. Narosa Publishing House Pvt Ltd, New Delhi

Riggs ED, Wang S, Singh RJ, Hymowitz T (1998) Possible transfer of resistance to *Heterodera glycines* from *Glycine tomentella* to *Glycine max*. J Nematol 30(4):547–552

Rooney WL, Aydin S (1999) Genetic control of a photoperiod-sensitive response in *Sorghum bicolor* (L.) Moench. Crop Sci 39:397–400

Rosenow DT, Clark LE (1995) Drought and lodging resistance for a quality sorghum crop. In: Proceedings of the 5th annual corn and sorghum industry research conference (Chicago, IL, 6–7 December 1995), American Seed Trade Association, Chicago, pp 82–97

Rosenow DT, Ejeta G, Clark LE, Gilbert ML, Henzell RG, Borrell AK, Muchow RC (1996) Breeding for pre- and post-flowering drought stress resistance in sorghum. In: Rosenow DT, Yohe JM (eds) Proceedings of the international conference on genetic improvement of sorghum and pearl millet (Lubbock, TX, 22–27 September 1996), ICRISAT, Lubbock, India, pp 400–411

Sabadin PK, Malosetti M, Boer M et al (2012) Studying the genetic basis of drought tolerance in sorghum by managed stress trials and adjustments for phenological and plant height differences. Theor Appl Genet 124(8):1389–1402

Sagnard F, Deu M, Dembe'le D'k et al (2011) Genetic diversity, structure, gene flow and evolutionary relationships within the *Sorghum bicolor* wild–weedy–crop complex in a western African region. Theor Appl Genet doi. doi:10.1007/s00122-011-1662-0

Sanchez AC, Subudhi PK, Rosenow DT, Nguyen HT (2002) Mapping QTLs associated with drought resistance in sorghum (*Sorghum bicolor* L. Moench). Plant Mol Bio 48:713–726

Sangduen N, Hanna WW (1984) Chromosome and fertility studies on reciprocal crosses between two species of autotetraploid sorghum *Sorghum bicolor* (L.) Moench and *S. halepense* (L.) Pers. J Hered 75:293–296

Sattari M, Kathiresan A, Gregorio GB, Hernandez JE, Nas TM, Virmani SS (2007) Development and use of a two-genemarker-aided selection system for fertility restorer genes in rice. Euphytica 153:35–42

Sávoly S (1921) A seprőczirok termesztése és a czirokseprő készítése házilag. Frankli-Társulat, Press, Boca Raton

Schecter Y, de Wet JMJ (1975) Comparative electrophoresis and isozyme analysis of seed proteins from cultivated races of sorghum. Amer J Bot 62:254–261

Schertz KF, Sotomayor-Rios A, Torres-Cardona S (1989) Cytoplasmic-nuclear male sterility: opportunities inbreeding and genetics. Proc Grain Sorghum Res Utility Conf 16:175–186

Schnable PS, Wise RP (1996) The molecular basis of cytoplasmic male sterility and fertility restoration. Trends in Plant Sci 3(5):175–182

Sears ER (1993) Use of radiation to transfer alien chromosome segments to wheat. Crop Sci 33:897–901

Sharma HC (1993) Host plant resistance to insects in sorghum and its role in integrated pest management. Crop Prot 12:11–34

Sharma DC, Forsberg RA (1977) Spontaneous and induced interspecific gene transfer for crown rust resistance in *Avena*. Crop Sci 17:855–860

Sharma HC, Franzmann BA (2001) Host-plant preference and oviposition responses of the sorghum midge, *Stenodiplosis sorghicola* (Coquillett) (Dipt., Cecidomyiidae) towards wild relatives of sorghum. J Appl Ent 125:109–114

Sharma HC, Hariprasad KV (2002) Flowering events in sorghum in relation to expression of resistance to sorghum midge, Stenodiplosis sorghicola. Euphytica 127:411–419

Sharma HC, Taneja SL, Leuschner K, Nwanze KF (1992) Techniques to screen sorghum for resistance to insects. Information bulletin no. 32. International Crops Research Institute for the Semi-Arid Tropics, Patancheru 502 324, Andhra Pradesh, India, p 48

Shi AN, Leath S, Murphy JP (1998) A major gene for powdery mildew resistance transferred to common wheat from wild einkorn wheat. Phytopath 88:144–147

Singh RJ (2003) Plant cytogenetics. CRC Press, Boca Raton

Singh BU, Sharma HC, Rao KV (2011) Mechanisms and genetic diversity for host plant resistance to spotted stem borer, *Chilo partellus* in sorghum *Sorghum bicolor*. JApplEntomoldoi.doi:10.1111/j.1439-0418.2011.01647

Singh VK, Singh A, Singh SP et al (2012) Incorporation of blast resistance into 'PRR78', an elite Basmati rice restorer line, through marker assisted backcross breeding. Field Crop Res 128:8–16

Smith CW, Frederiksen RA (2000) Sorghum: origin, history, technology and production. John Wiley & Sons, Inc, USA ISBN ISBN 0-471-24237-3

Snowden JD (1936) The cultivated races of sorghum. Adlard, London, pp 1–274

Snowden JD (1955) The wild fodder sorghums of the section Eu-sorghum. J Linn Soc Lond 55:191

Sonah H, Deshmukh RK, Singh VP, Gupta DK, Singh NK (2010) Genomic resources in horticultural crops: status, utility and challenges. Biotech Adv 29:199–209

Stafford HA (1969) Changes in phenolic compounds and related enzymes in young plants of sorghum. Phytochem 8:743–752

Stalker HT, Harlan JR, de Wet JMJ (1977) Observations on introgression of Tripsacum into maize. Amer J Bot 64:1162–1169

Stamova BS, Chetelat RT (2000) Inheritance and genetic mapping of cucumber mosaic virus resistance introgressed from *Lycopersocon chilense* into tomato. Theor Appl Genet 101:527–537

Stebbins GL Jr (1950) Variation and evolution in plants. Columbia University Press, New York

Steele KA, Price AH, Shashidhar HE, Witcombe JR (2006) Marker-assisted selection to introgress rice QTLs controlling root traits into an Indian upland rice variety. Theor Appl Genet 112:208–221

Stemler ABL, Harlan JR, de Wet JMJ (1975) Evolutionary history of cultivated sorghums (*Sorghum bicolor* [Linn.] Moench) of Ethiopia. Bull Torrey Bot Club 102:325–333

Stemler ABL, Harlan JR, de Wet JMJ (1977) The sorghums of Ethiopia. Econ Bot 31:446–450

Stephens JC, Holland PF (1954) Cytoplasmic male sterility for hybrid sorghum seed production. J Agron 46:20–23

Subudhi PK, Rosenow DT, Nguyen HT (2000) Quantitative trait loci for the stay-green trait in sorghum (*Sorghum bicolor* (L.) Moench) consistency across genetic backgrounds and environments. Theor Appl Genet 101:733–741

Sun Y, Suksayretrup K, Kirkham MB, Liang GH (1991) Pollen tube growth in reciprocal interspecific pollinations of *Sorghum bicolor* and *S. versicolor*. Plant Breed 107:197–202

Takebe I, Labib G, Melchers G (1971) Regeneration of whole plant from isolated mesophyll protoplasts of tobacco. Naturwissenschaften 58:318–320

Tang H, Liang GH (1988) The genomic relationship between cultivated sorghum [*Sorghum bicolor* (L.) Moench] and johnsongrass [*S. halepense* (L.) Pers.]: a re-evaluation. Theor Appl Genet 76:277–284

Tanksley SD (1993) Mapping polygenes. Ann Rev Genet 27:205–233

Tao Y, Manners JM, Ludlow MM, Henzell RG (1993) DNA polymorphism in grain sorghum (*Sorghum bicolor* L. Moench). Theor Appl Genet 86:679–688

Tao YZ, Henzell RG, Jordan DR, Butler DG, Kelly AM, McIntyre CL (2000) Identification of genomic regions associated with stay green in sorghum by testing RILs in multiple environments. Theor Appl Genet 100:1225–1232

Tao YZ, Hardy A, Drenth J et al (2003) Identifications of two different mechanisms for sorghum midge resistance through QTL mapping. Theor App Genet 107:116–122

Tenkouano A (1995) Current trends in regional integration of sorghum research in Western and Central Africa. In: SICNA and ICSIRAT (eds) Int. Sorghum Millets Newsl 36:41–44

Tesso T, Kapran I, Grenier C et al (2008) The potential for crop-to-wild gene flow in sorghum in Ethiopia and Niger: a geographic survey. Crop Sci 48:1425–1431

Thomas B, Prue VD (1997) Photoperiodism in plants, 2nd edn. Academic, London

Thomas R, Venkatraman TS (1930) Sugarcane-Sorghum hybrids. Agric J India 25:114–118

Thornsberry JM, Goodman MM, Doebley J, Kresovich S, Nielsen D (2001) Dwarf polymorphisms associate with variation in flowering time. Nat Genet 28:286–289

Thurber CS, Ma JM, Higgins RH, Brown PJ (2013) Retrospective genomic analysis of sorghum adaptation to temperate-zone grain production. Genome Bio 14:68–80

Tovar MJ, Romero MP, Girona J, Motilva MJ (2002) L-Phenylalanine ammonialyase and concentration of phenolics in developing olive (*Olea europaea* L. cv. Arbequina) fruit grown under different irrigation regimes. J Sci Food Agric 82:892–898

Trouche G, Da S, Pale G, Sohoro A, Oue'Draogo O, Gosso G (2001) Evaluation participative de nouvelles variétés de sorgho. In: Hocde' H, Lanc¸on J, Trouche G (eds) Sélection participative. CIRAD, Montpellier, pp 36–55

Tuinstra MR, Grote EM, Goldsbrough PB, Ejeta G (1996) Identification of quantitative trait loci associated with drought tolerance in sorghum. Crop Sci 36:1337–1344

Tuinstra MR, Grote EM, Goldsbrough PB, Ejeta G (1997) Genetic analysis of post-flowering drought tolerance and components of grain development in *Sorghum bicolor* (L.) Moench. Mol Breed 3:439–448

Tuinstra MR, Ejeta G, Peter G (1998) Evaluation of near-isogenic sorghum lines contrasting for QTL markers associated with drought tolerance. Crop Sci 38:835–842

USDA ARS (2007) National Genetic Resources Program. Germplasm Resources Information Network (GRIN) [Online Database]

Vaksman M, Traoé S, Niangado O (1996) Le photopériodisme des sorghos africains. Agriculture et Dévélopment 9:13–18

Van Orsouw NJ, Hogers RCJ, Janssen A et al (2007) Complexity reduction of polymorphic sequences (CRoPSTM): a novel approach for large-scale polymorphism discovery in complex genomes. PLoS One 2, e1172

Van K, Rastogi K, Kim KH, Lee SH (2013) Next-generation sequencing technology for crop improvement. SABRAO J Breed Genet 45(1):84–99

Varshney RK, Nayak SN, May GD, Jackson SA (2009) Next-generation sequencing technologies and their implications for crop genetics and breeding. Trend Biotechnol 27:522–530

Vavilov NI (1992) Origin and geography of cultivated plants. Translated by D. Love Cambridge University Press, Cambridge, 498 pp

Vida G, Gal M, Uhrin A et al (2009) Molecular markers for the identification ofresistance genes and marker-assisted selection in breedingwheat for leaf rust resistance. Euphytica 170:67–76

Vigouroux Y, Mitchell S, Matsuoka Y et al (2005) An analysis of genetic diversity across the maize genome using microsatellites. Genetics 169:1617–1630

Vishnu-Mittre A (1974) The beginning of agriculture, paleobotanical evidence in India. In: Hutchinson JB (ed) Evolutionary studies in world crops: diversity and change in the Indian subcontinent. Cambridge University Press, Cambridge, pp 3–30

von Klercker J (1892) Eine Methode zur isolier lebender protoplasten. Ofvers Ventensk Akad For handle 49:463–474

Waara S, Glimelius K (1995) The potential of somatic hybridization in crop breeding. Euphytica 85:217–233

Waniska RD, Venkatesha RT, Chandrashekar A et al (2001) Antifungal proteins and other mechanisms in the control of sorghum stalk rot and grain mold. J Agric Food Chem 49:4732–4742

Washburn RS, Martin JH (1933) An economic study of broomcorn production. U.S. Department of Agriculture, Technical bulletin No. 347

Weir TL, Park SW, Vivanco JM (2004) Biochemical and physiological mechanisms mediated by allelochemicals. Curr Opin Plant Biol 7:472–479

Wendorf F, Close AE, Schild R, Wasylikowa K, Housley RA, Harlan JR, Królik H (1992) Saharan exploitation of plants 8,000 years B.P. Nature 359:721–724

Weston LA, Nimbal CI, Jeandet P (1999) Allelopathic potential of grain sorghum [*Sorghum bicolor* (L.) Moench] and related species. In: Inderjit KMMD, Foy CL (eds) Principles and practices in plant ecology. Allelochemical interactions. CRC Press, Boca Raton, pp 467–477

Wilde F, Korzun V, Ebmeyer E, Geiger HH, Miedaner T (2007) Comparison of phenotypic and marker-based selection for *Fusarium* head blight resistance and DON content in spring wheat. Mol Breed 19:357–370

Wooten DR (2001) The use of *Sorghum propinquum* to enhance agronomic traits in sorghum. M.S. Thesis. Texas A&M University, College Station, Texas

Wright S (1931) Evolution in Mendelian populations. Genetics 16:97

Wu TP (1990) *Sorghum macrospermum* and its relationship to the cultivated species *S. bicolor*. Cytologia (Tokyo) 55:141–151

Xia X (2009) Assessing substitution saturation with DAMBE. In: Lemey P, Salemi M, Vandamme AM (eds) The phylogenetic handbook: a practical approach to phylogenetic analysis and hypothesis testing. Cambridge University Press, Cambridge

Xu Y, Crouch JH (2008) Marker-assisted selection in plant breeding: from publications to practice. Crop Sci 48:391–407

Xu GW, Magill CW, Schertz KF, Hart GE (1994) A RFLP linkage map of *Sorghum bicolor* (L.) Moench. Theor Appl Genet 89:139–145

Xu W, Subudhi PK, Crasta OR, Rosenow DT, Mullet JE, Nguyen HT (2000) Molecular mapping of QTLs conferring stay-green in grain sorghum (*Sorghum bicolor* L. Moench). Genome 43:461–469

Xu Y, Lu Y, Xie C, Gao S, Wan J, Prasanna BM (2012) Whole-genome strategies for marker-assisted plant breeding. Mol Breed 29:833–854

Yabe S, Ohsawa R, Iwata H (2013) Potential of genomic selection for mass selection breeding in annual allogamous crops. Crop Sci 53(1):95–105

Yemets AI, Blume YB (2003) Microprotoplasts as an efficient technique for chromosome transfer of single chromosomes between incompatible plant species. Cyto Genet 37:38–48

Yim KO, Bayer DE (1997) Rhizome expression in a selected cross in the *Sorghum* genus. Euphytica 94:253–256

Zhu YL, Song QJ, Hyten DL, Van Tassell CP, Matukumalli LK (2003) Single-nucleotide polymorphisms in soybean. Genetics 163:1123–1134

Zimmerman U, Scheurich P (1981) High frequency fusion of plant protoplasts by electric fields. Planta 151:26–32

Zizumbo-Villarreal D, Colunga-Garcia Marin P, Payro de la Cruz E, Delgado-Valerio P, Gepts P (2005) Population structure and evolutionary dynamics of wild–weedy–domesticated complexes of common bean in a Mesoamerican region. Crop Sci 45:1073–1083

Zwick MS, Islam-Faridi MN, Zhang HB et al (2000) Distribution and sequence analysis of the centromere-associated repetitive element CEN38 of *Sorghum bicolor* (Poaceae). Am J Bot 87:1757–1764

Pearl Millet

8

H.P. Yadav, S.K. Gupta, B.S. Rajpurohit, and Nisha Pareek

Abstract

Pearl millet [*Pennisetum glaucum* (L). R. Br.] is cultivated on more than 8 million ha in India, ranking third after rice and wheat in acreage. It is an important source of staple food for human beings and valuable source of fodder for livestock in marginal environments. A significant portion of pearl millet grain is also used for non-food purposes such as poultry and cattle feed and alcohol extraction. Pearl millet improvement research in India is coordinated through the All India Coordinated Pearl Millet Improvement Project (AICPMIP) under the aegis of Indian Council of Agricultural Research (ICAR). Enormous accomplishments have been made in pearl millet improvement during the last 25 years. This chapter presents an overview of achievements in genetic improvement, cultural practices, disease and insect-pest management, seed production and value addition.

Germplasm from diverse genetic backgrounds and breeding materials with adequate disease resistance have been utilized in hybrid parental line breeding programmes leading towards development of hybrids with good adaptation to diverse production environments. During the last 25 years, a total of 115 improved cultivars were released, which provided a wider cultivar choice to farmers in various agro-ecological regions.

Agronomic research led to the establishment of detailed recommendations for agroecology specific pearl millet zones which made it possible to harness the yield potential of high-yielding hybrids and varieties.

H.P. Yadav (✉) • B.S. Rajpurohit • N. Pareek
All India Coordinated Research Project on Pearl Millet, Jodhpur, Agriculture University, Jodhpur, Rajasthan, India
e-mail: hpyadav2008@gmail.com

S.K. Gupta
International Crops Research Institute for the Semi-Arid Tropics,
Patancheru-502 324, Hyderabad, India

M. Singh, S. Kumar (eds.), *Broadening the Genetic Base of Grain Cereals*,
DOI 10.1007/978-81-322-3613-9_8

High-yielding hybrids and open-pollinated varieties (OPVs) have been widely adopted by Indian farmers. Currently, nearly 65 % of pearl millet area is under improved cultivars, mainly hybrids. A genetic linkage map of pearl millet has been developed, and quantitative trait loci (QTL) have been identified for traits of economic importance to facilitate molecular marker-assisted selection. Several processing technologies have been standardized to popularize pearl millet-based traditional and health food products. In the future climate change scenario, pearl millet being a highly climate resilient crop will play a greater role in providing food and nutritional security.

Keywords
Pearl millet • Evolution • Gene pool • Genetic diversity • Wide hybridization

8.1 Introduction

Pearl millet [*Pennisetum glaucum* (L.) R. Br.] is a highly cross-pollinated crop due to protogynous habit of flowering. Being a C_4 plant, it has a very high photosynthetic efficiency and dry matter production capacity. The crop adapts well to the driest and marginal soils than most other cereals. It is cultivated in the marginal environments where it encounters frequent drought, high temperatures, low and erratic rainfall and infertile soils with poor water-holding capacity. Pearl millet is grown as rainy season crop in north, while in south and northwest India, it may be grown in two or even three seasons a year. Pearl millet owing to its high drought tolerance potential occupies unique position in rainy season (*kharif*) crops. It also provides good quality fodder to cattle in the arid and semi-arid tropical regions and recognized as valuable forage crop because of its robust and fast growth habit. Its stalk is used for fuel and thatching.

In India, pearl millet is a primary source of dietary energy (360 k cal/kg) for rural population and the fourth most important cereal after rice, wheat and sorghum. It is a rich source of protein, calcium, phosphorous and iron. Pearl millet grain contains fairly high amounts of thiamine, riboflavin and niacin. Pearl millet grain is also used for non-food purpose such as poultry feed, cattle feed and alcohol extraction (Basavaraj et al. 2010). The exploitation of heterosis in pearl millet was considered easy with its protogynous flowering and high outcrossing rates (Rao et al. 1951; Chavan et al. 1955). However, the usual method of developing chance hybrids by taking advantage of the natural crossing was not successful in India in obtaining yield and uniformity of the desired level to make such hybrids a viable alternative to local landrace varieties. The availability and knowledge of cytoplasmic-nuclear male sterility (CMS), the development of CMS lines and their maintainers and restorers made it possible to produce the seed of commercial single-cross F_1 grain hybrids in India (Athwal 1966). Such hybrids had significant grain-yield advantage over popular open-pollinated varieties (OPVs). Mahadevappa and Ponnaiya (1966) recorded 4–157 % more grains yield in hybrids with CMS lines over CO I, the then OPV check. Moreover, these single-cross hybrids were phenotypically much more uniform than OPVs. At this stage, the prospect of hybrid breeding appeared bright (Rachie et al. 1967). In India, CMS has been exploited to realize grain-yield heterosis on farmers' fields. In fact, pearl millet grain hybrids using a CMS line from the USA were first developed in India in the mid-1960s (Athwal and Rachie 1963). Pearl millet hybrid using the same male-sterile line was also released for forage production in the southern coastal plain region of the USA in 1972 (Burton 1977).

Pearl millet grain hybrid intended for use in animal feed has been released in the same region. Significant heterosis has been demonstrated among variety/cross hybrids in experimental plots in Western Africa (Quendeba et al. 1993) where a large area of pearl millet is grown in a wide range of environments. At present about 70 % of pearl millet area in India is under improved cultivars, mainly hybrids. Following the adoption of high-yielding and disease-resistant cultivars, pearl millet productivity has gone up from 539 kg/ha during 1986–1990 to 1198 kg/ha during 2012–2013 registering a 73 % improvement, which is highest among all food crops.

8.2 Origin

Pearl millet originated in tropical Western Africa some 4000 years ago. From there, it differentiated into two races: *globosum* race that moved to the western side and the *typhoides* race that reached Eastern Africa and spread to India and southern Africa some 2000–3000 years ago. The evolution of pearl millet under the pressures of drought and high temperatures imparted the ability to tolerate drought, nutrient-deprived soil and extremes of temperature more effectively than other cereals like wheat and rice.

8.3 Distribution, Production and Productivity

Pearl millet accounts for about 50 % of the total area under all millets in the world. Pearl millet is cultivated on about 32 m ha in more than 30 countries of four continents, viz., Asia, Africa, North America and Australia. Pearl millet is cultivated on about 14 m ha in Africa and on about 12 m ha in Asia. Among all pearl millet growing countries, India has the largest area (>8 million ha) with 8.5 million tons of production. The developing countries in Asia and Africa contribute about 93 % of total millet production in the world. Asia alone contributes 43 % of world millet production. Recently pearl millet cultivation has expanded to 4–5 m ha area in Brazil. The states growing pearl millet in India are Rajasthan, Maharashtra, Gujarat, Uttar Pradesh, Haryana, Tamil Nadu, Andhra Pradesh and Karnataka, though the first four states account for >90 % of pearl millet acreage in the country.

Based on the wide range of environmental conditions under which pearl millet is grown in India and considering requirements for local adaptation, the whole pearl millet area has been divided into three zones, viz., A_1, A and B (Fig. 8.1). Zone A_1 is composed of parts of Rajasthan, Gujarat and Haryana receiving less than 400 mm annual rainfall. This zone accounts for about 4 m ha area, is highly drought prone and has light sandy soils with high temperatures. Zone A is composed of the remaining parts of the states of Rajasthan, Gujarat and Haryana and the entire pearl millet growing areas of other northern states like Uttar Pradesh, northern Madhya Pradesh, Punjab and Delhi. This zone has sandy loam soils and an annual rainfall of greater than 400 mm. Irrigation facilities are also available in some areas. Zone B is comprised of the southern states of Maharashtra, Karnataka, Tamil Nadu and Andhra Pradesh with rainfall greater than 400 mm, heavy soils and mild temperature conditions. The B zone has shorter days compared to A_1 and A zones. At present, about 75 % of the pearl millet is grown in A and A_1 zones, while B-zone accounts for the remaining area. This concept of zones was validated based on a 3-year hybrid trial study which indicated the existence of two distinct mega-environments corresponding to the AICPMIP's A- and B-zones (Gupta et al. 2013).

Most of pearl millet in India is grown during rainy (kharif) season (June–September) but is also cultivated during summer (February–May) in Gujarat, Rajasthan and Uttar Pradesh and during post-rainy (rabi) season (November–February) at a small scale in Maharashtra and Gujarat (Mula et al. 2009). During kharif season, pearl millet is largely grown as rainfed crop except in some areas in Eastern Rajasthan, Southern Haryana and Western Uttar Pradesh where supplemental irrigation is provided in case of shortage of rainfall during the crop season.

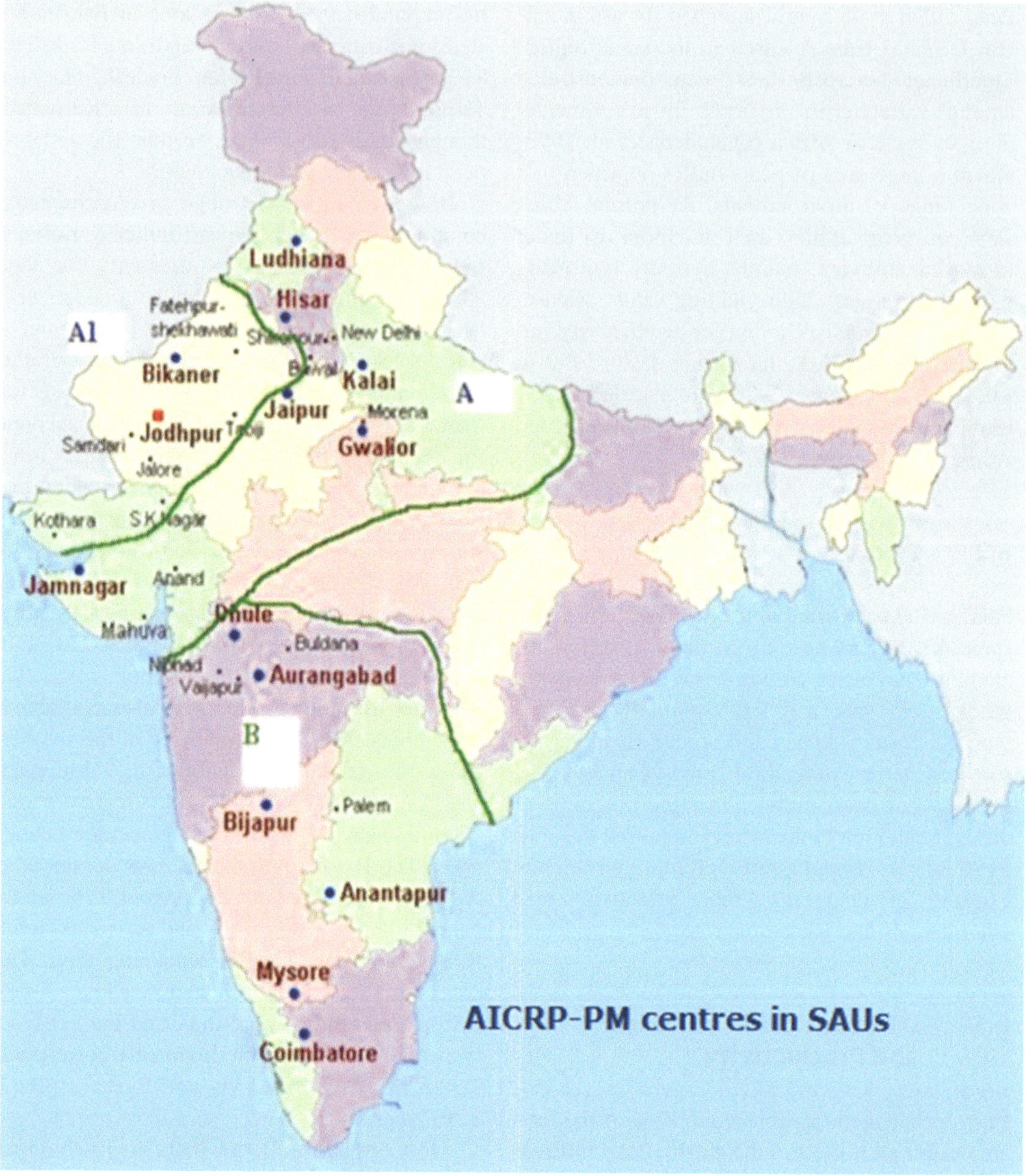

Fig. 8.1 Pearl millet growing zones of India

Summer season pearl millet is cultivated as an irrigated crop under high levels of agronomic management. Yields of more than 50 q/ha have been experienced in FLDs and 70 and 80 q/ha were achieved in national demonstrations with good management practices.

8.4 Research Prioritization

Pearl millet is cultivated under diverse agroecologies in different regions. The growing conditions vary from better-endowed environments with high-external input to highly drought-prone environments. This necessitates prioritization of research in relation to production constraints and differential requirement of various crop-growing regions.

8.4.1 Abiotic Constraints

Pearl millet is mostly (>92 %) cultivated under rainfed conditions in the arid and semi-arid regions of country where annual rainfall ranges from 150 to 750 mm, most of which is received during June to September (Harinarayana et al. 1999). Owing to its cultivation in rainfed systems, its cultivation is challenged by several abiotic stresses. Among abiotic constraints, drought affects this crop the most and is caused due to low and erratic distribution of rainfall. The coefficient of variation of annual rainfall ranges from 20 to 30 % leading to variable drought conditions within and between crop seasons. Hence, the development of pearl millet cultivars suitable for rainfed and unpredictable low-rainfall situations has been a priority area in crop management. Average rainy season maximum and minimum air temperatures (°C) in India are around 35 °C and 25 °C, respectively. The maximum air temperature around 43 °C is common in the beginning of rainy season crop. The soil surface temperatures during germination may reach 60–62 °C in the Indian arid zone. Formation of crust is also common in soil with high-silt contents. Both of these factors lead to the disturbed plant stand. Hence, cultural interventions have been explored to get improved emergence and adequate plant stand crop geometry, weed control and water management to optimize crop yields in drought-prone environments have also been worked out. Soils in the regions where pearl millet is cultivated are often infertile as they contain low amount of organic matter (0.05–0.40 %) because of low vegetation cover, coarse texture of soils and prevailing high temperatures (Kumar et al. 2009). Soils also contain low to medium levels of available phosphorous (10–25 kg/ha). Organic fertilizers are rarely applied resulting in extremely nutrient-depleted soils. Therefore, research on nutrient management has been a critical component of research in order to increase and stabilize the crop productivity. Pearl millet is also grown with supplemental irrigation with high input and management condition in small pockets scattered throughout the pearl millet growing regions. The environmental resources in such blocks are sufficient enough to support high levels of productivity to obtain a maximum return for the input supplied. Hence, maximization of yield under high crop management is also an important research priority.

Pearl millet cultivation has recently occupied large areas in irrigated summer season (February–June) crop in parts of Gujarat, Rajasthan and Uttar Pradesh where high temperatures (>42 °C) are of common occurrence during flowering. This crop faces high air temperatures during flowering leading to high reproductive sterility which finally leads to drastic yield reduction. However, a few hybrids from some of the seed companies (e.g. 86 M 11 and Proagro 9444), which specifically target this environment for hybrid development, have been found having good seed set and high grain yield of the order of 4–5 t/ha. Based on multilocation and multi-year screening in target ecology, large genetic variation for tolerance to heat at reproductive stage among pearl millet breeding lines and populations has been observed, and heat-tolerant lines have been identified. These include several maintainer lines (ICMB 92777, ICMB 05666, ICMB 00333, ICMB 01888, ICMB 02333 and ICMB 03555), improved populations (ICMV 82132, MC 94, ICTP 8202 and MC-Bulk) and germplasm accessions (IP 19799, IP 19877 and IP 19743) (Gupta et al. 2015a, b).

8.4.2 Biotic Constraints

Pearl millet production is confronted with relatively few biotic stresses as compared to other crops. Among the diseases, downy mildew (*Sclerospora graminicola*) is the most important constraint, especially on genetically uniform hybrids. Other diseases include smut (*Moesziomyces penicillariae*), rust (*Puccinia substriata* var. *indica*), blast (*Pyricularia grisea*) and ergot (*Claviceps fusiformis*). Development, screening and evaluation of germplasm, breeding material and hybrid parental lines for their reaction to various diseases are hence integral components of research. Although many insects have been reported in pearl millet, only a few of them are of some, albeit little, economic importance in causing losses to the crop. The insect-pest incidence on commercial cultivars and experimental test genotypes needs to be closely monitored during crop seasons.

8.4.3 Addressing Utilization Constraints

Enhancing the utilization of pearl millet as food is constrained by the property of its flour to acquire a rancid smell within 7–10 days of milling because of high concentration of lipids that contribute to the development of fat acidity, lipolytic activity and accumulation of peroxide of lipids in the meal during storage. Another constraint is the presence of various anti-nutrients like phytate and polyphenols which affect the nutritional quality and interfere with mineral bioavailability, inhibit proteolytic and amylolytic enzymes and reduce the protein and starch digestibility. Polyphenols also affect the nutritional quality and interfere with mineral bioavailability and protein digestibility by inhibiting the activity of several enzymes. Several processing techniques have been explored to circumvent these constraints and develop value-added food products with enhanced shelf life.

8.5 Genetic Improvement

8.5.1 Germplasm Utilization

Utilization of diverse germplasm is very critical to broaden the genetic base of cultivars. Hybrid parents in pearl millet, both seed parents (A-lines) and restorer parents (R-lines), have been bred utilizing germplasm for different geographical regions. A-lines have been bred utilizing significant germplasm of African origin, while locally adapted Indian material was used in breeding R-lines in most of public and private sector hybrid programmes. For instance, trends in utilization of genetic resources at ICRISAT suggest that a wide range of germplasm, composites, elite lines and crosses between these three groups have been used in the breeding of both maintainer (B-) and R-lines of hybrids (Table 8.1). The nature of germplasm, composites and elite lines that have gone into the development of B- and R-lines varied enormously. This was clearly reflected in the molecular characterization data of diversity among B- and R-lines developed at ICRISAT. In a recently conducted study at ICRISAT, genetic diversity pattern between two groups of parents was investigated, bred till 2004 (Set I) and developed during 2004–2010 (Set II). Combined analysis of 379 hybrid parents (current 166 parents and 213 previously developed hybrid parents) carried out using a set of highly polymorphic 28 SSRs detected 12.7 alleles per locus. Distance matrix differentiated these currents and previously developed hybrid parents into two separate clusters, indicating infusion of new genetic variability over time as reflected by the involvement of more

Table 8.1 Type of material used in the development of designated maintainer (*B*) and restorer (*R*) lines of pearl millet bred at ICRISAT during 1981–2004

Type of genetic material	No. of lines derived		Remarks
	B-lines	R-lines	
Germplasm	3	19	Inbreeding and selection directly from germplasm
Composites	9	11	Includes composites and open-pollinated varieties
Germplasm × elite line	21	8	Includes early-generation breeding lines derived from crosses between germplasm and elite lines
Composite × elite line	19	17	Includes early-generation breeding lines derived from crosses of composites and elite lines
Elite line × elite line	47	59	Includes crosses between elite lines from advanced generations

genotype-specific alleles. Also, the seed and restorer parents were found clearly separated from each other in both sets with few crossovers, indicating the existence of two diverse and broad-based pools in hybrid parents of pearl millet. Results suggested that newly developed lines were as much divergent when compared with previously developed lines, indicating that current ICRISAT pearl millet breeding programme was moving towards the development of diverse new hybrid parental lines Fig. 8.2.

Germplasm from different regions has also been used to develop a large number of composites (Rai and Anand Kumar 1994) which have been further utilized in the development of different open-pollinated varieties (OPVs) following recurrent selection. Many pearl millet varieties such as WCC75, ICMV 155, ICTP 8203, CZP 9802, Raj 171, ICMV 221, JBV 2 and JBV 3 were developed from germplasm.

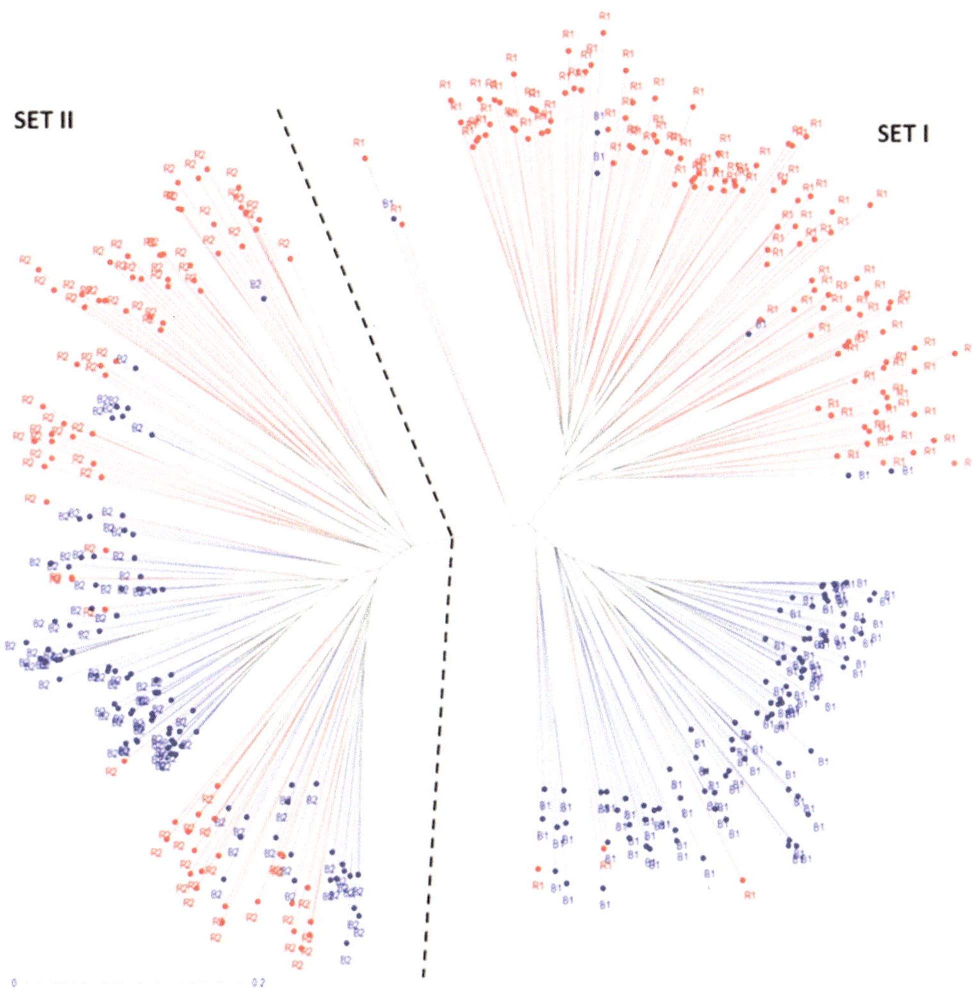

Fig. 8.2 Unweighted neighbour-joining tree based on a simple matching dissimilarity matrix for allele sizes detected by 28 simple sequence repeat primer pairs across 379 pearl millet hybrid parental lines (213 set I lines and 166 set II lines). Accessions are identified as "B" for seed parental lines and "R" as restorer lines. Suffix 1 and 2 with B- and R-lines represent set I and set II lines, respectively. B-lines are shown in *blue* and R-lines in *red colour* (Gupta et al. 2015b)

8.5.1.1 Development of Seed Parents

Considering the availability of a diverse range of restorers available in the Indian pearl millet breeding programmes and a serious lack in the number and diversity in seed parents, NARS and ICRISAT prioritized its research with much greater emphasis on seed parents' development.

8.5.1.2 Improving Male-Sterile Lines

In the breeding of seed parents (A-lines), high grain-yield potential of A-lines, both as lines per se and in hybrids (i.e. combining ability), is the most important consideration. Thus, high yield potential is the first target trait for which selection is made visually in un-replicated nurseries. High yield, coupled with other agronomic and farmer preferred traits, is targeted in A-line breeding. Some of the traits considered for all the environments include lodging resistance, compact panicles, good exertion and seed set, while regional preferred traits, like maturity, plant height (grain vs. dual purpose), tillering ability, seed colour and seed size, are specific to different agroecologies. Most of these agronomic traits have high heritability for which visual selection during advance generations is fairly effective. The A lines must also have complete and stable male sterility, and B-lines must have profuse pollen production ability across the seasons and sites. The d_2 dwarf plant height has emerged as the most dominant plant type concept in seed parent breeding. This has several operational advantages: (i) it provides the option for breeding hybrids of varying heights, (ii) it provides greater control on seed yield and quality by reducing the risk of lodging that can occur under high management seed production conditions, and (iii) it allows a much rapid detection and efficient rouging of off-types and pollen shedders in A lines. A large number of designated A lines bred represent considerable morphological diversity for the agronomic traits. In view of the increasingly important role of the stover, seed parent development programme is targeting medium and late maturity duration. New plant types such as A lines with long panicles (30–80 cm compared to standard normal of 10–20 cm), thick panicles (40–50 mm diameter compared normal 20–30 mm) and large seed size (17–20 g of 1000-seed mass compared to standard normal of 9–12 g) are being developed at ICRISAT and many All India Coordinated Pearl Millet Improvement Centres. Progenies are evaluated for downy mildew resistance during the generation advancement of breeding lines which runs concurrent to agronomic evaluation to ensure that B- and R-lines finally produced are resistant to this disease.

8.5.1.3 Genetic and Cytoplasmic Diversity

The male-sterile lines and pollinators being used to develop superior hybrids should be diverse in genetic and cytoplasmic nature.

8.5.1.4 Genetic Diversification of Male-Sterile Lines

Till date A_1 CMS has been the most stable source of male sterility in pearl millet hybrid breeding. This source continues to be the only one involved in almost all the commercial hybrids produced so far. It also continues by far to be the most extensively used source in seed parent breeding. As a consequence a large number of male-sterile lines developed with this source are now available, especially in India ICRISAT alone developed and disseminated more than 90 male-sterile lines of diverse genetic backgrounds and with diverse morphological characteristics during 1971–2013.

8.5.1.5 Cytoplasmic Diversification

Burton and Athwal (1967) studied the relationships between cytoplasms of Tift 23A, L66A and L67A. They crossed all the three CMS lines with each of their respective maintainers (B-lines) and also to some of the restorers (R-lines). Based on the fertility/sterility reaction as measured by seed set in A × B hybrids, genetic models were proposed for A-, B- and R-lines. Several other CMS sources were identified in the following years. For instance, Appadurai et al. (1982) developed a CMS line PT732A and showed its cytoplasm to be different from that of Tift 23A. Aken'Ova and Chheda (1981) identified male-sterile plants in a Nigerian pearl millet population (ex-Bornu), a gero millet. In later studies, Aken'Ova (1985) crossed Tift $23B_1$, Tift $238B_2$ and L67B (sterility maintainers of A_1, A_2 and A_3, respectively) with ex-Bornu male-sterile line. All the three maintained sterility in crosses with ex-Bornu source but not as well as did gero-B (the selfed progeny of an ex-Bornu line) which was an excellent maintainer for ex-Bornu source of CMS. Gero-B maintained sterility in Tift 23A and L67A sources but was a partial fertility restorer of the Tift $239A_2$. Tift 186 which was reported to be a maintainer for A_1, A_2 and A_3 sources restored fertility in ex-Bornu, confirming that male sterility identified in ex-Bornu was a new CMS source. Based on the similar field studies of male fertility restoration in hybrids, Marchais and Pernes (1985) identified a cytoplasm from an accession of *P. glaucum* subsp. *violaceum* from Senegal and showed its male sterility to be different from the A_1, A_2 and A_3 cytoplasms. However, in studies involving near-isonuclear lines, it was found that this cytoplasm had considerable resemblance with A_1 system (Talukdar et al. 1987). This CMS source was designated as Av. Hanna (1989) identified a cytoplasm from another accession of *P. glaucum* subsp. *monodii* (*violaceum*) from Senegal and showed its male sterility to be different from the A_1, A_2 and A_3 cytoplasms. This new cytoplasm has been designated as A_4. Rai and Hash (1990) observed significant effect of nuclear genetic background on the fertility restoration patterns of hybrids and suggested that isonuclear lines (the same nuclear genome transferred into different cytoplasmic backgrounds) should be used for reliable classification of CMS sources in pearl millet. Using isonuclear A-lines, Rai et al. (1996) showed that male sterility from the above two wild species sources are different from each other as well as from the A_1, A_2 and A_3 sources. Several other CMS sources identified at ICRISAT, in the Indian national programmes and elsewhere were shown either to be no different from the A_1 CMS system or had no more stable male sterility and hence were of little applied value. Among the other CMS sources, the PT732A source has been mostly used at the Tamil Nadu Agricultural University, and the A_2 and A_3 sources have been largely used in seed parent breeding at the Punjab Agricultural University. Both of these have had problems with high levels of pollen shedders in their A lines. Good maintainers of these three CMS sources are very infrequent in the breeding materials, limiting their utility in genetic diversification of A lines. Other CMS sources have either not proved better than the A_1 or have not been properly characterized. Research shows that among all the reported CMS sources, the A_4 and A_5 have more stable male sterility than the A_1 source (Rai et al. 2009). Restorers of the A_4 and A_{egp} occur in materials of diverse origin, indicating their greater immediate utility in breeding seed parents of grain hybrids. For breeding seed parents of forage hybrids, the A_5 CMS source may be the best one as virtually any inbred is a potential maintainer of it and hence can be converted into an A-line. Restorer sources of A_5 are hard to find (Rai 1995). However, its restorer genes have now been discovered, paving the way for the potential utility of this CMS source in breeding seed parents of grain hybrids as well. The use of the A_4 source in seed parent breeding has already begun in several hybrid breeding programmes in India and the USA. CMS lines were bred on A_1, A_2, A_3 and other sources and exploited for commercial production of hybrids.

8.5.1.6 Stability of CMS Systems

It was observed that all the male-sterile lines based on A_1 CMS system produce varying, albeit low, frequencies of pollen shedders, which is influenced by the genetic background of A-lines and the environments in which they are grown. At the same time, it had also been established that A_2 and A_3 CMS systems were unstable for male sterility (Rai et al. 1996). In order to assess the stability of alternative CMS sources, viz., A_{egp}, A_4 and A_5 vis-à-vis that of A_1, isonuclear A-lines with four cytoplasms (A_1, A_{egp}, A_4 and A_5) in each of the three diverse nuclear genetic backgrounds (81B, 5054B and ICMB 88004) were evaluated for pollen shedders in six environments created by three planting dates (rainy season, early summer and late summer season) for 2 years. The results showed that the frequency of pollen shedders was higher in A-line with the A_1 cytoplasm (0.0–2.5 %) as compared to the A_4

Table 8.2 Percent pollen shedders in isonuclear A-lines of pearl millet in the background of four cytoplasms

	Pollen shedders (%) in cytoplasm			
Genotype	A_1	A_{egp}	A_4	A_5
81B	0.3–0.6	0.0–0.1	0.0	0.0
5054B	0.1–2.5	0.0	0.0–0.3	0.0
ICMB 88004	0.0–0.6	0.0–0.1	0.0–0.1	0.0

cytoplasm (0.0–0.3 %) and the A_{egp} cytoplasm (0.0–0.1 %) (Table 8.2). There were no pollen shedders in A-lines with the A_5 cytoplasm, irrespective of their genetic backgrounds or the environments in which they were grown.

8.5.1.7 Prioritization of CMS Systems

Considering the key attributes of the CMS systems as mentioned above, the highest priority should be on breeding A-lines with the A_5 and A_4 CMS systems. A beginning has been made in this direction at ICRISAT and AICPMIP centres, as reflected in the increasing proportion of A-lines with the A_4 cytoplasm, followed by the A_5 cytoplasm (Table 8.3). Additional advantage is that the genetic background of male-sterile lines in A_4 and A_5 cytoplasms does not affect the fertility restoration of hybrids, whereas the genetic background of A_1 cytoplasm has significant effect on the fertility restoration (Gupta et al. 2010). Hence, ICRISAT focused on greater use of the A_4 and A_5 CMS systems for breeding a diverse range of A-lines. Thus, during the last 18 years (1996–2013), ICRISAT developed and disseminated 150 A-lines to the public and private sector breeding programmes, of which 64 are based on A_1 CMS system, 68 on A_4 CMS system and 16 on A_5 CMS system. A recent consultation meeting with hybrid parent users also suggested increase in cytoplasmic diversification and hence recommended greater use of the A_4 and A_5 CMS systems in A-line breeding (Rai et al. 2012). However, it would require a major shift in breeding efforts for restorer development as a high proportion of the restorers of currently most exploited A_1 CMS system in all the hybrid programmes in India and elsewhere fail to restore the fertility of hybrids made on A-lines based on the A_4 and A_5 CMS systems.

8.5.2 Restorer Parents

While higher frequency of maintainers is a positive attribute for A-line development, the implied low frequency of restorers makes it a negative attribute for restorer line (R-line) development. Thus, to make the A-lines of new CMS systems commercially attractive, greater efforts would be required to breed their restorers. Considering the medium to high frequency of restorers in a diverse range of populations mentioned earlier, inbreeding and selection can be usefully practiced to develop restorer lines of the A_1 and the A_4 CMS systems.

8.5.2.1 Restorer Lines

Restorer lines must produce profuse pollen that should remain viable at air temperatures as high as 42–44 °C. Also pollen parents must produce highly fertile hybrids, which confer some degree of protection from ergot and smut infection. Besides being able to produce high-yielding hybrids, the restorers should also be highly productive which is important from the viewpoint of seed production economy. It is desirable to breed pollinators of 150–180 cm height, but no shorter than the A-line with built-in attributes of panicle, maturity and tillering that will be preferred by farmers in the hybrids. Pollinators must have acceptable level of lodging resistance and should also possess adequate levels of resistance to various diseases.

8.5.2.2 Diversification of Restorers

So far, almost entire emphasis in restorer breeding has been on the utilization of the A_1 CMS system. Hence, excellent restorers of this CMS system are abundantly available. But there is a serious lack of the A_4 and A_5 restorers in elite agronomic genotypes. However, excellent genetic stocks of A_4 and A_5 restorers developed at ICRISAT are being used in backcross breeding for the development of restorer lines of A_4 and A_5 CMS systems. An efficient backcross breeding method for converting elite inbred lines into their A_4 and A_5 restorer versions has been developed. Also, moderate and low frequency of restorers of the A_4 and A_5 CMS systems, respectively, have

Table 8.3 Number of male-sterile (*A*-) lines developed in background of different CMS systems at AICPMIP centres and ICRISAT

Period	Number of designated A-lines				
	A_1 cytoplasm	A_4 cytoplasm	A_5 cytoplasm	Others	Total
AICPMIP centres					
1986–2011	157	30	3	25	215
ICRISAT					
1986–1995	26	1	0	0	27
1996–2004	32	31	4	1	68
2005–2013	32	37	12	0	81
Total	243	90	14	26	373

been found in most of the populations surveyed. Their frequency in these populations can be rapidly increased by recurrent selection as shown for the A_1-system restorers.

8.6 Cultivar Development

Earlier efforts in pearl millet improvement in India concentrated on the utilization of local germplasm material. Using simple mass selection, a few varieties were developed. The introduction of material in the 1960s from African countries yielded useful varieties for Indian conditions. Jamnagar Giant, Improved Ghana and Pusa Moti were developed by selection from African introductions. Since pearl millet is a highly cross-pollinated crop and displays a high degree of heterosis for grain and stover yields, attempts were made in the 1950s to exploit heterosis in hybrids by utilizing the protogynous nature of flowering to produce chance hybrids and to raise crop productivity. The chance hybrids, however, could not become popular due to their limited superiority over OPVs, narrow range of adaptation and lack of seed production programmes.

8.6.1 Single-Cross Hybrids

CMS is used for developing high-yielding single-cross hybrids (SCH) of pearl millet. This is achieved by identifying a specific combination of a male-sterile line (seed parent) and an inbred (male parent pollinator) resulting in a high-yielding fertile hybrid, the seed of which can be multiplied economically. Exploitation of heterosis became a reality with the discovery of cytoplasmic-nuclear male sterility and release of male-sterile lines Tift 23A and Tift 18A in the early 1960s at Tifton Georgia, USA. These lines were made available to Indian breeding programmes. The male-sterile line Tift 23A was extensively utilized because of its semi-dwarf stature, profuse tillering, uniform flowering and good combining ability. As a result, a few hybrids based on this line were released between 1965 and 1969. One of these hybrids (HB 3) became highly popular and was extensively cultivated because of its early maturity, bold grains and adaptation to drought.

The development of single-cross hybrids started with the availability of male-sterile line Tift 23A (Athwal 1966). The spectacular yield advances that were achieved (Athwal and Rachie 1963) stimulated several breeders to make and test new hybrid combinations. During the last 42 years (1965–2012), a large number of pearl millet hybrids were identified and released in India for general cultivation (Khairwal et al. 2007). Although the numbers of hybrids bred are many, only a few became popular among farmers. Some public sector hybrids HB 3 (Tift 23A × J 104), BJ 104 (5141A × J 104), BK 560 (5141A × K 560–230), MH 179 or ICMH 451 (81A × ICMP 451), Pusa 23 (841A × D 23), HHB 50 (81A × H 90-4-5) and HHB 67 (843A × H 77/833-2) and at least five private sector hybrids (including MBH 110, GK 1004, PB9444 and MLBH 104 and 86 M86) have been widely grown, whereas CJ 104 (5054A × J 104), PHB 14 (IIIA × PIB 228) and HHB 60

(81A × H77/833-2) have been restricted to specific areas. If we trace the pedigree of popular released pearl millet hybrids, we would recognize only a limited fraction of the available genetic resources that have been exploited so far. Almost all pearl millet hybrids in India are still based on Tift 23B. Similarly, a handful of pollinators, especially J 104, K 560 and H77/833-2, have been repeatedly used in the national programmes in India. CCS HAU, Hisar, released nine pearl millet hybrids by utilizing only few pollinators (H90/4-5, H77/833-2, G73-107, H77/29-2 and HBL-11) and four male-sterile lines that were related to Tift 23A (Khairwal et al. 2007). Intensive cultivation of hybrids based on a single male-sterile line, however, led to downy mildew epidemic in the mid-1970s (Govila et al. 1997).

Using diverse male-sterile lines and pollinators, a large number of hybrids have been released in India during the last 25 years by both public and private sectors. A total of 107 improved cultivars were released since 1986. Three-fourth of these was hybrids which showed that hybrid breeding has been a major priority in India in pearl millet. The number of releases over a period of 25 years indicates that, on an average, 3–4 hybrids are released each year for general cultivation for different agro-ecological zones. Additionally, private sector seed companies have marketed significant number of hybrids as truthfully labelled seed. This has enabled farmers to choose from a wide range of available cultivars with appropriate trait combinations that they consider fit to meet their requirement in different crop production environments of various states. This cultivation of a large number of hybrids also helped in providing buffering mechanism against diseases, insect pests and environmental vagaries. Since 1986, 53 hybrids from public sector and 35 from private sector have been notified for different regions. These hybrids include HHB 45, HHB 50, MH 179, Pusa 23, GHB 30, HHB 60, HHB 67, MLBH 104, MBH 110, Eknath 301, ICMH 356, Shradha, Saburi, JKBH 26, 7686, 7688, HHB 68, HHB 94, HHB 146, HHB 117, HHB 67 Improved, RHB 121, GHB 538, GHB 558, GK 1004, Proagro 9444, GHB32, ICMH451, MBH110, Pusa23, Pusa332, PHB108, ICMH356, PB9444, 86 M86 Nandi 5, JK26, RHB121 and HHB197, HHB 223, RHB 177, MHB 17 and 86 M64.

Experiencing recurrent problems of downy mildew in hybrids in the 1970s and 1980s, AICPMIP and ICRISAT responded by increasing the efforts to breed open-pollinated varieties (OPVs) and by strengthening the research to diversify the genetic base of seed parents. As a result, many OPVs like ICTP 8203, WCC 75, HC 4, HC 10, HC 20, ICMV 155, ICMV 221, CZP 9802 and Raj 171 were adopted by growers at a large scale. Contrary to hybrids, there is no risk of breakdown of resistance of OPVs to downy mildew. The OPVs are currently being released for risk-prone areas where replanting is a common practice.

8.7 Molecular Breeding

8.7.1 Genetic Linkage Map

Efforts towards molecular breeding started in pearl millet in the early 1990s with the development of a molecular marker-based genetic linkage map which largely comprised of RFLP loci (Liu et al. 1994). This linkage map was short (circa 300 cM), but it has now been expanded (Qi et al. 2004), and current genetic linkage map of pearl millet covers 1148 cM (Supriya et al. 2011), and SNP markers have also been included (Bertin et al. 2005; Sehgal et al. 2012). The most recently published, well-saturated genetic linkage map of pearl millet provides coverage with 321 marker loci (258 DArTs and 63 SSRs) distributed over 1148 CM (Supriya et al. 2011).

8.7.2 Quantitative Trait Loci Mapping and Marker-Assisted Selection

Initial map was followed by identification of QTLs using several different mapping populations. The target traits for pearl millet QTL mapping have been downy mildew resistance, grain and stover yield under favourable conditions, stover quality and iron and zinc concentrations in pearl millet grain.

8.7.3 Downy Mildew Resistance

Several putative QTLs have been identified that determine a significant proportion of downy mildew resistance in pearl millet (Hash et al. 1997). Downy mildew-resistant version of an early-maturing hybrid HHB 67 has been released as HHB 67 Improved for drought-prone areas in the states of Rajasthan, Haryana and Gujarat. It is suggested that resistance pyramiding conferred border or broader spectrum than expected based on the performance of parents involved. It is further recommended that host plant resistance deployed in genetically uniform hybrids be backstopped with appropriate management practices (crop and cultivar rotation and the use of appropriate prophylactic fungicidal seed dressings) to extend the useful economic life of this resistance (Hash et al. 1997, 1999; Witcombe and Hash 2000; Hash and Witcombe 2002).

8.7.4 Drought Tolerance

Because of intrinsic difficulties in breeding for drought adaptation by conventional breeding (Barker et al. 2005; Blum 1988), this field has become a prime focus for molecular marker-assisted breeding. Genetic mapping for drought tolerance has targeted terminal drought. Several major QTLs have been identified that have significant effects on pearl millet yield in drought-stress environments (Bidinger et al. 2005; Yadav et al. 2002, 2004). In addition, a number of other QTLs were detected that were associated with maintenance of grain-yield-determining component traits (Yadav et al. 2011). There are reports that the LG 2 QTL is also (not mentioned earlier) associated with salinity tolerance (Sharma et al. 2011), presumably as a result of its effects on transpiration rate.

8.7.5 Stover Quality

Efforts are under way to use marker-assisted breeding for enhancement of stover quality (Hash et al. 2003). Two QTLs one each on LG 2 and LG 6 have been identified which govern several fodder quality traits. It seems that the improved stover quality conferred is likely due to improved host plant resistance to blast (Nepolean et al. 2010).

8.8 Seed Production and Supply

In the last 25 years, great progress has been made in terms of seed production and processing by using high-yielding seed parents, adoption of improved production technologies by seed producers and modernization in seed processing and packaging technologies. Both public and private sectors are involved in pearl millet seed production and distribution. The major share in marketing of improved pearl millet seed comes from private sector because of involvement of hybrid cultivars. However, public sector also plays very vital role in providing seed of improved cultivars at reasonable price to the farmers.

8.8.1 Planning

Planning is a very important aspect of seed production. Need assessment is the first most important step in production planning to produce adequate quantity of all states and allocates seed production programme to various seed-producing organizations. Depending upon the sale projections of a particular hybrid and inventory stock of various classes of seed with acceptable quality, private companies also work out requirements for foundation and breeder seed. An example of the method for calculating quantity of different classes of seed and land requirement for pearl millet seed production is given in Table 8.4. Thus, it can be seen that total annually required quantity of breeder seed of A-, B- and R-lines is

Table 8.4 Area and seed requirement for various seed classes to produce 22,000 tons (enough to plant 5.5 million ha) of certified pearl millet hybrid seed

	Year 1 Season I		Year 1 Season II		Year 2 Season I	
	For *breeder seed* (BS) production		For *foundation seed* (FS) production		For *certified seed* (CS) production	
Parental line	Area (ha)	Nucleus seed quantity (kg)	Area (ha)	BS quantity (kg)	Area (ha)	FS quantity (kg)
A-line	0.352	1.408	88.0	352.0	22,000	88,000
B-line	0.088	0.352	22.0	88.0	–	–
R-line	0.088	0.352	22.0	88.0	5500	22,000
Total	0.528	2.112	132.0	528.0	27,500	110,000

The calculations assume 1000 kg/ha of seed yield in production plots, 4 kg/ha of seed rate and female/male row ratios of 4:1

528 kg to ultimately produce 22,000 tons of hybrid seed that is sufficient enough to plant an area of 5.5 m ha. The total area to produce required breeder seed is 0.53 ha only which is equal to a plot size of 73 m × 73 m. Such small seed quantity needed to produce breeder seed to meet the huge certified seed demand shows the massive seed production potential of pearl millet. Thus, only 110 tons of foundation seed of hybrid parental lines are required to fulfil the current requirement of certified hybrid seed, and 27500 ha area is needed to produce certified seed of hybrids. The whole seed production chain is completed in three seasons (1½ year).

The most critical component in foundation and certified hybrid seeds is to ensure adequate isolation distance for which planning is very critical. This requirement can be easily fulfilled through seed village concept to undertake seed production programme in nontraditional area in Telangana during summer season. Carry-over strategy is a very critical component of planning. Carry-over seed is produced as a buffer stock to insure against sudden demand or unforeseen shortfalls due to vagaries of nature. A general limit of carry-over seed in pearl millet is 100 % for breeder seed, 50 % for foundation seed and 20 % for certified seed.

8.8.2 Production Chain

Seed production and supply chain in India are very well developed for pearl millet and involve multiplication process of various classes (nucleus, breeder, foundation and certified). Commercial seed production chain starts from breeder seed, which is produced by AICPMIP centres, SAUs, ICRISAT and some private companies. Foundation seed of OPVs and parental lines of hybrids is produced by the National Seed Corporation (NSC), State Farms Corporation of India, State Seed Corporation (SSCs) and some seed companies. Certified seed of released and notified cultivars is mainly produced by SSCs and NSC following established procedure of seed production which starts from compiling of indents by the Government of India that is passed on to the Assistant Director General (Seeds), Indian Council of Agricultural Research, who in turn, passes on the indents to the Project Coordinator (Pearl millet). The Project Coordinator allots the breeder seed production of OPVs and hybrid parental lines to different originating centres and monitors the production programme. The breeder seed produced is then passed on to the indenting agencies to enable them to use it in foundation seed production. Private sector is involved in large-scale seed production of public bred hybrids as well as their own proprietary hybrids. The multiplication and distribution of certified seed of public hybrids like Pusa 23, HHB 67, HHB 67 Improved, HHB 197 and OPV (ICTP 8203) by the private sector is one example contributing towards the common goal of farmers' benefit. Private sector also produces large quantities of truthfully labelled seed which need not be certified by the seed certification agencies. The quantity of such seed produced and marketed by the private sector is yet to be assessed.

8.8.3 Production Calendar

The certified seed production programme is organized primarily in farmers' fields by various seed corporations and private companies and is undertaken in South India. Almost 80 % of total seed is produced during summer (January–April) in Telangana. After harvesting and processing, the seed is then dispatched to the target area. This cycle of seed production reduces the overhead cost as no seed storage is required for long. Some seed is produced during the rainy season in Andhra Pradesh and Karnataka to meet any deficit. However, rainy season seed production sometimes is risky due to pollen wash and ends up with lower seed yields. Seed quality is also not as good as that obtained from summer production.

8.8.4 Production Technology

Substantial refinement in seed production technology has taken place in the recent past. Planting time, planting ratio of male and female parental lines of hybrids and agronomic and fertilizer management have been worked out to maximize the yields in production plots. In certified seed production plot of hybrids, 8:2 ratio of A- and R-lines is followed to economize the seed production. Seed yield, often in the range of 1000–1200 kg/ha, is easily obtained as seed production is undertaken under high agronomic management. However, the yield of parental lines is determined by their genetic potential. Yield levels in seed production plots are also influenced by climate, soil conditions and crop management.

8.8.5 Seed Quantity

There has been no shortfall in the supply of indented breeder seed in any of the variety or hybrid parental lines. Each year, on an average, 20,000 metric tons of hybrid seed is produced that is sufficient to cover more than 60 % of total pearl millet area, which is highest among crops in India in which hybrids are preferred as choice of cultivar types. There is still a great scope to future further improve the quality and seed availability of improved cultivars.

8.9 Alternative Uses and Value Addition

Unfermented bread *(chapatti)* is the most common food made from pearl millet in rural households, followed by other traditional products like thin porridge (*khichdi*) and thick porridge (fermented and unfermented). Recently, technologies for various processing treatments, such as milling, malting, blanching, acid treatment, dry heating and fermentation, which reduce antinutritional factors and increase the digestibility and shelf life of various alternative food products such as unleavened flat bread (*roti/chapati*), porridges, noodles, bakery products and extruded and weaning food products, have been developed and tested at the laboratory scale.

Pearl millet is gluten-free and, hence, has a good chance of being commercialized for the food-based management of this problem owing to its low-glycemic index; pearl millet offers unique food for diabetic patients (Table 8.5). Pearl millet is rich in oil, and linoleic acid accounts for 4 % of the total fatty acids in this oil, giving it a higher percentage of n-3 fatty acids as compared to maize in which linoleic acid accounts for only 0.9 % of the total fatty acids

Table 8.5 Health value of pearl millet-based diabetic products

Product	Glycemic index	
	Control (wheat flour)	Pearl millet-based products
Biscuit	72.7	58.1
Chapati	69.4	48.0
Dhokla	68.4	38.0
Instant *idli*	69.8	52.1
Pasta	71.3	54.1

Source: Mani et al. (1993)

and, hence, is highly deficient in *n*-3 fatty acids. The *n*-3 fatty acids play an important role in many physiological functions, including platelet aggregation, cholesterol accumulation and the immune system. Pearl millet in poultry feed can have a significant effect on the fatty acid composition of eggs and, consequently, on human health. In a poultry feeding trial, it was observed that eggs produced from layers fed on a pearl millet-based diet had lower *n*-6 fatty acids and higher *n*-3 fatty acids and, thus, led to lower *n*-6/*n*-3 fatty acid ratios than those fed on corn-based diets (Table 8.6). These eggs are of special health value, especially for those prone to high levels of low-density lipoproteins (LDL).

8.10 Research and Development Partnerships

Research and development are essential components of an integrated programme required to make an impact. The organizations dealing with these two components may differ in their mandate, organizational structure, operational domain and the nature and magnitude of resources, but many of them may have common goal of making an impact on the target populations and environments. It is these shared goals and ambitions to succeed that drive the research and development organization towards building the partnership and pursuing it further for its improvement, adaption to changing conditions and successful maintenance over time. The AICPMIP has joined hands with international and national organizations broadly in three categories in pearl millet improvement research.

Table 8.6 Cereal grains and egg composition of *n*-6 and *n*-3 fatty acids

	Diet		
Fatty acid	Corn	Corn + pearl millet	Pearl millet
Diet composition of fatty acid (% of total fatty acids)			
Total *n*-6	59.3	47.0	40.0
Total *n*-3	2.4	2.5	3.3
n-6/*n*-3 ratio	25.2	19.0	12.8
Egg composition of fatty acid (mg/g yolk)			
Total *n*-6	66.8	55.6	47.3
Total *n*-3	5.1	5.5	5.7
n-6/*n*-3 ratio	13.1	10.1	8.3

Modified from Collins et al. (1997)

8.10.1 Partnership with International Research Centres

The foremost among international research centres (IRCs) is ICRISAT, which has a global mandate for pearl millet improvement, especially targeting the arid and semi-arid tropical regions of Asia and sub-Saharan Africa. Having the largest collection of more than 21,000 well-characterized germplasm accessions, ICRISAT has played a significant role in strengthening pearl millet improvement research through dissemination of large number and diverse range of breeding lines and hybrid parents. ICRISAT organizes Pearl Millet Scientists Field Day biennially and breeders from public and private sector select breeding materials of their choice (Table 8.7). ICRISAT is also involved in joint evaluation of materials in multilocational trials, generation of strategic research information and building research capacity through training. These activities are largely undertaken under the umbrella of ICAR-ICRISAT research partnership projects.

Table 8.7 Pearl millet breeding lines selected and seed samples supplied from the Scientists Field Day Selections at ICRISAT, Patancheru, India

Description	Field day year	Public	Private	Total
Number of participants	2012	22	38	60
	2014	31	38	69
Number of lines selected	2012	712	2484	2782
	2014	1302	1668	2523
Number of samples supplied	2012	1304	4941	6245
	2014	2677	1339	4016

8.11 Crop Management

Agronomic research conducted in AICPMIP included both the research station experiments and farm-level extension trials and has led to the establishment of detailed recommendations for specific/individual pearl millet growing zones with respect to the time of sowing, seed rate, weed management, fertilizer application including bio-fertilizers, intercropping, cropping sequence/rotation systems and moisture conservation techniques. This has been reviewed by several workers (Singh 1985; De and Gautam 1987; Bhatnagar et al. 1998).

Agronomic research in pearl millet can be divided in two broad areas: (1) intensive management in areas where moisture is generally adequate and (2) low-input management in areas where moisture is the major production constraint. Intensive management focuses on the increased plant population (1,75,000–2,20,000 plants/ha) achieved by maintaining 45 cm distance between the rows and 10–15 cm between plants within rows and high nutrient application. However, a lower plant population of 1,20,000/ha should be maintained in drier zones with rainfall <400 mm. Response to applied inorganic nitrogen fertilizer as high as 90–140 kg/ha has been reported from experimental trials in high rainfall areas, but farmer recommendations are in the range of generally 40–80 kg/ha in various agroclimatic zones. Inoculation of seed with *bio-fertilizers* has been reported to improve the availability of nitrogen and phosphorus to the extent of 10–20 kg/ha. The recommended dose of phosphorous for pearl millet is 20–40 kg/ha. Though weed management by chemical means has been standardized but has found little application on farmers' field. Applying atrazine at 0.5 kg/ha as pre-emergence combined with one hand weeding has also given good results.

Agronomic research for low-input arid areas has focused mainly on cropping system with legumes and on the moisture conservation techniques as pearl millet in drier areas is traditionally grown in mixture or in rotation with legumes and pulses to obtain stability in production and maintain soil fertility. Most suitable cropping systems have been worked out for diverse regions. Conservation of moisture through various techniques forms an important recommendation in dry regions. These techniques include wide-spaced crop and the use of mulch either through manipulating top soil or by organic means and to grow short duration cultivars to avoid moisture-stress situation. Though, adoptions of agronomic recommendations are mainly confined to better rainfall/assured moisture available areas, but the need to adopt improved agronomic practices is more in drier pearl millet growing areas to sustain the productivity. There exists a considerable potential for further adoption of modern inputs mainly nitrogen fertilizer and new cultivars in drier areas.

8.12 Adoption of Improved Technologies and Impact

Following the adoption of high-yielding and disease-resistant cultivars and production technology, pearl millet productivity has been consistently increasing since 1986. During the last 25 years, the productivity has gone up from 539 kg/ha during 1986–1990 to 1198 kg/ha during 2012–2013 registering a 73 % improvement, which is highest among all food crops (Table 8.8). The extent of improvement in pearl millet productivity has resulted in 45 % increase in its grain production from 5.83 million tons to 8.74 million tons, in spite of 18 % decline in crop area from 10.7 million ha to 7.3 million ha.

Table 8.8 Five-year means for grain yield and percent improvement in yield over average yield of 1986–1990 of principal food crops in India during 1986–2010

Period	Grain yield (kg/ha)					Improvement (%) in yield over average yield of 1986–1990				
	Rice	Wheat	Sorghum	Maize	Pearl millet	Rice	Wheat	Sorghum	Maize	Pearl millet
1986–1990	1622	2113	744	1371	539	–	–	–	–	–
1991–1995	1818	2429	827	1564	620	12	15	11	14	15
1996–2000	1918	2648	825	1768	733	18	25	10	29	36
2001–2005	1997	2661	784	1913	856	23	26	5	40	59
2006–2010	2161	2812	962	2124	932	33	33	29	55	73

Source: DAC, Government of India as on 7 February, 2010 available at http://www.agricrop.nic.in

References

Aken'Ova ME (1985) Confirmation of a new source of cytoplasmic genic male-sterility in bulrush millet [*Pennisetum americanum* (L) Leeke]. Euphytica 34:669–672

Aken'Ova ME, Cheeda HR (1981) A new source of cytoplasmic genic male sterility in pearl millet. Crop Sci 21:984–985

Appadurai R, Raveendran TS, Nagarajan C (1982) A new male sterility system in pearl millet. Indian J Agri Sci 52:832–834

Athwal DS (1966) Current plant breeding research with special reference to *Pennisetum*. Indian J Genet Plant Breed 26A:73–85

Athwal DS, Rachie KO (1963) Potentialities and future breeding for the improvement of bajra. Indian J Genet Plant Breed 23:155–157

Barker T, Campos H, Cooper M et al (2005) Improving drought tolerance in maize. Plant Breed Rev 25:173–226

Basavaraj G, Parthasarathy Rao P, Bhagavatula S, Ahmed W (2010) Availability and utilization of pearl millet in India. J SAT Agri Res 8:1–6

Bertin I, Zhu JH, Gale MD (2005) SSCP-SNP in pearl millet – a new marker system for comparative genetics. Theor Appl Genet 110:1467–1472

Bhatnagar SK, Yadav OP, Gautam RC (1998) Research achievements in pearl millet (*Pennisetum glaucum*). Indian J Agri Sci 68:423–430

Bidinger FR, Serraj R, Rizvi SMH, Howarth C, Yadav RS, Hash CT (2005) Field evaluation of drought tolerance QTL effects on phenotype and adaptation in pearl millet [*Pennisetum glaucum* (L.) R. Br.] topcross hybrids. Field Crop Res 94:14–32

Blum A (1988) Plant breeding for stress environments. CRC Press, Boca Raton

Burton GW (1977) Registeration of Gahi 3 pearl millet. Crop Sci 17:345–346

Burton GW, Athwal DS (1967) Two additional sources of cytoplasmic male sterility in pearl millet and their relationship to Tift 23A. Crop Sci 7:209–211

Chavan VM, Patil JA, Chowdhary BB (1955) Hybrid bajra in Bombay state. Poona Agric Coll Mag 46:14–150

Collins VP, Cantor AH, Pescatore AJ, Straw ML, Ford MJ (1997) Pearl millet in layer diets enhances egg yolk *n*-3 fatty acids. Poult Sci 76:326–330

De R, Gautam RC (1987) Management practices to increase and stabilize pearl millet production in India. In: Proceedings of the international pearl millet workshop, 7–11 April 1986, Ed. JR Witcombe ICRISAT Centre. Patancheru, Hyderabad, India, pp 247–254

Govila OP, Rai KN, Chopra KR, Andrews DJ, Stegmier WD (1997) Breeding pearl millet hybrids for developing countries: Indian experience. In: Proceedings of international conference on genetic improvement of sorghum and pearl millet held from 22–27 September 1996 at Lubbock, Texas, USA, pp 97–118

Gupta SK, Rai KN, Kumar SM (2010) Effect of genetic background on fertility restoration of pearl millet hybrids based on three diverse cytoplasmic-nuclear male-sterility systems. J SAT Agri Res 8:1–4

Gupta SK, Rathore A, Yadav OP, Rai KN, Khairwal IS, Rajpurohit BS, Das RR (2013) Identifying mega-environments and essential test locations for pearl millet cultivar selection in India. Crop Sci 53(6):2444–2453

Gupta SK, Rai KN, Singh P, Ameta VL, Gupta SK, Jayalekha AK, Mahala RS, Pareek S, Swami ML, Verma YS (2015a) Seed set variability under high temperatures during flowering period in pearl millet (Pennisetum glaucum L. (R.) Br.). Field Crop Res 171:41–53. ISSN 0378–4290

Gupta SK, Nepolean T, Sankar SM, Rathore A, Das RR, Rai KN, Hash CT (2015b) Patterns of molecular diversity in current and previously developed hybrid parents of pearl millet [pennisetum glaucum (L.) R. Br.]. Am J Plant Sci 6(11):1697–1712. ISSN 2158–2742

Hanna WW (1989) Characteristics and stability of a new cytoplasmic-nuclear male-sterile source in pearl millet. Crop Sci 29(6):1457–1459

Harinarayana G, Anand Kumar K, Andrews DJ (1999) Pearl millet in global agriculture. In: Khairlwal IS, Rai KN, Andrews DJ, Harinarayana G (eds) Pearl millet breeding. Oxford & IBH Publishing, New Delhi, pp 480–506

Hash CT, Witcombe JR (2002) Gene management and breeding for downy mildew resistance. In: Leslie JF (ed) Sorghum and millets pathology. Iowa State Press, Ames, pp 27–36

Hash CT, Witcombe JR, Thakur RP, Bhatanagar SK, Singh SD, Wilson JP (1997) Breeding for pearl millet disease resistance. In: Proceedings of international conference on genetic improvement of sorghum and pearl millet held from 22–27 September 1996 at Lubbock, Texas, USA, pp 337–373

Hash CT, Singh SD, Thakur RP, Talukder BS (1999) Breeding for disease resistance. In: Khairwal IS, Rai KN, Andrews DJ, Harinarayana G (eds) Pearl millet breeding. Oxford & IBH, New Delhi, pp 337–379

Hash CT, Bhasker Raj AG, Lindup S et al (2003) Opportunities for marker-assisted selection (MAS) to improve the feed quality of crop resistance in pearl millet and sorghum. Field Crop Res 84:79–88

Khairwal IS, Rai KN, Diwakar B, Sharma YK, Rajpurohit BS, Nirwan B, Bhattacharjee R (2007) Pearl millet: crop management and seed production manual. In: International Crop Research Institute for the semi-arid tropics, Patancheru 502 324, Andhra Pradesh, India, p 108

Kumar P, Tarafder JC, Painuli DK, Raina P, Singh MP, Beniwal RK, Soni ML, Kumar M, Santra P, Shamsudin M (2009) Variability in arid soils characteristics. In: Kar A, Garg BK, Singh MP, Kathju S (eds) Trends in arid zone research in India. Central Arid Zone Research Institute, Jodhpur, pp 78–112

Liu CJ, Witcombe JR, Pittaway TS, Nash M, Hash CT, Busso CS, Gale MD (1994) An RFLP-based genetic map of pearl millet (*Pennisetum glaucum*). Theor Appl Genet 89:481–487

Mani UV, Prabhu BM, Damle SS, Mani I (1993) Glycemic index of some commonly consumed foods in western India. Asia Pacific J Clin Nut 2(3):111–114

Marchais L, Pernes J (1985) Genetic divergence between wild and cultivated pearl millets (*Pennisetum typhoides*). I. Male sterility. Z Pflanzenzuchtg 95:103–112

Mula RP, Rai KN, Dangaria CJ, Kulkarni MP (2009) Pearl millet as a post-rainy cool season crop: case studies from Gujarat and Maharashtra, India. J SAT Agri Res 7:1–7

Nepolean T, Hash CT, Blummel M et al (2010) Marker-assisted backcrossing (MABC) to improve pearl millet stover quality traits simultaneously improves blast resistance. In: National symposium on genomic and crop improvement: relevance and reservations, Feb 25–27, 2010, Acharya NG Ranga Agricultural University, Hyderabad, p 162

Qi X, Pittaway TS, Lindup S et al (2004) An integrated genetic map and a new set of simple sequence repeat markers for pearl millet, *Pennisetum glaucum*. Theor Appl Genet 109:1485–1493

Quendeba B, Ejeta G, Nyquist WE, Hanna WW, Anand Kumar K (1993) Heterosis and combining ability among African pearl millet landraces. Crop Sci 33:735–739

Rachie KO, Singh A, Bakshi JS (1967) Development of hybrid grain millet (*Pennisetum typhoides* S & H) for India. Western Branch meeting, American Society of Agronomy, Los Cruces, New Mexico, USA.

Rai KN (1995) A new cytoplasmic-nuclear male sterility system in pearl millet. Plant Breed 114:445–447

Rai KN, Anand Kumar K (1994) Pearl millet improvement at ICRISAT-an update. Int Sorghum Millets Newsl 35:1–29

Rai KN, Hash CT (1990) Fertility restoration in male-sterile × maintainer hybrids and pearl millet. Crop Sci 30:889–892

Rai KN, Virk DS, Harinarayana G, Rao AS (1996) Stability of male-sterile source and fertility restoration of their hybrids in pearl millet. Plant Breed 115:494–500

Rai KN, Khairwal IS, Dangaria CJ, Singh AK, Rao AS (2009) Seed parents breeding efficiency of three diverse cytoplasmic nuclear male-sterility systems in pearl millet. Euphytica 165:495–507

Rai KN, Yadav OP, Gupta SK, Mahala RS, Gupta SK (2012) Emerging research priorities in pearl millet. J SAT Agri Res 10:1–5

Rao PK, Nambiar AK, Madhava Menon P (1951) Maximization of production by cultivation of hybrid strains with special reference to cumbu (pearl millet). Madras Agri J 38:95–100

Sehgal D, Rajaram V, Armstead IP, Vadez V, Yadav YP, Hash CT, Yadav RS (2012) Integration of gene-based markers in pearl millet genetic map for identification of candidate gene underlying drought tolerance quantitative trait loci. BMC Plant Bio 12(1):1–13

Sharma PC, Sehgal D, Singh G, Yadav RS (2011) A major terminal drought tolerance QTL of pearl millet is also associated with reduced salt uptake and enhanced growth under salt stress. Mol Breed 27:207–222

Singh RP (1985) Pearl millet. In: Balasubramaniam V, Venkateswarlu J (eds) Efficient management of dry land crops. Central Research Institute for Dry Land Agriculture, Hyderabad, pp 51–62

Supriya A, Senthilvel S, Nepolean T et al (2011) Development of molecular linkage map of pearl millet integrating DArT and SSR markers. Theor Appl Genet 123:239–250

Talukdar BS, Rai KN, Rao AM, Witcombe JR (1987) Heterosis and fertility restoration in ICRISAT pollinators. Millet Newsl 6:7–8

Witcombe JR, Hash CT (2000) Resistance gene development strategies in cereal hybrids using marker-assisted selection: gene pyramiding, three-way hybrids, and synthetic parent populations. Euphytica 112:175–186

Yadav RS, Hash CT, Bidinger FR, Cavan GP, Howarth CJ (2002) Quantitative traits loci associated with traits determining grain and stover yield in pearl millet under terminal drought stress conditions. Theor Appl Genet 104:67–83

Yadav RS, Hash CT, Bidinger FR, Devos KM, Howarth CJ (2004) Genomic regions associated with grain yield and aspects of post-flowering drought tolerance in pearl millet across stress environments and testers background. Euphytica 136:265–277

Yadav RS, Sehgal D, Vadez V (2011) Using genetic mapping and genomics approaches in understanding and improving drought tolerance in pearl millet. J Exp Bot 62:397–408

Finger Millet

9

Salej Sood and B. Kalyana Babu

Abstract

Finger millet (*Eleusine coracana* (L.) Gaertn.) is an important traditional crop of Asia and Africa. It has very high adaptation and can be grown from sea coast to high hills. It originated in the Ethiopian highlands of Africa and was introduced into India. It is mainly cultivated for food, fodder and medicinal uses. *Eleusine* is a small genus with about nine species and is distributed over the tropical and subtropical parts of Africa, Asia and South America. There are indications that wild and weedy ancestor *E. africana* in Africa has contributed to the cultivated gene pool diversity. Moreover, the African germplasm appears to possess a higher level of resistance to blast. Despite its recognised value as an important staple crop, particularly for poor people in arid and semiarid areas, finger millet has been neglected in mainstream crop improvement research. Identification of molecular markers/QTLs for important agro-morphological traits can accelerate the crop improvement of finger millet with more emphasis on introgression of genes into locally well-adopted genotypes for development of stress-tolerant, high-yielding varieties with improved nutritional properties. Till now, little efforts have been made to enhance finger millet genotypes at molecular level. Combination of conventional and molecular breeding techniques will allow for rapid development of high-yielding finger millet varieties with double the current yield levels.

Keywords

Finger millet • Evolution • Crop wild relatives • Genetic diversity • Wide hybridisation

S. Sood (✉) • B.K. Babu
ICAR – Vivekananda Institute of Hill Agriculture,
Almora 263601, India
e-mail: salej1plp@gmail.com

M. Singh, S. Kumar (eds.), *Broadening the Genetic Base of Grain Cereals*,
DOI 10.1007/978-81-322-3613-9_9

9.1 Introduction

Finger millet (*Eleusine coracana* (L.) Gaertn.) is an annual plant widely grown as cereal in arid and semiarid areas of Asia and Africa. It is a highly adaptable crop and even grown in higher elevations up to 2400 m above mean sea level in the Himalayas. Finger millet is originally native to the Ethiopian highlands and was introduced into India approximately 4000 years ago. It is the most important minor millet in the tropics and grown in more than 25 countries in Africa (Eastern and Southern) and Asia (from Near East to Far East) and accounts for 12% of the global millet area. It is a cereal crop cultivated for food and for making beer (Phillips 1974), as well as fodder and medicinal uses (Bisht and Mukai 2002; Oduori 2005). It is a hardy crop that can be grown in very diverse environments and has an excellent food value. It is a potential nutritious crop for the increasing world population, particularly in arid and semiarid regions where it is usually ranked third in cereal production, after sorghum and pearl millet (Bisht and Mukai 2002). Finger millet serves as a subsistence and food security crop especially important for its nutritive and cultural value (Oduori 2005; Verma 2009).

Botanical name: *Eleusine coracana* (L.) Gaertn. (Hilu and de Wet 1976a).
Tribe: Eragrosteae.
Subfamily: Chloridiodeae.
Family: Poaceae.
Common name: Various common names of finger millet in different countries have been reported and are listed in Table 9.1.

Table 9.1 Common names of finger millet in different countries

Country		Common name
Afrikaans and Dutch	:	*Vogel gierst*
Arabic	:	*Taibbon*
England	:	*Finger millet, African millet, koracan*
French	:	*Petit millet, eleusine cultivée, coracan, koracan*
German	:	*Fingerhirse* (horse)
Swahili	:	*Wimbi, ulezi*
Ethiopia	:	*Dapussa, tokusa, bavankiya*
India	:	*Ragi*
Kenya	:	*Limbi, mugumbi*
Malawi	:	*Hawere, lipoke, usanje, khawke, malesi, mulirubi, lupodo, mawe*
Nepal	:	*Koddo*
Sudan	:	*Tailabon (Arabic), ceyut (Bari)*
Tanzania	:	*Mwirubi, mbege*
Uganda	:	*Bulo*
Zambia	:	*Kambale, lupoke, mawele, majolothi, awale, bule*
Zimbabwe	:	*Rapoka, zviyo, niera, mazhovole, rukweza, uphoka, poha*

The precise estimate of area and production at the global level is not available except for India (Table 9.2). The global annual planting area of finger millet is estimated to be around 4–4.5 million ha, with a total production of 5 million t of grains, of which India alone produces about 2.2 million t and Africa about 2 million t. The rest comes from other countries in South Asia. The important finger millet-growing countries in Eastern and Southern Africa include the subhumid regions of Ethiopia, Kenya, Malawi, Tanzania, Uganda, Zaire, Zambia and Zimbabwe. Similarly in South Asia, the crop is largely grown in India, Nepal and, to some extent, in Bhutan and Sri Lanka. Finger millet is reported to be grown in both China and Japan to a limited extent (https://www.croptrust.org/sites/default/files/documents/files/Finger%20Millet-final.pdf).

The grain of finger millet produces or releases fine aroma when cooked or roasted and is known to have many health-promoting qualities. The grain can be stored for years without any insect damage, which makes it a valuable crop particularly for famine-prone areas. Although grown under dry lands, it provides an assured harvest, thus making it indispensable in specific ecosystems. The crop provides food grain and straw which are valued animal feed, especially in the rainfed areas and hills. Finger millet is considered as a coarse grain because of its fibrous and tough outer layer that irritates the tongue and not readily accepted by people accustomed for the

Table 9.2 Area production and productivity of finger millet in India

State	1950–1951			2008–2009		
	Area (000 ha)	Production (000 t)	Productivity (kg/ha)	Area (000 ha)	Production (000 t)	Productivity (kg/ha)
Andhra Pradesh	273.0	242.0	885	50.0	52.0	1040
Bihar	194	94.0	481	11.4	9.3	816
Gujarat	60.0	55.0	916	19.0	20.0	1053
Jharkhand	–	–	–	12.1	8.5	702
Karnataka	800.0	441.0	550	841.0	1394.0	1658
Maharashtra	193.0	134.0	694	126.0	125.0	992
Orissa	123.0	69.0	562	65.7	41.0	624
Tamil Nadu	318.0	273.0	861	90.1	170.0	1887
Uttarakhand	–	–	–	135.0	193.0	1130
West Bengal	12.0	6.0	519	12.7	14.7	1157
India	2199.0	1427.0	649	1381.4	2039.9	1552

Source: Ashok et al. (2011)

consumption of wheat and rice. Apart from palatability, it lacks gluten characteristic of wheat and hence cannot be used for the preparation of chapattis or baked products. It remains food of choice for lower socio-economic groups and traditional consumers, because of its coarse texture and intense colour of seed coat. Finger millet contains a large proportion of carbohydrates and thus provides bulk of energy in diets. It is also rich in proteins and sulphur-containing amino acids. Because of low glycaemic index with high fibre content, it is recommended for diabetic patients. Apart from the major nutrients, it also contains iron and calcium, which are deficient in most Indian and African women. High calcium, high soluble fibre, low fat and high diastatic power of malted grains render finger millet unique place among food grains. Its consumption has proved to be very effective in controlling blood glucose level of diabetics. Consumption of finger millet also prevents constipation and cholesterol (http://www.ccrp.org/sites/default/files/finger_millet___year_2___development_of_high-yielding_disease-resistant_and_drought-tolerant_finger_millet_genotypes.pdf).

The Green Revolution has contributed tremendously towards making India self-sufficient in food production. However, the higher-yielding varieties also require higher inputs to reach their potential, a cost that cannot be afforded by the small and marginal farmers. Approximately 70 % of India's working force relies on agriculture for their livelihood, and a significant portion of them are subsistence farmers. These farmers are extremely vulnerable to the effects of adverse conditions on crop yields as they barely produce enough food to feed their families. Therefore, efforts need to be directed towards such indigenous crops which can play a major role in achieving nutritional security as these are rich sources of energy and micronutrients.

9.2 Origin and Domestication

The origin of finger millet is argued as Indian (De Candolle 1886; Burkill 1935; Cobley 1956; Dixit et al. 1987) or independently as African and Indian (Vavilov 1951) or African (Mehra 1963b; Chennaveeraiah and Hiremath 1974). The cultivated *E. coracana* is a tetraploid ($2n = 4x = 36$) and exhibits morphological similarity to both *E. indica* ($2n = 18$) and *E. africana* ($2n = 36$). Earlier, it was thought that cultivated *E. coracana* originated from *E. indica*, which is widely distributed, from Africa eastwards to Java. The cytological evidence indicates that *E. indica* has contributed one of the genomes (AA) to the cultivated *E. coracana*. The species *E. africana* ($2n = 36$), which is also a tetraploid, exhibits great similarity in morphological features with *E. coracana* that are genetically related, and gene flow occurs between them in nature, suggesting that *E. coracana* possibly originated from *E. africana* through selection and further mutation that resulted into larger grains (Chennaveeraiah and Hiremath 1974; Hilu and de Wet 1976a) (Fig. 9.1).

Hilu (1988), after analysing the chloroplast genome, showed that *E. indica* is also the maternal genome donor to *E. africana*. The RAPD and isozyme analyses also proved close affinity between *E. indica*, *E. africana* and *E. coracana* (Hilu 1995; Werth et al. 1994). The comparison of rDNA sites on the chromosomes of diploid and polyploid species of *Eleusine* also suggested that *E. indica* and *E. floccifolia* may have contributed two genomes to *E. africana* (Bisht and Mukai 2000). The number of 18S-5.8S-26S and 5S rDNA sites and location were similar on two pairs of chromosomes of *E. africana* and one pair of chromosomes of *E. indica* and *E. floccifolia* (Bisht and Mukai 2000). The genomic in situ hybridisation results suggested that *E. indica* and *E. floccifolia* are two genome (A and B) donors to the polyploid species *E. coracana* (Bisht and Mukai 2001). Later, Neves et al. (2005) refuted *E. floccifolia* as B-genome donor. Recent studies suggest *E. indica* and *E. tristachya* as the A-genome donors, with a differential degree of relatedness to *E. coracana*. McGISH supports *E. indica* as the primary A-genome parent and *E. tristachya* (or its extinct sister or ancestor) as the secondary A-genome parent for the derivation of finger millet (*E. coracana*). The most likely scenario is that the B-genome donor is extinct (Liu et al. 2014).

The archaeological findings of finger millet from Ethiopia date to about the third millennium BC (Hilu et al. 1979). The two distinct races of finger millet recognised are the African highland race and Afro-Asiatic lowland race. The African highland race is considered to be derived from *E. africana* under cultivation, and this gave rise to the African lowland race which later migrated to India and developed as the Afro-Asiatic lowland race (Mehra 1962; Purseglove 1976). This migration of finger millet to the Indian subcontinent is likely to have occurred around 3000 BC. Studies of the patterns of variability in African and Asian finger millets have indicated relatively larger diversity in African germplasm compared to Indian collections, supporting the view that Africa could be the primary centre of origin. The

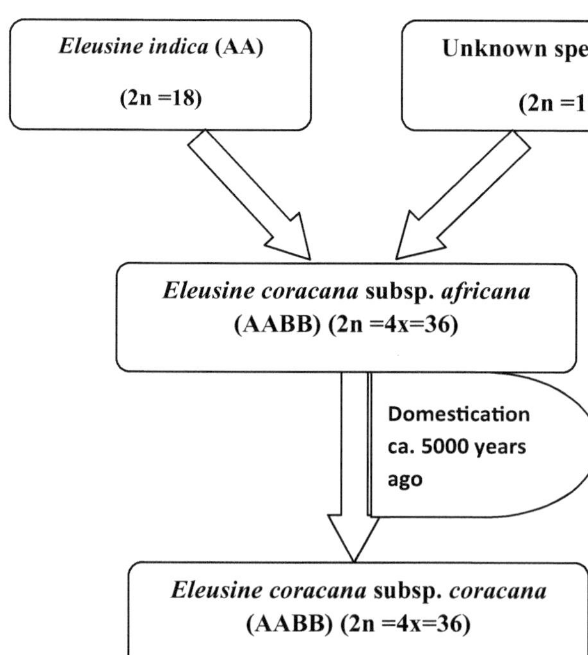

Fig. 9.1 Origin and evolution of finger millet

long history of finger millet cultivation in the Indian subcontinent, i.e. for more than 5000 years, accompanied by human selection, has resulted in the generation of a large diversity in landraces and local cultivars in India. Close study of various characters in Indian germplasm has revealed that economically important characters, such as finger length, finger width, finger number, grain yield, ear weight, total biomass and leaf number, possess large variability indicating India as the secondary centre of diversity (Naik et al. 1993). There are a number of crops for which researchers have suggested that certain domestication traits might have evolved after the incipient crop was transferred to areas where the wild relatives were absent. Such a pattern, if widespread, would support the importance of reproductive isolation in domestication. Finger millet is one example among them, which has been suggested to have been domesticated initially in the East African Highlands, yet the earliest known remains date to 2000 BC in India, and it only shows up later in the paleorecord of East Africa (Fuller 2006). Similar patterns have also been suggested for pearl millet and sorghum (*Sorghum bicolor*) domestication (Haaland 1995, 1999; Fuller 2007), although this remains controversial. However, there are too few such crops to allow quantification of this potential 'allopatry effect'.

9.2.1 Morphological Description

Finger millet is a tufted annual crop, growing to a height of 30–150 cm and maturing in 75–160 days. The leaf blades are shiny, strongly keeled and difficult to break and are 22–50 cm long and 0.6–1.0 cm wide. The leaves and culms are typically green in colour. It has an exceptionally strong root system, and it is difficult to pull out of the ground by hand (Van Wyk and Van Oudtshoorn 1999). The culms and the leaf sheaths are prominently flattened. The ligule is a fringed membrane. There are two subspecies of finger millet, the wild form (*E. coracana* subsp. *africana*) and a cultivated form derived from it (*E. coracana* subsp. *coracana*). Wild African finger millet (*E. coracana* subsp. *africana*) is similar to Indian goosegrass (*E. indica*) and may be confused with it, but the latter has smaller spikelets and oblong, not rounded, grains. The grains of *Eleusine* are unusual in that the outer layer (pericarp) is not fused and can be easily removed from the seed coat (testa) (Van Wyk and Gericke 2000). The panicle consists of a group of digitally arranged spikes often referred to as fingers (Fig. 9.2). The spikelets are made up of 4–10 florets arranged serially on the finger. The spikelets are 5–8 mm long and 3–4 mm wide. The spikelets do not disarticulate (break apart at the joints) at maturity.

Fig. 9.2 Finger millet crop in fields at sub-mountainous Himalayan region, India; (**a**) closed-ear type (**b**) open-ear type

All florets are perfect flowers with the exception of terminal ones which may sometimes be infertile. The grain is oblong to round and oval, reddish-brown to white in colour with the grains' surface finely corrugated. While seed of wild finger millet relatives shatter easily when mature, cultivated forms lost this trait during domestication, and some but not all cultivars completely lack the ability to disperse seed naturally through shedding (de wet et al. 1984). By and large, seeds of cultivated finger millet are nondormant and germinate within a few weeks, although several landraces show some dormancy (Kulkarni and Basavaraju 1976). Wild species have dormant seeds that remain viable in soil for 2 years or longer (Chuah et al. 2004) and germinate during crop season.

9.2.2 Growth and Development

Finger millet has the highly efficient C_4 photosynthetic pathway in common with maize, sorghum and sugarcane. The plants are adaptable to a wide range of environmental and climatic conditions. It tolerates salinity better than most of the cereals (CAB 2003) and has been reported to contain acid-producing, nitrogen-fixing bacteria (*Acetobacter diazotrophicus*) in roots, rhizosphere and stems (van Wyk and Gericke 2000). Typically as a tropical and rainfed crop, it is best suited for dry farming. It can be grown in a wide range of climates because of its short and flexible growing season. It is mainly produced within 20°N and 20°S latitude. Finger millet thrives well at higher altitudes than most other tropical cereals. It is grown at intermediate elevations, between 500 and 2400 m above sea level. However, at very high altitudes, long-duration varieties may take longer time to enter into the reproductive phase. Annual precipitation tolerance ranges from 300 to 4000 mm. It requires moderate rainfall (500–1000 mm) which should be well distributed during the growing season, but with an absence of prolonged droughts. Dry weather is required for drying the grains at harvest. The crop tolerates a cooler climate than other millets. It is best grown in regions with average temperature ranging from 11 to 28 °C and is reported to tolerate a soil pH of 5.0–8.2 (Duke 1978, 1979). Areas with low precipitation and low relative humidity during seed ripening and maturation are best for regeneration. However, it thrives under hot conditions also. It can be grown in the areas where temperature is as high as 35 °C. In southern states of India and some African countries, the crop grows best where the average maximum temperature exceeds 27 °C and the average minimum does not fall below 18 °C (Vara Prasad and Staggenborg 2011). The crop matures in 90–140 days (AICSMIP 2014).

The crop is grown on a variety of soils. In the tropics, it is mainly produced on reddish-brown lateritic soils with good drainage but reasonable water-holding capacity. It grows best on free-draining sandy loam soils with well-distributed rainfall during the growing season. Its thick adventitious root system makes it ideally suited to these conditions (McMaster 1962). The absence, however, of deep roots to draw on subsoil moisture makes the crop extremely sensitive to dry spells when the topsoil dries out rapidly.

It can also tolerate some waterlogging, though excessive moisture or stagnation of water limits the growth and yield potential. It seems to have more ability to utilise rock phosphate than other cereals (Neshamba 2010).

Although a C_4 crop, it grows slowly and has a poor ability to compete with weeds. Finger millet volunteers are common in fields planted in earlier seasons with the crop. Feral finger millet plants are rather uncommon as the crop is a poor competitor to weeds unless there is human intervention, i.e. weeding. This problem has often been observed in finger millet crop varietal trials where wild-type plants grow within the plot every year even after careful quality check in maintenance breeding.

9.2.3 Crop Management

In India and Africa, finger millet is generally grown as a pure continuous crop, or it is intercropped or rotated with soya bean, pigeon pea, cowpea, groundnut or horse gram. In India, it is generally rotated with pulses or other cereals. Finger millet+pigeon pea in 8:2 or finger millet+field bean in 8:1 has become popular in growing areas of the crop. Finger millet+soya bean (90+10% mixtures) intercropping is a common system in hilly areas (Ashok et al. 2011). There are two main seasons in India, rainy and post-rainy season. In high rainfall zones, sowing is done with early varieties during June in order to harvest the crop by September or October. Most of the other areas are sown with late varieties between July and August, in order to harvest the crop by November or December. A late-sown crop is normally affected by water stress during seedling stage, which affects the yield. There is a minimum land preparation required for planting finger millet in both Africa and India. Higher yields are obtained in cultivated fields if the soil is worked to a fine filth with a disc harrow. Finger millet can be broadcasted or sown in rows. Row spacing permits easy weeding and higher yields. Seeds are sown thinly in rows 25 cm apart and the plants are thinned to 7.5–10 cm between each plant in a row. Although in most regions fertilisers are not applied for finger millet cultivation, the crop responds well to fertilisers. Application of 40–60 kg N/ha was found sufficient for a sole cropping system in sandy clay and clay loam soils of medium fertility in southern India; application of N increased grain yields by about 50%. In Uganda, a combined application of N and P, or manure and P, resulted in higher yields. Application of N alone favours initiation of flowers and physiological maturity of the crop by 1 and 2 weeks, respectively. Growth of finger millet on sterile, phosphorus-deficient soils could be improved by inoculation with vesicular-arbuscular mycorrhizal fungus. Application of 20 kg P2O5 and 20 kg K2O/ha is also recommended for enhancing productivity (Ashok et al. 2011; Vara Prasad and Staggenborg 2011).

In India, seeds are traditionally broadcasted by hand or drilled using seed drills. The seed rate is 5–12 kg/ha depending on the method of establishment. As millets do not germinate in cool and wet soil, it might be necessary to delay planting until soils are thoroughly warmed up. Two to three inter-cultivations are needed for in-line sown cropping systems. In broadcasted systems, weeding should be done manually.

Transplanted crop of finger millet is taken in irrigated areas in India where the nursery is raised and transplanted after 30 days of sowing in moist fields. Yield levels of more than 5 tonnes/ha have been recorded in research trials under irrigated conditions. SRI method of planting in finger millet has also been introduced in certain states of India under organic cultivation. The nursery is sown in 40 m^2 raised beds for 1 acre area using 500 g seeds. The 20–25-day-old plants are transplanted in field at a spacing of 10 × 10 inches. Three intercultural operations are carried out on the 15th, 25th and 40th day after transplanting. The results were encouraging as the plants had thick stem, more number of fingers, deeper roots and higher yield (2.5–3.75 t/ha in comparison to 1.0 t/ha in conventional method) – if the lines are related, then check the tense (http://sri.ciifad.cornell.edu/aboutsri/othercrops/fingermillet/In_SFMI_Pradan.pdf). However, more studies need to be conducted on SRI method for its economic benefits before recommendations.

9.3 Gene Pool

It is a small genus that includes about 9–12 species (Phillips 1972; Willis 1973; Hilu and de Wet 1976b; Hilu 1981) and is distributed in the tropical and subtropical parts of Africa, Asia and South America (Phillips 1972). East Africa is considered the centre of diversity for the genus, and eight species, namely, *E. africana*, *E. coracana*, *E. kigeziensis*, *E. indica*, *E. floccifolia*, *E. intermedia*, *E. multiflora* and *E. jaegeri*, occur in this region (Mehra 1963a; Phillips 1972). The genus includes diploid as well as polyploidy species and has three basic chromosome numbers, x=8, 9 and 10. On the basis of growth habit of the species, the genus is divided into two separate groups, annual and perennial (Table 9.3). Within each group, the differences between the species are often small, and among the annuals, in particular, introgression is a frequent phenomenon

Table 9.3 Chromosome numbers, growth habit and geographical distribution of Eleusine species

Species	Common name	2n chromosome numbers and genome formula	Growth habit	Geographical distribution
E. indica (L.) Gaertn.	Yard grass, wire grass, goose grass	2n=2x=18, AA	Annual	Japan, India, North America, Tanzania, Belgium, Hungary
E. tristachya (Lam.) Lam.	Three-spike grass	2n=2x=18 AA	Annual	South America
E. multiflora Hochst. ex A. Rich.		2n=2x=16 CC	Annual	Kenya, Tanzania
E. floccifolia (Forssk.) Spreng.		2n=2x=18 BB	Perennial	Ethiopia, Somalia, Kenya, Yemen
E. intermedia (Chiov.) S. M. Phillips		2n=2x=18 AB	Perennial	Ethiopia
E. jaegeri Pilger	Manyatta grass	2n=2x=20 DD	Perennial	Tanzania
E. kigeziensis S. M. Phillips		2n=4x=38, AADD	Perennial	Burundi
E. coracana subsp. *africana* (Kennedy-O'Byrne) Hilu and de Wet	Wild finger millet	2n=4x=36, AABB	Annual	Malawi, Kenya, Rhodesia, Tanzania
E. coracana (L.) Gaertn. subsp. *coracana*	Finger millet, bird's foot millet, African millet, ragi	2n=4x=36, AABB	Annual	India, Nepal, Uganda, Kenya, Ethiopia

Adapted from Bisht and Mukai (2002); genome designations are according to Bisht and Mukai 2002

with the occurrence of intermediates (Mehra 1962; Phillips 1972). Two groups also differ in spikelet characters. In annuals, the lower glume is 1–3 nerved with a winged keel, the upper glume is 3–7 nerved with a thickened to narrowly winged keel, the lemmas have 1–3 subsidiary nerves close to each side of the central nerve forming a conspicuous thickened keel and the paleas are winged on the keels. In contrast, perennials usually have simple spikelets with one nerved unwinged glume, three nerved lemmas without subsidiary nerves and unwinged paleas (Phillips 1972).

The taxonomic, phylogenetic and biosystematic relationships among the species in the genus *Eleusine* are not conclusively resolved yet. The boundaries of primary, secondary and tertiary gene pool are also not distinguishable. However, based on the available literature, the gene pool has been categorised into three classes. The primary gene pool includes the domesticated and wild forms of finger millet. Diploid progenitor wild species constitute the secondary gene pool, and the tertiary gene pool includes other *Eleusine* species (Fig. 9.3). It is now evident that cultivated finger millet is highly variable within its primary centre of origin in Africa and secondary centre in the Indian subcontinent. Both wild finger millet (*Eleusine coracana* subspecies *africana*) and the cultivated finger millet (*Eleusine coracana* subspecies *coracana*) are important from the point of germplasm collection and conservation and form the primary gene pool. Wild finger millet (subspecies *africana*) is native to Africa but has migrated to several warmer parts of Asia and America. Natural hybridisation between wild and cultivated finger millets has resulted in hybrid derivatives and in companion weeds of the crop in Africa. This has led to the generation of new diversity of intermediate forms (https://www.croptrust.org/sites/default/files/documents/files/Finger%20Millet-final.pdf). The weedy and wild forms are restricted to the closely related annual species. The diploid wild species *E. indica*, *E. floccifolia* and *E. tristachya* form the secondary gene pool, and the tertiary gene pool comprises species *E. intermedia*, *E. jaegeri*, *E. kigeziensis*, *E. multiflora* and *E. semisterlis (E. compressa)*.

Fig. 9.3 Eleusine gene pool along with species

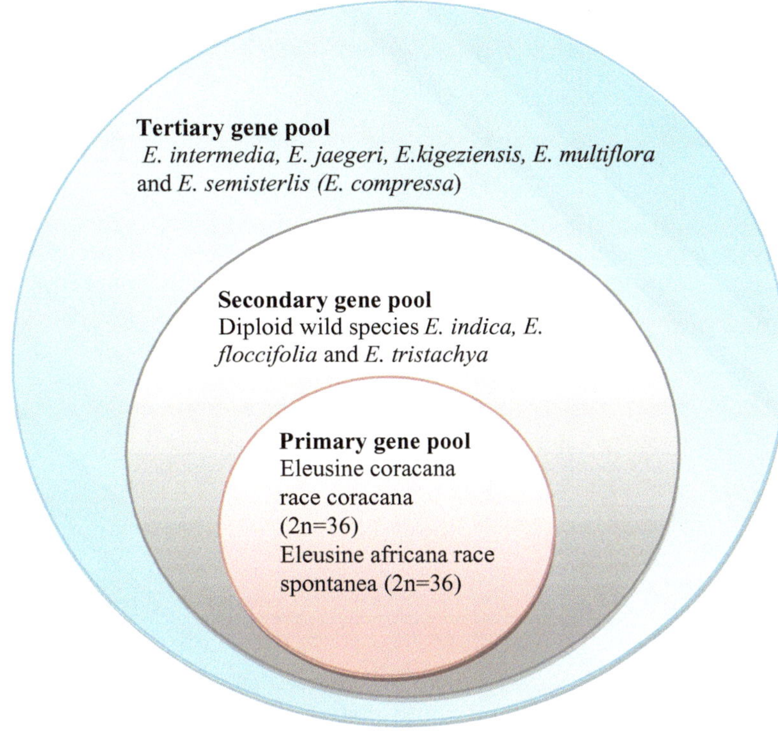

9.4 Germplasm Collection

The largest collection of finger millet crop is maintained by Indian gene bank, National Bureau of Plant Genetic Resources, New Delhi, which maintains 10,507 accessions of finger millet under long-term conservation. Most of these collections are indigenous in nature with 117 accessions of exotic origin. These Indian collections also include 6 wild relatives, 154 advanced improved varieties and 64 breeding/research material. Most of the Indian collections (58 %) are also being maintained as medium-term storage at the All India Coordinated Small Millets Improvement Project (AICSMIP) centre at Bangalore, which is also being used as a working collection, and meet the demands of national researchers. The International Crops Research Institute for the Semi-Arid Tropics (ICRISAT), Patancheru, India, maintains global collections comprising a total of 5957 accessions of which 105 are of wild species, 5665 landraces, 137 of improved cultivars and 50 breeding/research materials. Most of these collections are of exotic or non-Indian origin (4585 accessions). The Agricultural Research Station of the USDA in Griffin, Georgia, maintains 766 accessions of which 17 are of wild relatives (*E. floccifolia*, *E. indica*, *E. jaegeri*, *E. multiflora* and *E. tristachya*). These accessions belong to 11 countries which include Ethiopia (2), India (646), Kenya (3), Nepal (4), Pakistan (2), South Africa (1), Tanzania (1), Uganda (54), Zaire (3), Zambia (1) and Zimbabwe (2). The other countries holding large collections from South Asia include Nepal (877), Sri Lanka (393) and Bhutan (84). Among African countries, the largest collections are reported from gene banks based in Kenya (1902), Zimbabwe (1158), Uganda (1155), Zambia (497), Tanzania (293), Malawi (145), Eritrea (120), Burundi (113), Ethiopia (71), Nigeria (20) and South Africa (17). Other countries reported to conserve finger millet collections are China (300), the Russian Federation (110) and Vietnam (52). A total of 25,707 finger millet accessions have been recorded under various sources (Table 9.4). Landraces represent the largest proportion (%) of accessions of finger millet

Table 9.4 Germplasm collection details in finger millet

Country	Total number of accessions	Origin of germplasm		Landraces	Wild relatives	Advanced/improved varieties	Breeding/research material	Others
		Indigenous	Exotic					
Bhutan	84	84		84				
Burundi	113	113		113				
China	300	255	45	300				
Eritrea	120	120		120				
Ethiopia	71	71		67	4			
India	1,0507	1,0549	117	1,0180				
Kenya	2920	1899	1021	1813	6	154	64	103
Malawi	145	145		145	1071		36	
Nepal	877	877		877				
Nigeria	20	20		19	1			
Russian Federation	110	110		51		9	50	
South Africa	17	17		17				
Sri Lanka	393	335	58	344	3	46		
Tanzania	293	293		283	10			
Uganda	1155	1155		1155				
USA	766		766	719	17			30
Vietnam	52	52		52				
Zambia	497	497		497				
Zimbabwe	1158	1158		1157	1			
ICRISAT (Global)	5957	1372	4585	5665	105	137	50	
Total	2,5555	1,6598	6592	2,3658	1218	346	200	133

Source: https://www.croptrust.org/sites/default/files/documents/files/Finger%20Millet-final.pdf

germplasm conserved in gene banks worldwide. Only 5 % are wild relatives (1274), fewer than 2 % are advanced improved varieties (347), less than 0.8 % are breeding/research materials (200) and 0.6 % are of unknown description (154) (https://www.croptrust.org/sites/default/files/documents/files/Finger%20Millet-final.pdf).

9.5 Taxonomy

Based on inflorescence compactness and shape, finger millet germplasm is classified into races and subraces. Species *E. coracana* consists of two subspecies *africana* and *coracana*. Subspecies *africana* consists of two wild races *africana* and *spontanea*; and subspecies *coracana* consists of four cultivated races, namely, *elongata*, *plana*, *compacta* and *vulgaris*. Each of these races is further classified into subraces. The race *elongata* has three subraces: *laxa*, *reclusa* and *sparsa*. Race *plana* also consists of three subraces: *seriata*, *confundere* and *grandigluma*. Race *compacta* has no subrace, while the race *vulgaris* has four subraces, *liliacea*, *stellata*, *incurvata* and *digitata* (Prasada Rao et al. 1993).

9.5.1 Subspecies *africana*

It is a tufted annual, with slender and geniculately ascending culms that branch at the lower nodes. Flowering culms are up to 135 cm tall, with leaf blades up to 36 cm long and 10 mm wide. Inflorescence branches are 8–17 cm long and rarely more than 5 mm wide, with the spikelets arranged in two rows on one side of the rachis. Spikelets have four to nine flowers and are 5–8 mm long. Glumes are shorter than spikelet and lanceolate – oblong in profiles, rarely over 5 mm long and narrowly winged along the keel.

E. coracana is predominately self-fertilised. Subspecies *africana*, however, crosses occasionally with subsp. *coracana* to produce fully fertile hybrids. Derivatives of such crosses are aggressive colonisers and are grouped under the race *spontanea* (de Wet et al. 1984).

9.5.2 Subspecies *coracana*

It includes all cultivated finger millets. Plants are annual, tufted and erect or with geniculately ascending culms that are up to 165 cm high and sometimes root from the lower nodes.

Race *vulgaris* This race is commonly found in Africa and Asia. The inflorescence fingers are reflexed in subrace *liliacea*, twisted in subrace *stellata*, incurved in subrace *incurvata* giving fist-like appearance and top curved in subrace *digitata*.

Race *plana* This race is characterised by large spikelets that are arranged in two, almost even rows along the rachis, giving the inflorescence branch a flat ribbon-like appearance. In the subrace *seriata*, the spikelets are serially arranged giving a typical ribbon-like appearance; in *confundere*, the fertile florets are numerous and almost surround the rachis at maturity thus giving a compact appearance to the panicle; and in *grandigluma* are characterised by large pointed glumes, which are several times longer than the spikelets.

Race *compacta* Members of this race are commonly referred to as cockscomb finger millets in both Africa and India. Spikelets are composed of nine or more florets, sometimes with the inflorescence axis divided at the base, ascending and incurved at the tip to form a fist-like inflorescence.

Race *elongata* This race is morphologically the most distinct of the four races of finger millet (Prasada Rao et al. 1993). It is characterised by long slender inflorescence branches, digitately arranged, spreading and curved outward at the time of maturity. Subrace *laxa* has long open fingers with spikelets arranged in narrow rows on inflorescence branches, closely resembling wild *africana*, while subrace *reclusa* has short open fingers without any curving out of finger branches. Subrace *sparsa* also has open fingers, but spikelets are arranged in clusters on the inflorescence branch with naked space in between (Bharathi 2011).

9.6 Genetic Improvement

9.6.1 Floral Morphology

The florets are covered by two large barren leaves each being enclosed between a pair of scales known as palea. The florets are in the axil of lower flowering glumes known as lemma, which had small appendage. Near the base of the ovary, two little scales, the lodicules, are present. The androecium consisted of three hypogynous stamens. Gynoecium is bicarpellary and unilocular with superior ovary and single ovule. The androecium almost surrounds the stigmas, which ensure self-pollination. The filament is very short (0.7 mm), while anthers (0.98 mm) are bigger than filaments. Feathery branched stigma is of 0.83 mm in length. Dodake and Dhonukshe (1998) studied the pollination system in finger millet. Their study revealed that when stigmas come out of the lemma, it was covered with the thick cloud of pollen dust due to association of elongation of style and filaments with anthers bursting that gives very little chance for cross-pollination. The anthers dehisced when still inside the palea. Soon after the dehiscence of anthers, the flower was closed with no traces of stigma, only empty dehisced anthers hanging out from the closed flowers at low humidity and high temperature. At high humidity and low temperature, both anthers and stigma were outside from the closed flowers. This behavioural sequence in finger millet predisposes it into cleistogamous and chasmogamous species. In each spike, spikelet open from the top to downward while in each spikelet floret opens from bottom to top and one floret in the spikelet opens per day. The maximum number of flowers opens on the third day. It takes 5–7 days to complete flowering. Anthesis is known to take place from 1 am to 4 am and ends by 11 am (Ratnakar et al. 2009). As soon as glume and palea begin to gap, the stigma and anthers emerge almost concurrently. The dehiscence of anthers occurred prior to opening of florets. The sticky stigma and anthers attained the same height inside the flower at the time of dehiscence. The anthers dehisce and pollinate their own stigmas. The pollen remains viable for 20 min, while receptivity of stigma is up to 5 h. Finger millet is a highly self-pollinating crop, and estimates of natural crossing generally have not exceeded 1 % (Rachie and Peters 1977). Its pollen are dispersed through wind (Phillips and Renvoize 1974; de wet et al. 1984).

9.6.2 Hybridisation

Due to small floret size, crossing, emasculation and hybridisation are difficult in finger millet. The contact method of hybridisation is therefore an easy choice for breeders (Ayyangar 1934; Ravikumar 1988). In contact method, the panicles at appropriate hybridisation stage are tied together by intertwining the fingers of a male panicle inside the female panicle. The tied panicles are covered with butter paper bag for protection and exclusion of external pollen pollination. The lines having dominant traits like stem or panicle pigmentation are generally used as male plants for identification of hybrid plants. Only very few hybrid seeds are recovered in this method, and all the seeds from female panicles need to be grown for identification of hybrid plants in the field. Growing all seeds for identification of few hybrid plants in this method requires more resources including time, space and labour. Therefore, hot water emasculation method (Rao and Rao 1962; Raj et al. 1984) is a better alternative for hybridisation in finger millet. The temperature requirement of water for emasculation may vary with the location and growing conditions. We at ICAR-VPKAS, Almora, Uttarakhand (79°39′ E and 29°35′ altitude 1250 m amsl), have standardised the temperature and duration for effective emasculation in finger millet. The female panicles in an appropriate stage are immersed in hot water at a temperature of 48 °C for 5 min. Thereafter, air-dried-treated panicles are tied with male panicle in the same fashion as is done in contact method. In this method, only few seeds set in the female panicle, most of which give rise to hybrid plants. This saves both time and resources.

Genetic male sterility is also reported in finger millet. It carries the ms_1 allele and was developed through chemical mutagen treatment (1.5% aqueous solution of ethyl methane sulfonate for 6 h at 25 °C), and this genetic male-sterile line was named as 'INFM 95001' (Gupta et al. 1997). However, it could not be utilised for hybridisation in finger millet due to maintenance problem in genetic male sterility. Gametocides have also been used to induce male sterility and produce F_1 hybrids in finger millet (Reddy et al. 1983).

9.7 Gene Flow

Based on the information available from Africa and Indian germplasm collections of finger millet, it is now widely recognised that the highlands of Africa, especially Ethiopia/Uganda, are the primary centre of origin and the Indian subcontinent is the secondary centre. The crop has been in cultivation on both continents for more than 5000 years, but separated. This spatial isolation led to the emergence of two distinctly different gene pools both genetically and morphologically. There have been very few systematic studies comparing the diversity found in the African and Indian collections. A large proportion of Indian accessions were pigmented compared to those from Africa. However, African germplasm presented much larger variation for inflorescence colour compared to Indian germplasm. The African and Indian germplasm too differs for inflorescence characters, with the majority of Indian accessions having semi-compact or compact ears (race *vulgaris*) and African accessions showing diverse ear types ranging from open to fist shaped (race *plana* and *compacta*). African accessions also exhibited higher intensity of finger branching compared to Indian accessions (Naik et al. 1993). Among quantitative traits, the Indian germplasm exhibited higher variability for characters associated with grain and fodder yield. This included characters such as finger length, finger width, grain yield potential, ear weight, total biomass and fodder weight and leaf number. For these characters, the mean values in Indian accessions were higher than in African accessions. On the other hand, African germplasm exhibited large variation, with mean values skewing towards late maturity, tallness, stout plant stature, long broad flag leaf, long narrow finger, higher number of spikelets, more florets per spikelet, small and long glumes, poor threshability and lower harvest index (Naik et al. 1993). There are indications that wild and weedy ancestor *E. africana*, which has introgressed fairly easily with cultivated finger millet, has contributed to the cultivated gene pool diversity generated in Africa. Furthermore, the African germplasm appears to possess a higher level of resistance to blast.

Systematic finger millet improvement work has been ongoing in India since the early 1950s. The most significant aspect of finger millet improvement in India has been the use of African germplasm in hybridisation programmes (Gowda et al. 1986). Exotic germplasm, particularly African finger millet germplasm, was introduced into India during the 1970s. The African germplasm was superior to Indian germplasm in rainy (*kharif*) and post-rainy (*rabi*) seasons, while the performance was comparable during summer (Swaminath 1979). The African genotypes have also many desirable characters like high initial vigour, large ears, high grain density and thick robust stems with broad dark-green leaves (Seetharam 1982a). Following hybridisation among indigenous, exotic and between indigenous and exotic collections, pedigree selection was adopted. Early (less than 100 days), mid-late (100–110 days) and late (more than 110 days) selections were made in Indian x Indian, Indian x exotic or exotic x Indian and exotic x exotic crosses. The Indian x Indian derivatives had intermediate productivity. The Indo-African crosses have provided the real backbone for breaking the grain yield barriers in the improvement of finger millet. They helped in increasing finger millet productivity by more than 50% in Karnataka State (Seetharam 1982b) and by 60% in Tamil Nadu (Nagarajan and Raveendran 1983).

The intercrossing of African germplasm with the Indian germplasm has revealed a potential for using the two isolated germplasm pools. These efforts resulted in the generation of breeding material, and subsequent selection resulted in the development and release of a series of highly productive varieties, more recently combined with blast resistance for cultivation. More than a dozen varieties of finger millet have been released at the national level (PR202, RAU8, KM13, PES400, VL149, VL146, GPU28, GPU45, GPU26, GPU66, GPU67, VR708, BM9-1, VL 347, VL 352) along with other varieties released for cultivation at the state level (AKP2, RATNAGARI, Kalyani, Gujarat Nagali 2, Gujarat Nagali 3, MR1, MR6, L5, HR911, Indaf5, Indaf8, Chilka, Bhariaba, Subra, Co12, C013, Co14, VL315, VL324, PRM 1, PRM2, etc.). In the finger millet breeding programme in India, blast resistance is one of the priority objectives, and many stable sources of resistance have been identified and used. Many sources of resistance have come from African germplasm. Several initiatives are underway to enhance the use of finger millet germplasm through deployment of core and mini-core approaches and identification of trait-specific germplasm.

Pollen-mediated gene flow and introgression from cultivated finger millet to its wild relatives are likely where the plants occur in sympatry. Although finger millet is a self-fertilising crop, spontaneous weedy hybrids are found in and around finger millet fields. Unfortunately, there is no information available on role of pollinators including wind and insects, the distance pollen can travel and outcrossing rates at varying distances from the pollen source. Studies are needed to investigate the rate of natural outcrossing between finger millet and its wild relatives, the extent and main direction of introgression and the fate of introgressed alleles in the later generations.

Seed exchange is another form of gene flow and introgression between finger millet and its wild relatives that needs to be considered. In countries like India, farmers often exchange seed both within a village and with farmers in neighbouring villages. These informal seed exchange systems facilitate gene flow not only within the crop but also between domesticated and wild finger millet.

9.8 Traits of Importance

Breeders desire superior genotypes with wide adaptation especially for farmers in low-resource areas like the case for finger millet. The stable genotypes which perform well under stress and low-input conditions are desirable under smallholder farmers' conditions for sustainable crop production.

9.8.1 Grain Yield

High grain yield is the first and foremost consideration in the improvement of any crop plant. Unlike major crops, the yield levels in finger millet have not increased, and still the productivity of the crop is around 1.5 t/ha in India and even lesser in Africa. There is need to identify and introgress yield component traits for enhancing the productivity levels of the crop. The breeding efforts should be intensified towards improvement and pyramiding of component traits directly affecting yield like panicle size, number of fingers, number of effective tillers, density of spikelets on the finger, etc.

9.8.2 Lodging Tolerance

Finger millet cultivars generally tend to lodge under high fertility and moisture conditions. Taller genotypes with height more than 120 cm have higher lodging susceptibility in comparison to dwarf genotypes. However, the farmers require both food and fodder from finger millet; hence, ideal plant height should be around 90–110 cm. Selection of genotypes with reduced plant height and incorporation of dwarfing and semidwarfing genes is one of the important breeding objectives in finger millet improvement. Similarly, genotypes with profuse basal tillering, reduced number of nodal tillers and longer and more number of fingers will result in enhanced productivity of the crop.

9.8.3 Head Shape

Among the yield components of finger millet, head shapes play a vital role in yield improvement and acceptability among farmers (Baniya et al. 2003). Some studies have indicated that open-headed or short- and/or early-maturing genotypes are more susceptible to disease than incurving types. Genotypes with compact head shapes are reported to be reasonably resistant to head infections, but more susceptible to leaf infections (Esele and Odelle 1995). Progenies from crosses with disease-resistant parents have also been reported to show lower susceptibility followed by crosses from sterile (Esele 1993). Bua and Adipala (1995) also showed that head shapes may be important in determining resistance to blast disease. These are also important because farmers' preferences for different head shapes have been reported for various reasons, yet Oduori (2008) found a negative correlation between head shape and lodging, stating that open-headed genotypes were more prone to lodging than the fist-headed genotypes probably due to open heads offering resistance to wind. He further reported higher susceptibility of open-headed genotypes to head blast, as head blast increased with tendency to open headedness. Compact or fisted head shape was associated with higher grain yield, less shattering, blast disease tolerance and limited bird damage, making it a highly preferred trait.

9.8.4 Maturity

Since finger millet is grown in diverse habitats under different cropping sequences. There is need to identify different maturity groups suitable for the regional requirements. In higher hills, early- and medium-maturing genotypes perform well and fit into the cropping sequence of the area, late-maturing cultivars find a place in plains where the growing season is longer. Early-maturing genotypes (maturing in <100 days) are preferred in dry semiarid regions and hills where the growing season is short.

9.8.5 White Grain Genotypes for Industry Purpose

White finger millet genotypes are known to contain high protein and are low tannins as compared to brown-seeded genotypes. The tannins although have protective functions against fungi and pests, but impart astringent taste to the grains and reduce their palatability and nutritional quality. Thus, development of high-protein and low-tannin varieties is required. This will have advantage in bakery products, where the dark colour will not appear in the finished products.

9.8.6 Nutritional Traits

Breeding for improved grain quality is another major objective in finger millet. Finger millet is known for very high calcium content in its grains. Besides, it also contains high amount of polyphenols and dietary fibre and is source of essential amino acids. Nutritionally, its importance is well recognised because of its high content of calcium (0.38 %), dietary fibre (18 %) and phenolic compounds (0.3–3 %). The growing public awareness for nutrition and health care substantiates the potential of phytochemicals such as polyphenols and dietary fibre on their health-beneficial properties. Hence, finger millet provides a better option as nutraceutical and therapeutic foods. It is also recognised for their health-beneficial effects, such as antidiabetic, antitumorigenic and atherosclerogenic effects and antioxidant and antimicrobial properties.

9.8.7 Disease Resistance

The major biotic constraint in finger millet production is blast disease (*Pyricularia grisea*). There is no reported economic loss due to leaf blast; however, finger and neck blast can cause heavy losses in crop yield in susceptible varieties. The disease is quite severe in rainy season crop at all the growth stages. The losses are more if the disease appears in the nursery and on the ears affecting the neck and fingers. In recent years, brown spot caused by *Drechslera nodulosum* is gaining importance. Its damage could be severe if the crop suffers from drought or nutritional deficiency. Other diseases affecting the crop are foot rot (*Sclerotium rolfsii*), downy mildew or green ear (*Sclerospora macrospora*) and leaf spot (*Cercospora* sp.). Host plant resistance is the major strategy for disease control in the crop, and identification of varieties with blast resistance should be a priority in finger millet breeding programme.

9.8.8 Wide Adaptability

Andrews (1993) suggested growing the materials in sufficient test environments to evaluate for superior stable entries of finger millet so as to increase production, while Misra et al. (2009) used multi-environment trials (METs) in the identification of stable and productive environments and those that best suit particular finger millet genotypes in India to maximise its production. In addition, METs can also be used to select homogeneous subgroups of environments and agro-management conditions or mega environments that minimise genotype-environment interaction for recommendations in finger millet (Misra et al. 2009; Samonte et al. 2005). At national levels, these could constitute agro-ecological zones. The use of METs in environments which differ in altitude, latitude, photoperiod, temperature, rainfall, soil type and disease incidence allows the expression of high yield potential. Choice of selection sites is therefore particularly relevant in the case of production areas with variable levels of abiotic stresses (Ramagosa and Fox 1993).

9.8.9 Fodder Quality

Finger millet has the potential to be an important forage crop in dry land regions. Finger millet has superior forage quality in terms of nutritive value, mineral nutrients, lower lignin and superior starch content compared to corn or sorghum. However, its biomass yield is lower than corn and sorghum in good soils, but has the potential for greater yields in poor soils with limited resources (Vara Prasad et al. 2009).

E. coracana thrives under hot conditions, but also tolerates cool climates. Hence, the study was conducted in Northern Ireland for evaluating the finger millet as a forage crop. Problems of photoperiod sensitivity were encountered in the crop as observed in Davis, California, USA (Anonymous 1996). The plants remained growing throughout the summer without flowering until they were killed by frost in early October. This frost sensitivity of the crop indicated that finger millet can only be sown after the period of spring frosts and must be harvested before the autumn frosts (Mackey et al. 2007). Finger millet husk, a by-product from brewing as spent grain, has been reported to be a source of fibre as well as a good source of protein and is especially used in household poultry feeding (Obilana and Manyasa 2002).

9.9 Landraces in Crop Improvement

In finger millet and many other crops, landraces have not been bred as varieties but adapted to the local conditions of environment and inputs where they are cultivated (Evans 1996) and have been the mainstay of agricultural systems in many developing countries (Juma 1989). Several workers have found and reported genetic variability for traits related to high yield and tolerance to various stresses. For instance, Oduori (2008) evaluated finger millet landraces in Kenya and found variability for agronomic traits among 310 accessions, whereas Bezaweletaw et al. (2006) had similar findings on evaluating 66 accessions of which 64 were landraces from Ethiopia. Apart from genetic variability, breeders can find materials adapted to marginal environments within landraces. Despite the research already done, the role of landraces contributing germplasm to breeding programmes has not been fully appreciated (Hill et al. 1998). This has been due to inadequate testing and lack of genetic information about the landraces. Information on genetic studies could be useful in designing breeding programmes that will best exploit economically important and yield-related traits while developing new varieties.

9.10 Uses of Wild Relatives in Crop Improvement

Breeding for improvement in finger millet has focused on genetic characterisation and use of *E. coracana* cultivars and landraces only. There are no specific records of use of *E. coracana* wild relatives for crop improvement. The wild species even have not been characterised for important traits including nutritional traits. There is no precise information on trait values of wild species for their inclusion in crop improvement programme.

9.11 Genomics and Recent Advancements in Marker Discovery

Despite its recognised value as an important staple crop, particularly for poor people in arid and semiarid areas, finger millet has been neglected in mainstream crop improvement research (Upadhyaya et al. 2006). Till now, very little efforts have been made to enhance finger millet genotypes from the Indian subcontinent at molecular level (Salimath et al. 1995). Although the traditional plant breeding based on phenotypic selection is very effective, it has suffered from several limitations for complex traits. The advances in omics technologies, i.e. comprehensive and integrated genomics and transcriptomics, and proteomic analysis can elucidate genetic architecture of plant genomes and the relationships between genotype and phenotype. The rapid advances in DNA sequencing technology have made whole-genome sequencing (WGS) both technically and economically feasible. More than 25 economically important plants' genomes have been sequenced (Hamilton and Buell 2012).

The major disadvantages of morphological and biochemical markers are that they are limited in number and are influenced by environmental factors or the developmental stages of the plant (Winter and Kahl 1995). In contrast to the instability of morphological and biochemical markers to different environmental conditions, molecular markers are stable and are more in number. The DNA markers have been used to evaluate genetic diversity in different crop species (Cooke 1995). Various molecular markers are being used for fingerprinting such as restriction fragment length polymorphism (RFLP) (Dubrail and Charcosset 1998), random amplified polymorphic DNA (RAPD) (Williams et al. 1990), microsatellites (Smith et al. 2000) and amplified fragment length polymorphism (AFLP). The SSR markers offer many advantages such as higher frequency of polymorphism, rapidity, technical simplicity, use of fluorescence, requirement of only a few nanograms of DNA, compatible for high-throughput genotyping and feasibility of automation (Semagn et al. 2006). Therefore, SSRs have been used to analyse the genetic relatedness in several crop species (Varshney et al. 2001). The higher resolution of molecular markers made them a valuable tool for fingerprinting and protection of breeder's rights, facilitating appropriate choice of parents for breeding programmes, analysing quantitative traits and detection of quantitative trait loci (QTLs), gene mapping, marker-assisted selection (MAS) and gene transfer and understanding evolutionary pathways and for the assessment of genetic diversity.

Analyses of nuclear DNA content in the species of *Eleusine* were first carried out by Hiremath and Salimath (1991) using Feulgen microspectrophotometry. Mysore and Baird (1997) determined the genomic sizes for *E. coracana* subsp. *coracana*, as 2C=3.36–3.87 pg (2n=36) (1 pg=980 Mbp). Gupta and Ranjekar (1981) found that approximately 39 to 49 % of the genome of finger millet consists of repetitive DNA sequences which intersperse with 18 % of single-copy DNA sequences of 1900 nucleotide pairs. Approximately 20 % of the repeated DNA sequences (4000–4200 nucleotide pairs) were involved in long-range interspersion pattern, while 60 % of the repeated DNA sequences (150–200 nucleotide pairs) were involved in short-period interspersion pattern. It is proposed that plants with haploid DNA content of more than 2.5 pg exhibited mostly the short-period interspersion pattern, while those with haploid DNA content of less than 2.5 pg showed diverse patterns of genome organisation.

9.11.1 Molecular Markers and Genetic Diversity Studies

Genetic diversity refers to the variations within the individual gene loci/among alleles of a gene, or gene combinations, between individual plants or between plant populations. Genetic diversity offers opportunities to utilise various genomic resources and technologies to manipulate desirable traits. A comprehensive exploration of potential genetic resources and exploitation of natural genetic variations can be a useful source of genomic information. Diversity of landraces can be explored for their desirable agro-morphological traits and can be further utilised to develop new varieties through molecular plant breeding approaches. This will pave the way towards the detection of novel and superior genotypes. There are very few reports on diversity studies of finger millet using molecular markers such as RAPDs, SSRs, isozymes and RFLPs.

Salimath et al. (1995) used RLFP (three restriction enzymes and eight different probes) as well as RAPD and ISSR markers among 22 accessions derived from five species. These workers suggested the use of ISSR markers to distinguish within and between species. RAPD markers offer many advantages such as higher frequency of polymorphism, rapidity, technical simplicity, use of fluorescence, requirement of only a few nanograms of DNA, no requirement of prior information of the DNA sequence and feasibility of automation (Subudhi and Huang 1999; Chalmers et al. 1992). RAPDs have also been used to study the genetic diversity and species relationships in the genus *Eleusine* (Hilu 1995; Salimath et al. 1995; Babu et al. 2007). Salimath et al. (1995) used RAPDs for the identification of 22 accessions belonging to 5 species of the genus *Eleusine*, and Fakrudin et al. (2004) analysed genetic diversity among 12 finger millet accessions using 37 RAPD primers. However, these two studies used only a limited number of finger millet accessions. There is little information on the extent of genetic diversity at molecular level among finger millet accessions adapted to various states of the peninsular region of India. Babu et al. (2007) fingerprinted 32 finger millet genotypes using 50 random amplified polymorphic DNA (RAPD) markers. Among the 32 finger millet genotypes, GEC 182 and CO 12 were distantly related with a low similarity index of 0.315. These two accessions also differed considerably in days to flowering and grain weight. The genotype GEC 182 is early flowering and has bold grains, while CO 12 is late flowering and has smaller grains. They suggested these two accessions with higher diversity at molecular level, phenology and grain weight will act as ideal parents in hybridisation programme. Singh and Kumar (2010) used RAPD primers to generate genomic fingerprint profile which depicted about 25–40 % linkage distance and resulted in formation of two major groups. Polymorphism range shown by RAPD primers was 71–90 %, while the range of total loci scored was from 7 to 10. The genetic relatedness of three varieties of finger millet with varying seed coat colour, namely, PRM-1(brown), PRM-701(white) and PRM-801 (golden), was assessed with ten RAPD and ten ISSR markers (Gupta et al. 2010). The RAPD markers generated better polymorphism than ISSR markers in their study. The RAPD and ISSR molecular marker-based analysis revealed maximum similarity among PRM-701 and PRM-801 which has been further confirmed by morphological, physiological and enzymatic (nitrate reductase and glutamine synthetase) characterisation.

Kumari and Pande (2010) conducted experiments to characterise the genetic diversity using 12 genotypes of finger millet including two of the same variety (VL-149) but from different regions. The study helped in identifying the accessions in a quick and reproducible manner and studying their relatedness. Panwar et al. (2010) studied genetic relationships among 52 genotypes collected from different districts of Uttarakhand, investigated by using RAPD, SSR and cytochrome P450 gene-based markers. Mean polymorphic information content (PIC) for each of these marker systems (0.351 for RAPD, 0.505 for SSR and 0.406 for cyt P450 gene-based markers) suggested that all the marker systems were effective in determining polymorphisms. Kumar et al. (2012) studied similar 52 genotypes using SSR, RAPD and protein profiling analysis to investigate the variation in protein content. In this study, although SSR, RAPD and SDS-PAGE profiles were reproducible and generated several bands, the banding patterns observed with protein profiling were almost similar and not discriminatory as observed with DNA fingerprinting. Thus, it can be concluded that RAPD and SSR markers could be used in differentiating a large number of *E. coracana* genotypes which could be useful in their characterisation. RAPDs are generally considered an unreliable fingerprinting technique (Salimath et al. 1995; Jones et al. 1997; Harris 1999). RAPDs are rarely used nowadays, especially considering the availability of far more reliable markers, such as microsatellites (SSRs) and ISSRs.

The ISSRs may be particularly useful for first assessment of genetic diversity in species where microsatellites have not yet been developed. Salimath et al. (1995) used ISSRs in the genus of finger millet and found the technique promising for analysis of plant diversity, including differentiation of accessions within the same species of *Eleusine*. The SSRs or microsatellites (Hancock 1999) are currently the most useful DNA markers for studies of genetic diversity and structure of natural and cultivated populations, mainly due to their high variability, abundance in the genomes, reliability, ease of use and affordability of the technique. For the first time in finger millet, genomic SSRs were developed by Dida et al. (2007). They isolated di- and trinucleotide SSRs (*Ugep*) from random genomic *Hin*dIII, *Pst*I and *Sal*I libraries of finger millet accession PI 321125 following hybridisation of 18,432 double-gridded colonies with mixtures of (AG)15/(AC)15, (AAG)10/(AAC)10/(ATC)10, (AGC)10/(AGG)10 and (CCG)10 oligonucleotides. A total of 96 SSR markers were identified. Bharathi (2011) used 20 genomic SSR markers on 959 accessions for genetic diversity analysis. The 20 SSR loci generated a total of 206 alleles ranging from 7 (UGEP11, UGEP56, UGEP104, UGEP) to 18 (UGEP5) with mean of 10.3 alleles per locus in reference collection. The highest and lowest gene diversity of the reference set was 0.852 (UGEP26) and 0.307 (UGEP31), respectively, with a mean value of 0.616.

9.11.2 Role of Functional Markers

The SSR markers also have disadvantages such as tedious, costly cloning and enrichment procedures required for their generation (Zane et al. 2002; Squirrell et al. 2003; Weising et al. 2005). Microsatellites developed from expressed sequence tags (ESTs), popularly known as EST-SSRs or genic microsatellites, represent functional molecular markers; putative function for majority of such markers can be deduced by database searches and other in silico approaches. Furthermore, EST-SSRs markers are expected to possess high interspecific transferability as they belong to relatively conserved genic regions of the genome. Up to July 2012, a total of 1956 EST sequences were available for finger millet in NCBI (GenBank) database. The mining of SSRs from EST sequences will be very useful for their ready use in diversity studies. Reddy et al. (2012) found that 324 ESTs had SSRs and developed 132 primer pairs out of the available ESTs (1927) by using the criteria as minimum of 5 for di-, tri-, tetra-, penta- and hexa- repeats. Reddy et al. (2012) found dimeric repeats as the most frequent among all the repeats. However, the results with the criterion were taken for in silico analysis of SSRs. Babu et al. (2014a) used two types of search criteria for identification of EST-SSRs. In the first case, the minimum number of repeats was six for dinucleotide, four for trinucleotide and three for tetra-, penta- and hexanucleotides. While in the second case, the minimum number of repeats was two instead of three for tetra, penta and hexanucleotides. Thirty-two EST sequences had more than two microsatellites, and 1357 sequences did not have any SSRs. The most frequent type of repeats was trimeric motif; however, the second place was occupied by dimeric motif followed by tetra, hexa and penta repeat motifs. The most common dimer repeat motif was GA and in case of trimeric SSRs, it was CGG. Zeng et al. (2010) found 55 % of the ESTs of Chinese medicinal plant (*Epimedium sagittatum* (Sieb. et Zucc.) Maxim) consisted of trimeric repeats. This showed that trimers are the most frequent motifs among crop species.

Functional SSR markers have been recently used for diversity studies in finger millet. Nirgude et al. (2014) used 36 EST-SSR primers for the *opaque2* modifiers and 20 anchored-SSR primers for calcium transporters and calmodulin genes for analysis of the genetic diversity of 103 finger millet genotypes for grain protein and calcium contents. The *opaque2* modifier-specific EST-SSRs differentiated the finger millet genotypes into high-, medium- and low-protein-containing genotypes. However, calcium-dependent candidate gene-based EST-SSRs broadly differentiate the genotypes based on the calcium content with a few exceptions. A significant negative correlation between calcium and protein content was observed. The orthologous regions for *o2* modifier genes have been identified in cereal crops such as rice, maize, sorghum and tef. To identify the o2 homologues in finger millet, Babu et al. (2014b) developed a set of 74 EST-SSR loci to investigate the genetic variation among the 190 finger millet genotypes for *o2* modifiers influencing tryptophan content. The functional SSR loci developed and broadly grouped the global collection of 190 finger millet genotypes into three major clusters (A, B and C) based on tryptophan content. Cluster A comprised genotypes that had medium levels of tryptophan content (0.70–0.85 %), while cluster B comprised genotypes containing high tryptophan (>0.85 %). Cluster C consisted mostly of the genotypes that had tryptophan values of less than 0.70 %, with a few exceptions.

9.11.3 Population Structure Analysis

The problem of analysing the structure of natural populations arises in many contexts and has attracted considerable attention. For example, methods for analysing population structure have been used in studies of human history, conservation genetics and domestication events and to correct for cryptic population stratification in genetic association studies. In the case of cereals, it is well studied. But in millets especially finger millet, only few studies have been conducted. Dida et al. (2008) studied genetic diversity and population structure of 9 finger millet accessions (*E. coracana* subsp. *coracana*) from 11 African and 5 Asian countries, plus 14 wild *E. coracana* subsp. *africana* lines collected from Uganda and Kenya with 45 genomic SSR markers distributed across the finger millet genome. Phylogenetic and population structure analyses showed that the *E. coracana* germplasm formed three largely distinct subpopulations, representing subsp. *africana*, subsp. *coracana* originating from Africa and subsp. *coracana* originating from Asia. Bharathi (2011) analysed the population structure of a large set of (nearly 990) finger millet accessions and grouped them into four subpopulations: subpopulation 1 and subpopulation 2 with mixture of East Africa and South Asian accessions, subpopulation 3 with maximum number of South Asia accessions and subpopulation 4 with mixture of all other geographical origins. There was good correspondence between the geographic patterns observed in the population structure. Although the population subgroups corresponded largely to geographic regions, there were some notable exceptions. Evidence of admixture between Indian and African germplasm was found in the Indian accessions, probably as a result of intercrossing between Indian and African germplasm, a breeding strategy that was adapted in India in the 1970s. Recently, Babu et al. (2014c) also conducted a similar study in 190 finger millet genotypes using the software STRUCTURE v. 2.3.4.

The global collection of finger millet genotypes belonged to India, and exotic accessions from nine different countries were evaluated for estimation of population structure using a panel of 46 SSR loci spread across all the chromosomes. The results obtained from phylogeny showed that there were at least two population groups, largely corresponding to their geographical origin (Fig. 9.4).

9.11.4 Linkage Map Studies

In the present next-generation sequencing era, a lot of progress has been made in several crops where sequencing has been done in 22 crops (Hamilton and Buell 2012). However, in crops like finger millet, still linkage maps are not available, which may be attributed to limited breeding efforts in finger millet. Crossing has long been impeded in the crop by the highly self-pollinating nature and small flower size, which has made hand emasculation difficult. Recently, Dida et al. (2007) put a very beginning effort in the construction of linkage maps. The genetic map incorporates homologous and heterologous RFLP, AFLP, EST and SSR markers and consists of 22 linkage groups, arranged in 9 homologous groups, plus 4 small linkage groups of unknown origin. Restriction fragment length polymorphism (RFLP), amplified fragment length polymorphism (AFLP), expressed sequence tag (EST) and simple sequence repeat (SSR) markers were used to generate a genetic map of the tetraploid finger millet (*Eleusine coracana* subsp. *coracana*) genome (2n=4x=36). The maps span 721 cM on the A-genome and 787 cM on the B-genome and covered all 18 finger millet chromosomes.

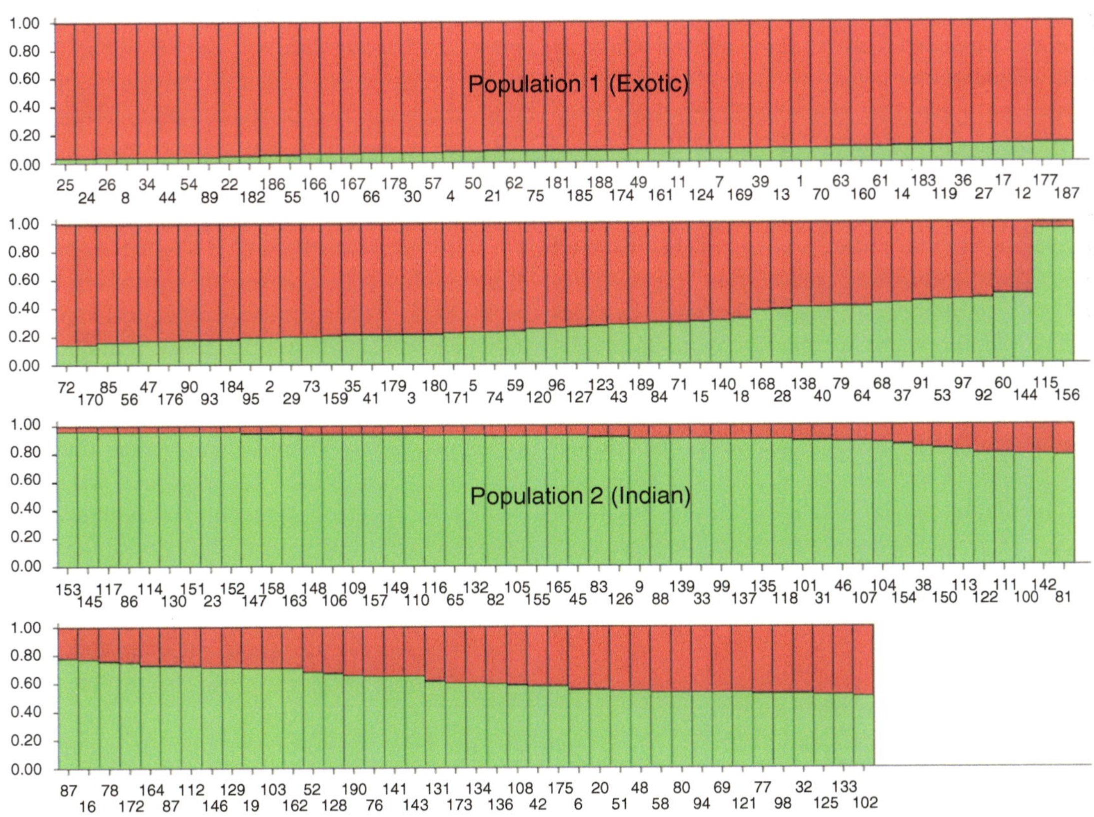

Fig. 9.4 The population structure of 190 finger millet genotypes obtained from STRUCTURE software using genomic SSR loci

9.11.5 Molecular Markers for Blast Resistance

In finger millet crop, blast disease caused by *Magnaporthe grisea* (anamorph *Pyricularia grisea*) has been identified as the highest priority constraint to finger millet production in Eastern Africa and India since most of the genotypes are highly susceptible (Nja et al. 2007). The causal organism of blast disease *Magnaporthe grisea* is also a causative agent of rice blast. Recently, functional marker-based molecular characterisation of resistance gene analogs encoding NBS-LRR disease resistance proteins in finger millet was done by Panwar et al. (2011). They described the analysis of a large collection of finger millet genotypes for association of NBS sequences with the blast disease resistance and susceptibility. The study established genetic relationships among the closely related genotypes and thus helped build a more complete picture of the molecular characterisation of finger millet genotypes. Singh and Kumar (2010) studied a total of 45 *Magnaporthe grisea* isolates using 15 RAPD primers to generate genomic fingerprint profile which depicted about 25–40 % linkage distance and resulted in formation of two major groups. Polymorphism range shown by RAPD primers was 71.40–90 %. Reddy et al. (2011) isolated resistance gene homologues from finger millet using degenerate oligonucleotide primers designed from the conserved regions of the nucleotide-binding site of the previously cloned plant disease resistance genes. Of the 107 clones sequenced, 41 showed homology to known R-genes and are denoted as EcRGHs (*Eleusine coracana* resistance gene homologues), while 11 showed homology to pollen signalling proteins (*PSiPs*) and are denoted as EcPSiPs (*Eleusine coracana* pollen signalling proteins).

9.11.6 Comparative Genomics

Comparative genomics is a powerful tool for genome analysis, with the objective of understanding the detailed process of evolution at the macro level and of translating DNA sequence data into proteins of known functions. With the availability of full-sequence information of the major cereal crops like rice and maize, it is now possible through comparative genomics to identify the genes influencing several important traits in finger millet. Srinivasachary et al. (2007) first generated a comparative map between rice and finger millet. Overall, colinearity is highly conserved between finger millet and rice. Three finger millet homologous groups, 2, 5 and 6, show colinearity with two rice chromosomes each, 2 and 10, 5 and 12 and 6 and 9, respectively. The finger millet-rice comparative maps demonstrate that, other than the rearrangements that are necessary to account for the difference in chromosome number between finger millet and rice, the finger millet genome has remained highly conserved since its divergence from a common ancestor with rice some 60 million years ago. Wang et al. (2005) made an attempt to identify the transferability of 210 SSR markers from major cereal crops (wheat, rice, maize and sorghum) to *E. coracana*, and more than half (57 %) of the SSR primers screened generated reproducible cross-species or cross-genus amplicons. They described that the transfer rate of SSR markers was correlated with the phylogenetic relationship or genetic relatedness.

The in silico analysis of NBS-LRR regions of finger millet with rice was done recently (Babu et al. 2014a). The sequences obtained from resistant and susceptible genotypes were subjected to BLASTN analysis to find the homologous sequences. The sequences from resistant genotypes showed high similarity with the rice blast *Pi* genes and NBS-LRR regions, while the sequences obtained from the susceptible genotypes did not show any similarity with the NBS-LRR regions. The nucleotide sequences were further analysed by conserved domain (CD) and resulted in identification of the NB-ARC (nucleotide-binding APAF-1, R proteins and CED-4) domain in sequences. In our earlier studies, we developed several EST-SSR markers for blast resistance and identified their positions on the finger millet and rice chromosome maps hypothetically (Fig. 9.5).

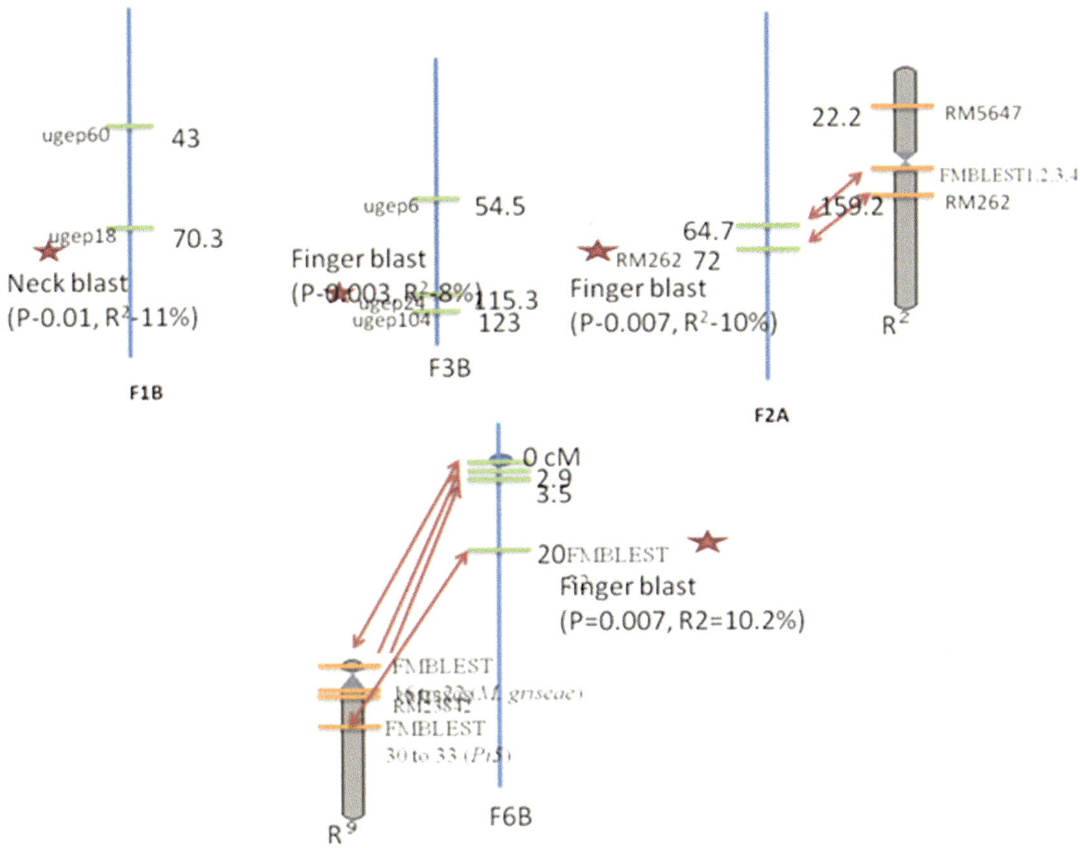

Fig. 9.5 The chromosomal location of the QTLs linked to the blast resistance in finger millet using genic and genomic SSRs (Babu et al. 2014d)

9.11.7 Association Mapping

The availability of plenty of landraces and diverse germplasm of finger millet have made it possible to dissect the highly useful and agronomically important alleles, through comparative genomics where genome sequence is not available. Molecular characterisation and identification of QTLs will facilitate marker-assisted breeding for improving the quality of finger millet. The goal of many plant researchers is to explain natural phenotypic variation in terms of simple changes in DNA sequence. Traditionally, linkage mapping has been the most commonly employed method to achieve this goal, where it involves generation of mapping populations by biparental crosses with known relatedness. By exploring the deeper population genealogy rather than the family pedigree, association mapping offers three advantages over linkage analysis: much higher mapping resolution, greater allele numbers and broader reference population. The terms linkage disequilibrium (LD) and association mapping have often been used interchangeably in literature. However, we feel that while association mapping refers to significant association of a molecular marker with a phenotypic trait, LD refers to non-random association between two markers or two genes/QTLs or between a gene/QTL and a marker locus. Thus, association mapping is actually one of the several uses of LD. Association mapping (AM) or association analysis detects and locates quantitative trait loci (QTLs) based on the strength of the correlation between mapped genetic markers and traits of unrelated individuals or natural population (Table 9.5).

Table 9.5 The details of the association mapping studies in finger millet

Trait	Markers	Reference
Finger blast	RM262, FMBLEST32, UGEP53	Babu et al. (2014d)
Neck blast	UGEP18	Babu et al. (2014d)
Leaf blast	FMBLEST35, RM23842, FMBLEST15	Babu et al. (2014d)
Tryptophan	OM5, FM8	Babu et al. (2014b)
Protein	FMO2EST1	Babu et al. (2014b)
Basal tiller number	UGEP81, UGEP8	Babu et al. (2014c) and Bharathi (2011)
Days to 50 % flowering	UGEP77, UGEP90, UGEP8, UGEP56	Babu et al. (2014c) and Bharathi (2011)
Flag leaf blade width	FM9, UGEP65	Bharathi (2011) and Babu et al. (2014c)
Plant height	FM9	Babu et al. (2014c)
Flag leaf blade length	UGEP56, UGEP8, UGEP31	Bharathi (2011)
Peduncle length	UGEP65, UGEP31	Bharathi (2011)

In finger millet very little progress has been made in identification of QTLs through association mapping. Bharathi (2011) carried out association analysis using phenotypic data of composite collection and reference set of three environments using 20 SSR markers and population structure data. Association analysis using phenotypic data on reference set data (300 accessions) and 20 SSR markers revealed 15 QTLs in environment 1 (E1), 22 QTLs in environment 2 (E2), 14 QTLs in environment 3 (E3) and 11 QTLs in combined analysis. QTL for days to 50 % flowering had consistent association with UGEP8 in LG3 (E2, E3 and combined for both composite collection and reference set) and UGEP56 in LG9 (E2 and E3 in composite collection and E1 in reference set). It indicated relatively tight linkage between the trait and marker. The loci UGEP56 (days to 50 % flowering and flag leaf blade length), UGEP8 (days to 50 % flowering, basal tiller number and flag leaf blade length) and UGEP65 (flag leaf blade width and peduncle length) had multi-trait association in E2 in the composite collection. Babu et al. (2014d) also studied association mapping of QTLs for 15 agro-morphological characters with 46 genomic SSRs. They reported identification of five markers tightly linked to QTLs of four traits at a significant threshold (P) level of 0.01 and 0.001. The QTL for basal tiller number was strongly associated with the locus UGEP81 at a P value of 0.001 by explaining the phenotypic variance (R2) of 10.8 %. The QTL for days to 50 % flowering was linked by two SSR loci UGEP77 and UGEP90 and explained 10 and 8.7 % of R2, respectively, at a P value of 0.01. The SSR marker, FM9 found to have strong association to two agro-morphological traits, flag leaf width (P 0.001, R2 14.1 %) and plant height (P 0.001, R2 11.2 %). Association mapping for *opaque2* modifiers *(Opm)* was done using 120 (74 genic and 46 genomic) SSR loci for identification of QTLs linked to *Opm* influencing tryptophan content and found two QTLs for tryptophan and one QTL for protein content. The QTLs for tryptophan content were associated with the genic marker OM5 at a P value of 0.009 and explained 11 % of phenotypic variance (R2) (Babu et al. 2014b). Association mapping for blast resistance was done using 104 SSR markers which identified four QTLs for finger blast and one QTL for neck blast resistance (Babu et al. 2014d). The genomic marker RM262 and genic marker FMBLEST32 were linked to finger blast disease at a P value of 0.007 and explained phenotypic variance (R2) of 10 % and 8 %, respectively. The genomic marker UGEP81 was associated to finger blast at a P value of 0.009 and explained 7.5 % of R2. The QTLs for neck blast were associated with the genomic SSR marker UGEP18 at a P value of 0.01, which explained 11 % of R2 (Babu et al. 2014d).

9.12 Constraints to Production

Finger millet production constraints can broadly be categorised as socio-economic, biotic, abiotic and policy considerations (low research consideration), some of which cannot be addressed by the scientific community alone.

9.12.1 Small Seed Size and Weed Control

The small seed size also contributes to complication in its cultivation (NRC 1996) as it necessitates planting in well-made and fine seed beds at higher plant densities (especially where it is planted by broadcasting). Subsequently, weeding becomes a problem further making cultivation of the crop labour-intensive. The difficulty in weeding is further complicated by wild relatives of the crop (for instance, *Eleusine indica*) that look like finger millet at the time of weeding. No effective chemical is available for weed control in finger millet. However, oxyfluorfen at 0.1 l a.i. per hectare or isoproturon at 0.5 kg a.i. per hectare as pre-emergent spray within 24 h of sowing and 2,4-D sodium salt at 0.75 kg a.i. per hectare as post-emergent spray around 20–25 days after sowing is recommended by All India Coordinated Small Millets Improvement Project, Bangalore, India (Ashok et al. 2011). Isoproturon pre-emergence application reduces the weed pressure; however, it does not fully control the weeds.

Chemical control is effective if the crop is sown in assured rainfall or irrigated areas. Since finger millet is grown in marginal areas as a rain-fed crop, the seed is broadcasted in the dry soils, and grains germinate whenever rains happen. The chemical weed control may not be feasible in finger millet.

9.12.2 Non-availability of Quality Seed of Improved Varieties

Poor seed chain is another bottleneck for reduction in acreage and production of this crop. The farmers still grow the susceptible low-yielding cultivars as no agency is willing to produce the seed of this crop. In India, only Karnataka State Seed Agency is producing sufficient quantity of quality seed of finger millet improved varieties. In recent years, the INSIMP programme was launched by the Government of India, where quality seed supply was also one of the objectives. However, the seed requirements under the programme could not be met in most of the states due to non-availability/non-production of quality seed.

9.12.3 Slow Initial Growth

Although finger millet is a C_4 crop, the initial growth is quite slow, whereas other C_4 crops grow rapidly. Due to slow initial growth, the crop is not able to compete with the weeds which overshadow the crop and starve the crop from nutrients and sunlight.

9.12.4 Biotic Factors

Among the biotic factors are weeds, pests and diseases. There is a high weed pressure, and the most important are wild species which come up along with the crop and are difficult to distinguish until flowering. This results in poor crop stand and yield of the crop. A wide range of fungal and bacterial diseases have been reported on finger millet (Ekwamu 1989), the most important of these diseases being blast caused by the fungus *Pyricularia grisea* (Sreenivasaprasad et al. 2005; NRC 1996; Bua and Adipala 1995). The disease is both economically significant and very destructive, causing over 50 % yield losses especially in wet seasons (Esele and Odelle 1995), but records of over 90 % yield losses also exist in Uganda (Ekwamu 1989). The crop is also affected by few insect pests. The principal pest problems in millet production are grasshoppers and army worms. Sometimes shoot fly, stem borers and chinch bugs or false chinch bugs may also cause economic damage, but these can be controlled by insecticides. Birds are also important pests, especially the notorious *Quelea quelea* and other small grain-feeding birds.

9.12.5 Finger Millet Blast

Finger millet blast disease caused by *Pyricularia grisea* affects finger millet at all stages of plant development, from seedling to grain formation. The pathogen has a wide host range that includes many graminaceous weeds, and therefore, initial sources of inoculum come from nearby weeds or cereal plants which act as collateral hosts (Takan et al. 2004). The fungus may also persist in crop debris which acts as potential inoculum reservoirs. The weeds also act as green bridges for finger millet blast. Depending on the severity of the infection, the disease may result in total inhibition of grain formation or production of shrivelled grain. Although the disease was first recorded in Uganda in 1933 (Esele 1993), there is still limited knowledge on its control. The cultivars grown therefore need to be improved for resistance to blast disease, and this is possible through hybridisation and selection.

The use of blast-resistant finger millet varieties could be key to an effective blast control programme as it would be compatible with low-cost input requirements of small-scale farmers who are the main growers of finger millet. It also poses no technical difficulties to the farmers. The effectiveness of this strategy is enhanced when resistance is available in adapted, productive germplasm (Seetharam and Ravikumar 1993). In India and to some extent in Africa, efforts have, over the years, been put in breeding for blast resistance in finger millet (Oduori 2008). The results, however, have been temporary with varieties currently succumbing to blast. This was probably because the breeding was done without clear and full understanding of the mechanisms of resistance operating in the available finger millet genotypes. There has also been slow breeding progress in recent times due to limited knowledge on genetics (combining ability) of finger millet blast resistance. Another problem has been existence of pathogen variability which probably leads to differences in aggressiveness, therefore presenting a challenge to identifying resistant genotypes in the field due to the interaction between host, parasite and environment. In addition, the challenges are further compounded by some varieties showing adequate resistance but exhibiting poor agronomic traits. Therefore, it is necessary to develop varieties that combine blast resistance with high grain yield potential which can be acceptable to farmers and consumers in the country and indeed the region. Crop damage by insects is minimal, but pests such as birds and *Striga* weed in Africa are a constant serious threat to the crop.

9.12.6 Abiotic Factors

Among the abiotic factors, the unpredictable rainfall patterns and poor soil fertility are the major limiting factors. Drought, high temperature and nutrient stresses further affect the crop production.

9.12.7 Hard Threshing

Mechanisation is lacking in the crop, and all operations are done manually right from sowing to harvesting. The panicles on maturity are hand-picked with the help of sickles at most of the places. After harvesting, the panicles are heaped for drying. The dried panicles are threshed with the help of sticks which again is a labour-intensive operation. This is followed by cleaning of grains. The mechanical harvesters and threshers are now available but are of limited use only because of unsuitability to small and marginal farmers. Drudgery involved in the crop production to processing is one of the major constraints in declining of the crop area.

9.12.8 No Availability of Market

The utmost important among various constraints is the non-availability of market for finger millet grains. These are considered as poor man's grains and no one likes to eat finger millet. There are no processing industries for finger millet, and neither any major development has happened in this area in recent past for popularisation of the crop. There is a need for value addition and incorporation of finger millet grains in breakfast cereals, baby products, bakery products, extrusion products, etc. This is again a major hindrance in the expansion of this crop as other commercial crops with easy package of practices are replacing finger millet.

9.12.9 Policy Issues

In terms of policy, finger millet is grossly neglected both nationally and internationally in terms of research, compared to crops like maize and rice (Oduori 2005), and this has resulted in use of unimproved, low-yielding, disease-susceptible cultivars which are also responsible for the low yields observed in finger millet.

Other constraints to finger millet production include poor incentives and marketing arrangements, low pricing, poor and inaccessible market channels, inaccessibility to credit facilities and inadequate improved processing and product development facilities at commercial levels (Mgonja et al. 2007).

There is urgent need to address the major constraints to make the crop more remunerative to the growers. These include:

1. Potential uses of millet in food, feed and industrial application
2. Improved national and regional trade of the commodity
3. Establishment of national stakeholder consultative groups to lobby for the subsector and coordinate production, processing and marketing of finger millet products
4. Intensify research on control and management of pests and diseases and drought stress and to develop technical packages for finger millet
5. Increase in labour productivity using simple planters, harvesters and threshers to be developed and introduced for ease in operations
6. Intensify dissemination and promotion of the crop training at various levels, information materials and demonstrations
7. Increasing awareness on the nutritional value and potential health benefits of the crop
8. Popularisation on the consumption of finger millet in diverse uses – porridge, bread, cakes and biscuits in schools, training institutions and school feeding programmes; and traditional dishes in the hotel industry

9.13 Conclusions

Finger millet is considered as wonderful millet because of its good storage quality and high nutritive and therapeutic value. With increasing awareness of good nutrition and healthy living, consumption of finger millet has increased recently. However, finger millet genetic studies lag considerably behind those of major cereal crops. Being a recent polyploidy, it has a very narrow genetic base. The genetic base of the crop can be widened using the wild relative gene pool. This strategy has been successfully used in other cereals where the wild species have been used as donors of novel genes particularly for disease resistance and quality traits. The finger millet germplasm and wild species have not been systematically characterised for various traits for their utilisation in breeding programme. Though recently little progress has been made in the application of molecular markers for finger millet improvement, still much needs to be done. Initially, there is a need to design the genomic SSR markers, EST-based SSR markers or transferability of close crops like rice. SSR markers will be required for development of linkage groups and QTL mapping studies. Comparative genomics also plays an important role in the identification of genes responsible for important traits. Identification of QTLs for important agro-morphological characters can greatly enhance the finger millet. At the same time, there should be more emphasis on introgression of genes into locally well-adopted germplasm of finger millet for development of biotic and abiotic stress-resistant varieties and high-yielding varieties with good amount of nutritional properties. Further, there is a need to sequence the full genome of finger millet which will help in addressing the major problems in finger millet crop improvement.

Unfortunately, finger millet has been branded as a food for the poor, and this negative label has contributed to the decline of the crop. However, its nutritional value and health benefits have resulted in the renaissance of the crop, which provides enormous scope for breeding improved cultivars for nutritional quality, best suited for developing newer products in finger millet so that good value addition could be achieved. Remarkable yield improvement was seen in wheat and rice yields in the last 60 years. This yield improvement was mainly due to introgression of semidwarfing genes. A similar strategy is required in finger millet also. Combination of conventional and molecular breeding techniques will allow for rapid development of high-yielding crop varieties in finger millet with double the current yield levels. Finger millet improvement requires collaborative efforts between breeders, biotechnologists and importantly funding agencies.

References

AICSMIP (2014) Annual report, Indian Council of Agricultural Research-all India coordinated small millets improvement project 2013–14, pp BR 49

Andrews DJ (1993) Principles for breeding small millets. In: Riley KW, Gupta SC, Seetharam A, Mushonga J (eds) Advances in small millets. International Science Publisher, New York, pp 411–423

Anonymous (1996) Lost crops of Africa, vol 1: grains. Board on Science and Technology for International Development, National Academy Press, Washington DC

Ashok EG, Dhananjaya BN, Gowda MVC et al (2011) Results of frontline demonstrations (2002–2010) and technologies for increasing small millets production in India, Project Coordinating Unit (Small Millets) ICAR. GKVK, Bangalore

Ayyangar GNR (1934) Recent work in the genetics of millets in India. Madras Agric J 22:16–26

Babu BK, Senthil N, Michael Gomez S, Biji KR, Rajendraprasad NS, Satheesh Kumar S, Chandra Babu R (2007) Assessment of genetic diversity among finger millet (Eleusine coracana (L.) Gaertn.) accessions using molecular markers. Genet Resour Crop Evol 54:399–404

Babu BK, Pandey D, Agrawal PK, Sood S, Kumar A (2014a) In-silico mining, type and frequency analysis of genic microsatellites of finger millet (Eleusine coracana (L.) Gaertn.): a comparative genomic analysis of NBS–LRR regions of finger millet with rice. Mol Biol Rep 41(5):3081–3090

Babu BK, Agrawal PK, Pandey D, Kumar A (2014b) Comparative genomics and association mapping approaches for opaque2 modifier genes in finger millet accessions using genic, genomic and candidate gene based SSR markers. Mol Breed 34:1261–1279

Babu BK, Agrawal PK, Pandey D, Jaiswal JP, Kumar A (2014c) Association mapping of agro-morphological characters among the global collection of finger millet genotypes using genomic SSR markers. Mol Biol Rep 41(8):5287–5297

Babu BK, Pandey D, Agrawal PK, Sood S, Chandrashekara C, Bhatt JC, Kumar A (2014d) Comparative genomics and association mapping approaches for blast resistant genes in finger millet using SSRs. PLoS One 9(6), e99182. doi:10.1371/journal.pone.0099182

Baniya BK, Subedi A, Rana RB, Tiwari RK, Chaudhary P (2003) Finger millet seed supply system in Kaski ecosite of Nepal. In: Sthapit BR, Upadhyaya MP, Baniya BK, Subedi A, Joshi BK (eds) On-farm management of agricultural biodiversity in Nepal, Proceedings of a national workshop, 24–26 April 2001, Lumle, Nepal. NARC/LIBIRD/IPGRI, Kathmandu, Nepal/Pokhara, Nepal/Rome, Italy, pp 171–175

Bezaweletaw K, Sripichitt P, Wongai W, Hongtrakul V (2006) Genetic variation, heritability and path-analysis in Ethiopian finger millet [Eleusine coracana (L.) Gaertn] landraces. Kasetsart J Nat Sci 40:322–334

Bharathi A (2011) Phenotypic and genotypic diversity of global finger millet (Eleusine coracana (L.) Gaertn.) composite collection. PhD thesis, Tamil Nadu Agricultural University, Coimbatore

Bisht MS, Mukai Y (2000) Mapping of rDNA on the chromosomes of *Eleusine* species by fluorescence *in situ* hybridization. Genes Genet Syst 75:343–348

Bisht MS, Mukai Y (2001) Genomic *in situ* hybridization identifies genome donor of finger millet (*Eleusine coracana*). Theor Appl Genet 102:825–832

Bisht MS, Mukai Y (2002) Genome organization and polyploid evolution in the genus Eleusine (Poaceae). Plant Syst Evol 233:243–258

Bua B, Adipala E (1995) Relationship between head blast severity and yield of finger millet. Int J Pest Manage 41:55–59

Burkill IH (1935) A dictionary of economic products of Malaya peninsula. Crown Agents of the Colonies, London

CAB International (2003) Crop protection compendium, 2005th edn. CAB International, Wallingford

Chalmers KJ, Waugh R, Sprent JI, Simons AJ, Powell W (1992) Detection of genetic variation between and within populations of *Gliricidia sepium* and *Gliricidia maculata* using RAPD markers. Genetics 69:465–475

Chennaveeraiah MS, Hiremath SC (1974) Genome analysis of Eleusine coracana (L.) Gaertn. Euphytica 23:489–495

Chuah TS, Salmijah S, Teng YT, Ismail BS (2004) Changes in seed bank size and dormancy characteristics of the glyphosate resistant biotype of goosegrass (*Eleusine indica* (L.) Gaertn.). Weed Biol Manage 4:114–121

Cobley LS (1956) An introduction to the botany of tropical crops. Longmans, Green and Co., London

Cooke RJ (1995) Variety identification of crop plants. In: Skerrit JH, Appels R (eds) New diagnostics in crop science, vol 13, Biotechnology in Agriculture No. CAB International, Wallingford, pp 33–63

De Candolle A (1886) Origin of cultivated plants. Hafner, NewYork

de Wet JMJ, Prasada Rao KE, Brink DG, Mengesha MH (1984) Systematic evolution of *Eleusine coracana* (Gramineae). Am J Bot 7:550–557

Dida MM, Srinivasachary RS, Bennetzen JL, Gale MD, Devos KM (2007) The genetic map of finger millet, Eleusine coracana. Theor Appl Genet 114:321–332

Dida MM, Wanyera N, Harrison Dunn MLN, Bennetzen JL, Devos KM (2008) Population structure and diversity in finger millet (Eleusine coracana) germplasm. Trop Plant Biol. doi:10.1007/s12042-008-9012-3

Dixit A, Dixit SS, Mittre V (1987) The occurrence of Eleusine africana Kennedy-O' Byrne in India and its significance in the origin of Eleusine coracana. Proc Indian Acad Sci, Plant Sci 85:1–10

Dodake SS, Dhonukshe BL (1998) Variability in floral structure and floral biology of finger millet (*Eleusine coracana* (L.) Gaertn.). Indian J Genet 58:107–112

Dubrail P, Charcosset A (1998) Genetic diversity within and among maize populations: a comparison between isozyme and nuclear RFLP loci. Theor Appl Genet 96:577–587

Duke JA (1978) The quest for tolerant germplasm. ASA Special Symposium 32, Crop tolerance to suboptimal land conditions. American Society of Agronomy, Madison, WI, pp 1–61

Duke JA (1979) Ecosystematic data on economic plants. Q J Crude Drug Res 17:91–110

Ekwamu A (1989) Host range, morphology and pathogenicity of the genus Pyricularia in Uganda. East Afr Agric Forest J 54:101–105

Esele JP (1993) The current status of research on finger millet blast disease (Pyricularia grisea) at Serere research station. In: Riley KW, Gupta SC, Seetharam A, Mushonga J (eds) Advances in small millets. International Science Publisher, New York, pp 467–468

Esele JPN, Odelle SE (1995) Progress in breeding for resistance to finger millet blast at Serere research station. In: Proceedings of the 8th EARSAM regional workshop on sorghum and millets, 30th Oct. – 5th Nov. 1992, Wad Medani, Sudan, pp 93–99

Evans LT (1996) Crop evolution, adaptation and yield. Cambridge University Press, Cambridge

Fakrudin B, Shashidhar HE, Kulkarni RS, Hittalmani S (2004) Genetic diversity assessment of finger millet, (Eleusine coracana (L) Gaertn.), germplasm through RAPD analysis. PGR Newsl 138:50–54

Fuller DQ (2006) Agricultural origins and frontiers in South Asia: a working synthesis. J World Prehist 20:1–86

Fuller DQ (2007) Contrasting patterns in crop domestication and domestication rates: recent archaeobotanical insights from the Old World. Ann Bot (Lond) 100:903–924

Gowda BTS, Seetharam A, Viswanath S, Sannegowda S (1986) Incorporating blast resistance to Indian elite finger millet cultivars from African cv. IE 1012. Sabrao J 18:119–120

Gupta VS, Ranjekar PK (1981) DNA sequence organization in finger millet (Eleusine coracana). J Biosci 3(4):417–430

Gupta SC, Muza FR, Andrews DJ (1997) Registration of INFM 95001 finger millet genetic male sterile line. Crop Sci 37:1409

Gupta R, Verma K, Joshi DC, Yadav D, Singh M (2010) Assessment of genetic relatedness among three varieties of finger millet with variable seed coat colour using RAPD and ISSR markers. Genet Eng Biotechnol J 2:1–9

Haaland R (1995) Sedentism, cultivation, and plant domestication in the Holocene Middle Nile region. J Field Archaeol 22:157–174

Haaland R (1999) The puzzle of the late emergence of domesticated sorghum in the Nile valley. In: Gosden C, Hather J (eds) The prehistory of food: appetites for change. Routledge, London/New York, pp 397–418

Hamilton JP, Buell CR (2012) Advances in plant genome sequencing. Plant J 70(1):177–190

Hancock JM (1999) Microsatellites and other simple sequences: genomic context and mutational mechanisms. In: Goldstein DB, Schlotterer C (eds) Microsatellites: evolution and applications, Oxford University Press, Oxford, pp 1–9

Harris SA (1999) RAPDs in systematics – a useful methodology? In: Hollingsworth PM, Bateman RM, Gornall RJ (eds) Molecular systematics and plant evolution. Taylor and Francis, London, pp 211–228

Hill J, Becker HC, Tigerstedt PMA (1998) Quantitative and ecological aspects of plant breeding. Chapman and Hall, London

Hilu KW (1981) Taxonomic status of the disputable Eleusine compressa (Gramineae). Kew Bull 36:559–562

Hilu KW (1988) Identification of the 'A' genome of finger millet using chloroplast DNA. Genetics 118:163–167

Hilu KW (1995) Evolution of finger millet: evidence from random amplified polymorphic DNA. Genome 38:232–238

Hilu KW, de Wet JMJ (1976a) Racial evolution of finger millet, Eleusine coracana. Am J Bot 63:1311–1318

Hilu KW, de Wet JMJ (1976b) Domestication of *Eleusine coracana*. Econ Bot 30:199–208

Hilu KW, deWet JMJ, Harlan JR (1979) Archaeobotany and the origin of finger millet. Am J Bot 66:330–333

Hiremath SC, Salimath SS (1991) Quantitative nuclear DNA changes in Eleusine (Gramineae). Plant Syst Evol 178:225–233

Jones N, Ougham H, Thomas H (1997) Markers and mapping: we are all geneticists now. New Phytol 137:165–177

Juma C (1989) The gene hunters: biotechnology and the scramble for seeds. Princeton University Press, Princeton

Kulkarni GN, Basavaraju V (1976) Firm seeds in the finger millet (*Eleusine coracana* (L.) Gaertn.). Curr Sci 45:425–426

Kumar A, Sharma N, Panwar P, Gupta AK (2012) Use of SSR, RAPD markers and protein profiles based analysis to differentiate Eleusine coracana genotypes differing in their protein content. Mol Biol Rep 39(4):4949–4960

Kumari K, Pande A (2010) Study of genetic diversity in finger millet (*Eleusine coracana* L. Gaertn) using RAPD markers. Afr J Biotech 9(29):4542–4549

Liu Q, Jiang B, Wen J, Peterson PM (2014) Low-copy nuclear gene and McGISH resolves polyploid history of *Eleusine coracana* and morphological character evolution in *Eleusine*. Turk J Bot 38:1–12

Mackey D, Brown A, Copeland R, Cooke L (2007) Evaluation of finger millet as a European forage crop: plant growth, weed control, and seed pelleting. In: Mgonja MA, Lenné JM, Manyasa E, Sreenivasaprasad S (eds) Finger millet blast management in East Africa. Creating opportunities for improving production and utilization of finger millet. Proceedings of the first international finger millet stakeholder workshop, Projects R8030 & R8445 UK Department for International Development – Crop Protection Programme held 13–14 September 2005 at Nairobi, Patancheru 502 324, Andhra Pradesh, India: International Crops Research Institute for the Semi-Arid Tropics, pp 85–89

McMaster DN (1962) A subsistence crop geography of Uganda. The world land use survey. Occasional papers no. 2. Geographical Publications Ltd., Bude, p 111

Mehra KL (1962) Natural hybridization between Eleusine coracana and E. africana in Uganda. J Indian Bot Soc 41:531–539

Mehra KL (1963a) Differentiation of cultivated and wild Eleusine species. Phyton 20:189–198

Mehra KL (1963b) Consideration of the African origin of *Eleusine coracana* (L.) Gaertn. Curr Sci 32:300–301

Mgonja MA, Lenné JM, Manyasa E, Sreenivasaprasad S (2007) Finger millet blast management in East Africa. Creating opportunities for improving production and utilization of finger millet. Proceedings of the first international finger millet stakeholder workshop, Projects R8030 & R8445 UK Department for International Development – Crop Protection Programme held 13–14 September 2005 at Nairobi, Patancheru 502 324, Andhra Pradesh, India: International Crops Research Institute for the Semi-Arid Tropics. 196 pp. ISBN: 978-92-9066-505-2

Misra RC, Das S, Patnaik MC (2009) AMMI model analysis of stability and adaptability of late duration finger millet (Eleusine coracana) genotypes. World Appl Sci J 6:1650–1654

Mysore KS, Baird V (1997) Nuclear DNA content in species of Eleusine (Gramineae): a critical re-evaluation using laser flow cytometry. Plant Syst Evol 207:1–11

Nagarajan C, Raveendran TS (1983) Germplasm mobilization and utilization in finger millet in Tamil Nadu. National Seminar on Finger Millet. Genetics and Breeding, 12–13 January 1983 UAS, Bangalore

Naik BJ, Shankare Gowda BT, Seetharam A (1993) Pattern of variability in relation to domestication of finger millet in Africa and India. In: Piley KW, Gupta SC, Seetharam A, Mushonga JN (eds) Advances in small millets. Oxford and IBH Publishing Co. Pvt. Ltd., New Delhi, pp 347–364

Neshamba SM (2010) Variability for drought tolerance in finger millet (*Eleusine coracana* (L.) Gaertn.) accessions from Zambia. M.Sc. thesis submitted to school of agricultural sciences of the University of Zambia, Lusaka, pp 7

Neves SS, Clark GS, Hilu KW, Baird WV (2005) Phylogeny of Eleusine (Poaceae: Chloridoideae) based on nuclear ITS and plastid trnT–trnF sequences. Mol Phylogenet Evol 35:395–419

Nirgude M, Kalyana Babu B, Shambhavi Y, Singh UM, Upadhyaya HD, Kumar A (2014) Development and molecular characterization of genic molecular markers for grain protein and calcium content in finger millet (Eleusine coracana (L.) Gaertn.). Mol Biol Rep 41(3):1189–1200

Nja MA, Lenne JM, Manyasa E, Sreenivasaprasad S (2007) Finger millet blast management in East Africa. Creating opportunities for improving production and utilization of finger millet. UK Department for International Development-Crop Programme, pp 196

NRC (National Research Council) (1996) Finger millet. In: Lost crops of Africa, vol 1: grains. National Academy of Sciences, Nairobi, pp 39–57

Obilana AB, Manyasa E (2002) Millets. In: Belton PS, Taylor JRN (eds) Pseudocereals and less common cereals. Grain properties and utilization potential. Springer, Berlin, pp 176–217

Oduori COA (2005) The importance and research status of finger millet in Africa. A paper presented at the McKnight foundation collaborative crop research programme on Tef and Finger millet: comparative genomics of the chloridoid cereals at the biosciences for East Africa (BECA) ILRI, Nairobi, Kenya

Oduori COA (2008) Breeding investigations of finger millet characteristics including blast disease and striga resistance in western Kenya. Unpublished PhD Thesis, University of Kwazulu Natal, Pietermaritzburg

Panwar P, Nath M, Kumar V, Kumar A (2010) Comparative evaluation of genetic diversity using RAPD, SSR and cytochrome P450 gene based markers with respect to calcium content in finger millet (Eleusine coracana L. Gaertn.). J Genet 89(2):121–133

Panwar P, Jha AK, Pandey PK, Gupta AK, Kumar A (2011) Functional markers based molecular characterization and cloning of resistance gene analogs encoding NBS-LRR disease resistance proteins in finger millet (Eleusine coracana). Mol Biol Rep 38:3427–3436

Phillips SM (1972) A survey of the Eleusine Gaertn. (Gramineae) in Africa. Kew Bull 27:251–270

Phillips SM (1974) Eleusine Gaertn. In: Polhill RM (ed) Flora of tropical East Africa. Crown Agents for Overseas Governments and Administrations, London, pp 260–267

Phillips SM, Renvoize SA (1974) Gramineae. In: Milne-Redhead E, Polhill RM (eds) Flora of tropical east Africa, Part 2 (AA Balkema on behalf of the East African Governments: Rotterdam, The Netherlands) pp 449

Prasada Rao KE, De Wet JMJ, Gopal Reddy V, Mengesha MH (1993) Diversity in the small millets collection at ICRISAT. In: Riley KW, Gupta SC, Seetharam A, Mushonga JN (eds) Advances in small millets. Oxford & IBH Publishing Co, New Delhi, pp 331–346

Purseglove JW (1976) Millets, Eleusine coracana, Pennisetum anzericantcm (Graminae). In: Simmonds NW (ed) Evolution of crop plants. Longman, London New York, pp 91–93

Rachie KO, Peters LV (1977) The Eleusines: a review of the world literature (International Crops Research Institute for the Semi-Arid Tropics, Hyderabad, – 500016 (A P) India) pp 179

Raj SM, Mahudeswaran K, Shanmuga Sundaram A (1984) Observations on the hot water technique of emasculation of ragi florets (Eleusine coracana Gaertn.). Madras Agric J 51:71

Ramagosa I, Fox PN (1993) Genotype x environment interaction and adaptation. In: Hayward MD, Bosemark NO, Ramagosa I (eds) Plant breeding: principles and prospects. Chapman and Hall, London, pp 373–390

Rao VR, Rao KVR (1962) Emasculation of ragi (Eleusine coracana Gaertn.) flowers by hot water treatment for hybridization work. Andhra Agric J 9:290–293

Ratnakar MS, Mallikarjuna NM, Naveen-Kumar KS, Jayarame Gowda MV, Mohan Kumar MV (2009) Cytomorphological studies in inter-specific hybrids of finger millet. Gregor Mendel Foundation Proceedings, pp 69–74

Ravikumar RL (1988) Genetic and biochemical basis of blast resistance in finger millet (*Eleusine coracana* Gaertn.). PhD Thesis, University of Agricultural Sciences, Bangalore

Reddy BGS, Pushpa G, Satyan BA, Prakash KS (1983) Preliminary studies on gibberellic acid induced male sterility in finger millet. National Seminar on Finger Millet Genetics and Breeding, UAS, Bangalore, 12–13 January 1983

Reddy BL, Srinivas Reddy D, Lakshmi Narasu M, Sivaramakrishnan S (2011) Characterization of disease resistance gene homologues isolated from finger millet (Eleusine coracana L. Gaertn). Mol Breed 27(3):315–328

Reddy BL, Lakshmi NM, Sivaramakrishnan S (2012) Identification and characterization of EST–SSRs in finger millet (Eleusine coracana (L.) Gaertn.). J Crop Sci Biotech 15(1):9–16

Salimath SS, Olivera ACD, Godwin ID, Bennetzen JL (1995) Assessment of genome origins and diversity in the genus Eleusine with DNA markers. Genome 38:757–763

Samonte SOPB, Wilson LT, McClung AM, Medley JC (2005) Targeting cultivars onto rice growing environments using AMMI and SREG GGE biplot analysis. Crop Sci 45:2414–2424

Seetharam A (1982a) Conservation of minor millets germplasm in India. MILWAI Newsl 1:14

Seetharam A (1982b) Finger millet improvement. Indian Farming 32(3):3–6

Seetharam A, Ravikumar RL (1993) Blast resistance in finger millet – its inheritance and biochemical nature. In: Riley KW, Gupta SC, Seetharam A, Moshanga J (eds) Recent advances in small millets. Proceedings of second international small millets workshop. Bulawayo (Zimbabwe). Oxford-IBH Publishing Company, New Delhi, pp 49–466

Semagn K, Bjørnstad A, Ndjiondjop MN (2006) Principles, requirements and prospects of genetic mapping in plants. Afr J Biotech 5(25):2569–2587

Singh Y, Kumar J (2010) Study of genomic fingerprints profile of Magnaporthe grisea from finger millet (Eleusine Coracana) by random amplified polymorphic DNA-polymerase chain reaction (RAPD-PCR). Afr J Biotech 9(46):7798–7804

Smith JSC, Kresovich S, Hopkins MS et al (2000) Genetic diversity among elite sorghum inbred lines assessed with simple sequence repeats. Crop Sci 40:226–232

Squirrell J, Hollingsworth PM, Woodhead M, Russell J, Lowe AJ, Gibby M, Powell W (2003) How much effort is required to isolate nuclear microsatellites from plants? Mol Ecol 12:1339–1348

Sreenivasaprasad S, Takan JP, Mgonja MA et al (2005) Enhancing finger millet production and utilization in East Africa through improved blast management and stakeholder connectivity. In: Harris D, Richards JI, Siverside P, Ward AF, Witcombe JR (eds) Pathways Out of poverty, aspects of applied biology 75. Association of Applied Biologists, Wellesbourne, pp 11–22

Srinivasachary DMM, Gale MD, Devos KM (2007) Comparative analyses reveal high levels of conserved colinearity between the finger millet and rice genomes. Theor Appl Genet 115:489–499

Subudhi PK, Huang N (1999) RAPD mapping in a doubled haploid population of rice (Oryza sativa L.). Hereditas 130:2–9

Swaminath MH (1979) Study on the genetic diversity in the African collection of finger millet (Eleusine coracana Gaertn.) and their interaction with environment (Abst.) Mysore J Agric Sci 13:225–226

Takan JP, Akello B, Esele JP et al (2004) Pathogen diversity and management of finger millet blast in East Africa: a summary of project activities and outputs. Int Sorghum Millets Newsl 45:66–69

Upadhyaya HD, Gowda CLL, Pundir RPS, Reddy VG, Singh S (2006) Development of core subset of finger millet germplasm using geographical origin and data on 14 quantitative traits. Genet Resour Crop Evol 53:679–685

Van Wyk BE, Gericke N (2000) People's plants: a guide to useful plants of southern Africa. Briza, Arcadia

Van Wyk E, Van Oudtshoorn F (1999) Guide to grasses of southern Africa. Briza Publications, Arcadia

Vara Prasad PV, Staggenborg SA (2011) Growth and production of sorghum and millets. http://www.eolss.net/sample-chapters/c10/e1-05a-14-00.pdf

Vara Prasad PV, Staggenborg S, Aiken R, Gowda P (2009) Comparative performance of finger millet for improving forage quality for dairy livestock in water-limited Ogallala Aquifer region. Ogallala Aquifer Program Final Report 2009 Kansas State University

Varshney RK, Kumar A, Balyan HS, Roy JK, Prasad M, Gupta PK (2001) Characterization of microsatellites and development of chromosome specific STMS markers in bread wheat. Plant Mol Biol Rep 18:5–16

Vavilov NI (1951) The origin, variation, immunity and breeding of cultivated plants. The Roland Press Co., New York

Verma V (2009) Textbook of economic botany. Ane Books, New Delhi

Wang ML, Barkley NL, Ju-Kyung Y, Dean R, Sorrells M, Paderson GA (2005) Transfer of simple sequence repeat (SSR) markers from major cereal crops to minor grass species for germplasm characterization and evaluation. Plant Genet Resour 3(1):45–57

Weising K, Nybom H, Wolff K, Kahl G (2005) Application of DNA fingerprinting in plant sciences. In: DNA Fingerprinting in plants-principles, methods, and applications. CRC Press, Boca Raton, pp 235–276

Werth CR, Hilu KW, Langner CA (1994) Isozyme of *Eleusine* (Gramineae) and the origin of finger millet. Am J Bot 81:1186–1197

Williams JGK, Kubelik AR, Livak KJ, Rafalski JA, Tingey SV (1990) DNA polymorphisms amplified by arbitrary primers are useful as genetic markers. Nucleic Acids Res 18:6531–6535

Willis JC (1973) A dictionary of the flowering plants and ferns, 8th edn. Cambridge University Press, Cambridge

Winter P, Kahl G (1995) Molecular marker technologies for plant improvement. World J Microbiol Biotech 11(4):438–448

Zane L, Bargelloni L, Patarnello T (2002) Strategies for microsatellite isolation: a review. Mol Ecol 11:1–16

Zeng S, Xiao G, Guo J, Fei Z, Xu Y, Bruce AR, Wang Y (2010) Development of a EST dataset and characterization of EST-SSRs in a traditional Chinese medicinal plant, Epimedium sagittatum (Sieb. Et Zucc.) Maxim. BMC Genomics 11:94–105

Foxtail and Barnyard Millets

10

Sandeep Kumar, N. Dikshit, Mohar Singh, and J.C. Rana

Abstract

Millets comprise group of crops which can play a major role in achieving the goal of nutritional security through diversification of diets in spite that these are generally grown in marginal and infertile lands. Compared to major cereals, millets provide carbohydrates, protein, minerals, dietary fibre, vitamins and various functional molecules to the people residing in Asian and African subcontinents. Among the small-grained millets, foxtail and barnyard millets are of short duration, drought resistant and widely grown crops. The availability of genome-wide sequences in foxtail millet has shifted the pace towards development of different kind of markers and linkage mapping. In contrary, little or no genomic work has been done in barnyard millet. But using comparative genomics from available foxtail sequences, different markers and QTLs can be identified in barnyard millet also. This chapter also covers the interspecific hybridisation, barriers to interspecific hybridisation and conventional and contemporary approaches used for the development of interspecific hybridisation. Further, in foxtail and barnyard millets, genetic-based broadening is required for agronomic traits particularly yield-related traits, resistance to biotic and abiotic stresses and nutritional traits.

Keywords

Genetic base • Broadening • Foxtail millet • Barnyard millet

S. Kumar
ICAR-National Bureau of Plant Genetic Resources, Pusa Campus, New Delhi 110 012, India

N. Dikshit (✉)
ICAR-National Bureau of Plant Genetic Resources, Regional Station, Dr PDKV Campus, Akola 444104, Maharashtra, India
e-mail: dikshitn@gmail.com

M. Singh
ICAR-National Bureau of Plant Genetic Resources Regional Station, Shimla, HP 171 004, India

J.C. Rana
Germplasm Evaluation Division, ICAR-National Bureau of Plant Genetic Resources, New Delhi 110 012, India

M. Singh, S. Kumar (eds.), *Broadening the Genetic Base of Grain Cereals*,
DOI 10.1007/978-81-322-3613-9_10

10.1 Introduction

Millets are small-grained cereals (annual grasses) and belongs to the family Poaceae. These are C4 plants and grown in marginal, drought prone, semiarid regions under subsistence farming conditions. Among the cereals, millets are considered important because of higher contents of minerals, protein, dietary fibre, resistant starch and polyphenols (Devi et al. 2011) and are important component of food security for smallholder farmers (Nelson et al. 2004; Naylor et al. 2004; Godfray et al. 2010; Varshney et al. 2010). Poor farmers with marginal and infertile lands grow small millets for their nutritional value and ability to tolerate extreme environmental conditions including salt stress, drought, temperature fluctuations and low-nutrient input conditions (Goron and Raizada 2015). Small millets are highly valued for their diverse benefits particularly for their low glycaemic index in comparison to major carbohydrate sources like wheat and rice (Hegde et al. 2005). Traits like higher tolerance to various biotic and abiotic stresses and long-term storability make the cultivation of small millets a good insurance against crop failure (Tsehaye et al. 2006; Reddy et al. 2011). In spite of vast potential of millets in mitigating nutritional deficiencies, the areas under millet cultivation has gone down in recent years.

Among the small-grained millets, foxtail and barnyard millets are used widely for food, fodder and brewing purposes. The two species varied with regard to chromosome number, genome size and breeding behaviour. Broadening the genetic base to suit the varying agro-climatic conditions needs to be studied. This chapter describes the status of crop gene pools, evolutionary relationships and systematics, levels of diversity in crop germplasm, limiting factors related to crop production, traits of importance for base broadening, interspecific hybridisation in crop species, barriers to interspecific hybridisation, molecular markers, genome mapping and genomics as an adjunct to breeding in two important small millet species, i.e. foxtail millet and barnyard millet.

Foxtail Millet [*Setaria italica* L. (P) Beauv.] is a diploid ($2n=2x=18$), considered among earlier domesticated crops and cultivated in 26 countries in the world. It is a suitable crop for cultivation under varying environmental conditions because of its short duration, tolerance to pests and diseases and higher efficiency of photosynthesis and for changing food habits of people due to very high nutritional value. It has a small genome of 490 million base pair (~400 Mb) with low repetitive DNA content (30 %). It is considered as a model species for detailed genetic and molecular studies as its genome structure is highly conserved in comparison to the ancestral grass lineage (Devos et al. 1998; Jayaraman et al. 2008). It is also used as a representative of several important biofuel crops due to its relatedness to switch grass, napier grass and pearl millet (Doust et al. 2009).

Barnyard Millet [*Echinochloa frumentacea* (Roxb.) Link.] is known for its early maturity, good yield, higher nutritional value and ability to tolerate the drought conditions. Its grain contains 55 % carbohydrate, 11.2 % protein, 3.9 % fat and 13.6 % crude fibre with significant amounts of both calcium and iron (Saleh et al. 2013). Also it is a nutritive fodder for animals. Low glycaemic index along with other valuable nutritional traits makes it an ideal food for management of type II diabetes (Ugare et al. 2014). Various food products such as baby foods, snacks and dietary foods are manufactured from barnyard millets (Ugare 2008; Surekha et al. 2013; Anju and Sarita 2010; Vijaykumar et al. 2010).

10.2 Crop Gene Pool, Evolutionary Relationships and Systematics

The genus *Setaria* (foxtail millet) is mainly grown in temperate regions and includes approximately 125 species worldwide. These species are further classified into sub-genera, out of which 74 belong to Africa (Hubbard 1915; Rominger 1962). The genus *Setaria* belongs to the tribe Paniceae, subfamily Panicoideae and family Poaceae (Pohl 1978). Based on genome organisation, three gene pools have been recognised. The primary gene pool includes *S. italica* and *S. viridis* with AA genome. Secondary gene pool includes *S. faberii* and *S. verticillata*, the weedy tetraploid species with AABB genome. The species in the tertiary gene pool comprise *S. grisebachii* from Mexico with CC genome and *S. queenslandica* (AAAA genome). Species *S. pumila* and *S. pallide-fusca* are also polyploid but lacks the AA genome (Willweber-Kishimoto 1962; Benabdelmouna and Darmency 2003). Foxtail millet and weedy green foxtail are morphologically and genetically allied and considered taxonomically to be conspecific (Harlan and de Wet 1971). Hence, the cultivated taxa are considered as *S. italica* ssp. *italica* and the spontaneous taxa as *S. italica* ssp. *viridis. S. italica* ssp. *italica is based on S. italica* including all cultivated foxtail millets. It differs consistently from ssp. *viridis* only in having lost the ability of natural seed dispersal. It depends on human beings for sowing in suitable habitats. It is morphologically variable and widely grown as cereal across temperate, subtropical and tropical Eurasia. Williams and Schreiber (1976) studied the phenetic relationships of green foxtail millet (*S. viridis* var. *viridis* (L.) Beauv.) and its allies by numerical and chemotaxonomic methods. Correlation coefficients and taxonomic distances based on quantitative data from 31 characters displayed close phenetic similarities among green foxtail, its varieties and foxtail millet (*S. italica* (L.) Beauv.). Prasad Rao et al. (1987) carried out discriminant function analysis among 1195 accessions of *S. italica* and recognised three cultivated races, i.e. *Moharia* from Europe and South Western Asia with 5–52 culms, each bearing several small, more or less erect inflorescences, cultivars in race *maxima* characterised by plants with 1–8 usually unbranched culms that bear large inflorescences (occurring in Transcaucasian Russia and the Far East). Race *C. indica* is intermediate in culm number (with average value 6.6) and inflorescence size between races *moharia* and *maxima* and is cultivated in Southern Asia. Based upon the inflorescence morphology, the genus has been classified into two species *S. pumila* and *S. italica. S. italica* is further classified into sub-species *viridis* and *italica*. Foxtail millet germplasm has been classified into three major races, namely, moharia, maxima and indica, which are further divided into 10 subraces, viz., aristata, fusiformis and glabra; compacta, spongiosa and asamense; and erecta, glabra, nana and profuse, respectively (Prasad Rao et al. 1987, 1993). Based upon the specific geographic regions, another report indicates four races of the genus including maxima (East Asia), moharia (Europe), indica (South Asia) and nana (Afghanistan and Lebanon) (Li et al. 1995).

Barnyard millet, *Echinochloa sp.* ($2n = 36, 54$) is a self-pollinating crop and belongs to the family *Poaceae* and is grown as cereal or fodder crop among the farmers of Southern and Eastern Asia. It is considered as the oldest domesticated and the fastest growing crop among all millets. The genus *Echinochloa* includes some 35 valid species spread over the world. Among these, *E. utilis* also known as the Japanese barnyard millet and *E. frumentacea* also known as the Indian or Sawa millet are the two domesticated species, and *E. utilis* is a temperate annual hexaploid ($2n = 6x = 54$) and is cultivated in Japan, Korea, etc. *E. frumentacea* ($2n = 36, 54$) is cultivated in India, Pakistan and Nepal. *E. colona* (L.) Link, the tropical grass, is the wild ancestor of *E. frumentacea*. Subspecies are further classified into races, viz., ssp. frumentacea (stolonifera, intermedia, robusta and laxa), ssp. *crus-galli* (crus-galli and macrocarpa) and ssp. *utilis* (utilis and intermedia), whereas, ssp. colona has no races. All members of the *Echinochloa* species complex are polyploid and *E. colona* and *E. crus-galli* are usually reported as hexaploids ($2n = 6x = 54$) (Prasad Rao et al. 1993; Upadhyaya et al. 2008). Wanous (1990) reported heterogeneity in the species. Total genome size has been estimated to be roughly 1.4 gigabases (Bennett et al. 1998, 2000).

10.3 Assessment of Gene Flow for Crop Improvement

Interspecific hybridisation between *S. viridis* and *S. italica* is common in nature (Li et al. 1942; Jusuf and Pernes 1985). Artificial crosses between these two species resulted in low yield, spindly shoot tissue, seed shedding and fertility losses associated with hybridisation and resistance to a variety of herbicides (Darmency and Pernes 1985, 1989; Wang et al. 1996; Wang and Darmency 1997). This highlights the scope for transferring agronomical valuable traits from wild to cultivated species for future crop improvement programme.

Both Japanese barnyard millet and Indian barnyard millet have overlapping morphological traits and can be differentiated on the basis of the presence/absence of awn, subtle spikelet differences and glume morphology (de Wet et al. 1983). Two species are genetically distinct and with sterile F_1 hybrids (Yabuno 1962; Hilu 1994).

10.4 Level of Diversity in Crop Germplasm

Foxtail millet has rich genetic diversity. Diverse genotypes have been characterised for morphological, physiological and biochemical traits and abiotic stresses. During 1997–2005, ICRISAT, Hyderabad, characterised global collection of foxtail millet germplasm (1535 accessions) for morphological traits and identified a core set of 155 accessions including 93 accessions of Indian origin. Athwal and Singh (1966) in a study of 32 varieties of foxtail millet reported high variability for yield and low variability for ear girth and days to flowering. Singh and Athwal (1966) observed significant interactions between genotypes and locations. The study also indicated that the genetic variation was influenced to a greater degree by interaction of genotypes with sowing dates and location than their interaction with season. Sandhu et al. (1974) in a study of 64 varieties for nine metric traits reported high variability for yield and secondary tillers per plant; intermediate for primary tillers per plant, height, ear length and ear circumference; and low for days to flowering, days to maturity and 1000-seed weight. Chinnaswamaiah (1975) studied the variability pattern through discriminant function technique among 48 varieties of foxtail millet and observed striking differences for all the traits excepting 1000-grain weight. Singh (1974) studied the utility of co-heritability, the correlation between pairs of characters and selection under four environments and concluded that selection indices constructed from data of late and abnormal sowing gave better efficiency than that of normal and favourable sowing conditions. Gill and Randhawa (1975) studied the heritabilities, interrelationships and genetic gains expected in 19 promising strains of foxtail millet. High heritability was recorded for days to heading and maturity, grain weight, plant height and tiller number compared to low heritability values for ear length, ear girth and grain yield. The expected genetic gains were high in tiller number, grain yield, 1000-grain weight, days to heading and plant height. Nagarajan and Prasad (1980) assessed the variability, correlation and path analysis in 50 foxtail millet accessions with diverse geographical origin. Higher GCV was recorded for straw yield, productive tiller and grains per branch, whereas high heritability values coupled with higher genetic advance was recorded for straw yield, grains per branch, productive tillers and number of branches. Vishwanatha et al. (1981) reported significant variability in 324 accessions for 13 characters. Moderate to high heritability values were recorded for all the characters studied except harvest index. Moderate heritability and high genetic advance were obtained for number of tillers per plant, grain density and panicle length indicating the additive gene action governing these characters and were reliable during selection. Godawat (1980) analysed the path coefficient analysis among 19 *tenai* varieties selected from advanced and initial foxtail millet trial and reported that 1000-grain weight and number of ears per plant are the two major characters that directly contribute to yield in foxtail millet. Ratnaswamy and Poonaiya (1963) studied the discriminant function for all the varieties and indicated that out of the three yield characters, viz., the number of productive tillers, the length of the main panicle and the yield of straw and the number of productive tillers exert maximum weight upon the grain yield of the plant. Hence, they suggested that the number of productive tillers might serve as an index in selection for yield in *S. italic*. Navale et al. (1984) mentioned the existence of quantitative variability for a number of tillers per plant, plant height, length of the panicle and yield in advanced generations of foxtail millet, SE 21–1. This genetic diversity in highly selected self-fertilising millet appears to be due to slow release of cryptic genetic variability through cytological homeostasis. Siles et al. (2004) studied heterosis in seven diverse parents along with 21 F_2S and 21 F_3S produced from biparental crosses and found that 18 out of 21 F_2S showed high parental heterosis for grain yield. They explained high heterosis for grain yield on the basis of existence of high amount of heterozygosity in hybrid over nonhybrid cultivars. Based on the significant correlations among estimates of additive and/or variety heterosis, they suggested that genes which control the plant height, yield and spike length were either the same or in coupling phase linkage. Nagarajan and Prasad (1980) studied the genetic divergence in 50 foxtail genotypes. The genotypes grouped into fifteen clusters whereas no relation was observed between genetic diversity and geographic regions. The clusters XV, XIV and XII were observed to be highly divergent from others. Superior genotypes in cluster IV for grain weight, cluster VIII for grain yield and cluster XIV for days to half bloom were reported. The study also indicated that the number of branches, grain yield panicle length and grains per branch contributed maximum to genetic divergence. Sheriff and Shivashankar (1992) assessed the genetic divergence among 225 foxtail genotypes through multivariate and canonical analysis. As per the Mahalanobis D^2 statistic, 33 clusters were formed. The intra- and intercluster distances ranged from 25.5 to 57.0 and 53.78 to 981.7, respectively. Both the Mahalanobis D^2 statistics and canonical analysis were in conformity as the contribution of characters for maximum divergence is concerned. The traits contributing significantly for the genetic diversity includes days to flowering, days to maturity, number of productive tillers, panicle weight, panicle length, panicle exertion, leaf angle and peduncle length. A total of 15 solitary clusters representing diverse genotypes in respect of yield and its components were observed. The study also revealed that genetic divergence has no relationship with geographic distance. Therefore, importance on genotypes rather than geographic diversity should be given while selecting parents for hybridisation programme. A large number of wild relatives of foxtail millet are conserved in national gene banks. Kumari et al. (2011) carried out genetic diversity analysis of 125 accessions from India as well as exotic collection and identified diversity-rich regions of southern states of India and the northern hills of Uttarakhand and Jammu Kashmir. Of the conserved wild relatives of the foxtail millet, China has the largest collection (26,670 accessions) followed by ICAR-National Bureau of Plant Genetic Resources, India (4330 accessions); ORSTOM-MONTP (3500 accessions); AICMMP, Bangalore (2512 accessions); ICRISAT (1535 accessions); NIAS (1299 accessions), North Central Regional Plant Introduction Station, Ames, USA (1000 accessions); IUT, France (850 accessions); Kenya Agriculture Research Institute (KARI) (772 accessions); Griffin, USA (762 accessions); and INFA, Mexico (350 accessions).

10.4.1 Barnyard Millet

The diversity in morphological and physiological traits is reported to be high in Japanese barnyard millet (Nozawa et al. 2006). Spectrum of variation is reported across landraces with respect to flowering time, inflorescence shape and spikelet pigmentation. Upadhyaya et al. (2014) recognised two races, viz., utilis and intermedia. Based on the length of flag leaves, peduncle, inflorescence racemes, plant height and basal tiller number, four morphological races, viz., laxa, robusta, intermedia and stolonifera, are recognised in Indian barnyard millet (de Wet et al. 1983). Species *E. frumentacea and E. colona* differ in inflorescence morphology, plant height, leaf length and width, flag leaf length and width, inflorescence length and width, number of racemes/inflorescence, raceme length, colour of inflorescence, days to maturity, inflorescence shape, inflorescence compactness, shape of lower raceme length of peduncle, arrangement of spikelet, degree of branching, degree of lodging, etc. Further, there is an urgent need to explore and collect race laxa which is not conserved in the ICRISAT gene bank and endemic to Sikkim state of India. In general, high variability has been observed in grain yield, number of productive tiller, straw yield and internodes length. These traits should be considered for selection point of view. Subramanyam and Rathore (1999) observed variation across genotypes in photosynthesis and yield-related traits. Gupta et al. (2009) tested 257 accessions for grain smut spores at anthesis and observed low susceptibility to grain smut (*Ustilago panici-frumentacei*) in advanced breeding lines in comparison to high variability in other accessions. Currently, India holds the largest collection of 1718 accessions at ICAR-National Bureau of Plant Genetic Resources followed by AICSMI, Bangalore (985 accessions); ICRISAT holds 743 accessions of both the species; VPKAS, Almora maintains more than 300 accessions, NIAS (156 accessions of Japanese barnyard millet) and Griffin, USA (232 accessions of Indian barnyard millet).

10.5 Limiting Factors Related to Crop Production

The principal factors limiting production of millets are declining area under cultivation due to switching over to remunerative crops like maize, soybean and horticulture crops, marginal and poor soils and low moisture conditions. Sometimes millets are grown under mixed cropping practices as sustenance agriculture without fertiliser. No adoption of improved varieties, tillage, weeding and interculture also resulted in low productivity.

10.6 Traits of Importance for Base Broadening

As very little attention is being given to millets by the researchers, these are commonly known as orphan or neglected crops (Wuthrich and Tadele 2012). These are generally grown under extreme environmental conditions such as areas with inadequate moisture or short-growing cycle and poor soil fertility (Baker 2003). Further, most of the genetic diversity is lost during the process of domestication. There is need for base broadening for the traits as discussed below.

10.6.1 Agronomic Traits

10.6.1.1 Yield-Related Traits

The major bottleneck affecting foxtail millet is its extremely low yield as it is mostly cultivated in marginal areas with poor moisture and fertility conditions. There is a need to broaden the genetic base for yield-related characteristics and to develop varieties for different agro-climatic regions – hills, coasts and plains. Similarly, Nirmalakumari and Vetriventhan (2010) established that better yield potential could be attained in foxtail millet by exploiting germplasm with medium flowering time, intermediate panicle length and more number of productive tillers. In addition to this, inherent characteristics, such as susceptibility to lodging, also significantly affect the productivity.

In Barnyard millet, the problem of lodging can be solved by reducing plant height to about 120–130 cm Grain and fodder yield can be enhanced by developing varieties with more and longer spikes and higher number of basal tillers, respectively. The cultivated barnyard millet has a narrow genetic base particularly for seed size. Breeding for easy dehulling is also required for easier postharvest processing.

10.6.1.2 Herbicide Resistance

Breeding foxtail millet to improve herbicide resistance is desirable for production improvement. Major weeds are grass weeds like species belonging to the genera *Digitaria*, *Eleusine* and *Setaria* and also some broadleaf species like *Amaranthus* species. Herbicides cannot be applied because of either a lack of selectivity or toxicity onto the crop. The gene pool of wild species related to foxtail millet has already provided two herbicide-resistant germplasms. Green foxtail, *S. viridis* (L.) Beauv., a wild relative of foxtail millet and resistant to triazine has been used for interspecific hybridisation and development of triazine-resistant germplasm. This resistance was maternally inherited both in the weed and the crop (Darmency and Pernes 1985). Trifluralin-resistant foxtail millet controlled by recessive nuclear genes at two linked loci was obtained following similar crosses with resistant green foxtail (Wang et al. 1996). Similarly, more wild species after screening for herbicide resistance can be used for pre-breeding.

In barnyard millet also, wild species can be used for transfer the herbicide resistance against crop-specific weeds.

10.6.2 Abiotic Stresses

Millets are important agricultural crops for arid regions due to short life span and their resistance to salinity and drought conditions. Similar to maize and sorghum, millets follow the C4 photosynthesis system (Brutnell et al. 2010; Warner and Edwards 1988); hence, they prevent photorespiration and as a consequence efficiently utilise scarce moisture in the semiarid regions. But still, drought affects foxtail millet yield adversely. Kafi et al. (2009) studied the salt stress effect on various millets and observed that with increase in salinity of irrigation water from 1.5 to 9.5 dS/m, grain yield of foxtail millet decreased by 37.2 % which was higher than that of pearl millet, i.e. 24.5 %. So, in both the cases, base broadening is required to minimise the effect of these stresses on yield-related traits.

Similarly, in case of barnyard millet also, both drought and salinity caused significant losses in yield. So there is a need to develop the varieties with more tolerance to these stresses with less yield losses.

10.6.3 Biotic Stresses

Resistance to various diseases and pests is a major area where base broadening of cultivated gene pool is required. In Foxtail millet, blast, downy mildew, rust and shoot fly are major biotic stresses. In case of barnyard millet, *Ustilago* spp. causing agent of grain smut is the major threat, and yield losses up to 60.8 % have been reported (Jain et al. 1997). The Japanese and Indian barnyard millet (wild progenitor) are resistant to *Ustilago* spp. and could be used in resistance breeding programmes. Leaf blight is also a major constraint in some areas. Anthracnose, sheath blight (*Rhizoctonia solani*) and Cercospora leaf spot (*Cercospora fujimaculans*) have also been found to affect the crop (http://www.nilgs.affrc.go.jp/db/diseases/contents/edisease.html). Shoot fly (*Atherigona falcata*) and pink stem borer (*Sesamia inferens*) are the major insect-pests of barnyard millet (Jagadish et al. 2008).

10.6.4 Quality Traits

Foxtail millet grains contain higher seed protein (14–16 %) and crude fat (5–8 %) than finger millet (Ravindran 1991; Dwivedi et al. 2012; Dikshit et al. 2013). Its seed is generally rich in protein content, macro- and micro-minerals and vitamins (Rai 2002). Millet proteins like other cereals are deficient in lysine and tryptophan amino acids but contain significant amounts of sulphur-containing amino acids (methionine and cysteine), the essential amino acids. Millets also contain higher fat content in comparison to major cereals (Obilana and Manyasa 2002). Important feature of millets lies in their high dietary fibre content, which is recognised for its hypocholesterolemic and hypoglycaemic effects, apart from contributing to high satiety value. Upadhyaya et al. (2011) carried out evaluation of core collection (155 accessions) of foxtail millet for nutritional traits. In core set, protein content ranged from 10.7 to 18.5 %, calcium (90.3–288.7 $mgkg^{-1}$), iron (24.1–68.0 $mgkg^{-1}$) and zinc (33.6–74.2 $mgkg^{-1}$). Its bran oil (9.4 %) is a rich source of linoleic acid (66.5 %), an essential fatty acid (Liang et al. 2010; Dwivedi et al. 2012), and high fibre content (42.56 %) (Amadou et al. 2011). Its grains are used for different purposes, viz., as an energy source for pregnant and lactating women, to reduce blood sugar level in diabetics (Sema and Sarita 2002) and to fight type 2 diabetes and cardiovascular diseases (Choi et al. 2005). Foxtail millet has an anti-proliferative property that may be due to the presence of phenolic extracts which could be helpful in preventing the cancer initiation (Chandrasekara and Shahidi 2011; Rao et al. 2011).

Barnyard millet grain is a good source of protein (11.1–13.9 %) (Monteiro et al. 1987). Carbohydrates constitute the major grain component (65 %) which is mostly present in non-starchy polysaccharide form. Various benefits of its consumption include blood pressure control, prevention of constipation and slow glucose release during digestion into the blood stream. Krishnakumari and Thayumanavan (1997) reported that barnyard millet could be considered as an effective faecal bulking, hypoglycaemic and hypolipidaemic agent better than other minor millets and rice. Again, Krishnakumari and Thayumanavan (1998) reported that in comparison to other millets, barnyard millet reduced blood glucose and lipid levels more effectively. These findings are also supported by the observations of other workers (Arora and Srivastava 2002; Ugare et al. 2014). Watanabe (1999) isolated three antioxidants including two flavonoids and one serotonin derivative from grains of Japanese barnyard millet (cv. 'Kurohie'). So, besides nutritious grains, the presence of strong antioxidative components makes us consider barnyard millet as a functional food crop which also has led to an increase in its demand (Kim et al. 2011). Amadou et al. (2013) found that millet foods can enhance the viability or functionality of probiotics with significant health benefits. Further, its gluten-free status also makes it advantageous over other major cereals for people allergic to gluten (Hoshino et al. 2010).

Wild species in general had higher content of micronutrients and vitamins which could be used for biofortification of the cultivated species. Further, resistant starch, high fibre content and functional biomolecules are other biochemical components where base broadening can play a major role.

10.6.5 Utilisation of Germplasm

In case of foxtail millet, large germplasm collections including 46,070 accessions of cultivated and 906 accessions of wild species are available with rich phenotypic variations (http://apps3.fao.org/wiews/germplasm_query.html; Reddy et al. 2006). After China, India contains the second largest germplasm of this crop. Some efforts have been made in characterisation of foxtail millet germplasm and identification of promising accessions. Core collections of 155 accessions in foxtail millet have been developed which resulted into identification of accessions with traits of early maturity, high grain yield and differences in panicle shape, size or seed colour and with higher grain Fe, Zn and/or Ca concentrations (Upadhyaya et al. 2011). A core collection of 107 accessions of foxtail millet was developed and used for determining population structure and allele mining of important agronomic and abiotic stress-related traits (Lata et al. 2011). 'Super Early Maturation No. 2', a new germplasm of foxtail millet, has been developed by the Institute of Millet Crops, China, using the targeted-character gene-bank breeding method for which 18 early maturation millet germplasms were evaluated (Liu et al. 2006). The germplasm has several outstanding characteristics including super early maturity (63 days), high iron content (54.1 mgkg−1), high crude fat (6.24 %) and protein content (14.36 %).

In case of barnyard millet, 729 accessions of *E. frumentacea* (Roxb.) Link were evaluated for yield and yield attributes over years and a core set including 50 accessions was developed (Gowda et al. 2009).

In both foxtail and barnyard millets, detailed evaluation of the available germplasm conserved in national gene banks particularly the wild/weedy species is required to identify the trait specific germplasm particularly for high grain yield, insect-pest and disease resistance and quality parameters which can ultimately be used for genetic base broadening.

10.7 Interspecific Hybridisation in Crop Species

Foxtail millet is a self-pollinating crop (Siles et al. 2004), in which gene flow occurs through intra- or interspecific hybridisation. In *Setaria*, introgression of pollen between weedy species, i.e. interspecific hybridisation, is of rare occurrence with almost sterile progenies (Li et al. 1942; Poirier-Hamon and Pernes 1986). Interspecific hybrids have been reported between green foxtail *S. viridis* L. and foxtail millet (*S. italica* P. Beauv) (De Wet et al. 1979; Till-Bottraud et al. 1992). Natural outcrossing within foxtail millet and its wild progenitor S. *viridis* occurred at between 0 and 2.2 % (reported up to 4.0 %, if different panicles are bagged together), and interspecific hybridisations occurred partly in the same order of magnitude but was different according to the distances of the partners and the design of the experimental plots (Boudry et al. 1993).The wild relatives of *S. italica* have been found to donate valuable traits to the cultivated gene pool of foxtail millet, e.g. resistance to certain herbicides such as triazine, sethoxydim, dinitroaniline and trifluralin. Herbicide resistance genes have been successfully transferred from *S. viridis* to *S. italica* (Darmency and Pernes 1985; Wang et al. 1996; Wang and Darmency 1997; Darmency et al. 2011).

Induced tetraploidy has been used to stabilise the progeny of interspecific hybrids in foxtail millet. The diploid and tetraploid F_2 populations of the *S. italica* × *S. viridis* cross were compared, and further selection resulted into tetraploid F_2 population which flowered earlier and had larger seed weight than the diploid F_2 (Darmency et al. 1987). Ahanchede et al. (2004) induced autotetraploid material by colchicine treatment of a pure line of foxtail millet and an interspecific hybrid of foxtail millet and giant green foxtail. Compared with diploid material, tetraploids were smaller, flowered later and had a twofold reduced fertility.

Two-line hybrids in foxtail millet have also been developed using the photoperiod-sensitive male sterility (PMS) as a genetic tool (Rongbai et al. 2005; Guo et al. 2006).

In barnyard millet (*Echinochloa* species), gene flow mainly occurs through intraspecific hybridisation. Maun and Barrett (1986) reported a high degree of self-fertilisation, among barnyard millet populations. The information on interspecific hybridisation in case of barnyard millet is very meagre. Interspecific hybrids for early maturity and high yield were developed using *E. frumentacea* (early maturing) and *E. esculenta* (high yielding) but found sterile both ways (Mehta et al. 2005). Similarly, sterile hybrids were obtained when interspecific hybridisation was carried out between Japanese and Indian barnyard millet for grain smut resistance (Sood et al. 2014). The cultivated species of barnyard millet can be improved utilising valuable wild progenitors (Mandelbaum et al. 1995).

10.7.1 Concerns Related to Interspecific Gene Transfer

Ellstrand (2003) reported gene flow from cultivated to wild relatives in a number of field crops. In this regard, gene flow from genetically modified crops to their wild relatives is a major concern. Shi et al. (2008) observed gene flow from foxtail millet (*S. italica*) to *S. viridis*, its weedy relative for herbicide resistance. So strategies should be developed to prevent this kind of gene flow.

10.8 Barriers to Interspecific Hybridisation

10.8.1 Morphological Traits

Cross fertility rates, for most of the foxtail millet varieties, have been found very low, i.e. 0.0–1.4 % (Till-Bottraud et al. 1992). The panicle height differences, times of fertilisation floral morphology and anthesis behaviour make cross pollination difficult in foxtail millet species (Willweber-Kishimoto 1962; Baltensperger 1996).

10.8.2 Genus Complexity

In case of genus *Setaria*, large germplasm collections including 46,070 accessions of cultivated and 906 accessions of wild species are available (http://apps3.fao.org/wiews/germplasm_query.

html), having diverse phenotypic traits, varied life cycles, ploidy levels and breeding systems suggesting a very complex taxonomy. The genus and related species contains basic chromosome number of nine ($x=9$) (Avdulov 1931; Giussani et al. 2001). On the basis of genome organisation, three gene pools have been designated for the *Setaria* species. *S. italica* has a diploid genome ($2n=2x=18$) designated as genome A by Li et al. (1945). Like *S. italica*, *S. viridis* also carry a diploid A genome and has been confirmed by various means: cytogenetics and protein and DNA markers (Harlan and de Wet 1971; Wang et al. 1998; Benabdelmouna et al. 2001a, b). Both these species occupy the primary gene pool. Species with AABB genome ($2n=4x=36$), *S. faberi* (giant foxtail) and *S. verticillata* (bristly foxtail) which probably originated from diploid ($2n=2x=18$) species, *S. viridis* (green foxtail) and *S. adhaerens* (bristly grass), constitute the secondary gene pool (Benabdelmouna et al. 2001a, b). Tertiary gene pool includes remaining landraces of *Setaria* namely, *grisebachii*, *queenslandica*, *pumila* and *pallide-fusca*. Recently, poor hybridisation signals, produced using GISH by genome of *S. grisebachii* Fourn. ex Hemsl ($2n=2x=18$) with A and B genomes, led to its designation as C genome (Wang et al. 2009). In this genus, *S. queenslandica* has been reported as the only autotetraploid species with AAAA genome, whereas *S. pumila* (yellow foxtail) and *S. pallide-fusca* are also polyploid but without the AA genome (Benabdelmouna et al. 2001a, b; Benabdelmouna and Darmency 2003). An unknown genome type has also been identified in *S. glauca* using GISH (Benabdelmouna et al. 2001a, b).

10.8.3 Cytoplasmic Male Sterility

Cytoplasmic male sterility (CMS) is a maternally inherited phenotype and a significant genetic tool to aid crossing and exploitation of hybrid vigour in crop plants. In foxtail millet, no successful CMS line has been developed to commercially exploit hybrid vigour (Dwivedi et al. 2012). However, effort has been made by Zhi et al. (2007) to develop CMS lines using green and cultivated foxtail millet species but obtained male sterile products. But there are some studies which exploited the CMS. For example, genetic male sterility has been exploited in foxtail millet to produce hybrid cultivars namely Jigu 16 and Suanxi 28 × Zhangnong 10 (Cui et al. 1979; Du and Wang 1997). So we can say that more efforts are needed to exploit male sterility in order to facilitate the development of hybrid progenies so that new segregants could be generated and genetic recombinations could be enhanced in foxtail millet breeding schemes.

In case of barnyard millet, no information is available in genome organisation. Nirmalakumari and Vetriventhan (2009) found that traits like flowering in early hours, small size of flowers which opens very little and only for a short duration and short pollen viability make hybridisation very difficult in this crop.

10.9 Conventional and Contemporary Approaches of Interspecific Gene Transfer

10.9.1 Conventional Approaches

Breeding foxtail millet to improve herbicide resistance is desirable for increasing the production. Foxtail millet germplasm resistant to triazine has been developed through interspecific hybridisation using *S. viridis* (L.) Beauv., a wild relative. This resistance was maternally inherited both in the weed and the crop (Darmency and Pernes 1985). Interspecific hybrids were developed successfully for improved sethoxydim (herbicide) resistance through crossing between *S. italica* and *S. viridis*, resistant green foxtail millet (Wang et al. 1997). Some valuable traits like herbicide resistance to sethoxydim, dinitroaniline and trifluralin have been transferred to cultivated species utilising resistant wild relatives of *S. italica* (Darmency and Pernes 1985; Wang et al. 1996; Wang and Darmency 1997; Darmency et al. 2011).

In $F_{2:3}$ progenies of foxtail millet developed through interspecific hybridisation between cultivated and wild species, the number of branches differs with the density of plants, genotype and various edaphic factors including G X E interactions (Doust and Kellogg 2006).

The diploid and tetraploid F_2 populations of the *S. italica* x *S. viridis* cross have been compared, and results showed that tetraploid F_2 population flowered earlier and had larger seed weight than the diploid F_2 (Darmency et al. 1987). Ahanchede et al. (2004) induced tetraploidy using the wild genetic resources of foxtail millet to stabilise the progeny of interspecific hybrids with *S. viridis*, the putative progenitor of foxtail millet. Autotetraploid material was induced by colchicine treatment of a pure line of foxtail millet and a hybrid developed from cultivated and green foxtail. Compared with diploid material, tetraploids were smaller, flowered later and had a twofold reduced fertility (number of grain per cm of spike). Grain weight increased by 20 % with polyploidy, whereas total grain yield decreased by 46 %. This work also illustrated the difficulty of breeding polyploid lines of foxtail millet that are of agronomic use.

Yuan et al. (2008) used PMS as a genetic tool for development of interspecific hybrids in *S. italica* P. Beauv. and identified a photoperiod-sensitive male sterile line JG1S.

In India, Uttarakhand and Tamil Nadu states are the major states involved in the breeding barnyard millet, and most of the varieties have been developed through mass and pure line selection followed by hybridisation and pedigree selection. As mentioned earlier also, Mehta et al. (2005) developed interspecific hybrids for early maturity and high yield using *E. frumentacea* (early maturing) and *E. esculenta* (high yielding) but found the products sterile both ways. Similarly, efforts to transfer grain smut resistance from Japanese to Indian barnyard millet were failed because of sterile hybrids (Sood et al. 2014).

In *Echinochloa,* gamma irradiation has been used to develop mutants for various traits including grain yield, plant height, tillers per plant and head length (Mehra et al. 1985) and waxiness (Hoshino et al. 2010).

10.9.2 Contemporary Approaches

Other than conventional approaches, genetic improvement could be achieved by using modern techniques for gene transfer, i.e. genetic engineering or recombinant DNA technology. In transgenic development, transformation success largely depends on the efficiency of regeneration and transformation protocols. Wuthrich and Tadele (2012) reviewed the millet improvement including foxtail and barnyard millets through regeneration and transformation.

In case of foxtail millet, in vitro studies have been carried out with immature inflorescence, mature embryo, mature seed and shoot tip as explants, for somatic embryogenesis, using 2, 4-D alone or with KIN or BA as well as for plant regeneration using NAA with BA or KIN or 2, 4-D with KIN (Xu et al. 1984; Rao et al. 1988; Reddy and Vaidyanath 1990; Osuna-Avila et al. 1995; Liu et al. 2005a, b; Wang et al. 2011). Methods for callus initiation, somatic embryo formation and plantlet regeneration from explants have recently been elaborated in Yugu1, a foxtail millet genotype by Chen et al. (2011; http://genomicscience.energy.gov/pubs/2011abstracts/2011GSPmeetingUSDA-DOEFeedstocks.pdf).

In barnyard millet, somatic embryogenesis has been carried out from mature seeds as explants using 2, 4-D by Gupta et al. (2001). Earlier workers have made successful attempts of callus and plant regeneration from different *Echinochloa* species viz., *E. frumentacea, E. crus-galli* and *E. colona* (Tyagi et al. 1985; Sankhla et al. 1992; Samantaray et al. 1996, 2001; Rout et al. 1997; Gupta et al. 2001; Bobkov 2005).

In case of foxtail millet, immature inflorescence has been used as initial explants for transformation studies. Embryogenic callus as transformed explants were successfully obtained using A (LBA4404; EHA105) media and Zm13, PF128 promoters (Liu et al. 2005a, b; Wang et al. 2011). Fang et al. (2008) employed the protocol to transform foxtail millet with *Zm13*:*Si401*, where *Zm13* was a pollen specific maize promoter and the plasmid vector used was pBI-Zm13Si401. The authors reported that *Si401* overexpression in foxtail millet led to many cellular aberrations such as premature tapetum degeneration, pre-disposition of endothecium cells' fibrous bands and aborted pollen grains during later stages of anther development. The findings of this study can be utilised for the development of male sterile lines of economic importance in near future (Fang et al. 2008; Ceasar and Ignacimuthu 2009). In barnyard millet, mature seed and leaf segment were used as initial explants. The stable transformation was recovered with embryogenic callus and leaf segment as explants using MB media with ZmUbi, CaMV 35S, OsAct, RbcS and ppcA-L-Ft promoters, but only ubi promoter was found effective for GUS gene expression (Gupta et al. 2001). Although *Agrobacterium*-mediated transformation is widely applied in cereals (Schrawat and Lorz 2006), microprojectile bombardment remains the dominant method of transformation in millets despite its drawbacks, where target genome in integrated number of times with transgene. Development of foxtail millet transgenic plants through particle bombardment of pollen and inflorescence have also been reported but their transformation efficiency was found to be low (Wang et al. 2011). Liu et al. (2005a, b) developed foxtail millet transgenics through *Agrobacterium*-mediated transformation with 6.6 % efficiency. Again in 2007, Liu et al. (2007) found that in foxtail millet, *Agrobacterium*-mediated transformation is influenced by genotype, source of explant and inoculation time. Chen et al. (2011) successfully transformed Yugu1, a foxtail millet genotype using *Agrobacterium*-mediated transformation.

10.10 Molecular Markers, Genome Mapping and Genomics as an Adjunct to Breeding

Discovery of molecular markers have revolutionised the agriculture breeding programmes which can be used for transfer of desired traits through marker-assisted selection. These have numerous advantages over morphological and biochemical markers and can be applied in cases where tight linkage between genes and the effects of unknown factors makes conventional breeding tough. The progress made in foxtail and barnyard millets with respect to the molecular markers, QTL identification, development of genetic maps, genome sequencing and comparative genomics is discussed in details.

10.10.1 Foxtail Millet

A set of 98 intron-length polymorphic markers along with 147 genomic microsatellite markers from foxtail millet have been developed by Gupta et al. (2011, 2012). In the late 1990s, two genetic maps were developed first from an inter-varietal cross and the other through an interspecific hybridisation of foxtail and green millet using available foxtail millet markers, 111 rice and 5 wheat RFLP markers (Devos et al. 1998; Wang et al. 1998). Compared study of genetic maps of foxtail millet and rice revealed high colinearity in two genomes (Devos et al. 1998). Recombinant inbred lines generated through a cross between *S. italica* Yugu1× *S. viridis* A10 has been used to develop a high-density map (Doust et al. 2009). In foxtail millet, four QTLs each have been identified for basal branching, and axillary branching present single each on chromosomes I and V and double on chromosome III in first case and single each on chromosomes VI and IX and double on chromosome V in the later case, and are partly controlled by different loci (Doust et al. 2004). Few of the basal branching deviations were influenced by loci other than the ones responsible for axillary branching in the mapping population. They also reported that allele responsible for the augmented axillary branching was transmitted by *S. viridis* while that for improved tillering by *S. italica*. Liu et al. (2011) obtained high level of genetic diversity using 79 SSR markers among 128 Chinese foxtail millet accessions. Wang et al. (2010) Low genetic diversity was obtained in wild green foxtail in comparison to cultivated foxtail millet based upon the DNA sequence survey of nine loci. Two factors are considered responsible for low genetic diversity and high linkage disequilibrium (LD) in foxtail millet – a population bottleneck and gene flow from foxtail millet to green foxtail. A total of 218 genic and genomic microsatellite markers have been developed in foxtail millet from microsatellite-enriched genomic libraries (147), RAPD-enriched genomic sequences from the gene bank (45) and expressed sequence tag sequences (26) before the availability of its genome-wide sequences in public domain (Jia et al. 2007; Lin et al. 2011; Gupta et al. 2012). Foxtail millet (*S. italica* (L.) Beauv.) genetic map was developed covering the genome length of 2 082.5 cM using molecular markers from different species and foxtail millet (Wang et al. 2014).

Comparative genomics involving foxtail millet and other members of grass family have also been reported earlier such as between rice and foxtail millet (Devos et al. 1998); pearl millet, rice and foxtail millet (Devos et al. 2000); and maize and foxtail millet (Doust et al. 2004). Genome Institute of the Department of Energy, USA, and Beijing Genomics Institute, China, have jointly sequenced the genome of two foxtail millet accessions which is now available in public domain and provides opportunity for comparative genomics. Based upon the available genome sequences, specific genetic markers can be developed, validated and used for characterisation of germplasm, phylogenetics, development of genetic linkage maps, transferability studies and finally in crop improvement programmes of foxtail millet and related species. Some efforts have been made in this direction as mentioned below:

In silico mining of the foxtail millet genome sequence led to the identification of 28,342 microsatellite markers which showed about 79 % and 93 % transferability in switchgrass and pearl millet, respectively. A high-density physical map of foxtail millet was generated through physical mapping of 15,573 microsatellites markers (Pandey et al. 2013). The further use of markers validated in foxtail millet or other species would accelerate the comparative genomics, development of high-density linkage maps and fine mapping of genes/QTLs for various economic traits. Kumari et al. (2013) developed EST-derived SSR (eSSR) markers for the first time in foxtail millet. A set of 447 primer pairs were designed from 24,828 non-redundant ESTs, out of which 327 were mapped physically and from these 106 produced high level of cross-genera amplification. Comparative mapping showed high level of similarity between foxtail millet chromosomes with sorghum (~68 %), maize (~61 %) and rice (~42 %) chromosomes. Similarly, out of 5123 intron-length polymorphic (ILP) markers developed in foxtail millet, 4049 were physically mapped on 9 chromosomes, and considerable proportion of sequence-based orthology and syntenic relationship were obtained using these physically mapped markers between foxtail millet chromosomes and sorghum (50 %), maize (46 %), rice (21 %) and Brachypodium (21 %) chromosomes through in silico comparative mapping (Muthamilarasan et al. 2014). Zhang et al. (2014) designed 788 SSR primer pairs on the basis of comparison of sequences between *S. italica* and *S. viridis*, and based on their amplification efficiency and polymorphic performance in *S. faberii*, *S. verticillata*, *S. glauca* and *S. adhaerens* found that evolutionarily, foxtail millet is more distant from *S. glauca* and *S. adhaerens*.

10.10.1.1 Barnyard Millet

In barnyard millet, genetic variability generated through RAPD markers was found in consistency with morphological variability (Hilu 1994). Efforts have also been made to classify different species of this complex genus including *E. crus-galli* (L.) Link., *E. colona* (L.) Beauv. and *E. crus-pavonis* (Kunth.) using microsatellite markers (Danquah et al. 2002). Genetic diversity and population structure of 155 accessions developed using SSR markers revealed that accessions belonging to var. *crus-galli* or var. *formosensis* were more diverse than of var. *esculenta* (Nozawa et al. 2006). Further, Sood et al. (2014) tested hybridity of four F1 plants selected on the basis of morphological features through rice SSR markers. Among 40 SSR markers screened for parental polymorphism, 12 markers were polymorphic. Among these, two markers, namely, RM 433 and RM439, were used for hybrid testing, and all the four F1 plants were found to be true hybrids.

In barnyard millet, very less information is available on genomic resources. Crop-specific markers are not available for barnyard millet which is also affecting its crop improvement programme. So transferability of markers from related species and development of a genetic map can play a major role in barnyard millet improvement. As mentioned earlier also, availability of foxtail millet genome sequences in public domain has led to the development of different type of markers such as SSRs and ILPs which have also been applied successfully across genera (Kumari et al. 2013; Pandey et al. 2013; Muthamilarasan et al. 2014). Further, the markers developed and validated are available to researchers on web-based foxtail millet marker database (FmMDb) (Venkata Suresh et al. 2013). So transferability of the markers available in foxtail millet database and then comparative genome mapping using sequence-based markers will be highly useful in barnyard millet improvement (Yadav et al. 2014).

Conclusions

The foxtail millet is very complex with different ploidy levels. The morphological barriers, genome complexity, lack of efficient materials and crossing techniques make genetic studies very difficult, in foxtail millet. Similarly, little efforts have been made in barnyard millet breeding. Therefore, the development of breeding materials through pre-breeding, i.e. utilisation of wild species and efficient crossing techniques among species, is essential to achieve genetic and breeding objectives in foxtail and barnyard millets.

The availability of genome-wide sequences in foxtail millet has led to the development of different kinds of molecular markers. After validation, these markers can be used for genetic diversity analysis, development of linkage maps and QTLs' identification for important characters. The developed markers can be used in related species for various applications. In barnyard millet, very little work has been done in genomics. Comparative genomics can play an important role. The identified markers in foxtail millet can be utilised in barnyard millet for genetic diversity analysis, in evolutionary studies in linkage mapping.

Further, field evaluation of wild species for specific traits and identification of respective genes through markers developed in foxtail millet which can also be used in barnyard millet and further introgression of these traits to the cultivated gene pool through MAS will lead to the genetic base broadening.

References

Ahanchede A, Poirier-Hamon S, Darmency H (2004) Why no tetraploid cultivar of foxtail millet? Genet Resour Crop Evol 51:227–230

Amadou I, Amza T, Yong-Hui S, Guo-Wei L (2011) Chemical analysis and antioxidant properties of foxtail millet bran extracts. Songklanakarin J Sci Technol 33:509–515

Amadou I, Gounga ME, Gua-Wei L (2013) Millets nutritional composition, some health benefits and processing-A (review). Emir J Food Agric 25(7):501–508

Anju T, Sarita S (2010) Suitability of foxtail millet (*Setaria italica*) and barnyard millet (*Echinochloa frumentacea*) for development of Low Glycaemic Index biscuits. Malaysian J Nutr 16(3):361–368

Arora S, Srivastava S (2002) Suitability of millet based food products for diabetics. J Food Sci Technol 39:423–428

Athwal DS, Singh G (1966) Variability in Kangni-1. Adaptation, genotypic and phenotypic variability in four environments. Indian J Genet 26:142–152

Avdulov NP (1931) Karyo-systematische untersuchungen der familie Gramineen. Bull Appl Bot Genet Plant Breed Supp 44:1–428

Baker RD (2003) Millet production. Guide A-414. In: Cooperative Extension Service, College of Agriculture and Home Economics, New Mexico University, Las Cruces, p 6

Baltensperger DD (1996) Foxtail and proso millet. In: Janick J (ed) Progress in new crops. ASHS Press, Alexandria, pp 182–190

Benabdelmouna A, Darmency H (2003) Copia – like retrotransposons in the genus *Setaria*: sequence heterogeneity, species distribution and chromosomal organization. Plant Syst Evol 237:127–136

Benabdelmouna A, Abirached-Darmency M, Darmency H (2001a) Phylogenetic and genomic relationships in *Setaria italica* and its close relatives based on the molecular diversity 5S and 18S-5.8S-25S rDNA genes. Theor Appl Genet 103(5):668–677

Benabdelmouna A, Shi Y, Abirached-Darmency M, Darmency H (2001b) Genomic *in situ* hybridization (GISH) discriminates between the A and the B genomes in diploid and tetraploid *Setaria* species. Genome 44:685–690

Bennett MD, Leitch IJ, Hanson L (1998) DNA amounts in two samples of Angiuosperm weeds. Ann Bot 82(1):121–134

Bennett MD, Bhandol P, Leitch IJ (2000) Nuclear DNA amounts in angiosperms and their modern uses- 807 estimates. Ann Bot 86:859–909

Bobkov SV (2005) Long-term regeneration in callus culture of paisa (*Echinochloa frumentacea* Link.). Int Sorghum Millets Newsl 46:120–122

Boudry P, Morchen M, Saumitou-Laprade P, Vernet P, van Dijk H (1993) The origin and evolution of weed beets : consequences for the breeding and release of herbicide-resistent transgenic sugar beets. Theor Appl Genet 87:471–478

Brutnell TP, Wang L, Swartwood K et al (2010) *Setaria viridis*: a model for C4 photosythesis. Plant Cell 22(8):2537–2544

Ceasar SA, Ignacimuthu S (2009) Genetic engineering of millets: current status and future prospects. Biotechnol Lett 31(6):779–788

Chandrasekara A, Shahidi F (2011) Antiproliferative potential and DNA scission inhibitory activity of phenolics from whole millet grains. J Funct Foods 3:159–170

Chen X, Zale J, Chen F (2011) The regeneration and transformation of foxtail millet (*Setaria italica*), a model biofuel crop. USDA-DOE Plant Feedstock Genomics for Bioenergy Awardee Meeting. Available at: http://genomicscience.energy.gov/pubs/2011 abstracts/2011 GSPmeetingUSDA-DOEFeedstocks. pdf

Chinnaswamaiah B (1975) Formulation of selection indices through discriminant function technique for yield in *Setaria italica* Beauv. Mysore J Agric Sci 9(2):357–358

Choi YY, Osada K, Ito Y, Nagaswa T, Choi MR, Nishizawa N (2005) Effects of dietary protein of Korean foxtail millet on plasma adiponectin, HDL-cholesterol, and insulin levels in genetically type 2 diabetic mice. Biosci Biotechnol Biochem 69:31–37

Cui WS, Ma HX, Zhang DY (1979) The selection and utilization of "Suan Hsi 28" a male sterility strain of millet. Sci Agric Sin 12:43–46

Danquah EY, Hanley SJ, Brookes RC, Aldam C, Karp A (2002) Isolation and characterisation of microsatellites in *Echinochloa* (L.) Beau.spp. Mol Ecol Notes 2:54–56

Darmency H, Pernes J (1985) Use of wild *Setariaviridis* (L.) Beauv. to improve triazine resistance in cultivated *S. italica* (L.) by hybridization. Weed Res 25:175–179

Darmency H, Pernes J (1989) Agronomic performance of a triazine resistant foxtail millet (*Setaria italica* (L.) Beauv. Weed Res 29:147–150. doi:10.1111/j.1365-3180.1989.tb00853.x

Darmency H, Ouin C, Pernes J (1987) Breeding foxtail millet (*Setaria italica*) for quantitative traits after interspecific hybridization and polyploidization. Genome 29:453–456

Darmency H, Picard JC, Wang T (2011) Fitness costs linked to dinitroaniline resistance mutation in *Setaria*. Heredity 107:80–86

de Wet JMJ, Oestry-Stidd LL, Cubero JI (1979) Origins and evolution of foxtail millets (*Setaria italica*). J Agric Trad Bot Appl 26:53–56

de Wet JMJ, Rao KEP, Mangesha MH, Brink DE (1983) Domestication of sawa millet. Econ Bot 37:283–291. doi:10.1007/BF02858883

Devi PB, Vijayabharathi R, Sathyabhama S, Malleshi NG, Priyadarisini VB (2011) Health befits of finger millet (*Eleusine coracana* L.) polyphenols and dietary fiber: a review. J food Sci Technol. doi: 10.1007/s13197-011-0584-9

Devos KM, Wang ZM, Beales J, Sasaki T, Gale M (1998) Comparative genetic maps of foxtail millet (*Setaria italica*) and rice (*Orvza sativa*). Theor Appl Genet 96:63–68

Devos KM, Pittaway TS, Reynolds A, Gale MD (2000) Comparative mapping reveals a complex relationship between the pearl millet genome and those of foxtail millet and rice. Theor Appl Genet 100:190–198

Dikshit N, Devi P, Yadav S (2013) Identification of high protein lines in foxtail millet (*Setaria italica* (L.) P. Beauv.) germplasm. Ann Plant Physiol 27(2):81–85

Doust AN, Kellogg EA (2006) Effect of genotype and environment on branching in weedy green millet (*Setaria viridis*) and domesticated foxtail millet (*Setaria italica*) (Poaceae). Mol Ecol 15:1335–1349

Doust AN, Devos KM, Gadberry MD, Gale MD, Kellogg EA (2004) Genetic control of branching in foxtail millet. Proc Natl Acad Sci, USA 101:9045–9050

Doust AN, Kellogg EA, Devos KM, Bennetzen JF (2009) Foxtail millet: a sequence driven grass model system. Plant Physiol 149(1):137–141

Du RH, Wang TY (1997) Development and application of the two-line system for hybrid production in foxtail millet. In: Li Y (ed) Foxtail millet breeding. China Agricultural Press, Beijing, pp 591–616

Dwivedi S, Upadhyaya H, Senthivel S et al (2012) Millets: genetic and genomic resources. In: Janick J (ed) Plant breed review, vol 35. Wiley/Wiley-Blackwell, New York, pp 247–375

Ellstrand NC (2003) Dangerous liaisons? When cultivated plants mate with their wild relatives. Johns Hopkins University Press, Baltimore

Fang FQ, Qian Z, Ao GM, Yu JJ (2008) Co-suppression of *Si401*, a maize pollen specific *Zm401* homologous gene, results in aberrant anther development in foxtail millet. Euphytica 163:103–111

Gill AS, Randhawa AS (1975) Heritable variation and inter-relationship in foxtail millet. (*Setaria italica* Beauv). Madras Agric J 62:253–258

Giussani LM, Cota-Sánchez JH, Zuloaga FO, Kellogg EA (2001) A molecular phylogeny of the grass subfamily

Panicoideae (Poaceae) shows multiple origins of C4 photosynthesis. Am J Bot 88(11):1993–2012

Godawat SL (1980) A note on the path coefficient analysis in foxtail millet (*Setaria italica* (L.)P. Beauv.). Madras Agric J 67(10):690–692

Godfray H, Beddington J, Crute I (2010) Food security: the challenge of feeding 9 billion people. Science 327:812–818

Goron TL, Raizada MN (2015) Genetic diversity and genomic resources available for the small millet crops to accelerate a new green revolution. Front Plant Sci 6:157. doi:10.3389/fpls.2015.00157

Gowda JS, Bharathi S, Somu G, Krishnappa M, Rekha D (2009) Formation of core set in barnyard millet [*Echinochloa frumentacea* (Roxb.) Link] germplasm using data on twenty four morpho-agronomic traits. Int J Plant Sci 4:1–5

Guo RX, Sun DF, Tan ZB et al (2006) Two recessive genes controlling termophotoperiod-sensitive male sterility in wheat. Theor Appl Genet 112:1271–1276

Gupta P, Raghuvanshi S, Tyagi AK (2001) Assessment of the efficiency of various gene promoters via biolistics in leaf and regenerating seed callus of millets, *Eleusine coracana* and *Echinochloa crusgalli*. Plant Biotechnol 18(4):275–282

Gupta A, Mahajan V, Kumar M, Gupta HS (2009) Biodiversity in the barnyard millet (*Echinochloa frumentacea* Link., Poaceace) germplasm in India. Genet Resour Crop Evol 56:883–889

Gupta S, Kumari K, Das J, Lata C, Puranik S, Prasad M (2011) Development and utilization of novel intron length polymorphic markers in foxtail millet [*Setaria italica* (L.) P. Beauv.]. Genome 54:586–602

Gupta S, Kumari K, Sahu PP, Vidapu S, Prasad M (2012) Sequence based novel genomic microsatellite markers for robust genotyping purposes in foxtail millet [*Setaria italica* (L.) P. Beauv]. Plant Cell Rep 31:323–337

Harlan JR, de Wet JMJ (1971) Towards a rational taxonomy of cultivated plants. Taxon 20:509–517

Hegde PS, Rajasekaran NS, Chandra TS (2005) Effects of the antioxidant properties of millet species on oxidative stress and glycemic status in alloxan-induced rats. Nutr Res 25:1109–1120. doi:10.1016/j.nutres.2005.09.020

Hilu KW (1994) Evidence from RAPD markers in the evolution of *Echinochloa* millets (Poaceae). Plant Syst Evol 189:247–257

Hoshino T, Nakamura T, Seimiya Y et al (2010) Production of a full waxy line and analysis of waxy genes in the allohexaploid crop, Japanese Barnyard millet. Plant Breed 129:349–355

Hubbard FT (1915) A taxonomic study of *Setaria italica* and its immediate allies. Am J Bot 2:169–198

Jagadish PS, Mohapatra HK, Chakravarthy K, Srivastava N, Nangia N (2008) A compendium of insect pests of Finger Millet and other small millets. All India Coordinated Small Millets Improvement Project, GKVK, Bangalore, p 60

Jain AK, Jain SK, Yadava HS (1997) Assessment of yield losses due to grain smut in barnyard millet. Indian Phytopathol 50:49–52

Jayaraman A, Puranik S, Rai NK, Vidapu S, Sahu PP, Lata C, Prasad M (2008) cDNA- AFLP analysis reveals differential gene expression in response to salt stress in foxtail millet (*Setaria italica*). Mol Biotechnol 40:241–251

Jia XP, Shi YS, Song YC, Wang GY, Wang TY, Li Y (2007) Development of EST-SSR in foxtail millet (*Setaria italica*). J Genet Res Crop Evol 54:233–236

Jusuf M, Pernes J (1985) Genetic variability of foxtail millet (*Setaria italica* P. Beauv.): electrophoretic study of five isoenzyme systems. Theor Appl Genet 71:385–391

Kafi M, Zamani G, Ghoraishi SG (2009) Relative salt tolerance of south Khorasan millets. Desert 14:63–70

Kim JY, Jang KC, Park BR et al (2011) Physicochemical and antioxidative properties of selected barnyard millet (*Echinocloa utilis*) species in Korea. Food Sci Biotechnol 20:461–469

Krishnakumari S, Thayumanavan B (1997) Comparative study of resistance starch from minor millets on intestinal response of blood glucose, serum cholesterol and triglycerides in rats. J Sci Food Agric 73:296–302

Krishnakumari S, Thayumanavan B (1998) Characterization of starches of proso, foxtail, barnyard, kodo, and little millets. Plant Food Hum Nutr 53:47–56

Kumari R, Dikshit N, Sharma D, Bhat KV (2011) Analysis of molecular genetic diversity in a representative collection of foxtail millet [*Setaria italica* (L.) P. Beauv.] from different agro-ecological regions of India. Physiol Mol Biol Plants 17(4):363–374

Kumari K, Muthamilarasan M, Misra G et al (2013) Development of eSSR-markers in Setaria italica and their applicability in studying genetic diversity, cross-transferability and comparative mapping in millet and non-millet species. PLOS One 8(6):e67742

Lata C, Jha S, Dixit V, Sreenivasulu N, Prasad M (2011) Differential antioxidative responses to dehydration-induced oxidative stress in core set of foxtail millet cultivars [*Setaria italica* (L.)]. Protoplasma 248:817–828

Li CH, Pao WK, Li HW (1942) Interspecific crosses in Setaria II, cytological studies of interspecific hybrids involving : 1, S. *faberi* and S. *italica*, and 2, a three way cross, F_2 of *S. italica* x *S. viridis* and *S. faberi*. J Hered 33:351–355

Li HW, Li CH, Pao WK (1945) Cytological genetical studies of the interspecific cross of cultivated foxtail millet, *Setaria italica* (L.)Beauv., and *S. viridis* L. J Am Soc Agron 37:32–54

Li J, Wu SZ, Cao JS (1995) Cluster analysis of an international collection of foxtail millet (*Setaria italica* (L.) P. Beauv). Euphytica 83:79–85

Liang S, Yang G, Ma Y (2010) Chemical characteristics and fatty acid profile of foxtail millet bran oil. J Am Oil Chem Soc 87:63–67

Lin HS, Chiang CY, Chang SB, Kuoh CS (2011) Development of simple sequence repeats (SSR) markers in *Setaria italica* (Poaceae) and cross-amplification in related species. Int J Mol Sci 12:7835–7845

Liu YH, Yu JJ, Zhao Q, Ao GM (2005a) Genetic transformation of millet (*Setaria italica*) by *Agrobacterium*-mediated. Agric Biotechnol J 13:32–37

Liu ZL, Cheng RH, Huang WS, Qu ZF, Liu JX (2005b) Innovation of super-prematurity new millet idioplasm with sweet straw and multi-spike. Sci Agric Sin 38:17–21 (in Chinese)

Liu ZL, Cheng RH, Zhang FL, Xia XY, Shi ZH, Hou SL (2006) Analysis on genetic foundation and study on millet variety and its pedigree in Boreali-Sinica summer millets region. Acta Agric Bor Sin:103–109

Liu Y, Yu J, Ao G, Zhao Q (2007) Factors influencing *Agrobacterium* mediated transformation of foxtail millet (*Setaria italica*). Chin J Biochem Mol Biol 23:531–536

Liu ZL, Bai GH, Zhang DD, Zhu CS, Xia XY, Cheng RH, Shi ZG (2011) Genetic diversity and population structure of elite foxtail millet [*Setaria italica* (L.) P. Beauv.] germplasm in China. Crop Sci 51:1655–1663

Mandelbaum CI, Barbeau WE, Hilu KW (1995) Protein, calcium, and iron content of wild and cultivated species of Echinochloa. Plant Food Hum Nutr 47:101–108

Maun MA, Barrett SC (1986) The biology of Canadian weeds. 77. *Echinochloa crus-galli* (L.) Beauv. Can J Plant Sci 66:739–759

Mehra HS, Joshi HC, Chikara J (1985) Induced mutations in Japanese millet. Indian J Agric Sci 55:294–295

Mehta H, Tyagi PC, Mohapatra KP (2005) Genetic diversity in Barnyard millet (*Echinochloa frumentacea* Roxb.). Indian J Genet 65:293–295

Monteiro PV, Sudharshana L, Ramachandra G (1987) Japanese barnyard millet (Echinochloa frumentacea): protein content, quality and SDS-PAGE of protein fractions. J Sci Food Agric 43:17–25

Muthamilarasan M, Suresh BV, Pandey G, Kumari K, Parida SK, Prasad M (2014) Development of 5123 intr1-12.on-length polymorphic markers for large-scale genotyping applications in foxtail millet. DNA Res 21(1):41–52

Nagarajan K, Prasad MN (1980) Studies on genetic diversity in foxtail millet (*Setariaitalica* Beauv.). Madras Agric J 67(1):28–38

Navale PA, Harinarayana G, Mungse HB (1984) Intra population variability in variety of foxtail millet. J Maharashtra Agric Univ 9(3):154

Naylor RL, Falcon WP, Goodman RM, Jahn MM, Sengooba T, Tefera H, Nelson RJ (2004) Biotechnology in the developing world: a case for increased investments in orphan crops. Food Policy 29:15–44

Nelson RJ, Nayor RI, Jahn MM (2004) The role of genomics research in improvement of "orphan crops". Crop Sci 44:1901–1904

Nirmalakumari A, Vetriventhan M (2009) Phenotypic analysis of anther and pollen in diversified genotype of barnyard millet (*Echinochloa frumentacea*) floral characters. ICFAI Univ J Genet Evol 2:12–16

Nirmalakumari A, Vetriventhan M (2010) Characterization of foxtail millet germplasm collections for yield contributing traits. Electron J Plant Breed 1:140–147

Nozawa SM, Takahashi M, Nakai H, Sato YI (2006) Differences in SSR variations between Japanese barnyard millet (*Echinochloa esculenta*) and its wild relative *E. crus-galli*. Breed Sci 56:335–340

Obilana AB, Manyasa E (2002) Millets. In: Beltson PS, Taylor JRN (eds) Pseudocereals and less common cereals. Grain properties and utilization potential. Springer, Berlin, -177217

Osuna-Avila P, Nava-Cedillo A, Jofre-Garfias AE, Cabrera-Ponce JL (1995) Plant regeneration from shoot apex explants of foxtail millet. Plant Cell Tiss Organ Cult 40:33–35

Pandey G, Misra G, Kumari K, Gupta S, Parida SK, Chattopadhyay D, Prasad M (2013) Genome-wide development and use of microsatellite markers for large-scale genotyping applications in foxtail millet [*Setaria italica* (L.)]. DNA Res 20:197–207

Pohl RW (1978) How to know the grasses, 3rd edn. WCB McGraw Hill, Boston

Poirier-Hamon S, Pernes J (1986) Instabilite chromosomique dans les tissus somatiques des descendants d'un hybride interspecifique *Setaria verticillata* (P. Beauv.), *Setaria italica* (P. Beauv.). R Acad Sci 302:319–324

Prasad Rao KE, de Wet JMJ, Brink DE, Mengesha MH (1987) Intraspecific variation and systematic of cultivated *Setaria italica*, foxtail millet (Poaceae). Econ Bot 41:108–116

Prasad Rao KE, de Wet JMJ, Reddy VG, Mengesha MH (1993) Diversity in small millet collection at ICRISAT. In: Riley KW, Gupta SC, Seetharam A, Mushonga JN (eds) Advances in small millets. Oxford & IBH Publishing Co. Pvt. Ltd., New Delhi, pp 331–346

Rai M (2002) Nutritive cereals. In: Survey of Indian agriculture. The Hindu, Chennai, Tamil Nadu, India, pp 59–62

Rao AM, Kishor PBK, Reddy LA, Vaidyanath K (1988) Callus induction and high frequency plant regeneration in Italian millet (*Setaria italica*). Plant Cell Rep 7:557–559

Rao BR, Nagasampige MH, Ravikiran M (2011) Evaluation of nutraceutical properties of selected small millets. J Pharm Bioallied Sci 3:277–279

Ratnaswamy MC, Poonaiya BWX (1963) Discriminant function technique in selection for yield in (*Setaria italica* Beauv.), Italian millet. Madras Agric J 50:84

Ravindran G (1991) Studies on millet: proximate composition, mineral composition, and phytate and oxalate contents. Food Chem 39:99–107

Reddy LA, Vaidyanath K (1990) Callus formation and regeneration in two induced mutants of foxtail millet (*Setaria italica*). J Genet Breed 44:133–138

Reddy VG, Upadhyaya HD, Gowda CLL (2006) Characterisation of world's foxtail millet germplasm collections for morphological traits. J SAT Agric 2:1–3

Reddy INBI, Reddy DS, Narasu ML, Sivaramakrishnan S (2011) Characterisation of disease resistance gene homologues isolataed from finger millet (*Eleusine coracana* L. Gaertn.). Mol Breed 27:315–328. doi:10.1007/s11032-010-9455-1

Rominger JM (1962) Taxonomy of *Setaria* (Gramineae) in North America. Illinois Biol Monogr 29:1–132

Rongbai L, Pandey MP, Sharma P (2005) Inheritance of thermosensitive genic male sterility in rice (*Oryza sativa* L.). Curr Sci 88:1809–1815

Rout GR, Samantaray S, Das P (1997) Regeneration of a metal tolerant grass *Echinochloa colona* via somatic embryogenesis from suspension cultures. Biol Plant 40:17–23

Saleh ASM, Zhang Q, Chen J, Shen Q (2013) Millet grains: nutritional quality: processing and potential health benefits. Comp Rev Food Sci Food Saf 12(3):281–295

Samantaray S, Rout GR, Das P (1996) Regeneration of plants via somatic embryogenesis from leaf base and leaf tip segments of *Echinochloa colona*. Plant Cell Tiss Org Cult 47:119–125

Samantaray S, Rout GR, Das P (2001) Induction, selection and characterization of Cr and Ni-tolerant cell lines of *Echinochloa colona* (L.) Link *in vitro*. J Plant Physiol 158:1281–1290

Sandhu TS, Arora BS, Singh Y (1974) Inter-relationships between yield and yield components in foxtail millet. Indian J Agric Sci 44(9):563–566

Sankhla A, Davis TD, Sankhla D, Sankhla N, Upadhyaya A, Joshi S (1992) Influence of growth regulators on somatic embryogenesis, plantlet regeneration, and post-transplant survival of *Echinochloa frumentacea*. Plant Cell Rep 11:368–371

Schrawat AK, Lorz H (2006) *Agrobacterium*–mediated transformation of cereals: a promising approach crossing barriers. Plant Biotechnol J 4:575–603

Sema A, Sarita S (2002) Suitability of millet-based food products for diabetics. J Food Sci Technol 39:423–426

Sheriff RA, Shivashankar G (1992) Genetic divergence in foxtail millet (*Setariaitalica* B.). Indian J Genet 52(1):29–32

Shi Y, Wang T, Li Y, Darmency H (2008) Impact of transgene inheritance on the mitigation of gene flow between crops and their wild relatives: The example of foxtail millet. Genetics 180:969–975

Siles MM, Russell WK, Baltensperger DD et al (2004) Heterosis for grain yield and other agronomic traits in foxtail millet. Crop Sci 44:1960–1965

Singh G (1974) Variability in *Kangni*-3.Association between plant characters and discriminant function for varietal selection in four environments. Indian J Genet 34(3):411–416

Singh G, Athwal DS (1966) Variability in kangni 2.genotypic X environment interaction heritability and genetic advance. Indian J Genet 26:153–161

Sood S, Khulbe RK, Saini N, Gupta A, Agrawal PK (2014) Interspecific hybrid between *Echinochloa esculenta* (Japanese barnyard millet) and *E. frumentacea* (Indian barnyard millet) – a new avenue for genetic enhancement of barnyard millet. Electron J Plant Breed 5:248–253

Subramanyam D, Rathore VS (1999) Variation in photosynthetic traits in barnyard millet (*Echinochloa frumentacea*) genotypes. J Agron Crop Sci 183:199–203. doi:10.1046/j.1439.037x.1999.0034I.x

Surekha N, Naik RS, Mythri S, Devi R (2013) Barnyard millet (*Echinochloa frumentacea* Link.) cookies : development, value addition, consumer acceptability, nutritional and shelf life evaluation. IOSR J Environ Sci Toxicol Food Technol 7(3):1–10

Till-Bottraud I, Reboud X, Brabant P, Lefranc M, Rherissi B, Vedel F, Darmency H (1992) Outcrossing and hybridization in wild and cultivated foxtail millets: consequences for the release of transgenic crops. Theor Appl Genet 83:940–946

Tsehaye Y, Berg T, Tsegaye B, Tanto T (2006) Farmers' management of finger millet (*Eleusine coracana* L.) diversity in Tigray, Ethiopia and implications for on-farm conservation. Biodivers Conserv 15:4289–4308. doi:10.1007/s10531-005-3581-3

Tyagi AK, Bharal S, Rashid A, Maheshwari N (1985) Plant regeneration form tissue cultures initiated from immature inflorescences of a grass *Echinochloa colona* (L.) Link. Plant Cell Rep 4:115–117

Ugare R (2008) Health benefits, storage quality and value addition of barnyard millet (*Echinochloa frumentacea* Link.). Masters dissertation, University of Agricultural Sciences, Dharwad, India

Ugare R, Chimmad B, Naik R, Bharathi P, Itagi S (2014) Glycemic index and significance of barnyard millet (*Echinochloa frumentacae*) in type II diabetics. J Food Sci Technol 51:392–395

Upadhyaya HD, Pundir RS, Gowda CLL, Reddy VG, Singh S (2008) Establishing a core collection of foxtail millet to enhance the utilisation of the germplasm of an underutilized crop. Plant Genet Resour 7:177–184

Upadhyaya HD, Ravishankar CR, Narasimhudu Y et al (2011) Identification of trait-specific germplasm and developing a mini core collection for efficient use of foxtail millet genetic resources. Field Crop Res 124:459–467

Upadhyaya HD, Dwivedi SL, Singh SK, Singh S, Vetriventhan M, Sharma S (2014) Forming core collections in barnyard millet, Kodo and Little millet using morpho-agronomic descriptors. Crop Sci 54:2673–2682

Varshney RK, Glaszmann JC, Leung H, Ribaut JM (2010) More genomic resources for less-studied crops. Trends Biotechnol 28(9):452–460

Venkata Suresh B, Muthamilarasan M, Misra G, Prasad M (2013) FmMDb : a versatile database of foxtail millet markers for millets and bioenergy grasses research. PLoS One 8:e71418

Vijaykumar TP, Mohankumar JB, Srinivasan T (2010) Quality evaluation of noodles from millet flour blend incorporated composite flour. J Sci Ind Res 69:48–54

Vishwanatha JK, Mallanna KN, Aradhya KM, Gowda MVC, Shanker RU (1981) Genetic variability in world collection of germplasm of foxtail millet, *Setaria italica* Beauv. Mysore J Agric Sci 15(2):234–238

Wang T, Darmency H (1997) Inheritance of sethoxydim resistance in foxtail millet Setaria italica (L.) Beauv. Euphytica 94:69–73

Wang T, Fleury A, Ma J, Darmency H (1996) Genetic control of dinitroaniline resistance in foxtail millet (*Setaria italica*). J Hered 87:423–426

Wang T, Chen HB, Reboud X, Darmency H (1997) Pollen-mediated gene flow in an autogamous crop : foxtail millet (*Setaria italica*). Plant Breed 116:579–583

Wang ZM, Devos KM, Liu CJ, Wamg RQ, Gale MD (1998) Construction of RFLP-based maps of foxtail millet, *Setaria italica* (L.) P. Beauv. Theor Appl Genet 96(1):31–36

Wang YQ, Zhi H, Li W, Li HQ, Wang YF, Huang ZJ, Diao XM (2009) A novel genome of C and the first autotetraploid species in the *Setaria* genus identified by genomic in situ hybridization (GISH). Genet Resour Crop Evol 56(6):843–850

Wang C, Chen J, Zhi H et al (2010) Population genetics of foxtail millet and its wild ancestor. BMC Genet 11:90

Wang MZ, Pan YL, Li C (2011) Culturing of immature inflorescences and *Agrobacterium*-mediated transformation of foxtail millet (*Setaria italica*). Afr J Biotechnol 10:16466–16479

Wang ZI, Wang J, Yuan F, Du XF, Yang HQ, Guo EH (2014) Construction of genetic map of foxtail millet (*Setaria italica* (L.) Beauv.) using PCR-based molecular markers. Sci Agric Sin 47(17):3492–3500

Wanous MK (1990) Origin, taxonomy and ploidy of the millets and minor cereals. Plant Var Seeds 3:99–112

Warner DA, Edwards GE (1988) C_4 photosynthesis and leaf anatomy in diploid and autotetraploid *Pennisetum americanum* (pearl millet). Plant Sci 56:85–92

Watanabe M (1999) Antioxidative phenolic compounds from Japanese barnyard millet (*Echinochloa utilis*) grains. J Agric Food Chem 47:4500–4505

Williams RD, Schreiber MM (1976) Numerical and chemotaxonomy of green foxtail complex. Weed Sci 24:331–335

Willweber-Kishimoto E (1962) Interspecific relationships in the genus *Setaria*. Contrib Bio Kyoto Univerity.1–41

Wuthrich SP, Tadele Z (2012) Millet improvement through regeneration and transformation. Biotechnol Mol Biol Rev 7(2):48–61

Xu ZH, Wang DY, Yang LJ, Wei ZM (1984) Somatic embryogenesis and plant regeneration in cultured immature inflorescences of *Setariaitalica*. Plant Cell Rep 3:149–150

Yabuno T (1962) Cytotaxonomic studies on the two cultivated species and the wild relatives in the genus *Echinochloa*. Cytologia 27:296–305

Yadav C, Muthamilarasan M, Pandey G, Khan Y, Prasad M (2014) Development of novel microRNA-based genetic markers in foxtail millet for genotyping applications in related grass species. Mol Breed. doi: 10.1007/s11032-014-0137-9

Yuan AP, Hou AB, Zhang FY, Guo YD (2008) Inheritance and effects of the photoperiod sensitivity in foxtail millet [*Setaria italica*(P.) Beauv]. Hereditas 145:147–153

Zhang S, Tang C, Zhao Q et al (2014) Development of highly polymorphic simple sequence repeat markers using genome-wide microsatellite variant analysis in foxtail millet [*Setaria italica* (L.) P. Beauv.]. BMC Genet 15:78

Zhi H, Wang YQ, Li W, Wang YF, Li HQ, Lu P, Diao XM (2007) Development of CMS material from intraspecies hybridization between green foxtail and foxtail millet. J Plant Genet Resour 8:261–264